PSA 1990

VOLUME TWO

PSA 1990

PROCEEDINGS OF THE 1990
BIENNIAL MEETING
OF THE
PHILOSOPHY OF SCIENCE
ASSOCIATION

volume two

Symposium and Invited Papers

edited by

**ARTHUR FINE,
MICKY FORBES
&
LINDA WESSELS**

1991
Philosophy of Science Association
East Lansing, Michigan

Copyright © 1991 by the
Philosophy of Science Association

All Rights Reserved

No part of this book may be utilized, or reproduced in any form or by any means electronic or mechanical, without written permission from the publishers except in the case of brief quotations embodied in critical articles and reviews.

Library of Congress Catalog Card Number 72-624169

Cloth Edition: ISBN 0-917586-31-X

ISSN: 0270-8647

Manufactured in the United States of America

CONTENTS

Preface ix
Synopsis xi

Part I. Presidential Address

1. *The Road Since Structure* 3
   Thomas S. Kuhn, Massachusetts Institute of Technology

Part II. John S. Bell Session: State Vector Reduction

2. *A Tribute to John S. Bell* 17
   Abner Shimony, Boston University

3. *I. Dynamical Reduction Theories: Changing Quantum Theory so the Statevector Represents Reality* 19
   GianCarlo Ghirardi and Philip Pearle,
   University of Trieste and Hamilton College

4. *II. Elements of Physical Reality, Nonlocality and Stochasticity in Relativistic Dynamical Reduction Models* 35
   GianCarlo Ghirardi and Philip Pearle,
   University of Trieste and Hamilton College

5. *Desiderata for a Modified Quantum Dynamics* 49
   Abner Shimony, Boston University

Part III. Biology: The Non-Propositional Side

6. *Are Pictures Really Necessary?* 63
   The Case of Sewell Wright's "Adaptive Landscapes"
   Michael Ruse, University of Guelph

7. *Material Models in Biology* 79
   James R. Griesemer, University of California, Davis

8. *Mapping Ecologists' Ecologies of Knowledge* 95
   Peter J. Taylor, Cornell University

9. *Taming the Dimensions–Visualizations in Science* 111
   William C. Wimsatt, University of Chicago

Part IV. History of Philosophy of Science

10. *Conventionalism and the Origins of the Inertial Frame Concept* 139
    Robert DiSalle, University of Western Ontario

11. *Designation and Convention:*
    A Chapter of Early Logical Empiricism 149
    Thomas A. Ryckman, Northwestern University

12. *Hidden Agendas: Knowledge and Verification*   159
   Joia Lewis, University of San Diego

13. *Philosophical Interpretations of Relativity Theory: 1910-1930*   169
   Klaus Hentschel, University of Göttingen

Part V. Deduction From the Phenomena

14. *Newton's Classic Deductions from Phenomena*   183
   William Harper, University of Western Ontario

15. *Reasoning from Phenomena: Lessons from Newton*   197
   Jon Dorling, University of Amsterdam

16. *"From the Phenomena of Motions to the Forces of Nature": Hypothesis or Deduction?*   209
   Howard Stein, University of Chicago

Part VI. Science as Process

17. *Phylogenetic Analogies in the Conceptual Development of Science*   225
   Brent D. Mishler, Duke University

18. *The Function of Credit in Hull's Evolutionary Model of Science*   237
   Noretta Koertge, Indiana University

19. *The Evolution of Scientific Lineages*   245
   Michael Bradie, Bowling Green State University

20. *Conceptual Evolution: a Response*   255
   David L. Hull, Northwestern University

Part VII. Self-Organization, Selection and Evolution

21. *Form and Order in Evolutionary Biology: Stuart Kauffman's Transformation of Theoretical Biology*   267
   Richard M. Burian and Robert C. Richardson, Virginia Polytechnic Institute and State University and University of Cincinnati

22. *Self Organization and Adaptation in Insect Societies*   289
   Robert E. Page, Jr. and Sandra D. Mitchell, University of California, Davis and University of California, San Diego

23. *The Sciences of Complexity and "Origins of Order"*   299
   Stuart A. Kauffman, University of Pennsylvania

Part VIII.	**Statistical Inference and Theory Change**	
24.	*The Appraisal of Theories: Kuhn Meets Bayes* Wesley C. Salmon, University of Pittsburgh	325
25.	*Giving up Certainties* Henry E. Kyburg, Jr., University of Rochester	333
26.	*Belief Revision and Relevance* Peter Gärdenfors, University of Lund	349
Part IX.	**Mathematical and Physical Objects**	
27.	*Between Mathematics and Physics* Michael D. Resnik, University of North Carolina at Chapel Hill	369
28.	*Elementarity and Anti-Matter in Contemporary Physics: Comments on Michael D. Resnik's "Between Mathematics and Physics"* Susan C. Hale, California State University, Northridge	379
29.	*Problematic Objects between Mathematics and Mechanics* Emily R. Grosholz, The Pennsylvania State University	385
30.	*Structuralism and Conceptual Change in Mathematics* Christopher Menzel, Texas A&M University	397
Part X.	**Rudolf Carnap Centennial**	
31.	*The Unimportance of Semantics* Richard Creath, Arizona State University	405
Part XI.	**Implications of the Cognitive Sciences for Philosophy of Science**	
32.	*Implications of the Cognitive Sciences for the Philosophy of Science* Ronald N. Giere, University of Minnesota	419
33.	*Paradigms and Barriers* Howard Margolis, University of Chicago	431
34.	*Barriers and Models: Comments on Margolis and Giere* Nancy J. Nersessian, Princeton University	441
35.	*Some Twists in the Cognitive Turn* Steve Fuller, Virginia Polytechnic Institute and State University	445

Part XII. **Three Views of Experiment**

36. *Allan Franklin, Right or Wrong*   451
    Robert Ackermann, University of Massachusetts, Amherst

37. *Reason Enough? More on Parity-Violation Experiments and*   459
    Electroweak Gauge Theory
    Andy Pickering, University of Illinois at Urbana-Champaign

38. *Allan Franklin's Transcendental Physics*   471
    Michael Lynch, Boston University

39. *Do Mutants Have to be Slain, or Do They Die of Natural Causes?:*   487
    The Case of Atomic Parity Violation Experiments
    Allan Franklin, University of Colorado

Part XIII. **Computer Simulations in the Physical Sciences**

40. *Computer Simulations*   497
    Paul Humphreys, University of Virginia

41. *Computer Simulation in the Physical Sciences*   507
    Fritz Rohrlich, Syracuse University

42. *Computer Simulations, Idealizations and Approximations*   519
    Ronald Laymon, The Ohio State University

Part XIV. **Laws, Conditions and Determinism**

43. *Determinism in Deterministic Chaos*   537
    Roger Jones, University of Kentucky

44. *How Free are Initial Conditions?*   551
    Lawrence Sklar, University of Michigan

45. *Law Along the Frontier:*   565
    Differential Equations and Their Boundary Conditions
    Mark Wilson, The Ohio State University

PREFACE

This is the second of two volumes comprising the proceedings of the 1990 Biennial Meeting of the Philosophy of Science Association held in Minneapolis, Minnesota. The first volume, consisting of the program and the contributed papers, was published in advance of the meeting in October, 1990. The second volume consists of the presidential address, invited papers and symposia. Northwestern University provides generous support for the preparation of these volumes.

The Program Committee (consisting of Robert Brandon, Deborah Mayo, Alan Nelson, Thomas Nickles and John Norton) with Linda Wessels as its chair, arranged for the symposia and invited papers. Micky Forbes, Assistant Editor of these *Proceedings,* supervised the editing and processing of the papers to produce uniform, camera ready copy. The PSA Business Office saw the copy through to publication.

Thanks are due to the Program Committee and the authors. We are grateful to Northwestern University for financial support and to its Academic Computing Center for the use of their word processing facilities. Special thanks are due Wendy Ward, whose expertise and imagination in using these facilities shows up on every page of the volume.

Linda Wessels
History and Philosophy of Science
Indiana University
Bloomington

Arthur Fine and
Micky Forbes
Department of Philosophy
Northwestern University

Synopsis

The following brief summaries, arranged here alphabetically by author, provide an introduction to each of the papers in this volume.

1. *Allan Franklin, Right or Wrong.* **Robert Ackermann.** Franklin and Pickering agree that scientists in an experimental sequence, like the one to be discussed here, choose to accept certain experiments and their results as crucial, but disagree as to whether such choice can be justified in terms of an on-line estimate of evidential reliability. This paper suggests that it is possible to define a position between Franklin's Bayesian objectivism and Pickering's social constructivism. This position depends on considering the sequence of improvement in material technique and instrumentation as more important than any measure of reliability determined merely from such factors as evidential spread in relevant sequences, a factor that neither Franklin nor Pickering takes sufficiently into account.

2. *The Evolution of Scientific Lineages.* **Michael Bradie.** The fundamental dialectic of *Science as a Process* is the interaction between two narrative levels. At one level, the book is a historical narrative of one aspect of one ongoing problem in systematics. At the second level, Hull presents a theoretical model of the scientific process which draws heavily on invoked similarities between biological and scientific change. I first situate the model as one alternative among several which loosely fit under the umbrella of 'evolutionary epistemologies.' Second, I explore one of the implications of Hull's model, namely, that insofar as scientific theories are [parts of] "conceptual lineages," they are "conceptual individuals."

3. *Form and Order in Evolutionary Biology: Stuart Kauffman's Transformation of Theoretical Biology.* **Richard M. Burian and Robert C. Richardson.** The formal framework of Kauffman (1991) depicts the constraints of self-organization on the evolution of complex systems and the relation of self-organization to selection. We discuss his treatment of 'generic constraints' as sources of order (section 2) and the relation between adaptation and organization (section 3). We then raise a number of issues, including the role of adaptation in explaining order (section 4) and the limitations of formal approaches in explaining the distinctively biological (section 5). The principal question we pose is the relation of generic constraints on evolution to more specific local constraints, imposed, for example, by the characteristic materials out of which organisms are constructed, the accidental features characteristic of the Bauplan of a lineage, and the local vicissitudes of adaptation. We offer no answer to this large question.

4. *The Unimportance of Semantics.* **Richard Creath.** Philosophers often divide Carnap's work into syntactic, semantic, and later periods, but this disguises the importance of his early syntactical writing. In *Logical Syntax* Carnap is a thoroughgoing conventionalist and pragmatist. Once we see that, it is easier to see as well that these views were retained throughout the rest of his life, that the breaks between periods are not as important as the continuities, and that our understanding of such Carnapian notions as analyticity and probability needs reevaluation.

5. *Conventionalism and the Origins of the Inertial Frame Concept.* **Robert DiSalle.** This paper examines methodological issues that arose in the course of the development of the inertial frame concept in classical mechanics. In particular it examines the origins and motivations of the view that the equivalence of inertial frames leads to a kind of conventionalism. It begins by comparing the independent versions of the idea found in J. Thomson (1884) and L. Lange (1885); it then compares Lange's conventionalist claims with traditional geometrical conventionalism. It concludes by examining some implications for contemporary philosophy of space and time.

6. *Reasoning from Phenomena: Lessons from Newton.* **Jon Dorling.** I argue that Newtonian-style deduction-from-the-phenomena arguments should only carry conviction when they yield unexpectedly simple conclusions. That in that case they do establish higher rational probabilities for the theories they lead to than for any known or easily constructible rival theories. However I deny that such deductive justifications yield high absolute rational probabilities, and argue that the history of physics suggests that there are always other not-yet-known simpler theories with higher rational probabilities on all the original evidence, and that these later turn out closer to the truth. My analyses rely on the modern Solomonoff-Levin solution to the problem of constructing a mathematically and philosophically acceptable inductive logic.

7. *Do Mutants Have to be Slain, or Do They Die of Natural Causes?: The Cause of Atomic Parity Violation Experiments.* **Allan Franklin.** In this paper I will reexamine the history of the early experiments on atomic parity violation, presenting both Pickering's interpretation and an alternative explanation of my own. I argue that, contrary to Pickering, there were good reasons for the decision of the physics community. I will also explore some of the differences between my view of science and that proposed by the "strong programme" or social constructivist view in the sociology of science.

8. *Some Twists in the Cognitive Turn.* **Steve Fuller.** I argue that the recent "cognitive turn" in the philosophy of science does not challenge nearly as much of traditional philosophy of science as it proponents have claimed. However, the turn has forced philosophers to embody such hallowed abstractions as knowledge, theories, rationality, and concepts in flesh-and-blood human thinkers. While I welcome this newfound ontological awareness, I criticize four "mistaken identities" committed by two representative cognitivists, Howard Margolis and Ronald Giere. Generally speaking, the misidentifications turn on a fundamental naivete about the social dimension of thought.

9. *Belief Revision and Relevance.* **Peter Gärdenfors.** A general criterion for the theory of belief revision is that when we revise a state of belief by a sentence A, as much of the old information as possible should be retained in the revised state of belief. The motivating idea in this paper is that if a belief B is irrelevant to A, then B should still be believed in the revised state. The problem is that the traditional definition of statistical relevance suffers from some serious shortcomings and cannot be used as a tool for defining belief revision processes. In particular, the traditional definition violates the requirement that if A is irrelevant to C and B is irrelevant to C, then A&B is irrelevant to C. In order to circumvent these drawbacks, I develop an amended notion of relevance which has the desired properties. On the basis of the new definition, I outline how it can be used to simplify a construction of a belief revision method.

10. *I. Dynamical Reduction Theories: Changing Quantum Theory so the Statevector Represents Reality.* **GianCarlo Ghirardi and Philip Pearle.** The propositions, that what we see around us is real and that reality should be represented by the statevector, conflict with quantum theory. In quantum theory, the statevector can readily become a sum of states of comparable norm, each state representing a different reality. In this paper we present the Continuous Spontaneous Localization (CSL) theory, in which a modified Schrodinger equation, while scarcely affecting the dynamics of a microscopic system, rapidly "reduces" the statevector of a macroscopic system to a state appropriate for representing individual reality.

11. *II. Elements of Physical Reality, Nonlocality and Stochasticity in Relativistic Dynamical Reduction Models.* **GianCarlo Ghirardi and Philip Pearle.** The prob-

lem of getting a relativistic generalization of the CSL dynamical reduction model, which has been presented in part I, is discussed. In so doing we have the opportunity to introduce the idea of a stochastically invariant theory. The theoretical model we present, that satisfies this kind of invariance requirement, offers us the possibility to reconsider, from a new point of view, some conceptually relevant issues such as non-locality, the legitimacy of attributing elements of physical reality to physical systems and the problem of establishing causal relations between physical events.

12. *Implications of the Cognitive Sciences for the Philosophy of Science.* **Ronald N. Giere.** Does recent work in the cognitive sciences have any implications for theories or methods employed within the philosophy of science itself? It does if one takes a naturalistic approach in which understanding the nature of representations or judgments of representational success in science requires reference to the cognitive capacities or activities of individual scientists. Here I comment on recent contributions from three areas of the cognitive sciences represented respectively by Paul Churchland's neurocomputational perspective, Nancy Nersessian's cognitive-historical approach, and Paul Thagard's computational philosophy of science. The main general conclusion is that we need to replace traditional linguistic notions of representation in science.

13. *Material Models in Biology.* **James R. Griesemer.** Propositions alone are not constitutive of science. But is the "non-propositional" side of science theoretically superfluous: must philosophy of science consider it in order to adequately account for science? I explore the boundary between the propositional and non-propositional sides of biological theory, drawing on three cases: Grinnell's remnant models of faunas, Wright's path analysis, and Weismannism's role in the generalization of evolutionary theory. I propose a picture of material model-building in biology in which manipulated systems of material objects function as theoretical models. In each of the cases, material systems such as diagrams play important generative as well as presentational roles.

14. *Problematic Objects between Mathematics and Mechanics.* **Emily R. Grosholz.** The existence of mathematical objects may be explained in terms of their occurrence in problems. Especially interesting problems arise at the overlap of domains, and the items that intervene in them are hybrids sharing the characteristics of both domains in an ambiguous way. Euclid's geometry, and Leibniz' work at the intersection of geometry, algebra and mechanics in the late seventeenth century, provide instructive examples of such problems and items. The complex and yet still formal unity of these items calls into question certain tenets of Resnik's structuralism, and of the reductive projects of the logicists.

15. *Elementarity and Anti-Matter in Contemporary Physics: Comments on Michael D. Resnik's "Between Mathematics and Physics".* **Susan C. Hale.** I point out that conceptions of particles as mathematical, or quasi mathematical, entities have a longer history than Resnik notices. I argue that Resnik's attack on the distinction between mathematical and physical entities is not deep enough. The crucial problem for this distinction finds its locus in the numerical indeterminacy of elementary particles. This problem, traced by Heisenberg, emerges from the discovery of antimatter.

16. *Newton's Classic Deductions From Phenomena.* **William Harper.** I take Newton's arguments to inverse square centripetal forces from Kepler's harmonic and areal laws to be classic deductions from phenomena. I argue that the theorems backing up these inferences establish systematic dependencies that make the phenomena carry the objective information that the propositions inferred from them hold. A review of the data supporting Kepler's laws indicates that these phenomena are Whewellian colli-

gations—generalizations corresponding to the selection of a best fitting curve for an open-ended body of data. I argue that the information theoretic features of Newton's corrections of the Keplerian phenomena to account for perturbations introduced by universal gravitation show that these corrections do not undercut the inferences from the Keplerian phenomena. Finally, I suggest that all of Newton's impressive applications of Universal gravitation to account for motion phenomena show an attempt to deliver explanations that share these salient features of his classic deductions from phenomena.

17. *Philosophical Interpretations of Relativity Theory: 1910-1930*. **Klaus Hentschel.** The paper (given in the section on "Recent work in the History of Philosophy of Science) discusses the method and some of the results of the doctoral dissertation on philosophical interpretations of Einstein's special and general theories of relativity, submitted to the Dept. for History of Science, Univ. of Hamburg, in 1989, also published by Birkhauser, Basel, in 1990. It is claimed that many of the gross oversimplifications, misunderstandings and misinterpretations occurring in more than 2500 texts about the theories of relativity written by scientists, philosophers, and laymen contemporary to Einstein can in fact serve as a clue to a better understanding of the general process by which philosophical interpretations are formed. Another very important source for answering the question of how misinterpretations are formed are hitherto unpublished documents in the estates of physicists and philosophers of that time, including apart from Einstein himself: Bergson, Bridgman, Carnap, Cassirer, Metz, Meyerson, Petzoldt, Reichenbach, Schlick and Vaihinger.

18. *Conceptual Evolution: A Response.* **David L. Hull.** Each of the commentators on my *Science as a Process* has emphasized a different part of my book. Mishler concentrates on the relevant biology, Koertge expands upon the sociological mechanism I propose, while Bradie discusses biological and conceptual lineages as historical entities. I respond to these comments and criticisms, emphasizing the roles played by sequences of ancestor-descendant tokens in replication and ecological types in interaction. Hence, selection results from the alternation of genealogical tokens with ecological types in both biological and conceptual evolution.

19. *Computer Simulations.* **Paul Humphreys.** This article provides a survey of some of the reasons why computational approaches have become a permanent addition to the set of scientific methods. The reasons for this require us to represent the relation between theories and their applications in a different way than do the traditional logical accounts extant in the philosophical literature. A working definition of computer simulations is provided and some properties of simulations are explored by considering an example from quantum chemistry.

20. *Determinism in Deterministic Chaos.* **Roger Jones.** John Earman's *A Primer on Determinism* treats the doctrine of Laplacian determinism by a careful look at a considerable variety of physical theories. This paper enriches Earman's discussion of chaos theory by considering in some detail the analysis of dripping faucets due to Robert Shaw. Shaw's analysis exhibits in a nice way some of the techniques used in chaos theory and gives a feel for research in this area. The paper concentrates on the tension between the determinism inherent in any description involving differential equations and the in-practice (any practice) unpredictability resulting from the extreme sensitivity to initial conditions of the non-linear differential equations characteristic of chaos theory.

21. *The Sciences of Complexity and "Origins of Order"*. **Stuart A. Kauffman.** This article discusses my book, *Origins of Order: Self Organization and Selection in Evolution,* in the context of the emerging sciences of complexity. *Origins*, due out of Oxford University Press in early 1992, attempts to lay out a broadened theory of evo-

lution based on the marriage of unexpected and powerful properties of self organization which arises in complex systems, properties which may underlie the origin of life itself and the emergence of order in ontogeny, and the continuing action of natural selection. The three major themes are: 1) that such self organized properties lie to hand for selection's further molding; 2) hence that the order we see is not due to selection alone, but in part reflects the order selection has always acted upon; 3) and finally that the marriage of natural order and natural selection may inevitably lead living entitites to a novel organized state, lying on the edge between order and chaos, as the inevitable evolutionary attractor of selection for the capacity to adapt.

22. *The Function of Credit in Hull's Evolutionary Model of Science.* **Noretta Koertge.** This paper first argues that evolutionary models of conceptual development which are patterned on Darwinian selection are unlikely to solve the demarcation problem. The persistence of myths shows that in most social environments unfalsifiable ideas are more likely to survive than ones which can be subjected to empirical scrutiny. I then analyze Hull's claims about how the credit system operates in science and conclude with him that it can perform a surprising variety of functions. However I argue that the credit system must be constantly tempered by internalized norms which encapsulate the traditional ultimate aims of science.

23. *The Road Since Structure.* **Thomas S. Kuhn.** A highly condensed account of the author's present view of some philosophical problems unresolved in *The Structure of Scientific Revolutions*. The concept of incommensurability, now considerably developed, remains at center stage, but the evolutionary metaphor, introduced in the final pages of the book, now also plays a principal role.

24. *Giving up Certainties.* **Henry E. Kyburg, Jr.** One of the serious motivations for the development of non-monotonic logics is the fact that, however sure we may be of some set of facts, there can come a time at which at least some of them must be given up. A number of philosophical approaches have stemmed from the study of scientific inference, in which a law or theory, accepted on good evidence at one time, comes to be rejected on the basis of more evidence. These approaches are reviewed, and an alternative approach, whose key idea is the control of observational error for the purpose of predictive adequacy is developed.

25. *Computer Simulations, Idealizations and Approximations.* **Ronald Laymon.** It's uncontroversial that notions of idealization and approximation are central to understanding computer simulations and their rationale. What's not so clear is what exactly these notions come to. Two distinct forms of approximation will be distinguished and their features contrasted with those of idealizations. These distinctions will be refined and closely tied to computer simulations by means of Scott-Strachey denotational programming semantics. The use of this sort of semantics also provides a convenient format for argumentation in favor of several theses I shall propose concerning the role computer implemented approximations and idealizations play in fixing what the acceptance of an underlying scientific theory is or should be.

26. *Hidden Agendas: Knowledge and Verification.* **Joia Lewis.** A complete reading of the works of Moritz Schlick reveals an apparent vacillation between a preference for holistic, formalistic accounts of knowledge and a preference for atomist, foundational accounts. A clearer picture of Schlick's philosophical development emerges from an appreciation of what I consider to be two separate "agendas," each of them fully formed and present in his earliest writings. Schlick's conviction that the assumptions in each agenda were equally correct can explain a good deal about the construction of his mature theses and is, I believe, responsible for much of the criticism that his work has

received, both before and after his death in 1936. Making the two agendas explicit can also provide us with a better framework within which to judge Schlick's work. Finally, the story provides an instructive example of how one's firmest beliefs constrain and dictate one's choice of solutions and perhaps define the problems themselves.

27. *Allan Franklin's Transcendental Physics.* **Michael Lynch.** This paper was presented at a session on "Three views of experiment: Atomic parity violations," in which Allan Franklin's study of an episode in the recent history of particle physics was discussed and criticized. Franklin argues in favor of what he calls "the evidence model," a general claim to the effect that physicists' theory choices are based on valid experimental evidence. He contrasts his position to that of the social constructivists, who, according to him, insist that social and cognitive interests, and not the evidence, explains physicists' practical and theoretical judgments. My paper argues that Franklin miscasts the debate between experimental realism and social constructivism, because constructivists do not insist that evidence has no role whatsoever in experimental practice. My position draws lessons from Wittgenstein's later philosophy and ethnomethodological studies of scientific practices. The paper does not aim to support social constructivism against Franklin's arguments, so much as to suggest that the terms of the realist-constructivist debate provide a poor context for the examination of the temporal production of experiments and observations.

28. *Paradigms and Barriers.* **Howard Margolis.** In a forthcoming study I give an account of paradigm shifts as shifts in habits of mind. This paper summarizes the argument. Habits of mind, on this view, are what constitute a paradigm. Further, some particular habit of mind (the "barrier") is ordinarily critical for a Kuhnian revolution. A contrast is drawn between this view and the "gap" view that is ordinarily implicit in analysis of the nature of of paradigm shifts.

29. *Structuralism and Conceptual Change in Mathematics.* **Christopher Menzel.** I address Grosholz's critique of Resnik's mathematical structuralism and suggest that although Resnik's structuralism is not without its difficulties it survives Grosholz's attacks.

30. *Phylogenetic Analogies in the Conceptual Development of Science.* **Brent D. Mishler.** I address David Hull's theses about the process of science from the perspective of an evolutionary biologist, particularly emphasizing phylogenetic systematics (a.k.a. cladistics), an area that has figured prominently in Hull's work as a source of both sociological data and metatheory. The goal is to carefully explore analogies and disanalogies between scientific process and comparative biology. There do seem to be remarkable analogies (e.g., research groups as lineages, scientists as interactors in selection processes), indeed these lead to important insights that might not otherwise have been made, yet some possible analogies present novel problems: Are "memes" like genes or like traits? What is the nature of replication in science? It is argued that the primary need is for some rigorously worked-out case studies.

31. *Barriers and Models: Comments on Margolis and Giere.* **Nancy J. Nersessian.** Giere's assessment is that the cognitive sciences, especially cognitive psychology, have much to offer the philosophy of science as it attempts to develop theories of the growth, development, and change of scientific knowledge as human activities. Margolis produces a model of scientific change by drawing from recent work in the cognitive sciences and attempts to show how this model explains salient cases of conceptual change. While agreeing with Giere's assessment, I argue that Margolis provides the wrong model both for scientific change and for how the interaction between cognitive science and philosophy of science should proceed.

32. *Self Organization and Adaptation in Insect Societies.* **Robert E. Page, Jr. and Sandra D. Mitchell.** Division of labor and its associated phenomena have been viewed as prime examples of group-level adaptations. However, the adaptations are the result of the process of evolution by natural selection and thus require that groups of insects once existed and competed for reproduction, some of which had a heritable division of labor while others did not. We present models, based on those of Kauffman (1984) that demonstrate how division of labor may occur spontaneously among groups of mutually tolerant individuals. We propose that division of labor itself is not a product of natural selection but instead is a "typical" outcome of self organization.

33. *Reason Enough? More on Parity-Violation Experiments and Electroweak Gauge Theory.* **Andy Pickering.** I respond to Allan Franklin's critique of my account of the establishment of parity-violating neutral-current effects in atomic and high-energy physics as an instance of a more general 'rationalist' attack on 'constructivist' understandings of science. I argue that constructivism does not entail the denial of 'reason' in science, but I note that there are typically too many 'reasons' to be found for 'reason' to count as an explanation of why science changes as it does. I show, first, that there were many 'reasonable' but different ways of reasoning about the field of evidence at issue in this episode and, second, that Franklin's articulation of how theory-choice should proceed on the basis of evidence implies a vicious conservatism which is fortunately not to be found in the history of science.

34. *Between Mathematics and Physics.* **Michael D. Resnik.** Nothing has been more central philosophy of mathematics than the distinction between mathematical and physical objects. Yet consideration of quantum particles shows the inadequacy of the popular spacetime and causal characterizations of the distinction. It also raises problems for an assumption used recently by Field, Hellman and Horgan, namely, that the mathematical realm is metaphysically independent of the physical one.

35. *Computer Simulation in the Physical Sciences.* **Fritz Rohrlich.** Computer simulation is shown to be philosophically interesting because it introduces a qualitatively new methodology for theory construction in science different from the conventional two components of "theory" and "experiment and/or observation". This component is "experimentation with theoretical models." Two examples from the physical sciences are presented for the purpose of demonstration but it is claimed that the biological and social sciences permit similar theoretical model experiments. Furthermore, computer simulation permits theoretical models for the evolution of physical systems which use cellular automata rather than differential equations as their syntax. The great advantages of the former are indicated.

36. *Are Pictures Really Necessary? The Case of Sewell Wright's "Adaptive Landscapes".* **Michael Ruse.** Philosophical analyses of science tend to ignore illustrations, implicitly regarding them as theoretically dispensable. If challenged, it is suggested that such neglect is justifiable, because the use of illustrations only leads to faulty reasoning, and thus is the mark of bad or inadequate science. I take as an example one of the most famous illustrations in the history of evolutionary biology, and argue that the philosophers' scorn is without foundation. I take my conclusions to be support for a naturalistic approach to philosophy.

37. *Designation and Convention: A Chapter of Early Logical Empiricism.* **T.A. Ryckman.** An examination of Carnap's *Aufbau* in the context of Schlick's *Allgemeine Erkenntnislehre* of ten years earlier, suggests that Carnap's focus there on the sign-relation (Zeichenbeziehung) is an effort to retrieve a verificationist account of the meaning of individual scientific statements from the abyss of meaning-holism entailed by

Schlick's proposal that scientific concepts be implicitly defined. The *Aufbau*'s antipodal aspects, its reductive phenomenalism and quasi-Kantian concern with the constitution of objectivity, are seen as complementary moments of the marriage of empiricism and a new emphasis on scientific concepts as "free creations of the human mind".

38. *The Appraisal of Theories: Kuhn Meets Bayes.* **Wesley C. Salmon.** This paper claims that adoption of Bayes's theorem as the schema for the appraisal of scientific theories can greatly reduce the distance between Kuhnians and logical empiricists. It is argued that plausibility considerations, which Kuhn considered outside of the logic of science, can be construed as prior probabilities, which play an indispensable role in the logic of science. Problems concerning likelihoods, especially the likelihood on the "catchall," are also considered. Severe difficulties concerning the significance of this probability arise in the evaluation of individual theories, but they can be avoided by restricting our judgments to comparative assessments of competing theories.

39. *Desiderata for a Modified Quantum Dynamics.* **Abner Shimony.** If quantum mechanics is interpreted as an objective, complete, physical theory, applying to macroscopic as well as microscopic systems, then the linearity of quantum dynamics gives rise to the measurement problem and related problems, which cannot be solved without modifying the dynamics. Eight desiderata are proposed for a reasonable modified theory. They favor a stochastic modification rather than a deterministic non-linear one, but the spontaneous localization theories of Ghirardi et al. and Pearle are criticized. The intermittent fluorescence of a trapped atom irradiated by two laser beams suggests a stochastic theory in which the locus of stochasticity is interaction between a material system and the electromagnetic vacuum.

40. *How Free are Initial Conditions?* **Lawrence Sklar.** Those who think of some aspects of the world as "physically necessary" usually think of this kind of necessity as being confined to the general law of nature, initial conditions being "contingent." Tachyon theory and general relativity provide independent but related reasons for thinking that some initial states are, however, "impossible." And statistical mechanics seems to lead us to conclude that some initial conditions are, if not impossible, "highly improbable." We are then, led from these aspects of physics to wonder if initial conditions are always "freely specifiable" and in the domain of physical contingency.

41. *"From the Phenomena of Motions to the Forces of Nature": Hypothesis or Deduction?* **Howard Stein.** This paper examines Newton's argument from the phenomena to the law of universal gravitation—especially the question how such a result could have been obtained from the evidential base on which that argument rests. Its thesis is that the crucial step was a certain application of the third law of motion—one that could only be justified by appeal to the consequences of the resulting theory; and that the general concept of interaction embodied in Newton's use of the third law most probably evolved in the course of the very investigation that led to this theory.

42. *Mapping Ecologists' Ecologies of Knowledge.* **Peter J. Taylor.** Ecologists grapple with complex, changing situations. Historians, sociologists and philosophers studying the construction of science likewise attempt to account for (or discount) a wide variety of influences making up the scientists' "ecologies of knowledge." This paper introduces a graphic methodology, mapping, designed to assist researchers at both levels—in science and in science studies—to work with the complexity of their material. By analyzing the implications and limitations of mapping, I aim to contribute to an ecological approach to the philosophy of science.

43. *Law Along the Frontier: Differential Equations and Their Boundary Conditions.* **Mark Wilson.** Physicists often allow the "laws" of a discipline, formulated as partial differential equations, to be disobeyed along various surfaces, arrayed along the boundary and inside the medium under study. What kinds of considerations permit these lapses in the applicability of the equations? This paper surveys a variety of answers found in the physical literature.

44. *Taming the Dimensions—Visualizations in Science.* **William C. Wimsatt.** The role of pictures and visual modes of presentation of data in science is a topic of increasing interest to workers in artificial intelligence, problem solving, and scientists in all fields who must deal with large quantities of complex multidimensional data. Drawing on studies of animal motion, aerodynamics, morphological transformations, the history of linkage mapping, and the analysis of deterministic chaos, I focus on the strengths and limitations of our visual system, the analysis of problems particularly suited to visualization—the analysis of similarities and differences between complex objects, and problems making conjoint use of information from several complex images.

Part I

PRESIDENTIAL ADDRESS

The Road Since Structure

Thomas S. Kuhn

Massachusetts Institute of Technology

On this occasion, and in this place, I feel that I ought, and am probably expected, to look back at the things which have happened to the philosophy of science since I first began to take an interest in it over half a century ago. But I am both too much an outsider and too much a protagonist to undertake that assignment. Rather than attempt to situate the present state of philosophy of science with respect to its past — a subject on which I've little authority — I shall try to situate my present state in philosophy of science with respect to its own past — a subject on which, however imperfect, I'm probably the best authority there is.

As a number of you know, I'm at work on a book, and what I mean to attempt here is an exceedingly brief and dogmatic sketch of its main themes. I think of my project as a return, now underway for a decade, to the philosophical problems left over from the *Structure of Scientific Revolutions*. But it might better be described more generally, as a study of the problems raised by the transition to what's sometimes called the historical and sometimes (at least by Clark Glymour, speaking to me) just the "soft" philosophy of science. That's a transition for which I get far more credit, and also more blame, than I have coming to me. I was, if you will, present at the creation, and it wasn't very crowded. But others were present too: Paul Feyerabend and Russ Hanson, in particular, as well as Mary Hesse, Michael Polanyi, Stephen Toulmin, and a few more besides. Whatever a *Zeitgeist* is, we provided a striking illustration of its role in intellectual affairs.

Returning to my projected book, you will not be surprised to hear that the main targets at which it aims are such issues as rationality, relativism and, most particularly, realism and truth. But they're not primarily what the book is about, what occupies most space in it. That role is taken instead by incommensurability. No other aspect of *Structure* has concerned me so deeply in the thirty years since the book was written, and I emerge from those years feeling more strongly than ever that incommensurability has to be an essential component of any historical, developmental, or evolutionary view of scientific knowledge. Properly understood — something I've by no means always managed myself — incommensurability is far from being the threat to rational evaluation of truth claims that it has frequently seemed. Rather, it's what is needed, within a developmental perspective, to restore some badly needed bite to the whole notion of cognitive evaluation. It is needed, that is, to defend notions like truth and knowledge from, for example, the excesses of post-modernist movements like the strong program. Clearly, I can't hope to make all that out here: it's a project for a book. But I shall try, however sketchily, to describe the main elements of the position the book develops. I begin by saying something about what I now take incommensurability to be, and then attempt to sketch its relationship to questions of relativism, truth, and realism. In the book, the issue of rationality will figure, too, but there is no space here even to sketch its role.

Incommensurability is a notion that for me emerged from attempts to understand apparently nonsensical passages encountered in old scientific texts. Ordinarily they had been taken as evidence of the author's confused or mistaken beliefs. My experiences led me to suggest, instead, that those passages were being misread: the appearance of nonsense could be removed by recovering older meanings for some of the terms involved, meanings different from those subsequently current. During the years since, I've often spoken metaphorically of the process by which later meanings had been produced from earlier ones as a process of language change. And, more recently, I've spoken also of the historian's recovery of older meanings as a process of language learning rather like that undergone by the fictional anthropologist whom Quine misdescribes as a radical translator (Kuhn 1983a). The ability to learn a language does not, I've emphasized, guarantee the ability to translate into or out of it.

By now, however, the language metaphor seems to me far too inclusive. To the extent that I'm concerned with language and with meanings at all — an issue to which I'll shortly return — it is with the meanings of a restricted class of terms. Roughly speaking, they are taxonomic terms or kind terms, a widespread category that includes natural kinds, artifactual kinds, social kinds, and probably others. In English the class is coextensive, or nearly so, with the terms that by themselves or within appropriate phrases can take the indefinite article. These are primarily the count nouns together with the mass nouns, words which combine with count nouns in phrases that take the indefinite article. Some terms require still further tests hinging, for example, on permissible suffixes.

Terms of this sort have two essential properties. First, as already indicated, they are marked or labelled as kind terms by virtue of lexical characteristics like taking the indefinite article. Being a kind term is thus part of what the word means, part of what one must have in the head to use the word properly. Second — a limitation I sometimes refer to as the no-overlap principle — no two kind terms, no two terms with the kind label, may overlap in their referents unless they are related as species to genus. There are no dogs that are also cats, no gold rings that are also silver rings, and so on: that's what makes dogs, cats, silver, and gold each a kind. Therefore, if the members of a language community encounter a dog that's also a cat (or, more realistically, a creature like the duck-billed platypus), they cannot just enrich the set of category terms but must instead redesign a part of the taxonomy. Pace the causal theorists of reference, 'water' did not always refer to H_2O (Kuhn 1987; 1990, pp. 309-14).

Notice now that a lexical taxonomy of some sort must be in place before description of the world can begin. Shared taxonomic categories, at least in an area under discussion, are prerequisite to unproblematic communication, including the communication required for the evaluation of truth claims. If different speech communities have taxonomies that differ in some local area, then members of one of them can (and occasionally will) make statements that, though fully meaningful within that speech community, cannot in principle be articulated by members of the other. To bridge the gap between communities would require adding to one lexicon a kind-term that overlaps, shares a referent, with one that is already in place. It is that situation which the no-overlap principle precludes.

Incommensurability thus becomes a sort of untranslatability, localized to one or another area in which two lexical taxonomies differ. The differences which produce it are not any old differences, but ones that violate either the no-overlap condition, the kind-label condition, or else a restriction on hierarchical relations that I cannot spell out here. Violations of those sorts do not bar intercommunity understanding. Members of one community can acquire the taxonomy employed by members of another, as the historian does in learning to understand old texts. But the process which permits understanding produces bilinguals, not translators, and bilingualism has a cost, which will be particularly important to what follows. The bilingual must always remember within which community discourse is occurring. The use of one taxonomy to make statements to someone who uses the other places communication at risk.

Let me formulate these points in one more way, and then make a last remark about them. Given a lexical taxonomy, or what I'll mostly now call simply a lexicon, there are all sorts of different statements that can be made, and all sorts of theories that can be developed. Standard techniques will lead to some of these being accepted as true, others rejected as false. But there are also statements which could be made, theories which could be developed, within some other taxonomy but which cannot be made with this one and vice versa. The first volume of Lyons' *Semantics* (1977, pp. 237-8) contains a wonderfully simple example, which some of you will know: the impossibility of translating the English statement, "the cat sat on the mat", into French, because of the incommensurability between the French and English taxonomies for floor coverings. In each particular case for which the English statement is true, one can find a co-referential French statement, some using 'tapis', others 'paillasson,' still others 'carpette,' and so on. But there is no single French statement which refers to all and only the situations in which the English statement is true. In that sense, the English statement cannot be made in French. In a similar vein, I've elsewhere pointed out (Kuhn 1987, p. 8) that the content of the Copernican statement, "planets travel around the sun", cannot be expressed in a statement that invokes the celestial taxonomy of the Ptolemaic statement, "planets travel around the earth". The difference between the two statements is not simply one of fact. The term 'planet' appears as a kind term in both, and the two kinds overlap in membership without either's containing all the celestial bodies contained in the other. All of which is to say that there are episodes in scientific development which involve fundamental change in some taxonomic categories and which therefore confront later observers with problems like those the ethnologist encounters when trying to break into another culture.

A final remark will close this sketch of my current views on incommensurability. I have described those views as concerned with words and with *lexical* taxonomy, and I shall continue in that mode: the sorts of knowledge I deal with come in explicit verbal or related symbolic forms. But it may clarify what I have in mind to suggest that I might more appropriately speak of concepts than of words. What I have been calling a lexical taxonomy might, that is, better be called a conceptual scheme, where the "very notion" of a conceptual scheme is not that of a set of beliefs but of a particular operating mode of a mental module prerequisite to having beliefs, a mode that at once supplies and bounds the set of beliefs it is possible to conceive. Some such taxonomic module I take to be pre-linguistic and possessed by animals. Presumably it evolved originally for the sensory, most obviously for the visual, system. In the book I shall give reasons for supposing that it developed from a still more fundamental mechanism which enables individual living organisms to reidentify other substances by tracing their spatio-temporal trajectories.

I shall be coming back to incommensurability, but let me for now set it aside in order to sketch the developmental framework within which it functions. Since I must again move quickly and often cryptically, I begin by anticipating the direction in which I am headed. Basically, I shall be trying to sketch the form which I think any viable evolutionary epistemology has to take. I shall, that is, be returning to the evolutionary analogy introduced in the very last pages of the first edition of *Structure*, attempting both to clarify it and to push it further. During the thirty years since I first made that evolutionary move, theories of the evolution both of species and of knowledge have, of course, been transformed in ways I am only beginning to discover. I still have much to learn, but to date the fit seems extremely good.

I start from points familiar to many of you. When I first got involved, a generation ago, with the enterprise now often called historical philosophy of science, I and most of my coworkers thought history functioned as a source of empirical evidence. That evidence we found in historical case studies, which forced us to pay close attention to science as it really was. Now I think we overemphasized the empirical aspect of our enterprise (an evolutionary epistemology need not be a naturalized one). What has for me emerged as essential is not so much the details of historical cases as the perspective or the ideology that attention to historical cases brings with it. The historian, that is, always picks up a process already underway, its beginnings lost in earlier time. Beliefs are already in place; they provide the basis for the ongoing research whose results will in some cases change them; research in their absence is unimaginable though there has nevertheless been a long tradition of imagining it. For the historian, in short, no Archimedean platform is available for the pursuit of science other than the historically situated one already in place. If you approach science as an historian must, little observation of its actual practice is required to reach conclusions of this sort.

Such conclusions have by now been pretty generally accepted: I scarcely know a foundationalist any more. But for me, this way of abandoning foundationalism has a further consequence which, though widely discussed, is by no means widely or fully accepted. The discussions I have in mind usually proceed under the rubric of the rationality or relativity of truth claims, but these labels misdirect attention. Though both rationality and relativism are somehow implicated, what is fundamentally at stake is rather the correspondence theory of truth, the notion that the goal, when evaluating scientific laws or theories, is to determine whether or not they correspond to an external, mind-independent world. It is that notion, whether in an absolute or probabilistic form, that I'm persuaded must vanish together with foundationalism. What replaces it will still require a strong conception of truth, but not, except in the most trivial sense, correspondence truth.

Let me at least suggest what the argument involves. On the developmental view, scientific knowledge claims are necessarily evaluated from a moving, historically-situated, Archimedian platform. What requires evaluation cannot be an individual proposition embodying a knowledge claim in isolation: embracing a new knowledge claim typically requires adjustment of other beliefs as well. Nor is it the entire body of knowledge claims that would result if that proposition were accepted. Rather, what's to be evaluated is the desirability of a particular change-of-belief, a change which would alter the existing body of knowledge claims so as to incorporate, with minimum disruption, the new claim as well. Judgements of this sort are necessarily comparative: which of two bodies of knowledge — the original or the proposed alternative — is *better* for doing whatever it is that scientists do. And that is the case whether what scientists do is solve puzzles (my view), improve empirical adequacy (Bas van Frassen's), or increase the dominance of the ruling elite (in parody, the strong program's). I do, of course, have my own preference among these alternatives, and it makes a difference (Kuhn, 1983b). But no choice between them is relevant to what's presently at stake.

In comparative judgements of the kind just sketched, shared beliefs are left in place: they serve as the given for purposes of the current evaluation; they provide a replacement for the traditional Archimedean platform. The fact that they may — indeed probably will — later be at risk in some other evaluation is simply irrelevant. Nothing about the rationality of the outcome of the current evaluation depends upon their, in fact, being true or false. They are simply in place, part of the historical situation within which this evaluation is made. But if the actual truth value of the shared presumptions required for the evaluation is irrelevant, then the question of the truth or falsity of the changes made or rejected on the basis of that evaluation cannot arise either. A number of classic problems in philosophy of science — most obviously Duhemian holism — turn out on this view to be due not to the nature of scientific knowledge but to a misperception of what justification of belief is all about. Justification does not aim at a goal external to the historical situation but simply, in that situation, at improving the tools available for the job at hand.

To this point I have been trying to firm-up and extend the parallel between scientific and biological development suggested at the end of the first edition of *Structure*: scientific development must be seen as a process driven from behind, not pulled from ahead — as evolution from, rather than evolution towards. In making that suggestion, as elsewhere in the book, the parallel I had in mind was diachronic, involving the relation between older and more recent scientific beliefs about the same or overlapping ranges of natural phenomena. Now I want to suggest a second, less widely perceived parallel between Darwinian evolution and the evolution of knowledge, one that cuts a synchronic slice across the sciences rather than a diachronic slice containing one of them. Though I have in the past occasionally spoken of the incommensurability between the theories of contemporary scientific specialties, I've only in the last few years begun to see its significance to the parallels between biological evolution and scientific development. Those parallels have also been persuasively emphasized recently in a splendid article by Mario Biagioli of UCLA (1990). To both of us they seem extremely important, though we emphasize them for somewhat different reasons.

To indicate what is involved I must revert briefly to my old distinction between normal and revolutionary development. In *Structure* it was the distinction between those developments that simply add to knowledge, and those which require giving up part of what's been believed before. In the new book it will emerge as the distinction between developments which do and developments which do not require local taxonomic change. (The alteration permits a significantly more nuanced description of what goes on during revolutionary change than I've been able to provide before.) During this second sort of change, something else occurs that in *Structure* got mentioned only in passing. After a revolution there are usually (perhaps always) more cognitive specialties or fields of knowledge than there were before. Either a new branch has split off from the parent trunk, as scientific specialties have repeatedly split off in the past from philosophy and from medicine. Or else a new specialty has been born at an area of apparent overlap between two preexisting specialties, as occurred, for example, in the cases of physical chemistry and molecular biology. At the time of its occurrence this second sort of split is often hailed as a reunification of the sciences, as was the case in the episodes just mentioned. As time goes on, however, one notices that the new shoot seldom or never gets assimilated to either of its parents. Instead, it becomes one more separate specialty, gradually acquiring its own new specialists' journals, a new professional society, and often also new university chairs, laboratories, and even departments. Over time a diagram of the evolution of scientific fields, specialties, and sub-specialties comes to look strikingly like a layman's diagram for a biological evolutionary tree. Each of these fields has a distinct lexicon, though the differences are local, occuring only here and there. There is no *lingua franca* capable of expressing, in its entirety, the content of them all or even of any pair.

With much reluctance I have increasingly come to feel that this process of specialization, with its consequent limitation on communication and community, is inescapable, a consequence of first principles. Specialization and the narrowing of the range of expertise now look to me like the necessary price of increasingly powerful cognitive tools. What's involved is the same sort of development of special tools for special functions that's apparent also in technological practice. And, if that is the case, then a couple of additional parallels between biological evolution and the evolution of knowledge come to seem especially consequential. First, revolutions, which produce new divisions between fields in scientific development, are much like episodes of speciation in biological evolution. The biological parallel to revolutionary change is not mutation, as I thought for many years, but speciation. And the problems presented by speciation (e.g., the difficulty in identifying an episode of speciation until some time after it has occurred, and the impossibility, even then, of dating the time of its occurrence) are very similar to those presented by revolutionary change and by the emergence and individuation of new scientific specialties.

The second parallel between biological and scientific development, to which I return again in the concluding section, concerns the unit which undergoes speciation (not to be confused with a unit of selection). In the biological case, it is a reproductively isolated population, a unit whose members collectively embody the gene pool which ensures both the population's self-perpetuation and its continuing isolation. In the scientific case, the unit is a community of intercommunicating specialists, a unit whose members share a lexicon that provides the basis for both the conduct and the evaluation of their research and which simultaneously, by barring full communication with those outside the group, maintains their isolation from practitioners of other specialties.

To anyone who values the unity of knowledge, this aspect of specialization — lexical or taxonomic divergence, with consequent limitations on communication — is a condition to be deplored. But such unity may be in principle an unattainable goal, and its energetic pursuit might well place the growth of knowledge at risk. Lexical diversity and the principled limit it imposes on communication may be the isolating mechanism required for the development of knowledge. Very likely it is the specialization consequent on lexical diversity that permits the sciences, viewed collectively, to solve the puzzles posed by a wider range of natural phenomena than a lexically-homogenous science could achieve.

Though I greet the thought with mixed feelings, I am increasingly persuaded that the limited range of possible partners for fruitful intercourse is the essential precondition for what is known as progress in both biological development and the development of knowledge. When I suggested earlier that incommensurability, properly understood, could reveal the source of the cognitive bite and authority of the sciences, its role as an isolating mechanism was prerequisite to the topic I had principally in mind, the one to which I now turn.

This reference to 'intercourse', for which I shall henceforth substitute the term 'discourse', bring me back to problems concerning truth, and thus to the locus of the newly restored bite. I said earlier that we must learn to get along without anything at all like a correspondence theory of truth. But something like a redundancy theory of truth is badly needed to replace it, something that will introduce minimal laws of logic (in particular, the law of non-contradiction) and make adhering to them a precondition for the rationality of evaluations (Horwich 1990). On this view, as I wish to employ it, the essential function of the concept of truth is to require choice between acceptance and rejection of a statement or a theory in the face of evidence shared by all. Let me try briefly to sketch what I have in mind.

Ian Hacking, in an attempt (1982) to denature the apparent relativism associated with incommensurability, spoke of the way in which new "styles" introduce into science new candidates for true/false. Since that time, I've been gradually realizing (the reformulation is still in process) that some of my own central points are far better made without speaking of statements as themselves being true or as being false. Instead, the evaluation of a putatively scientific statement should be conceived as comprising two seldom-separated parts. First, determine the status of the statement: is it a candidate for true/false? To that question, as you'll shortly see, the answer is lexicon-dependent. And second, supposing a positive answer to the first, is the statement rationally assertable? To that question, given a lexicon, the answer is properly found by something like the normal rules of evidence.

In this reformulation, to declare a statement a candidate for true/false is to accept it as a counter in a language game whose rules forbid asserting both a statement and its contrary. A person who breaks that rule declares him or herself outside the game. If one nevertheless tries to continue play, then discourse breaks down; the integrity of the language community is threatened. Similar, though more problematic, rules apply, not simply to contrary statements, but more generally to logically incompatible ones. There are, of course, language games without the rule of non-contradiction and its relatives: poetry and mystical discourse, for example. And there are also, even within the declarative-statement game, recognized ways of bracketing the rule, permitting and even exploiting the use of contradiction. Metaphor and other tropes are the most obvious examples; more central for present purposes are the historian's restatements of past beliefs. (Though the originals were candidates for true/false, the historian's later restatements — made by a bilingual speaking the language of one culture to the members of another — are not.) But in the sciences and in many more ordinary community activities, such bracketing devices are parasitic on normal discourse. And these activities — the ones that presuppose normal adherence to the rules of the true/false game — are an essential ingredient of the glue that binds communities together. In one form or another, the rules of the true/false game are thus universals for all human communities. But the result of applying those rules varies from one speech community to the next. In discussion between members of communities with differently structured lexicons, assertability and evidence play the same role for both only in areas (there are always a great many) where the two lexicons are congruent.

Where the lexicons of the parties to discourse differ, a given string of words will sometimes make different statements for each. A statement may be a candidate for truth/falsity with one lexicon without having that status in the others. And even when it does, the two statements will not be the same: though identically phrased, strong evidence for one need not be evidence for the other. Communication breakdowns are then inevitable, and it is to avoid them that the bilingual is forced to remember at all times which lexicon is in play, which community the discourse is occurring within.

These breakdowns in communication do, of course, occur: they're a significant characteristic of the episodes *Structure* referred to as 'crises'. I take them to be the crucial symptoms of the speciation-like process through which new disciplines emerge, each with it own lexicon, and each with its own area of knowledge. It is by these divisions, I've been suggesting, that knowledge grows. And it's the need to maintain discourse, to keep the game of declarative statements going, that forces these divisions and the fragmentation of knowledge that results.

I close with some brief and tentative remarks about what emerges from this position as the relationship between the lexicon — the shared taxonomy of a speech community — and the world the members of that community jointly inhabit. Clearly it cannot be the one Putnam (1977, pp. 123-38) has called metaphysical realism. Insofar as the structure of the world can be experienced and the experience communicated, it is constrained by the structure of the lexicon of the community which inhabits it. Doubtless some aspects of that lexical structure are biologically determined, the products of a shared phylogeny. But, at least among advanced creatures (and not just those linguistically endowed), significant aspects are determined also by education, by the process of socialization, that is, which initiates neophytes into the community of their parents and peers. Creatures with the same biological endowment may experience the world through lexicons that are here and there very differently structured, and in those areas they will be unable to communicate all of their experiences across the lexical divide. Though individuals may belong to several interrelated communities (thus, be multilinguals), they experience aspects of the world differently as they move from one to the next.

Remarks like these suggest that the world is somehow mind-dependent, perhaps an invention or construction of the creatures which inhabit it, and in recent years such suggestions have been widely pursued. But the metaphors of invention, construction, and mind-dependence are in two respects grossly misleading. First, the world is not invented or constructed. The creatures to whom this responsibility is imputed, in fact, find the world already in place, its rudiments at their birth and its increasingly full actuality during their educational socialization, a socialization in which examples of the way the world is play an essential part. That world, furthermore, has been experientially given, in part to the new inhabitants directly, and in part indirectly, by inheritance, embodying the experience of their forebears. As such, it is entirely solid: not in the least respectful of an observer's wishes and desires; quite capable of providing decisive evidence against invented hypotheses which fail to match its behavior. Creatures born into it must take it as they find it. They can, of course, interact with it, altering both it and themselves in the process, and the populated world thus altered is the one that will be found in place by the generation which follows. The point closely parallels the one made earlier about the nature of evaluation seen from a developmental perspective: there, what required evaluation was not belief but change in some aspects of belief, the rest held fixed in the process; here, what people can effect or invent is not the world but changes in some aspects of it, the balance remaining as before. In both cases, too, the changes that can be made are not introduced at will. Most proposal for change are rejected on the evidence; the nature of those that remain can rarely be foreseen; and the consequences of accepting one or another of them often prove to be undesired.

Can a world that alters with time and from one community to the next correspond to what is generally referred to as "the real world"? I do not see how its right to that title can be denied. It provides the environment, the stage, for all individual and social life. On such life it places rigid constraints; continued existence depends on adaptation to them; and in the modern world scientific activity has become a primary tool for adaptation. What more can reasonably be asked of a real world?

In the penultimate sentence, above. the word 'adaptation' is clearly problematic. Can the members of a group properly be said to adapt to an environment which they are constantly adjusting to fit their needs? Is it the creatures who adapt to the world or does the world adapt to the creatures? Doesn't this whole way of talking imply a mutual plasticity incompatible with the rigidity of the constraints that make the world real and that made it appropriate to describe the creatures as adapted to it? These difficulties are genuine, but they necessarily inhere in any and all descriptions of undirected evolutionary processes. The identical problem is, for example, currently the subject of much discussion in evolutionary biology. On the one hand the evolutionary process gives rise to creatures more and more closely adapted to a narrower and narrower biological niche. On the other, the niche to which they are adapted is recognizable only in retrospect, with its population in place: it has no existence independent of the community which is adapted to it. (Lewontin 1978.) What actually evolves, therefore, is creatures and niches together: what creates the tensions inherent in talk of adaptation is the need, if discussion and analysis are to be possible, to draw a line between the creatures within the niche, on the one hand, and their "external" environment, on the other.

Niches may not seem to be worlds, but the difference is one of viewpoint. Niches are where *other* creatures live. We see them from outside and thus in physical interaction with their inhabitants. But the inhabitants of a niche see it from inside and their interactions with it are, to them, intentionally mediated through something like a mental representation. Biologically, that is, a niche is the world of the group which inhabits it, thus constituting it a niche. Conceptually, the world is *our* representation of *our* niche, the residence of the particular human community with whose members we are currently interacting.

The world-constitutive role assigned here to intentionality and mental representations recurs to a theme characteristic of my viewpoint throughout its long development: compare my earlier recourse to gestalt switches, seeing as understanding, and so on. This is the aspect of my work that, more than any other, has suggested that I took the world to be mind-dependent. But the metaphor of a mind-dependent world — like its cousin, the constructed or invented world — proves to be deeply misleading. It is groups and group-practices that constitute worlds (and are constituted by them). And the practice-in-the-world of some of those groups *is* science. The primary unit through which the sciences develop is thus, as previously stressed, the group, and groups do not have minds. Under the unfortunate title, "Are species individuals?", contemporary biological theory offers a significant parallel (Hull, 1976, provides an especially useful introduction to the literature). In one sense the procreating organisms which perpetuate a species are the units whose practice permits evolution to occur. But to understand the outcome of that process one must see the evolutionary unit (not to be confused with a unit of selection) as the gene pool shared by those organisms, the organisms which carry the gene pool serving only as the parts which, through bi-sexual reproduction, exchange genes within the population. Cognitive evolution depends, similarly, upon the exchange, through discourse, of statements within a community. Though the units which exchange those statements are individual scientists, understanding the advance of knowledge, the outcome of their practice, depends upon seeing them as atoms constitutive of a larger whole, the community of practitioners of some scientific specialty.

The primacy of the community over its members is reflected also in the theory of the lexicon, the unit which embodies the shared conceptual or taxonomic structure that holds the community together and simultaneously isolates it from other groups. Conceive the lexicon as a module within the head of an individual group member. It can then be shown (though not here) that what characterizes members of the group is possession not of identical lexicons, but of mutually congruent ones, of lexicons with the same structure. The lexical structure which characterizes a group is more abstract than, different in kind from, the individual lexicons or mental modules which embody it. And it is only that structure, not its various individual embodiments, that members of the community must share. The mechanics of taxonomizing are in this respect like its function: neither can be fully understood except as grounded within the community it serves.

By now it may be clear that the position I'm developing is a sort of post-Darwinian Kantianism. Like the Kantian categories, the lexicon supplies preconditions of possible experience. But lexical categories, unlike their Kantian forebears, can and do change, both with time and with the passage from one community to another. None of those changes, of course, is ever vast. Whether the communities in question are displaced in time or in conceptual space, their lexical structures must overlap in major ways or there could be no bridgeheads permitting a member of one to acquire the lexicon of the other. Nor, in the absence of major overlap, would it be possible for the members of a single community to evaluate proposed new theories when their acceptance required lexical change. Small changes, however, can have large-scale effects. The Copernican Revolution provides especially well-known illustrations.

Underlying all these processes of differentiation and change, there must, of course, be something permanent, fixed, and stable. But, like Kant's *Ding an sich*, it is ineffable, undescribable, undiscussible. Located outside of space and time, this Kantian source of stability is the whole from which have been fabricated both creatures and their niches, both the "internal" and the "external" worlds. Experience and description are possible only with the described and describer separated, and the lexical structure which marks that separation can do so in different ways, each resulting in a different, though never wholly different, form of life. Some ways are better suited to some purposes, some to others. But none is to be accepted as true or rejected as false; none gives privileged access to a real, as against an invented, world. The ways of being-in-the-world which a lexicon provides are not candidates for true/false.

References

Biagioli, M. (1990), "The Anthropology of Incommensurability," *Studies in History and Philosophy of Science* 21: 183-209.

Hacking, I. (1982), "Language, Truth and Reason," in *Rationality and Relativism,* M. Hollis and S. Lukes (eds.). Cambridge: MIT Press, pp. 49-66.

Horwich, P. (1990), *Truth.* Oxford: Blackwell.

Hull, D.I. (1976), "Are Species Really Individual?", *Systematic Zoology* 25:174-191.

Kuhn, T.S. (1983a), "Commensurability, Comparability, Communicability," *PSA 1982, Volume Two.* East Lansing: Philosophy of Science Association, pp. 669-688.

_____. (1983b), "Rationality and Theory Choice," *Journal of Philosophy,* 80: 563-570.

_____. (1987), "What are Scientific Revolutions?" in *The Probabilistic Revolution, Volume 1: Ideas in History,* L. Krüger, L.J. Daston, and M. Heidelberger (eds.). Cambridge: MIT Press, pp. 7-22.

_____. (1990), "Dubbing and Redubbing: the Vulnerabiltity of Rigid Designation," in *Scientific Theories,* Minnesota Studies in the Philosophy of Science, XIV, C.W. Savage (ed.). Minneapolis: University of Minnesota Press, pp. 298-318,

Lyons, J. (1977), *Semantics, Volume I.* Cambridge: Cambridge University Press.

Lewontin, R.C. (1978), "Adaptation," *Scientific American* 239: 212-30.

Putnam, H. (1978), *Meaning and the Moral Sciences.* London: Routledge.

Part II

JOHN S. BELL SESSION:
STATE VECTOR REDUCTION

A TRIBUTE TO JOHN S. BELL

Among the people whom I have met personally there was no one whom I revered more than John Bell. He was a scientist of great intellectual power, as I know first hand by studying his papers on the foundations of quantum mechanics and hearing his lectures on the subject. From others whose judgment I trust comes the testimony that his work on quantum field theory and elementary particle theory was also deep and original. Furthermore, although he was not a professional philosopher, he was deeply concerned with philosophical questions concerning the nature of the physical world and the character of human knowledge. He was aware of the philosophical implications of his own scientific work, and his explicit philosophical comments were penetrating, pungent, judicious, and wonderfully sensible. What was most remarkable about John Bell was the fact that his character was as great as his intellect. He was simple, direct, free from self-aggrandizement, intensely dedicated to finding the truth, and completely honest. He was on occasion a sharp critic, as the other speakers have testified, but his sharpness never arose from a lapse in generosity or from a desire to gain the upper hand, but rather from his impatience with arguments that in his opinion obscured the truth. Bell's Theorem, for which he is most famous, was more a triumph of character than of intellect. It was not difficult to prove mathematically. The difficult thing about it was the realization of what was understood and what was not understood in the discussions of hidden variables. Bell's honesty about his own understanding provided the impetus for his formulation and proof of the theorem. It is very sad to realize that we shall have no further revelations and instruction from him.

Abner Shimony

I. Dynamical Reduction Theories: Changing Quantum Theory so the Statevector Represents Reality[1]

GianCarlo Ghirardi and Philip Pearle

University of Trieste and Hamilton College

We dedicate these papers to the memory of John Bell, whose contributions to, support for, and encouragement of the research program described here has meant more than words can say to those involved in it.

1. Describing Reality

In Schrödinger's "cat paradox" example, a nucleus which has a 50% probability of decaying within an hour is coupled to a cat by a "hellish contraption" which, if it detects the decay, will kill the cat. If we take the point of view that what we see around us is real, what occurs in reality is one of the two following evolutions :

$$\text{cat alive} \rightarrow \text{cat alive} \quad \text{OR} \quad \text{cat alive} \rightarrow \text{cat dead} \quad (1.1a, b)$$

($\rightarrow$ means "evolves in an hour into").

According to Schrödinger's equation, the evolution of the statevector describing this situation is

$$|\text{cat alive}> \rightarrow \frac{1}{\sqrt{2}} |\text{cat alive}> + \frac{1}{\sqrt{2}} |\text{cat dead}> \quad (1.2)$$

The right hand side of Eq. (1.2) does not correspond to either the reality on the right hand side of (1.1a) nor to the reality on the right hand side of (1.1b). This has led to various interpretations of just what it is that the right hand side of (1.2) represents (e.g. human knowledge, an infinite ensemble, many worlds, etc.), none of which will be discussed here.

Instead, we will pursue the proposition that the statevector ought to describe the cat. This leads us to modify the Schrödinger equation so that "statevector reduction" takes place, i.e. one of the following two evolutions is obtained:

$$|\text{cat alive}> \rightarrow .99... |\text{cat alive}> + .00... |\text{cat dead}> \quad (1.3a)$$
$$|\text{cat alive}> \rightarrow .00... |\text{cat alive}> + .99... |\text{cat dead}> \quad (1.3b)$$

(.99... is "almost one" and .00... is "almost zero.") We identify (1.3a) as corresponding to the reality (1.1a), and likewise (1.3b) describes the reality (1.1b). The part of the statevector in (1.3a) or (1.3b) that is multiplied by "almost zero" we shall call a "tail."

How can we justify representing a live (or dead) cat by a statevector with a tail? First we must say what the states | cat alive > and | cat dead > are. Their associated wavefunctions describe the particles of the cat as being *mostly* localized in appropriate locations around the center of mass of the live (or dead) cat. We say *mostly* localized because the wavefunctions of these states are small but nonvanishing at arbitrarily large distances from the live (or dead) cat's center of mass: that is the nature of atomic wavefunctions. In spite of this, what allows us to identify the state as e.g. | cat alive > is that the probability distribution of the eigenvalues of an operator A, representing an observable property such as the position or velocity of a part of the cat, is sharply peaked at values consistent with our visual knowledge of a live (or dead) cat as a localized moving (or still) object. But it is immediately clear that this is also true for states with sufficiently small tails such as those in Eqs. (1.3). Thus the tail is just another tiny bit of a wavefunction to add to the rest of the tiny bit of wavefunction far from the cat's center of mass which, while it is part of the representation of reality, has no observational significance (for more on this issue, see II, sections 4-6).

As has been mentioned, when looking at a cat, what allows us to distinguish between its live and dead states is the location of particles that make up the cat. Ghirardi, Rimini and Weber (1985, 1986) suggested as a possible fundamental physical behavior that nature also uses the location of the cat's particles to distinguish between the cat's states, and they introduced a mechanism to implement statevector reduction such as (1.3) according to this criteria (sections 8, 9).

But, even if a tail causes no interpretive problems, why is the tail there? It *is* possible to construct modified Schrödinger equations that describe statevector reduction, with the tail vanishing in a finite time (Pearle 1976, 1979, 1982, 1984b, 1985, 1986a, 1986b). However, these equations are less satisfactory than that of the Continuous Spontaneous Localization (CSL) theory to be described here. For example, they are nonlinear in the statevector whereas the CSL equation is linear (Pearle 1989). Moreover, as shown by Gisin (1984a, 1984b, 1989), they allow superluminal communication, whereas the CSL theory does not. In short, the "best" reduction theory which has been constructed has tails in it.

What is meant by "best" theory? The criteria involve aesthetics (e.g. following Dirac's dictum, to try to get beauty into the equations) and principle (e.g. no superluminal communication), as well as agreement with all tested predictions of quantum theory. One must use these to provide guidance because, although unexplained experimental results have historically provided guidance and motivation for the construction of new theories, they are lacking in the present case. Here the motivation is provided by the perception that the difficulties in finding a satisfactory theory of measurement to accompany quantum theory are a clue that quantum theory should be modified and, following Einstein (1949), that what is needed is a description of objective properties of individual systems, i.e. of reality.

2. What Determines Events?

Since the modified Schrödinger equation's solution must sometimes be one result (e.g. "cat alive") and sometimes another (e.g. "cat dead"), there must be extra variables in the equation which help determine the outcome. In the first theory of this type, due to Bohm and Bub (1966), these were certain Wiener-Siegel "hidden variables" which, as in the Pilot Wave theory of de Broglie-Bohm, provide a more specific description of the physical state of the sytem under consideration than just the statevector. However, the point of view taken here is that the statevector *is* the complete specification of the physical state of the system, and that the extra variables are associated with additional degrees of freedom in nature. After a brief flirtation with random phases (Pearle 1976), the present theories have settled on randomly fluctuating functions of time as the extra variables (Pearle 1979).

In particular, for mathematical simplicity, these functions have been taken to be the derivative of Brownian motion paths, $w(t)=dB(t)/dt$. So we may imagine that some jiggly Brownian paths $B(t)$ lead to "cat alive," while other such paths lead to "cat dead." As we shall see, in the CSL theory the above paths do not exhaust the full set of paths, but the remaining paths have vanishing probability, so they are not of interest.

True Brownian motion $B_0(t)$ is defined by giving not only the paths, but also their probabilities. It can be conveniently characterized as the result of a diffusion process, with diffusion constant λ and with zero drift (i.e. vanishing average velocity over all paths). Then $w_0(t) \equiv dB_0(t)/dt$ is called white noise, and its probability density distribution (which we shall call the "raw" distribution) can be written (in functional integral form) as:

$$P_{raw}(w_0) = C\exp[-(1/2\lambda)\int_{-\infty}^{\infty} w_0^2(t)dt)] \qquad (2.1)$$

This Gaussian probability distribution of $w_0(t)$ at every instant t is completely characterized by the moments

$$<< w_0(t) >> = 0, \quad << w_0(t)w_0(t') >> = \lambda\delta(t-t') \qquad (2.2a, b)$$

($<< F >>$ is the ensemble average of F using P_{raw}). In the CSL theory, the Brownian paths $B(t)$ *look* the same as the paths $B_0(t)$ of true Brownian motion, but they have different probabilities because they are the result of a diffusion process with nonzero drift. Since Eq. (2.2a) does not hold for $w(t)$, the relevant probability distribution of $w(t)$ is not that of Eq. (2.1). However, the diffusion behavior of $w(t)$ is still governed by Eq. (2.2b).

What physical quantity does $w(t)$ represent? It has been suggested by Karolyhazy (1966, 1990) (see also Karolyhazy, Frenkel and Lukacz (1982, 1986), Frenkel (1990) and Penrose (1986)) that fluctuations of the metric may be responsible for statevector reduction. However, we shall treat the fluctuations $w(t)$ phenomenologically, without speculating further about their physical nature.

3. Predictions

How can fluctuations produce the right predictions, i.e. the predictions of quantum theory? For example, for Schrödinger's cat experiment, the statevector (1.3a) (or (1.3b)) must comprise 50% of the solutions of the modified Schrödinger equation. It turns out that this can be achieved remarkably simply, if the reduction dynamics of the modified Schrödinger equation is analogous to the so-called Gambler's Ruin game (Pearle,1982).

Consider two gamblers, A and D, each of whom starts the game with $50. They toss a coin: if it turns up heads, gambler D gives a dollar to gambler A, while if it turns up tails, the dollar goes the other way. During the game, the amount of money possessed by either gambler fluctuates, but one of the gamblers eventually wins. (In a slight modification of this game, necessary to make a more precise analogy to CSL, a gambler down to his last dollar wagers half of it, and if he loses he wagers half of the remainder, etc. Thus the game never actually ends, although after enough time one can clearly identify the winner.) Moreover, if the game is repeated many times, each gambler wins 50% of the games.

Analogously, consider the initial statevector (1.2), in which the squared amplitudes of the states | cat alive > and | cat dead > start with value 1/2 each. The fluctuating function of time corresponds to the coin toss. The squared amplitudes fluctuate during the reduction "game," but one squared amplitude eventually "wins" (almost reaches 1). Moreover, each amplitude wins for 50% of the statevectors in the ensemble.

Crucial to the desired result is that the game be fair or, in mathematician's words, be a Martingale. Suppose we are considering the statevector

$$|\psi,t> = \alpha(t)|a> + \beta(t)|b> \qquad (3.1)$$

which is expressed in terms of the two orthonormal states | a >, | b > which are the eventual outcomes of the game. The desired result, which agrees with the predictions of quantum theory, is that as t•,

$$r(t) \equiv \frac{|\alpha(t)|^2}{|\alpha(t)|^2 + |\beta(t)|^2} \to 1 \text{ for } |\alpha(0)|^2 \text{ of the games.} \qquad (3.2)$$

(and also for the exchange $\alpha \leftrightarrow \beta$). The Martingale condition is that

$$<<r(t)>> = <<r(0)>> \qquad (3.3)$$

i.e., the ensemble expectation value of the ratio r(t) does not change with time. But this implies (3.2), as follows. First, $<<r(0)>> = |\alpha(0)|^2$ (since at t=0 the statevector (3.1) is precisely known and is normalized to 1). Second, $<<r(\infty)>> = 1 \text{x} \text{Prob}[r(\infty)=1]$ (as there is sure to be a winner, then r(∞) either equals 1 or 0, so $<<r(\infty)>> = 1 \text{x} \text{Prob}[r(\infty)=1] + 0 \text{x} \text{Prob}[r(\infty)=0]$). Thus, by Eq. (3.3), evaluated at t=∞, $\text{Prob}[r(\infty)=1] = |\alpha(0)|^2$.

In the CSL theory, the squared amplitudes (more precisely, the ratios r(t)) may be thought of as continuously playing the Gambler's Ruin game[2]. Following the proposal (to be described in sections 8, 9) of Ghirardi, Rimini and Weber (1986) for the physical mechanism underlying reduction, the *rate* at which the game is played turns out to increase as more particles described by state | a > come to occupy distinctly different locations (i.e. when their separation is of the order of $\approx 10^{-5}$ cm. or more) from the same particles described by state | b >. For a macroscopic object containing $\approx 10^{23}$ particles placed in such a superposition of macroscopically spacially distinct states, the reduction time (i.e. the time in which the "almost one" form (1.3a) or (1.3b) is achieved) is quite small, of the order of a microsecond. Thus, the ill-defined instantaneous reduction *rule* of quantum theory is replaced by the well-defined short-time dynamical *evolution* of CSL theory, but the predicted frequencies of outcomes are the same.

When the states | a > and | b > differ only in the location of a microscopic object, say a single particle, the reduction time is quite long, of the order of 10^{16} secs. $\approx 10^8$ years. This means that the reduction process only slightly interferes with the normal quantum evolution of such a particle. However, the statevector evolution *is* different from that predicted by quantum theory. This makes it possible to think of experiments that could distinguish between a dynamical reduction theory like CSL and quantum theory.

As an example, consider the two-slit neutron interference pattern observed by Zeilinger et al. (1984), which agrees with the prediction of quantum theory to $\approx 1\%$ accuracy. According to CSL, once the two neutron packets of identical amplitudes leave the two slits, they occupy positions in space separated by more than 10^{-5} cm., so the amplitudes begin playing the Gambler's Ruin game. By the time the packets reach the detector, they are not quite equal in amplitude. The interference pattern created by many such neutrons is slightly "washed out" compared to the pattern predicted by quantum theory, i.e. the separation between maxima and adjacent minima is not quite as large. The experimental results can be interpreted as showing, if such a reduction mechanism exists, that the characteristic reduction time for the neutron in this experiment must be greater than ≈ 8 secs. (Pearle 1984a, Zeilinger 1986). Since the theory's suggested characteristic time is $\approx 10^{16}$ secs., the results of this experiment are consistent both with CSL and with quantum theory. But this illustrates how an experiment of high accuracy (all that is needed is an improvement by a factor of 10^{15}!) can determine whether reduction dynamics really takes place.

4. Dynamical Mechanisms

Before presenting the CSL dynamical equation, it is useful to discuss some of its important features. Chief among these is that it is linear in the statevector, just like the ordinary Schrödinger equation.

But-but-but you may say, there is a theorem which says that a linear equation *can't* produce statevector reduction. Here is the theorem. Statevector reduction requires

$$\alpha |a\rangle + \beta |b\rangle \rightarrow |a'\rangle \text{ with probability } |\alpha|^2$$
$$\text{OR} \quad \alpha |a\rangle + \beta |b\rangle \rightarrow |b'\rangle \text{ with probability } |\beta|^2 \quad (4.1a, b)$$

where, in a measurement situation, $|a\rangle$ and $|b\rangle$ represent the microscopic system and an untriggered apparatus while $|a'\rangle$ and $|b'\rangle$ represent the evolved microscopic system and the triggered apparatus. In particular, as a special case of (4.1),

$$|a\rangle \rightarrow |a'\rangle \text{ with probability } 1$$
$$\text{AND} \quad |b\rangle \rightarrow |b'\rangle \text{ with probability } 1 \quad (4.2a, b)$$

But if a linear evolution produces (4.2), then that same linear evolution produces

$$\alpha |a\rangle + \beta |b\rangle \rightarrow \alpha |a'\rangle + \beta |b'\rangle \quad \text{with probability } 1 \quad (4.3)$$

which is certainly not the same as (4.1).

So, how does the CSL equation overcome this theorem? First, it violates one of the theorem's tacit premises, that a statevector's norm does not change during the linear evolution. Undoubtedly you took for granted that $\langle a'|a'\rangle = \langle b'|b'\rangle = 1$ when looking at Eqs. (4.2). But, in CSL, Eqs. (4.2) are replaced by

$$|a\rangle \rightarrow |\mathbf{a'}\rangle \quad \text{OR} \quad |a\rangle \rightarrow |a'\rangle \quad \text{with probability } 1 \quad (4.4a,b')$$
$$|b\rangle \rightarrow |b'\rangle \quad \text{OR} \quad |b\rangle \rightarrow |\mathbf{b'}\rangle \quad \text{with probability } 1 \quad (4.4a',b)$$

(the notation employed is $\langle \mathbf{a'}|\mathbf{a'}\rangle \gg \langle b'|b'\rangle$, $|a'\rangle = |\mathbf{a'}\rangle/\langle \mathbf{a'}|\mathbf{a'}\rangle^{1/2}$, $|b'\rangle = |\mathbf{b'}\rangle/\langle \mathbf{b'}|\mathbf{b'}\rangle^{1/2}$, etc.).

Moreover, each fluctuation, in governing its respective linear evolution of $|\psi,t\rangle$, is in one of two classes. Class A fluctuations evolve (4.4a, a') and class B fluctuations evolve (4.4b, b'). Therefore, Eq. (4.3) is replaced by

Class A: $\quad \alpha|a\rangle + \beta|b\rangle \rightarrow \alpha|\mathbf{a'}\rangle + \beta|b'\rangle \quad$ with probability $|\alpha|^2 \quad$ (4.5a)
Class B: $\quad \alpha|a\rangle + \beta|b\rangle \rightarrow \alpha|a'\rangle + \beta|\mathbf{b'}\rangle \quad$ with probability $|\beta|^2 \quad$ (4.5b)

Thus we see how a linear evolution which changes the norm of a statevector can produce reduced statevectors, provided we are comfortable with the idea of tails. For, since the operation is linear, we are permitted to normalize our statevector at any time without affecting its subsequent evolution, and when the right hand side of (4.5a) ((4.5b)) is normalized, the result is essentially $|a'\rangle$ ($|b'\rangle$).

But, more is involved than just linear evolution. In Eqs. (4.5) it is indicated that the fluctuations in class A (B) occur with probability $|\alpha|^2$ ($|\beta|^2$). What determines the probability of a fluctuation $w(t)$ and the statevector $|\psi,t\rangle$ that it evolves from $|\psi,0\rangle$? The linear Schrödinger equation is supplemented by a *probability rule* :

$$P(w) = P_{raw}(w) < \psi,t | \psi,t > \qquad (4.6)$$

That is, the probability of a fluctuation depends in a nonlinear way on the statevector it is evolving. Of course, for Eq. (4.6) to have any meaning, it is necessary that the probabilities $P(w)dw$ sum to 1, so Eq. (4.6) implies that

$$\ll <\psi,t|\psi,t> \gg = 1 \qquad (4.7)$$

Thus the consistency of the probability rule is bound up with the CSL dynamical equation: it must be constructed so that Eq. (4.7) is satisfied.

We close by noting the similarity of the structure of CSL theory to quantum theory, in that both have a linear evolution equation and a probability rule that is quadratic in the statevector. Moreover, in quantum theory, statevectors all have constant norm (which is conveniently taken equal to 1), but those statevectors which differ by only a phase factor are equivalent, while in CSL theory, the norm of a statevector can change, but those statevectors which differ by *any* normalization factor are equivalent.

5. Simplest Example

We now improve the qualitative discussion of CSL dynamics given in the previous section by presenting a simple quantitative example, whose *modified Schrödinger equation*[3] is (Ghirardi, Pearle and Rimini 1990):

$$d|\psi,t\rangle/dt = -iH|\psi,t\rangle + (Aw(t) - \lambda A^2)|\psi,t\rangle \qquad (5.1)$$

For simplicity, we will set the Hamiltonian H equal to zero. A is a Hermitian operator with two eigenvectors $|a\rangle$, $|b\rangle$ with corresponding eigenvalues a, b: we will see that reduction takes place to one of these two eigenvectors.

If the initial statevector is

$$|\psi,0\rangle = \alpha|a\rangle + \beta|b\rangle \qquad (5.2)$$

then the solution to Eq. (5.1) is easily seen to be

$$|\psi,t\rangle = \alpha e^{[aB(t)-\lambda a^2 t]}|a\rangle + \beta e^{[bB(t)-\lambda b^2 t]}|b\rangle \qquad (5.3)$$

Since in this case the solution depends only upon B(t) (and not upon the value of B at any earlier time), the *probability rule* (4.6) gives the probability density of B at time t:

$$P(B) = [(2\pi\lambda t)^{-1/2} \exp{-B^2/2\lambda t}] \langle \psi,t | \psi,t \rangle \qquad (5.4)$$

From Eqs. (5.3), (5.4) we can construct the following table:

CASE	$\alpha|a\rangle + \beta|b\rangle \to \approx$	$P \to \approx$
A: $B(t) = 2\lambda at \pm O(\lambda t)^{1/2}$	$e^{\lambda a^2 t}[\alpha|a\rangle + \beta\Delta|b\rangle]$	$|\alpha|^2 + (|\beta|^2-|\alpha|^2)\Delta^2$
B: $B(t) = 2\lambda bt \pm O(\lambda t)^{1/2}$	$e^{\lambda b^2 t}[\beta|b\rangle + \alpha\Delta|a\rangle]$	$|\beta|^2 + (|\alpha|^2-|\beta|^2)\Delta^2$
C: any other $B(t)$	$?|a\rangle + ?|b\rangle$	0, so who cares?

($\Delta \equiv e^{-\lambda t(a-b)^2}$, ? is an unspecified number, and $\to \approx$ means that the limit for large t is exact for $B(t) = 2\lambda at$ or $2\lambda bt$, but approximate elsewhere in the specified ranges of B.) We see, for example from case A, that B(t) fluctuates so as to be in a range about $2\lambda at$ with probability approaching $|a|^2$. The associated (normalized) statevector approaches $|a\rangle$, with the exponent $\approx \lambda t(a-b)^2$ governing the tail's decrease: *the reduction rate goes faster as the eigenvalue difference increases.* The norm of the statevector grows exponentially with exponent $\approx \lambda t a^2$. Thus, all of the properties of the solutions discussed in section 4 are displayed here.

It is useful also to display the the density matrix, which contains all experimental information. When the normalized pure density matrix $|\psi,t\rangle\langle\psi,t|/\langle\psi,t|\psi,t\rangle$ is multiplied by its probability (5.4) and integrated over all B values, the resulting expression for the density matrix is simply

$$D(t) = \langle\langle\,|\psi,t\rangle\langle\psi,t|\,\rangle\rangle \qquad (5.5)$$

Putting Eq. (5.3) into Eq. (5.5) we obtain the matrix elements of D(t):

$$\langle a|D(t)|a\rangle = |a|^2, \quad \langle b|D(t)|b\rangle = |b|^2, \qquad (5.6a, b)$$
$$\langle a|D(t)|b\rangle = \alpha\beta^* \exp{-(\lambda t/2)(a-b)^2} \qquad (5.6c)$$

which shows how, because the eigenstates of A have different eigenvalues, the initially pure density matrix decays to a mixture due to the reduction process.

It is worth emphasizing that the density matrix behavior in Eq. (5.6) can be achieved without statevector reduction. For this reason, such behavior by itself should not be regarded as sufficient for resolving the quantum measurement problem (Ghirardi, Rimini and Weber 1987). For example, it can readily be shown (Stapp 1989, Ghirardi and Rimini 1990, Ghirardi, Grassi and Pearle 1990a) that the Schrödinger equation

$$d|\psi,t\rangle/dt = -iAw_0(t)|\psi,t\rangle \qquad (5.7)$$

with the usual raw probability distribution (2.1) for $w_0(t)$ (and the statevector $|\psi,t\rangle$ it generates) leads to the same density matrix (5.6). However, the solution to Eq. (5.7),

$$|\psi,t\rangle = \alpha e^{-iaB_0(t)}|a\rangle + \beta e^{-ibB_0(t)}|b\rangle \tag{5.8}$$

remains forever in a superposition of states $|a\rangle$, $|b\rangle$ with unchanging squared amplitudes $|\alpha|^2$, $|\beta|^2$. The decay (5.6c) of the off-diagonal density matrix elements occurs in this case because of the random behavior of the phase factors multiplying each state. Even though there is no experimental distinction between the density matrix evolutions caused by Eqs. (5.1) and (5.7), there is a big conceptual difference. Bell (1990) has strongly criticized, as not knowing the difference between *and* and *or*, those who consider that they have solved the quantum measurement problem by replacing the right hand side of (5.8) by $|a\rangle$ *or* $|b\rangle$ when the off-diagonal density matrix elements have become so small as to be experimentally undetectable, even though each statevector is, as in Eq. (5.8), the sum of $|a\rangle$ *and* $|b\rangle$ with coefficients α, β of comparable magnitude.

6. Nonlinear Equation

There are two other mathematically equivalent ways to present the dynamical equation and probability rule.

It can be shown (Ghirardi, Pearle and Rimini 1990) that the probability rule (4.6) is equivalent to

$$w(t) \equiv dB(t)/dt = w_0(t) + 2\lambda \langle \psi,t|A|\psi,t\rangle/\langle \psi,t|\psi,t\rangle \tag{6.1}$$

The CSL theory is now expressed as two equations, (5.1) and (6.1), which describe the mutual interaction of $|\psi,t\rangle$ and $B(t)$, a form which is suggestive for future development of the theory.

If Eq. (6.1) is substituted into Eq. (5.1), we obtain an equation for $|\psi,t\rangle$, which may then be recast as an equation for the normalized statevector $|\phi,t\rangle \equiv |\psi,t\rangle/\langle\psi,t|\psi,t\rangle^{1/2}$:

$$d|\phi,t\rangle/dt = -iH|\phi,t\rangle \\ + [(A-\langle\phi,t|A|\phi,t\rangle)w_0(t) - \lambda(A-\langle\phi,t|A|\phi,t\rangle)^2]|\phi,t\rangle \\ + \lambda[\langle\phi,t|A^2|\phi,t\rangle - (\langle\phi,t|A|\phi,t\rangle)^2]|\phi,t\rangle \tag{6.2}$$

(This equation was arrived at independently by Gisin (1989). Diosi (1988) considered a special case while modelling the Brownian motion of a quantum particle, and Belavkin (1989) arrived at this as well as the linear equation also from a different perspective, the modelling of a continuous nondemolition measurement.). In Eq. (6.2), the probability of $w_0(t)$ (and the statevector $|\phi,t\rangle$ it generates) is given by the usual raw expression (2.1).

Thus, the linear equation (5.1) for the unnormalized statevector $|\psi,t\rangle$ (which we shall call the "raw" equation) plus the probability rule (4.6) (which we shall call the "cooking procedure") is mathematically equivalent to a nonlinear equation (6.2) for the normalized statevector $|\phi,t\rangle$ (which we shall call the "cooked equation"). (Gatarek and Gisin (1990) have recently given a rigorous discussion of their equivalence.) However, there is a conceptual difference between these two schemes. In the raw scheme, the fluctuating quantity $w(t)$ we see in the world depends upon the state of the world $|\psi,t\rangle$. In the cooked case, the fluctuating quantity is $w_0(t)$, and it does not depend upon the state of the world. In II, section 13 this distinction will be utilized to show that the raw scheme can be, and the cooked scheme cannot be, the basis of a special relativistically invariant theory (Ghirardi, Grassi and Pearle 1990b).

7. Many-Operator Generalization

We need to generalize the dynamical equation (5.1) to incorporate many operators $A_1, A_2, \ldots$ (with each operator having possibly many nondegenerate eigenvalues), each of which commutes with all the others. Then these operators have joint eigenvectors:

$$A_n | a_1, a_2, \ldots \rangle = a_n | a_1, a_2, \ldots \rangle, \quad A_n | b_1, b_2, \ldots \rangle = b_n | b_1, b_2, \ldots \rangle, \text{ etc.} \quad (7.1)$$

and the reduction takes place to one of these eigenvectors.

The dynamical equation (5.1) is replaced by

$$d|\psi,t\rangle/dt = -iH|\psi,t\rangle + \sum (A_n w_n(t) - \lambda A_n^2)|\psi,t\rangle, \quad (7.2)$$

where the $w_n(t)$ are statistically independent random functions. If we set $H=0$ and repeat the calculations as in section 5, in this case the off-diagonal density matrix expression (5.6c) is replaced by

$$\langle a_1, a_2, \ldots | D(t) | b_1, b_2, \ldots \rangle = \alpha\beta^* \exp[-(\lambda t/2) \sum (a_n - b_n)^2] \quad (7.3)$$

From Eq. (7.3) we observe that a difference $|a_n - b_n| \neq 0$ in eigenvalues of just *one* operator A_n causes reduction, but *the reduction proceeds more rapidly* when the eigenvalues of *more* operators differ. This property will play an important role in the CSL theory, to whose ideas we now turn.

8. Choice of Reduction Mechanism

A point of view which has had wide acceptance has been that, in a sense, every measurement (objective event) is a position measurement. It appears that every experiment seems to involve, at some point, a macroscopic collection of particles (which we may dub "a pointer") which occupy distinctly different positions directly correlated to the distinctly different outcomes of the experiment. If the experiment involves our sense perceptions, we remark that they too require, inside of our bodies, different spatial separations of particles which are correlated to different sensations (Aicardi, Borsellino, Ghirardi and Grassi 1991).

This suggests exploration of the hypothesis (Ghirardi, Rimini and Weber 1985), introduced in section 1, that nature chooses which event is to occur based upon discrimination of the spatial location of the involved particles.

This idea can be implemented within the framework already discussed by choosing each operator $A_n \rightarrow A_z$ of section 7 to represent essentially the particle number density in a volume $\approx (10^{-5} \text{ cm.})^3$ around the spatial point z (Pearle 1989). The consequence of this is that a statevector in a superposition of states describing particles in different places will reduce - and with more particles and bigger differences, the faster will the reduction take place.

9. Spontaneous Localization (SL) Theory

The CSL theory incorporates the physical ideas of the SL theory of Ghirardi, Rimini and Weber (1985, 1986, 1987, 1988) (see also Benatti, Ghirardi, Rimini and Weber 1987, Ghirardi and Rimini 1990) that preceded it. The SL theory is most easily explained in terms of an example, of a single particle whose wavefunction consists of two wavepackets, each of width less than $\approx 10^{-5}$ cm., which are spatially separated by a distance greater than $\approx 10^{-5}$ cm.

The GRW idea (see also Bell (1987)) is that this wavefunction evolves according to the ordinary Schrödinger equation, unless it is suddenly "hit," i.e. multiplied by a Gaussian. This instantaneously changes the wavefunction:

$$\psi(x) \rightarrow N^{-1}\psi(x)\exp(-(\alpha/2)(x-z)^2) \tag{9.1}$$

The constant N is chosen so that the wavefunction after the hit is normalized to 1. The width of the Gaussian is $\alpha^{-1/2} \approx 10^{-5}$ cm. The center of the Gaussian z is located randomly, such that the probability of a hit with the center lying in the interval z to z+dz is $\lambda N^2 dz$, where $\lambda^{-1} \approx 10^{16}$ sec. is chosen as the characteristic time for reduction. Since N, and therefore the probability of a hit, is small if the center of the hit lies far from both packets, a hit is most likely to occur with its center within one of the packets. This scarcely affects the packet that is hit, but the other packet, which is multiplied by the tail of the Gaussian, is effectively wiped out.

The evolution equation of the density matrix for a particle (with Hamiltonian set equal to zero) subjected to this hitting process is

$$\partial < x \mid D(t) \mid x' > /\partial t = -\lambda\{1 - \exp(-(\alpha/4)(x-x')^2)\}< x \mid D(t) \mid x' > \tag{9.2}$$

from which can be seen that the density matrix elements decay at a rate $\propto \lambda$ if $|x-x'|>>\alpha^{-1/2}$, but change negligibly over the time scale λ^{-1} if $|x-x'|<<\alpha^{-1/2}$.

For a wavefunction describing n particles comprising a solid which is in a superposition of two states, with the center of mass of the states respectively at x and x', it was shown by GRW that the theory naturally produces a density matrix evolution which is described by Eq. (9.2) as well, except that λ is replaced by $n\lambda$. Thus we see, as stated in section 3, how a single particle wavefunction is only affected over a long time scale λ^{-1}, but the many particle wavefunction is reduced rapidly, on the time scale $(n\lambda)^{-1}$. This increased reduction rate occurs because, for such a many-particle entangled state, the hit of a *single* particle reduces the *whole* wavefunction: since the hit of each particle is an independent event, *some* particle in the n-particle wavefunction is likely to be hit in characteristic time $(n\lambda)^{-1}$.

The SL theory does not provide a modified Schrödinger equation, but that is remedied in the CSL theory. Moreover, the SL hitting process does not maintain the symmetry of a many-particle wavefunction, but that is also remedied in the CSL theory. In both theories, the reduction process will narrow a wavepacket over time and thus increase the particle's energy. Ghirardi, Rimini and Weber (1986) (see also Ghirardi and Rimini 1990) showed that their parameter (α, λ) choices are such that this energy increase is at present unobservable. One may take the position that this energy increase actually exists in nature, or one might regard this energy increase as an undesirable feature of the theory, and use its elimination as a guideline for further development of this evolving set of ideas.

10. Continuous Spontaneous Localization (CSL) Theory

The CSL modified Schrödinger equation for the wavefunction of a single particle moving in one dimension is (Pearle 1989):

$$\partial\psi(x,t)/\partial t = -iH\psi(x,t) + [w(x,t) - \lambda]\psi(x,t) \qquad (10.1)$$

We obtain Eq. (10.1) directly from Eq. (7.2) by replacing the operators A_n by functions $A_z(X)$ of the position operator X:

$$A_n \to A_z(X) \equiv (\alpha/\pi)^{-1/4} \exp(-(\alpha/2)(X-z)^2) \qquad (10.2)$$

whose action on an eigenstate $|x>$ of position is to give a value $\approx (\alpha/\pi)^{1/4}$ for $|x-z| \ll \alpha^{-1/2}$ and a value ≈ 0 for $|x-z| \gg \alpha^{-1/2}$: thus $A_z(X)$ distinguishes between "nearness" and "farness" relative to z. The counterterm $-\lambda \Sigma A_n^2$ in Eq. (7.2) becomes $-\lambda \int A_z(X)^2 dz = -\lambda$ in Eq. (10.1). The random function w(x,t) in Eq. (10.1) is obtained from Eq. (7.2) by replacing $w_n(t)$ there by independent random functions $w_z(t)$ at each point z of space:

$$w(x,t) = \int dz A_z(x) w_z(t) \qquad (10.3a)$$

$$<< w_z(t) >> = 0, \quad << w_z(t) w_{z'}(t') >> = \lambda \delta(t-t') \delta(z-z') \qquad (10.3b)$$

The moments of w(x,t) with respect to the raw probability distribution follow from Eqs. (10.3):

$$<< w(x,t) >> = 0, \quad << w(x,t) w(x',t') >> = \lambda \delta(t-t') \exp(-(\alpha/4)(x-x')^2) \qquad (10.4a,b)$$

Thus, while w(x,t) fluctuates as much positively as negatively (Eq. (10.4a)), its fluctuations tend to be correlated over a distance $\alpha^{-1/2}$ (Eq. (10.4b)).

How does Eq. (10.1) work? It essentially describes a wavefunction undergoing a continuous succession of "little hits." To see this in more detail, we rewrite Eq. (10.1) (with H=0) as

$$\psi(x,t) + w(x,t) dt \psi(x,t) - \lambda dt \psi(x,t) = \psi(x,t+dt) \qquad (10.5)$$

Let $\psi(x,t)$ be the two-packet wavefunction described in the last section. To this wavefunction at time t is added two terms. One term $w(x,t)dt\psi(x,t)$ can be thought of as most likely multiplying the wavefunction by a tiny positive Gaussian which has its center within one of the packets, so as to increase the size of that packet (since the probability rule (4.6) favors fluctuations which enhance the norm of the wavefunction). The other term $-\lambda dt\psi(x,t)$ contributes a tiny decrease in the size of both packets (indeed, the whole wavefunction would exponentially decay if this were the only operating term). The net effect is an increase in the size of the hit packet (the fluctuation's effect is larger than the decay) while the other packet declines due to the decay. The ratios r(t) (see Eq. (3.2)) of squared packet amplitudes fluctuate up and down, playing the Gambler's Ruin game, as one packet eventually predominates over the other with time scale λ^{-1}.

The density matrix evolution equation that follows from the CSL Eq. (10.1) turns out to be identical to that of the SL theory (Eq. (9.2) when H=0), even though the two processes are quite different as far as the evolution of individual statevectors is concerned. However, it can be shown (Ghirardi, Pearle and Rimini 1990, Nicrosini and Rimini 1990) that there always is a hitting model which, in an appropriate limit of infinite hitting frequency, coincides in all respects with any CSL model.

11. CSL Theory for Many Particles

The generalization of Eq. (10.1) to many particles moving in three dimensions is (Pearle 1989)

$$d|\psi,t>/dt = -iH|\psi,t> + [\int dxn(x)w(x,t) - \lambda << \int dx \int dx' n(x) n(x') \Phi(x-x')>>]|\psi,t> \quad (11.1)$$

In Eq. (11.1), $n(x) = a^\dagger(x)a(x)$ is the particle number density operator, $a^\dagger(x)$ is the creation operator for a particle at the position x, w(x,t) is white noise defined by the obvious generalization of Eqs. (10.4) to three dimensions, and $\Phi(x-x') \equiv \exp-\alpha(x-x')^2/4$. Another way of writing Eq. (11.1), which makes the comparison with the expression $\Sigma A_n w_n(t)$ in Eq. (7.2) more transparent, is to replace $\int dx n(x)w(x,t)$ in Eq. (11.1) by $\int dz n_\alpha(z) w_z(t)$ where $n_\alpha(z) \equiv \int dx a^\dagger(x) a(x) A_z(x)$. From the Gaussian expression (10.2) for $A_z(x)$, we see that the operators $n_\alpha(z)$, to whose joint eigenstates the reduction takes place, correspond to roughly the number of particles in a spherical volume $\alpha^{-3/2} \approx (10^{-5} \text{ cm.})^3$ about z, for all z.

To illustrate how Eq. (11.1) works, consider an object such as Schrödinger's cat's tail in a superposition of two states, one upright (state $|\uparrow>$ with initial amplitude α), the other starting to droop a little (state $|\downarrow>$ with initial amplitude β). The density matrix between these two states has the form (7.3)

$$<\uparrow|D(t)|\downarrow> = \alpha\beta^* e^{-\Gamma t} \text{ where } \Gamma \equiv (\lambda/2) \int dz [n\uparrow(z) - n\downarrow(z)]^2 \quad (11.4)$$

where $n\uparrow(z)$, $n\downarrow(z)$ are the eigenvalues of $n(z)$ in these two states. The reduction rate integral (11.4) is calculated to be

$$\Gamma \approx \lambda \alpha^{-3/2} \rho N_u \quad (11.5)$$

where ρ is the density of particles in the tail. N_u is the number of particles in the tail which are "uncovered" in the process of taking two precisely identical overlapping upright tails and moving one of them to the drooping position. Thus, as more particles in the tail are moved, the reduction rate (11.5) increases, achieving its maximum value when the tails do not overlap and N_u equals the number of particles in the tail.

12. Concluding Remarks

The theory presented here has been awarded the sobriquet "serious" by Bell (1990), in that "apparatus" is not "separated off from the rest of the world into black boxes, as if it were not made of atoms and not ruled by quantum mechanics." The statevector represents reality, and the associated wavefunction $\Psi(x_1, x_2,...)$ in configuration space has a special status. It is the projection of the statevector onto the preferred basis $a^\dagger(x_1) a^\dagger(x_2)..|0>$ of eigenstates of the number-in-a-Gaussian-volume operators $n_\alpha(z)$. These localized states are those to which the reduction mechanism continuously strives to reduce the statevector. $|\Psi(x_1, x_2,..)|^2$ may be thought of as representing the density of *stuff* (Bell 1990) of which the world is made.

If the statevector of the world can be written as the direct product of the statevector of a particular particle with the rest of the world, it is not unreasonable to think of the wavefunction of that particle $\psi(x)$ as representing the *stuff* of that particle, and so sustain Schrödinger's original picture of the meaning of the wavefunction. Thus one is, in these circumstances, free to think of a particle as *really* spread out over space. Feynman's (1965) "the *only* mystery" of quantum theory, the two-slit interference pattern, is explained by the picture of *stuff* that *really* goes through *both* slits. Whereas Schrödinger abandoned his picture, when faced with ordinary quantum theory's implication that microscopic objects could be spread out over space, he might not have done so had he known of CSL theory, where the *stuff* of a micro-object collapses to a small region when its position is measured by a macro-object whose *stuff* is (almost) always (mostly) localized.

Notes

[1] One of us (P. P.) would like to acknowledge the hospitality of Prof. Abdus Salam, the International Atomic Energy Agency and UNESCO, as well as financial support from Hamilton College, the Istituto Nazionale di Fisica Nucleare, and the Consorzio per l'Incremento degli Studi e delle Richerche in Fisica dell'Università di Trieste.

[2] Although w(t), which causes the amplitudes to fluctuate, has nonzero drift so that it is not a Martingale, its effect can be expressed in terms of white noise $w_0(t)$ which is a Martingale, and which makes the ratio r(t) a Martingale.

[3] This, and the rest of the stochastic differential equations in this paper are Stratonovich equations, which simply means for us that we may perform calculus manipulations as if B(t) were an ordinary function.

References

Aicardi, F., Borsellino, A., Ghirardi, G. C., and Grassi, R. (1991),"Dynamical Models For State-Vector Reduction: Do They Ensure That Measurements Have Outcomes?", to be published in *Foundations of Physics Letters*.

Belavkin, V. P. (1989), *Physics Letters A* 140: 355-358.

Bell, J. S. (1987), *Schrodinger-Centenary celebration of a polymath*, C. W. Kilmister (ed.). Cambridge: Cambridge University Press.

_ _ _ _ _ . (1990), *Sixty-Two Years of Uncertainty*, Miller, A. (ed.). New York: Plenum, pp. 17-32.

Benatti F., Ghirardi, G. C., Rimini, A. and Weber, T. (1986), *Nuovo Cimento* 100B: 27-41.

Bohm, D. and Bub, J. (1966), *Reviews of Modern Physics* 38: 453-469.

Diosi, L. (1988), *Physics Letters* A132: 233-236.

Einstein, A. (1949), *"Reply to Criticism"* in *Albert Einstein: Philosopher-Scientist*, P.A. Schilpp (ed.), LaSalle, Ill.: Open Court.

Feynman, R.P., Leighton, R.B. and Sands, M. (1965), *The Feynman Lectures on Physics, Vol III*. Mass: Addison Wesley, p.1-1.

Frenkel, A. (1990), *Foundations of Physics* 20: 159-188.

Gatarek, D. and Gisin, N. (1990), "Continuous Quantum Jumps and Infinite-Dimensional Stochastic Equations": preprint.

Ghirardi, G.C., Rimini, A. and Weber, T. (1985), *Quantum Probability and Applications II*, Accardi, L. and von Waldenfels,W. (eds.). Berlin: Springe pp. 223-232.

_____. (1986), *Physical Review* D34: 470-491.

_____. (1987), *Physical Review* D36: 3287-3289.

_____. (1988), *Foundations of Physics* 18: 1-27.

Ghirardi, G. C., Pearle, P. and Rimini, A. (1990), *Physical Review* A42: 78-89.

Ghirardi, G. C., and Rimini, A. (1990), *Sixty-Two Years of Uncertainty*, Miller,A. (ed.). New York: Plenum, pp. 167-191.

Ghirardi, G. C., Grassi, R. and Pearle, P. (1990a), *Foundations of Physics* 20: 1271-1316.

_____.(1990b), "Relativistic Dynamical Reduction Models and Locality", to be published in the *Proceedings of the Symposium on the Foundations of Modern Physics 1990*, Lahti, P. and Mittelstaedt, P. (eds.). Singapore, World Scientific.

Gisin, N. (1984a), *Physical Review Letters* 52: 1657-1660.

_____. (1984b), *Physical Review Letters* 53: 1776.

_____. (1989), *Helvetica Physica Acta* 62: 363-371.

Karolyhazy, F. (1966), *Il Nuovo Cimento* 42A: 390-402.

_____. (1990), *Sixty-Two Years of Uncertainty*, Miller, A. (ed.). New York: Plenum, pp. 215-231.

Karolyhazy, F., Frenkel, A. and Lukacz, B. (1982), *Physics as Natural Philosophy*, Shimony, A. and Feshbach, H. (eds.). Cambridge: M.I.T., pp. 204-239.

_____. (1986), *Quantum Concepts in Space and Time*, Penrose, R. and Isham, C. J. (eds.). Oxford: Clarendon, pp.109-128.

Nicrosini, O. and Rimini, A. (1990), *Foundations of Physics* 20: 1317-1328.

Pearle, P. (1976), *Physical Review* D13: 857-868.

_____. (1979), *International Journal of Theoretical Physics* 48: 489-518.

_____. (1982), *Foundations of Physics* 12: 249-263.

_____. (1984a), *Physical Review* D29: 235-240.

_____. (1984b), *The Wave-Particle Dualism*, Diner, S. et. al (eds.). Dordrecht: Reidel, pp. 457-483.

_____. (1985), *Journal of Statistical Physics* 41: 719-727.

_____. (1986a), *Quantum Concepts in Space and Time*, Penrose, R. and Isham, C. J. (eds.). Oxford: Clarendon, pp. 84-108.

_____. (1986b), *New Techniques in Quantum Measurement Theory*, Greenberger, D. M. (ed.). New York: N.Y. Acad. of Sci., pp. 539-552.

_____. (1989), *Physical Review* A39: 2277-2289.

_____. (1990), *Sixty-Two Years of Uncertainty*, Miller, A. (ed.). New York: Plenum, pp. 193-214.

Penrose, R. (1986), *Quantum Concepts in Space and Time*, Penrose, R. and Isham, C. J. (eds.). Oxford: Clarendon, pp. 129-146.

Stapp, H. (1989), "Noise-Induced Reduction of Wave Packets": preprint LBL-26968.

Zeilinger, A., Gaehler, R., Shull, C. G. and Treimer,W. (1984), *Symposium on Neutron Scattering*, Faber, J. Jr. (ed.). New York: Amer. Inst. of Phys.

Zeilinger, A. (1986), *Quantum Concepts in Space and Time*, Penrose, R. and Isham, C. J. (eds.). Oxford: Clarendon, pp. 16-27.

II. Elements of Physical Reality, Nonlocality and Stochasticity in Relativistic Dynamical Reduction Models[1]

GianCarlo Ghirardi and Philip Pearle

University of Trieste and Hamilton College

In this part we will consider recent attempts to get a relativistic CSL theory. The problem of getting such a generalization, or at least of making plausible that it exists, is of great interest. J. Bell (1990), after having expressed his dissatisfaction with the fundamental lack of precision of the standard formulation of quantum mechanics and the opinion that the only available acceptable alternatives are the Pilot Wave and the Spontaneous Localization schemes has stated: *The big question, in my opinion, is which, if either, of these two precise pictures can be redeveloped in a Lorentz invariant way.* The point of view of this paper is that what is most interesting is not a detailed exposition of the formal aspects of the generalization, but some crucial conceptual points which have come into focus in this attempt. First, it is interesting to discuss the idea of stochastic relativistic invariance (Sections 1, 2). Next we reconsider quantum nonlocality and the related issue of tails in a nonrelativistic context (Sections 3-6) before turning to the fundamental problem of attributing objective properties to physical systems in the light of the relativistic formulation (Sections 7-12). We will also see how, taking into account basic conditions which have to be imposed in connection with the cause-effect relation, the two versions (raw and cooked schemes) of nonrelativistic CSL which have been presented in Part I acquire a completely different conceptual status and, as a consequence, one of them (the cooked scheme) has to be abandoned (Section 13).

1. Stochastic Galilean Invariance

In the previous paper (hereafter referred to as I), the CSL theory was presented. The dynamical equation (I.11.1)

$$d|\psi,t>/dt = -iH|\psi,t> + [\int dx a^\dagger(x)a(x)w(x,t) - \lambda \int dx \int dx' \Phi(x-x') a^\dagger(x)a(x)a^\dagger(x')a(x')]|\psi,t> \quad (1.1)$$

($\Phi(x-x') = \exp[-\alpha(x-x')^2/4]$, $a^\dagger(x)$ is the creation operator for a particle at the position x) describes the evolution of the statevector $|\psi,t>$ which represents physical reality. This equation says that, in addition to evolution under the usual hermitian hamiltonian H, the statevector is subjected to a new physical process in which a randomly fluctuating field w(x,t) "reduces" the statevector, i.e. rapidly turns a superposition of different macroscopic states into a single macroscopic state. A second equation

$$P(w) = P_0(w) < \psi,t \mid \psi,t> \qquad (1.2)$$

where $P_0(w)$ is the gaussian probability distribution characterized by the two moments

$$<<w(x,t)>> = 0, \quad <<w(x,t)w(x',t')>> = \lambda\delta(t-t')\Phi(x-x') \qquad (1.3)$$

which assigns the probability of occurrence to each fluctuation w(x,t) and thereby to its associated statevector $\mid \psi,t>$ completes the specification of the theory.

That this theory is invariant under transformations from one Galilean frame to another may be surmised from the manifestly Galilean invariant form of the density matrix evolution equation

$$\frac{dD(t)}{dt} = -i[H,D(t)] - \frac{\lambda}{2}\int dx \int dx' \Phi(x-x')[a\dagger(x)a(x),[a\dagger(x')a(x'),D(t)]]. \qquad (1.4)$$

However, it is not so obvious, when looking at the dynamical equation (1.1), that the theory is Galilean invariant. Indeed, the fluctuating field w(x,t) is space and time dependent, so the dynamical equation (1.1) *does* change as one time or space translates, rotates or boosts to new Galilean frames. However, the *ensemble* of equations (1.1) corresponding to the ensemble of fluctuations {w(x,t)} is identical in each Galilean frame, since Eqs. (1.2), (1.3) are Galilean invariant (Pearle 1990). It is this that makes the theory stochastic Galilean invariant, by which we mean that the ensemble of predicted results of experiments is frame-independent: after all, that (and not the invariance of any individual equation) is the bottom line.

One last important point should be made. Given a statevector $\mid \psi,T>$ at time T, the ensemble of equations evolves it into an ensemble of statevectors for any time t>T. However, for t<T, there is no ensemble that can evolve into $\mid \psi,T>$, and so for this time range there is no ensemble description of the system (i.e. there is actually (Ghirardi, Rimini and Weber 1988) only Galilean semigroup invariance). This is a physically reasonable restriction when $\mid \psi,T>$ is not precisely known. It is the usual case considered in the quantum theory of measurement, where $\mid \psi,T>$ represents a system prepared by an apparatus that is unspecified, so there is no hope of tracking the statevector backwards in time. (If the statevector and the fluctuation w(x,t) for t<T *are* precisely known, then one can retrodict $\mid \psi,t>$, but then one has a unique description for which the notion of Galilean invariance is problematic.)

2. Stochastic Relativistic Invariance

Construction of a Lorentz invariant theory along the same lines as the Galilean invariant theory meets an immediate obstacle. If a statevector $\mid \psi,\sigma>$ is defined on a hyperplane σ, a description on a boosted hyperplane σ* cannot be achieved. This is because part of σ* is earlier in time than σ so, as in the Galilean case, an appropriate ensemble cannot be defined.

A way out of this difficulty is provided by the Tomonaga-Schwinger formalism. This is a relativistically invariant evolution equation for a statevector defined on a succession of spacelike *hypersurfaces*. We can use it to describe the evolution of a single statevector defined on a hypersurface σ into an ensemble of statevectors described on a hypersurface σ*, such that the succession of hypersurfaces lying between σ and σ* move everywhere forward in time. The evolution equation is

$$\delta| \psi,\sigma >/\delta\sigma(x) = [L_I(x)w(x) - \lambda L_I^2(x)]| \psi,\sigma > \qquad (2.1)$$

Eq. (2.1) is expressed in the interaction picture, where $L_I(x)$ is a Lorentz scalar function of the field operators which evolve according to the usual hamiltonian, $[L_I(x),L_I(x')]=0$ for x spacelike with respect to x', and w(x) is white noise in all four variables x=(x,t):

$$<< w(x) >> = 0, \quad << w(x)w(x') >> = \lambda\delta(t-t')\delta(x-x') \qquad (2.2)$$

The CSL formalism based on Eq. (2.1) is relativistically invariant in that an individual in any Lorentz frame can use it to describe the evolution of a statevector defined on an initial hypersurface σ to an ensemble of statevectors on any hypersurface σ^* that is everywhere forward in time from σ, and an individual in any other Lorentz frame has an identical description (Ghirardi, Grassi and Pearle 1990a).

How does Eq. (2.1) work? Suppose that $L_I(x) = \phi(x)$ is a scalar field that is coupled to a fermion field. Suppose that | ψ,σ > represents a single fermion in a superposition describing two displaced packets. Then, since a virtual scalar field surrounds a fermion, the field amplitude $\phi(x,t)$ at each point x is also in a superposition of contributions from each packet. The reduction mechanism acts on the field amplitude enhancing one field configuration and decreasing the other. Since the fermion is attached to its field configuration, the reduction "brings the fermion along with it", effectively reducing the statevector to a single fermion packet (Pearle, 1990).

However, this model has an unsatisfactory feature. In order to achieve relativistic invariance, it was found necessary in Eq. (2.1) to locally couple the scalar field to the white noise, i.e., there is no nonlocal coupling through a Gaussian as in the nonrelativistic theory. This has the unfortunate consequence that there is an infinite energy production rate per unit volume, instead of the small finite amount as in the nonrelativistic theory. This new kind of ultraviolet divergence occurs because the white noise, in acting to narrow the field amplitude, does so on an arbitrarily short distance scale, creating scalar particles of arbitrarily short wavelength. This behavior can be seen from the density matrix equation corresponding to Eq. (2.1), where σ=t is a succession of parallel hyperplanes:

$$\frac{dD(t)}{dt} = -\frac{\lambda}{2}\int_{-\infty}^{\infty}dx) [\phi(x,t),[\phi(x,t),D(t)]] \qquad (2.3)$$

from which we find, for E≡TrHD, that

$$\frac{dE(t)}{dt} = \frac{\lambda}{2}\int_{-\infty}^{\infty}dx)\delta(0) \qquad (2.4)$$

using $[\phi(x),H]=i\pi(x)$ and $[\phi(x),\pi(x)]=i\delta(0)$.

Although, for the above reasons, we are not yet in possession of a physically completely satisfactory relativistic model, it is still possible to explore some of the consequences of the relativistic reduction behavior of a CSL theory based upon Eq. (2.1), and that is the subject of the remainder of this paper.

3. Properties of a System

Standard quantum mechanics, from the knowledge of the state of a system, allows us to make probabilistic predictions about the outcomes of all possible prospective measurements on that system. Consider a system described by the state vector |Ψ>.

Also consider an observable A with two eigenvalues U and D and corresponding eigenstates |U> and |D>. The probability $P(A=U|\Psi)$ of getting the result U is :

$$P(A=U|\Psi)=|<U|\Psi>|^2 \tag{3.1}$$

In the case in which $P(A=U|\Psi)=1$, we state, following Einstein, Podolsky and Rosen (1935), that the system has the property U. Analogous statements hold concerning property D.

Entangled states of composite systems occur. It is important to recall the implications concerning the possibility of attributing properties to the individual constituents. To this purpose let us consider a system containing two constituents and let us denote by |iU> and |iD>, i=1,2, two orthogonal states in the Hilbert spaces of the constituents. Suppose the composite system is in the entangled state:

$$|\Psi>=\frac{1}{\sqrt{2}}\ [|1U>|2D>-|1D>|2U>] \tag{3.2}$$

As is well known in such a case, even though the system as a whole has properties, the individual constituents have completely lost their individuality. In particular, contrary to what happens in the case of factorized states, one cannot attribute to them the properties corresponding to definite eigenvalues of a complete set of commuting observables.

4. The Property of Being Localized

Among the properties we are interested in, a particular role is played by those which have a local character. To discuss this, let us consider the case of a particle in one dimension in a state |Ψ> which is a linear superposition, with equal weights, of two states |h> and |t>. In this and the next two sections, we restrict the discussion to nonrelativistic quantum theory. In this representation, the states |h> and |t> are associated to the wave functions h(x) and t(x), which are assumed to be different from zero only in two disjoint intervals h and t, respectively. Let us put I=h∪t. According to the previous remarks we can say that, while $P(x \in I|\Psi)=1$, $P(x \in h|\Psi)=P(x \in t|\Psi)=1/2$. The ensuing impossibility, of attributing to the system the property of being at a given place, is a fact that nature has compelled us to accept in the case of microscopic systems. This same fact, however, becomes conceptually embarrassing when the specifications $x \in h$, i.e., the system has the property of being h(ere), and $x \in t$, i.e. the system has the property of being t(here), refer to a macroscopic object. CSL is a theory which is designed just to overcome such a difficulty.

To prepare the ground for the analysis of the following sections, we have to discuss in more detail the property of being at a given place. To this purpose we remark that, in quantum mechanics, a wave function in the coordinate representation never has compact support. To be more precise, even if it vanishes identically outside a certain interval Δ at a certain time, as a consequence of the Schrödinger evolution it is different from zero on the whole x-axis at any subsequent time. With reference to this fact let us consider a particle in one dimension which is described, in the coordinate representation, by the Gaussian wave function

$$\Psi(x)=\exp[-\frac{x^2}{\delta^2}\ ] \tag{4.1}$$

For such a state, the probability of finding the particle in any finite interval (no matter how large) is smaller than 1 and so we are not entitled to assert that the particle has the property of being within it. However, let us consider e.g. the interval $I=(-\Delta,\Delta)$ with $\Delta= k\delta$, and let us evaluate $P(x\in I|\Psi)$. We have

$$P(x\in I|\Psi) =\text{erf}[k]\cong 1-e^{-100}, \text{ for } k=10. \tag{4.2}$$

We can now raise the question: are we allowed to state that the particle has the property of being in the interval I? If one strictly sticks to the previous definition about possessed properties, the answer is no; however we stress that the opposite attitude is actually assumed in all discussions about quantum processes. The natural reason for being compelled to do so derives from the fact that, in measurement processes, the outcomes are always correlated to differing displacements of a "pointer", i.e. a macroscopic number of particles, and the probability of a pointer being in a certain interval never equals 1. If one is to be able to attribute definite properties to such a system at all, one is unavoidably led to enlarge the conditions allowing such attribution to the case in which the probability of getting the considered outcome in a prospective measurement is extremely close to one.

Why did we spend so much time in discussing a seemingly obvious fact? The reason for this is that it is extremely important to stress that the persistence, within CSL, of the tails of wave functions, as well as the position we will take about the attribution of a property to a system (i.e. to relate it to the fact that a definite outcome is "almost absolutely certain") are not specific features (or, as somebody could be inclined to think, difficulties) of CSL theory. They generally occur even within standard quantum mechanics with the postulate of wave packet reduction.

5. Tails in "Measurement Like" Processes

In particular, it is important to make clear the conceptual differences between the problem of the persistence of tails in CSL and a similar issue in the so called "measurement problem" of the standard theory. To raise this issue, suppose each of two states of a microsystem, |M1> and |M2>, can trigger an apparatus initially in its ready state |AR>, thereby inducing the pointer of the apparatus to occupy a position around H or a position around T respectively. Then, when the apparatus is triggered by the linear superposition with equal weights of the two previously considered states, one has as the final state

$$|\Psi> = \frac{1}{\sqrt{2}} [\,|M1>|H>+|M2>|\,T>] \tag{5.1}$$

In the above equation the states |H> and |T> are supposed to be Gaussian functions of width $d<<<<|H-T|$, describing the position X of the centre of mass of the macroscopic pointer. With reference to state (5.1) we remark that, first of all, since it is entangled it does not allow us to speak of individual properties, neither of the system nor of the apparatus. Concerning the apparatus, one notices that, if (5.1) holds, then $P(X\in [H-kD, H+kD]|\Psi)\cong 1/2$. Suppose now one performs an ideal position measurement aimed at ascertaining whether the pointer is in the interval [H-kD,H+kD] and further suppose that wave packet reduction occurs. When the answer yes is obtained in the measurement, the resulting state is

$$|\Psi^*> = \frac{1}{\sqrt{2}} [\,|M1>|H^*>+|M2>|\,T^*>] \tag{5.2}$$

In Eq. (5.2), the coordinate representation of |H*> and |T*> are obtained from those of |H> and |T> by multiplying them by the characteristic function of the interval [H-kD,H+kD]. The state is still entangled; furthermore, since wave packets spread instantaneously, |H*> and |T*> evolve immediately into states |H*'> and |T*'> whose space support covers the whole X-axis. However since the norm $\| |M1\rangle|H*'\rangle \|$ of the first term is much larger than that of the second one, one has:

$$P(X \in H|\Psi *'\rangle) = P(M=M1|\Psi *'\rangle) \cong 1-10^{-100} \quad (5.3)$$

and one legitimately asserts that "the pointer points at H", and also that "the system has the property M1".

6. Tails in CSL

We are now ready to discuss in greater detail the problem of the tails of the wave function within CSL. As explicitly shown in Part I, CSL forbids the persistence of linear superposition of states in which a macroscopic number of particles are in different (with respect to the localization distance) positions. When CSL describes the case just considered, the unique, universal dynamical principle on which it is based leads, in an extremely short time, to a state

$$|\Psi\#\rangle = \frac{1}{\sqrt{2}} [|M1\rangle|H\#\rangle + |M2\rangle|T\#\rangle] \quad (6.1)$$

This state is essentially the same as (5.2), i.e. the one obtained by the adoption of wave packet reduction. The only difference is that, instead of multiplying the wavefunctions $h(x)$ and $t(x)$ corresponding to the states $|H\rangle$ and $|T\rangle$ by the characteristic function of the interval [H-kD,H+kD], CSL effectively multiplies them by a Gaussian of width $1/\sqrt{\alpha}$, centered around H. Therefore, the state $|\Psi\#\rangle$ is extremely well localized within an interval $1/\sqrt{\alpha}$ around the point H, just as are the states $|H*\rangle$ and $|T*\rangle$ which immediately evolve from $|H*\rangle$ and $|T*\rangle$. Again, since $\| |M1\rangle|H\#\rangle \| \gg \| |M2\rangle|T\#\rangle \|$

$$P(x \in H|\Psi\#\rangle) = P(M=M1|\Psi\#\rangle) \cong 1-10^{-100} \quad (6.2)$$

and one can legitimately assert that "the pointer is around H" and that "the system has the property M1".

In the above discussion we have made use of the standard "probability density" interpretation of quantum mechanics within the context of CSL. As already mentioned, however, the CSL scheme, due to its fundamental feature of leading (in extremely short times) to well localized positions (e.g. of the centre of mass of an almost rigid body) seems to allow one to interpret the wave function itself as representing, as suggested by J. Bell (1990), *"the density of stuff of which the universe is made "*. We are well aware that taking such a position requires a detailed analysis and a clarification of many open problems (as J. Bell himself pointed out in a letter to one of us). However it is worth noticing that it becomes more natural to correlate our perceptions about the pointer being at H or at T to an overwhelming unbalance of a "density of stuff" between two regions (possibly in the 3N-dimensional configuration space of the macroscopic body) than to the unbalance of a *probability density* as is done within the standard interpretation.

7. Chance and Determinism

We specify that, from now on, we will consider exclusively the individual level of description (as opposed to the ensemble level of description) of physical processes.

The first important point which needs to be made precise in order to discuss nonlocality in CSL is the distinction between *chance* and *determinism*. In order to do this let us explore, within the standard quantum scheme, the following experiment. We consider, at time t, the system of a free spin-1/2 particle in the eigenstate |x+⟩ of σ_x belonging to the eigenvalue +1. Expressing such a state as the linear superposition of σ_z eigenstates we have, in the usual notation,

$$P(\sigma_z=+1|x+)=P(\sigma_z=-1|x+)=1/2 \qquad (7.1)$$

Suppose that at time $t_1>t$ a measurement of σ_z is performed and the result +1 is obtained. In accordance with the conceptual structure of the theory, which does not allow one to account for the result in terms of some (hidden) parameter, we state that the outcome +1 is due to chance. If we decide to repeat the measurement at a subsequent time $t_2>t_1$, since the state of the system is now |z+⟩ we have:

$$P(\sigma_z=+1|z+)=1 \qquad (7.2)$$

so that we can foresee in advance that the outcome will surely be +1. Correspondingly, this second outcome will be said to be deterministic.

8. Local Properties and Changes of Reference Frames

Let us consider now the system of two free spin-1/2 particles propagating in opposite directions along the x-axis, in the singlet spin state. In Fig. 1 we have drawn the space-time diagram of the process, and we have denoted by 1 and 2 the world lines of the two particles, respectively. Let us denote by $|\Psi(t_1)\rangle$ the state vector at time t_1 in the reference frame of the Figure. With the usual notation for the probability of an outcome, denoting by $\sigma_z(2)$ the spin z-component for particle 2, we have:

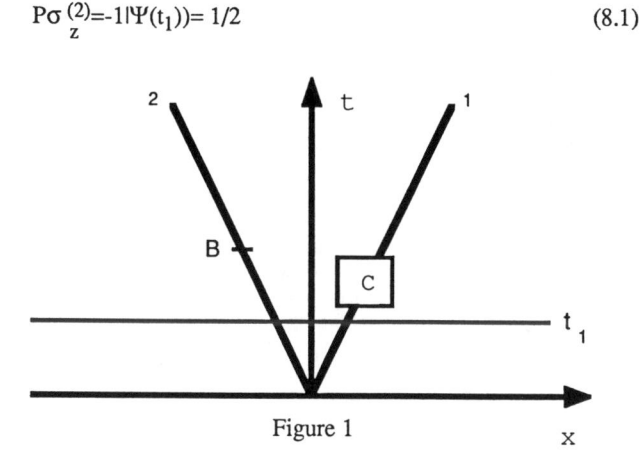

$$P_{\sigma_z^{(2)}=-1}|\Psi(t_1))= 1/2 \qquad (8.1)$$

Figure 1

Suppose now that, in the region C of the figure, there is an apparatus devised to measure the spin z-component of particle 1, and suppose that the outcome +1 is found. Suppose also that wave packet reduction occurs instantaneously. Then, at all times subsequent to the measurement time, and in particular at time t_B which characterizes the t=const. hyperplane going through a point B which is space-like with respect to C, we have

$$P(\sigma_z^{(2)}=-1|\Psi(t_B))= 1 \qquad (8.2)$$

This is one of the aspects (Redhead 1987) of quantum nonlocality: an observable that has no property at all can be given a local property by actions performed at a distance from the property's location.

Since the two events C and B can be arbitrarily far apart in space, even for an observer O' moving with a (correspondingly) arbitrarily small velocity with respect to the observer O of the figure, the time order of C and B can be reversed. One can then raise the question: how does wave packet reduction take place for O'? If the answer is the standard one, i.e. instantaneously at the time $t_C{}'$ which O' attributes to the event C, then observer O' will attribute to the system at his time $t_B{}'$ a statevector $|\Psi'(t_B')\rangle$ which essentially equals $|\Psi(t_1)\rangle$ (because of the small relative velocity of the O and O' reference frames) so that we have (Aharonov and Albert 1980, 1981):

$$P(\sigma_z^{(2)}=-1|\Psi'(t_B'))= 1/2 \qquad (8.3)$$

By comparing Eqs. (8.2) and (8.3) we see that the possibility of attributing a local property to the system at an objective space-time point is ambiguous since it depends on the reference frame. It should be noted that this ambiguity would not be present if B was an event in the future of the event C.

It is interesting to compare the situation just discussed with the situation occurring when there are two measuring apparata, one at C and the other at B. In the reference frame of the figure, i.e. for observer O, the fact that the apparatus B records that $\sigma_z(2)=-1$ is deterministic, while for O' it is due to chance.

9. Local Properties in Relativistic Reduction Models

Within the framework of relativistic CSL discussed in section 2, let us consider a local observable

$$A_I(\sigma) = \int_\sigma f_\alpha(x) F[\Phi_I(x), \partial_\mu \Phi_I(x)] dx) \qquad (9.1)$$

where $f_\alpha(x)$ is a function having compact support on the space-like surface σ. Let us consider now another space-like surface σ^* also containing the support α of $f_\alpha(x)$. We denote by $|\Psi(\sigma)\rangle$ the initial state vector, i.e. the one on the surface σ, and by $|\Psi(\sigma^*)\rangle$ the statevector it evolves into on the surface σ^*, according to the Tomonaga Schwinger evolution equation (2.1) with a particular fluctuating background field. Contrary to what happens in standard quantum field theory, the probability associated to a given eigenvalue of $A_I(\sigma)$ depends on the surface (Ghirardi, Grassi and Pearle, 1990a), i.e. it changes in going from σ to σ^*, and can be (extremely close to) 1 on one hypersurface and (appreciably) different from 1 on the other. Therefore the previously discussed ambiguity about the possibility of attributing definite properties to local observables is present also in relativistic CSL, at the individual level. It should be remarked that no ambiguity is present at the ensemble level, i.e. when the stochastic average of $\langle\Psi|A_I(\sigma)|\Psi\rangle$ is taken it does not change in going from σ to σ^*.

10. Ambiguity about Microproperties

To enlarge upon this issue let us consider two local observables A_1 and A_2 of a particle corresponding to the yes-no experiment ascertaining whether it is around H or T. Suppose in the space-time region around C (see Figure 2) there is an apparatus devised to measure A$_1$. Suppose that "initially" (i.e. on the space-like hyperplane σ$_0$ of the figure) the state of the particle plus apparatus is

$$|\Psi(\sigma_0)\rangle = \frac{1}{\sqrt{2}} \ [|H\rangle + |T\rangle] \ | A1, ready\rangle \qquad (10.1)$$

Figure 2

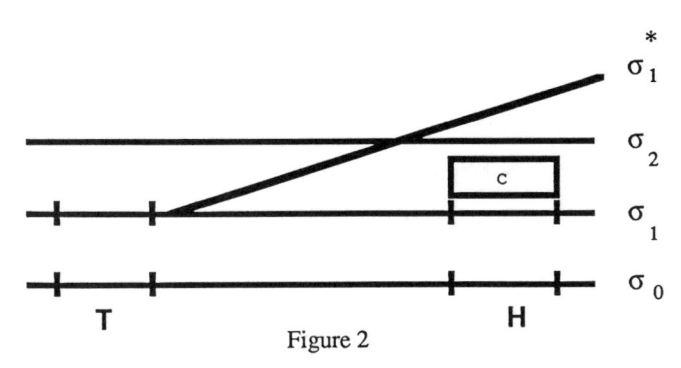

The same state appears on the t=const. hyperplane σ$_1$. Following the evolution from one such hyperplane to the subsequent ones, at time t$_C$ when the region C is crossed, the system interacts with the apparatus. The CSL dynamics leads, according to the specific realization of the fluctuating field which we assume to be one leading to the result that the particle has been detected by the apparatus, to the state $|\Psi(\sigma_2)\rangle$ on the hyperplane σ$_2$:

$$|\Psi(\sigma_2)\rangle = |H\rangle \ |A1, records \ Yes\rangle \qquad (10.2)$$

(Here and in subsequent equations we are discarding the tail for simplicity: see section 12 for an essential comment on this point). According to the evolution equation, however, the same state is associated to the space-like surface σ$_1$* obtained from σ$_1$ by tilting it as indicated in the figure. As usual, let us denote by x the position of the particle. It is worth noticing that, if one consider the evolution from σ$_1$ to σ$_2$, the probability P(x∈ T|$\Psi(\sigma_1)$)=1/2 changes to P(x∈ T| $\Psi(\sigma_2)$)=0. One would therefore be tempted to say that, due to the system-apparatus interaction at C, a property emerged at σ$_2$∩T i.e., in the space region T at the time at which the reduction became effective (i.e. immediately after t$_C$). If, on the contrary, one looks at the evolution from σ$_1$ to σ$_1$*, where the time associated with the space region T is earlier than t$_C$, this implies the same change (from 1/2 to 0) of the probability which, were the same line of reasoning followed, would lead one to say that the property emerged at σ$_1$*∩T=σ$_1$∩T. One then realizes that the theory gives rise to an ambiguity in the attribution of a property at an objective space-time point, which is the exact analogue of the ambiguity discussed in Section. 8 in connection with changes of reference frames.

11. Ambiguities about Macroproperties

We now discuss the analogous physical process when we assume that two measuring apparata are present, devised to measure the local observables A$_1$ and A$_2$ at the same time, in the reference frame characterized by horizontal lines in figure 3. We suppose that the specific realization of the fluctuating field leads to the record: the particle is within the interval H. In the figure we have indicated 4 regions: C1, B1 and C2, B2. B1 and C1 represent the regions where the system apparatus interactions take place with the ensuing triggering of the apparatus. B2 and C2 represent the regions in which the fluctuating field leads to the reduction. Let us write the states associated to the t=const. hyperplanes σ_0, σ_1 and σ_2:

$$|\Psi(\sigma_0)\rangle = \frac{1}{\sqrt{2}} [|H\rangle + |T\rangle] \, |A_1, \text{ready}\rangle |A_2, \text{ready}\rangle \quad (11.1a)$$

$$|\Psi(\sigma_1)\rangle = \frac{1}{\sqrt{2}} [|H\rangle| A_1, \text{records Yes}\rangle|A_2, \text{records No}\rangle$$
$$+ |T\rangle| A_1, \text{records No}\rangle|A_2, \text{records Yes}\rangle] \quad (11.1b)$$

$$|\Psi(\sigma_2)\rangle = |H\rangle| A_1, \text{records Yes}\rangle|A_2, \text{records No}\rangle \quad (11.1c)$$

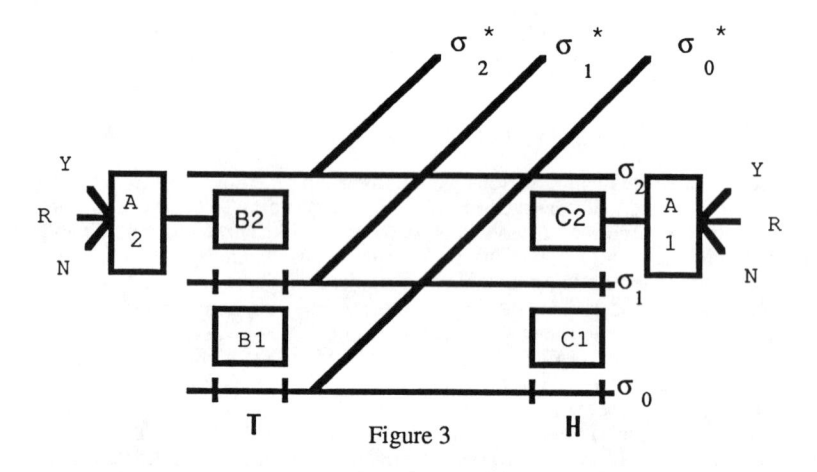

Let us also write the states which are associated to the tilted surfaces σ_0*, σ_1* and σ_2*:

$$|\Psi(\sigma_0*)\rangle = [|H\rangle | A_1, \text{records Yes}\rangle |A_2, \text{ready}\rangle \quad (11.2a)$$

$$|\Psi(\sigma_1*)\rangle = |H\rangle | A_1, \text{records Yes}\rangle |A_2, \text{records No}\rangle \quad (11.2b)$$

$$|\Psi(\sigma_2*)\rangle = |H\rangle | A_1, \text{records Yes}\rangle |A_2, \text{records No}\rangle \quad (11.2c)$$

Figure 3

Comparing these equations we see from (11.1a) and (11.2a) that, on σ_0 and σ_0* and "earlier" hypersurfaces, the state of the apparatus A_2 corresponds to its having the macroscopic property of "pointing at ready", independently of which hypersurface we consider. However, on σ_1 and σ_1* there is an ambiguity about a macroscopic property of the apparatus. In fact, while Eq. (11.1b) does not allow one to attribute a definite position to the A_2 pointer on σ_1, according to (11.2b) the pointer "points at No" on σ_1*, even though both hypersurfaces coincide in the spatial region around T where the A2 pointer is located. However, it should be remarked that, on σ_2 and σ_2* and "later" hypersurfaces, such an ambiguity is no longer present.

Conclusion: according to relativistic CSL there are ambiguities about local properties of physical systems. However, in the case of macro-objects such ambiguities last only for the extremely short characteristic reduction time of the theory.

12. Properties and Macro-objectivism

To be rigourous, the statements made in the previous Section about macrosystems "having definite properties" which have been based on equations like (11.2c) are not strictly correct. In fact, in writing down such a state we have disregarded the tail (of extremely small norm) which, as previously discussed, unavoidably accompanies it. We are then led to adopt the following modified criterion for the attribution of objective local properties to physical systems: *Consider a local observable A with compact support α on a space-like surface and one of its eigenvalues, say a. We state that the physical system has the objective property A=a iff the probability of getting the result a, as a consequence of a system-apparatus interaction, is extremely close to one on any space-like surface containing α.*

Using this criterion and taking into account the features of CSL we can then state that:
- Macroscopic systems practically always have definite macroproperties. This shows, as anticipated, that relativistic CSL allows one to take a macro-objectivist position about natural phenomena.
- No objective local property for a micro-object can emerge as a consequence of a measurement occurring in a space-like separated region. Such property can only emerge after the time necessary for a light signal to propagate from the apparatus to the macro-object has elapsed.
- It is impossible to raise the question whether the result of a measurement, which is space-like separated from another measurement, is due to chance or is determined by the other measurement.
- In any case, no faster than light physical effect can occur.

13. Counterfactuals and Stochastic Features of the Two Versions of CSL

As remarked in Sect. 6 of Part I, there are two mathematically equivalent formulations of nonrelativistic CSL, one being based on the "raw" linear equation (5.1) plus the probability rule (4.6), the other based on the "cooked" nonlinear equation (6.2). We will now analyze their stochastic features by resorting to counterfactual arguments (Ghirardi, Grassi and Pearle 1990b). For this purpose we have to specify the possible worlds we will consider. We will limit our considerations to "worlds" for which the "laws of nature" embodied in the specific CSL version under consideration are assumed to hold. Since we will always deal with situations involving microsystems interacting with macroscopic measuring devices, we will assume that the switching on and off of the system-apparatus interactions are the free variables whose different choices characterize the actual and the alternative worlds.

Within the nonrelativistic context we consider the physical system of a spin-1/2 particle in the state |x+> and two apparata M1 and M2 devised to sequentially measure σ_z. In the actual world, at time t_1, the system-apparatus interaction with coupling constant g_1 triggers M1 and the realization of the fluctuating field is assumed to lead to the result $\sigma_z=+1$. At time t_2 an interaction with coupling constant g_2 triggers M2, and the CSL dynamics leads M2 to record again the result $\sigma_z=+1$. We can now raise the counterfactual question: *For the considered individual case, if we would not have performed the first measurement, would the result of the second measurement have still been* $\sigma_z=+1$?

We can argue as follows. Within the nonlinear equation scheme the specific fluctuation which occurs in the individual process under consideration does not depend on g_1 being equal or different from zero, so we can assert that the same fluctuation (whichever one it is) occurs in the alternative world in which $g_1=0$. Then, from a conceptual point of view we could, in principle, solve the evolution equation with this potential and give a definite answer to the counterfactual question. At this point one has to remark (Ghirardi, Grassi, and Pearle 1990b) that there are fluctuations, among all possible ones which, when substituted in the evolution equation with both g_1 and g_2 different from zero, lead to the result $\sigma_z=+1$ in both measurements but, on the contrary, when used in the same equation with $g_1=0$ give the result $\sigma_z=-1$ in the second measurement (actually such fluctuations have probability 1/4). For such a set of fluctuations, there is an answer to the counterfactual question and it is obviously: No!

If we confine our attention to this class of fluctuations (their identification is conceptually possible although, of course, in practice there is no way to separate them from the rest of the fluctuations) we are allowed to state that there is a cause-effect relation between the switching on of g_1 and the record by M2. In fact, if we denote by a, b the two events:

a- switching on of g_1

b- registering $\sigma_z=+1$ by M2

then a can be made to happen at will; if a then b; if a does not happen then b does not happen either.

The above analysis can be repeated for an EPR-Bohm like set up, leading to the conclusion that, for an appropriate subset of all possible occurrences of the stochastic fluctuations, one is allowed to assert that a measurement at a given space-time point "causes" the outcome of a measurement taking place in a space-like separated region. (Of course, due to the above remark about the impossibility of experimentally isolating this subset of fluctuations from the rest, there is no possibility of superluminal communication.)

In the case of the raw plus cooking scheme the situation is completely different. In particular, within such a formulation, the probability of occurrence of a fluctuation depends, through the cooking procedure, on the overall physical situation and, in particular, on whether or not g_1 vanishes. There is then no possibility of relating the occurrence of the fluctuation in the alternative world to the one of the actual world. As a consequence there is no possibility of identifying, even conceptually, occurrences for which the considered counterfactual question admits an answer, and specifically the negative answer allowing the establishment of a cause-effect relation.

When one comes to discuss the same problem within relativistic CSL one would meet the same difficulty if one could consider the nonlinear equation scheme. Actually, as discussed at the beginning of this paper, in looking for a relativistic dynamics leading to reductions at the individual level, we have felt compelled to use the linear Tomonaga-Schwinger equation plus cooking approach. The formal counterpart of the conceptual difficulties we have just outlined consists in the fact that, if one tries to generalize the nonlinear equation approach to a relativistic scheme utilizing the

Tomonaga-Schwinger equation, one is led to consider a nonintegrable evolution equation.

We can then conclude by pointing out that the use of counterfactual arguments shows that the two schemes have a conceptually different stochastic structure. In particular if one requires that it must not be possible, even conceptually, to accept cause-effect relations between space-like separated macroscopic events (the records of the apparata), one is compelled to choose the linear equation plus cooking scheme and reject the nonlinear scheme. This latter scheme, at any rate, does not allow a relativistic generalization, since it leads to a nonintegrable dynamical equation.

Note

[1] We are deeply indebted to Dr. R. Grassi for useful discussions and stimulating suggestions. This paper II is largely based on research work to which her contribution has been quite relevant. One of us (P. P.) would like to acknowledge the hospitality of Prof. Abdus Salam, the International Atomic Energy Agency and UNESCO, as well as financial support from Hamilton College, the Istituto Nazionale di Fisica Nucleare, and the Consorzio per l'"Incremento degli Studi e delle Richerche in Fisica dell'Università di Trieste.

References

Aharonov, Y., and Albert, D.Z. (1980), *Physical Review* D21, 3316-3324.

_____. (1981), *Physical Review* D24, 359-370.

Bell, J.S. (1990), in*Sixty-Two Years of Uncertainty*, Millert, A. (ed.). New York: Plenum, pp. 17-32

Einstein, A., Podolsky, B., and Rosen, N. (1935), *Physical Review,* 47: 777-780.

Ghirardi. G.C., Rimini, A., and Weber, T. (1988), in *XVII International Colloquium on Group Theoretical Methods in Physics* . Singapore: World Scientific, pp. 648-651.

Ghirardi, G.C., Grassi, R., and Pearle, P. (1990a), *Foundations of Physics* 20: 1271-1316.

_____. (1990b), "Relativistic Dynamical Reduction Models and Locality", to be published in the *Proceedings of the Symposium on the Foundations of Modern Physics 1990*, Lahti, P. and Mittelstaedt, P. (eds). Singapore: World Scientific.

Pearle, P. (1990), in *Sixty-Two Years of Uncertainty*, Miller, A. (ed.). New York: Plenum, pp. 193-214.

Redhead, M. (1987), *Incompleteness, Nonlocality, and Realism: A Prolegomenon to the Philosophy of Quantum Mechanics*. Oxford, Clarendon Press.

Desiderata for a Modified Quantum Dynamics[1]

Abner Shimony

Boston University

1. The Motivation for Modifying Quantum Dynamics

A cluster of problems — the "quantum mechanical measurement problem," the "problem of the reduction of the wave packet," the "problem of the actualization of potentialities," and the "Schrödinger Cat problem" — are raised by standard quantum dynamics when certain assumptions are made about the interpretation of the quantum mechanical formalism. Investigators who are unwilling to abandon these assumptions will be motivated to propose modifications of the quantum formalism. Among these, many (including Professor Ghirardi and Professor Pearle) have felt that the most promising locus of modification is quantum dynamics, and in their presentations (this Volume) they have suggested stochastic modifications of the standard deterministic and linear evolution of the quantum state. Others who have followed this avenue of investigation are F. Károlyházy, A. Frenkel, and B. Lukács (Károlyházy et al. 1982), N. Gisin (1984, 1989), A. Rimini and T. Weber (in Ghirardi et al. 1986), L. Diósi (1988, 1989), and J.S. Bell (1987, pp.201-12). At a workshop at Amherst College in June 1990 Bell remarked that the stochastic modification of quantum dynamics is the most important new idea in the field of foundations of quantum mechanics during his professional lifetime. My own attitude is somewhat cautious and exploratory. The stochastic modification of quantum dynamics ought to be examined intensively, but the possibility should be kept in mind that it may fail, in which case the aforementioned assumptions about the interpretation of quantum mechanics will have to be reassessed. Many, and perhaps all, of the investigators listed above share this exploratory and empiricistic attitude.

It will be useful for later discussion to review briefly the standard dynamics of quantum mechanics and to state its implications for a schematized formulation of the measurement process. According to quantum mechanics, the state of a physical system is represented by a normalized vector of an appropriate Hilbert space (the representation being many-one, for two normalized vectors which are complex scalar multiples of each other represent the same state). If the system is closed, then there is a family of time-dependent unitary (and hence linear) operators $U(t)$, with the property $U(t_1)U(t_2)=U(t_1+t_2)$, such that if $s(0)$ is a vector representing the state of the system at time 0, then the state at an arbitrary time t is represented by $s(t)$, where

$$s(t) = U(t)s(0). \tag{1}$$

Eq. (1), which is slightly more general than the familiar time-dependent Schrödinger equation, is the fundamental dynamical principle of non-relativistic quantum mechanics. Its relativistic counterpart, the Tomonaga-Schwinger equation, will not be needed for our purposes. What is most important for the problem of measurement is the linearity of $U(t)$:

$$U(t)(c_1 s_1 + c_2 s_2 + ... + c_n s_n) = c_1 U(t) s_1 + c_2 U(t) s_2 + ... + c_n U(t) s_n, \tag{2}$$

for any vectors $s_1, s_2,...,s_n$ and any scalars $c_1, c_2,...,c_n$.

Suppose now that an object of interest and an apparatus employed to measure some property of the object together constitute a closed physical system (perturbations from the rest of the universe being negligible). Then by the dynamical principle of quantum mechanics there is a family of unitary operators $U(t)$ governing the temporal evolution of the states of object-plus-apparatus, as in Eqs. (1) and (2). The apparatus is to serve the purpose of revealing the value of a property of the object which is represented by the hermitian operator A, where

$$As_i = a_i s_i \quad (a_i \neq a_j \text{ if } i \neq j) \tag{3}$$

for some basis s_i in the object's Hilbert space. Then there must be some vector v_o in the Hilbert space of the apparatus (representing a "neutral" apparatus state) such that for some time t the vector $U(t)(s_i \otimes v_o)$ is an eigenvector of an operator representing a property of the apparatus, with an eigenvalue b_i from which one can infer a_i. A highly idealized version of this schema of measurement is one in which for each i there is a normalized vector v_i in the Hilbert space of the apparatus such that

$$U(t)(s_i \otimes v_o) = s_i \otimes v_i \tag{4}$$

and

$$Bv_i = b_i v_i \quad (b_i \neq b_j \text{ if } i \neq j). \tag{5}$$

In general, however, the initial state of the object will not be represented by a single one of the eigenvectors of A, but by a superposition of the form

$$s(0) = c_1 s_1 + ... + c_n s_n, \tag{6}$$

with the sum of the absolute squares of the scalar coefficients c_i being unity and with more than one of them non-zero. Then the state of object-plus-apparatus at time t is represented by

$$U(t)((c_1 s_1 + ... + c_n s_n) \otimes v_o) = c_1 U(t)(s_1 \otimes v_o) + ... + c_n U(t)(s_n \otimes v_o), \tag{7}$$

which is a superposition of n vectors, each representing a state in which the property of the apparatus has a different value b_i. It is at this point that the problems mentioned in the first paragraph are revealed. The purpose of a measurement is to obtain information about a property of an object (typically a microscopic object which cannot be directly scrutinized) by means of a correlation established between that property and a property of the apparatus. But in the state represented in Eq. (7) the apparatus property does not have a definite value, and hence the purpose of the measurement has not been achieved. Thus the "measurement problem" is posed. Furthermore, Eq.

(7) shows that the peculiar indefiniteness of a physical property is not confined to microscopic objects (as in Eq. (6)), but is manifested by a property of a macroscopic apparatus on account of the linearity of the dynamical evolution of object-plus-apparatus. In particular, when the notorious experimental arrangement of Schrödinger (1935) is analyzed quantum mechanically, then in the final stage of the experiment it is indefinite whether the cat is alive or dead — the "Schrödinger cat problem." These problems arising in the context of physical measurement may be considered to be special cases of a more general problem of "the actualization of potentialities," for it is obscure how actual events — such as the emission or absorption of photons, or the replication of a macro-molecule, or the firing of a neuron — can occur if quantum dynamics typically gives rise to states in which these events are merely potential because of the indefiniteness of relevant properties.

The following assumptions concerning the interpretation of the quantum mechanical formalism have the consequence of making the foregoing problems so serious that it is difficult to envisage their solution without some modification of the formalism itself. The assumptions themselves are strongly supported by physical and philosophical considerations, and therefore a high price would be paid by sacrificing one of them in order to hedge standard quantum mechanics against modifications.

(i) The quantum state of a physical system is an objective characterization of it, and not merely a compendium of the observer's knowledge of it, nor merely an intellectual instrument for making predictions concerning observational outcomes.

(ii) The objective characterization of a physical system by its quantum state is complete, so that an ensemble of systems described by the same quantum state is homogeneous, without any differentiations stemming from differences in"hidden variables."

(iii) Quantum mechanics is the correct framework theory for all physical systems, macroscopic as well as microscopic, and hence it specifically applies to measuring apparatuses.

(iv) At the conclusion of the physical stages of a measurement (and hence, specifically, before the mind of an observer is affected), a definite result occurs from among all those possible outcomes (potentialities) compatible with the initial state of the object.

I shall very briefly point out how these assumptions preclude some of the proposals that have been made for solving the problem of measurement and related problems. Assumption (i) stands in the way of an instrumentalist interpretation of the quantum mechanical formalism. Such an interpretation could accommodate an expression of the form of Eq. (7), with many terms corresponding to different observational outcomes, just as well as a characterization of the final state of object-plus-apparatus by a single term; either expression would merely be an instrument for anticipating an observational outcome or the probabilities of various outcomes. Some arguments against such an instrumentalist interpretation are given in Shimony (1989), along with references to other discussions. Assumption (ii) rejects a hidden variables interpretation of quantum mechanics, according to which the indefiniteness of the value a_i in Eq. (6) and of b_i in Eq. (7) applies only to ensembles and not to individual members of the ensembles. The main consideration in favor of Assumption (ii) is the incompatibility proved by Bell (1987, pp.14-21 and 29-39) between quantum mechanics and local hidden variables theories, but Bell himself emphasizes that there is still an option of non-local hidden variables theories, which he does not regard as completely repugnant (1987, pp.173-80). Assumption (iii) rules out all variants of the Copenhagen interpretation, which rejects the impasse of Eq. (7) by rejecting the application of quantum mechanics to the apparatus of measurement. In favor of this Assumption is the immense success of the general physical program of understanding macroscopic systems in terms of microscopic parts, conjoined with the immense success of quantum mechanics in the microscopic domain. Wigner (1971, pp.14-15) emphasizes particularly that we have no theory at present for dealing with the interaction of a quantum system and a classical system. Finally, Assumption (iv) precludes the "many-worlds" interpretation, in which all terms on the right hand side of Eq. (7) are considered to be equally real. The conceptual difficulties of this point of view, which effectively denies the distinction between actuality and potentiality, has been analyzed by many writers, for example Bell (1987, pp.93-100) and Stein (1984).

Henceforth in this paper I shall not question Assumptions (i)-(iv), even though, as stated in the first paragraph, an eventual re-assessment is not ruled out. Given these Assumptions, however, one finds Eq. (7) intolerable as a description of the final physical stage of the measurement process. There must be a further stage in which a selection is made from the superposition, and that further stage must be physical. A modification of quantum dynamics is thereby required.

2. Proposed Desiderata

   a. *The proposed modification of quantum dynamics should not be restricted to situations of measurement, for such a restriction would inject an anthropocentric element into fundamental physical theory.* This desideratum would preclude von Neumann's (1955, p. 351) postulate of a special process of reduction occurring when a physical variable is measured, unless that postulate could be shown to be a special case of a more general dynamical law. All the authors mentioned in the first paragraph are committed to satisfying this desideratum.

   b. *The modified dynamics must agree very well with quantum dynamics in the domain of sucessful application of the latter.* This desideratum is primarily the demand of experimental adequacy of a proposed new theory, for if standard quantum dynamics makes very accurate predictions of such phenomena as resonance and beats, then the new theory would have to agree closely with quantum dynamics in order to fit these phenomena. One may anticipate, however, some additional content in this desideratum: that the modified dynamics be related to standard quantum dynamics in a systematic way, by some limiting principle, which would be analogous to the "correspondence principle" relating quantum to classical mechanics.

   c. *If the proposed modified dynamics is applied to a measurement situation, it should predict definite outcomes in a "short" time, where the vague word "short" is made quantitative by the known reaction time of the experimental apparatus.* This desideratum strongly favors a stochastic modification of quantum dynamics over a deterministic non-linear modification. If the composite system object-plus-apparatus is governed by a non-linear dynamical equation, then one could not preserve Eq. (7), in which a final superposition mirrors the initial superposition of eigenvectors of the object variable A; and one could easily imagine a continuous dwindling away of all coefficients except one, which asymptotically would approach absolute value unity. But it is very difficult to construct a plausible dynamical equation for which this asymptotic behavior occurs in a finite time interval. Difficult is not impossible, however, and Pearle (1985) actually succeeded in constructing such an equation (but it is tailored to the measurement of a particular object variable, contrary to desideratum a). The non-deterministic "jumps" of a stochastic dynamical theory — whether they are sporadic and finite, as in the Spontaneous Localization theory of Ghirardi, Rimini, and Weber (1986), or infinitesimal, as in the Continuous Spontaneous Localization theory of Pearle (1989) — are a promising means for achieving definite measurement outcomes rapidly.

d. *If a stochastic dynamical theory is used to account for the outcome of a measurement, it should not permit excessive indefiniteness of the outcome, where "excessive" is defined by considerations of sensory discrimination.* This desideratum tolerates outcomes in which the apparatus variable does not have a sharp point value, but it does not tolerate "tails" which are so broad that different parts of the range of the variable can be discriminated by the senses, even if very low probability amplitude is assigned to the tail. The reason for this intolerance is implicit in Assumption (iv) of Section 1. If registration on the consciousness of the observer of the measurement outcome is more precise than the "tail" indicates, then the physical part of the measurement process would not yield a satisfactory reduction of the initial superposition, and a part of the task of reducing the superposition would thereby be assigned to the mind. For this reason, I do not share the acquiescence to broad "tails" that Pearle advocates (1990, pp.203-4), with the concurrence of Bell and Penrose (*ibid.*, p.213, footnote 30).

e. *The modified dynamics should be Lorentz invariant.* This desideratum has not been achieved by any of the proposed stochastic theories, and it evidently will be very difficult to satisfy. A discussion both of the difficulties and of some progress towards solving them is given by Pearle (1990, pp.204-12) and Fleming (1989).

f. *The modified dynamics should not lose the "peaceful coexistence" with special relativity that standard quantum mechanics possesses — that is, the impossibility of capitalizing upon the entanglement of the state of spatially separated systems to send a superluminal message.* Gisin (1989) has shown that a large class of non-linear deterministic modifications of quantum dynamics violate this desideratum. His argument provides a consideration supplementary to desideratum c for preferring stochastic theories.

g. *The modified dynamics should preclude the gestation of Schrödinger's cat, and in general the occurrence — even for a brief time — of states of a system in which a macroscopic variable is indefinite.* This desideratum is less strongly entrenched than the others discussed so far, because one could presumably achieve agreement with our failure to observe such states by supposing that they are highly unstable and decay very rapidly into states where macroscopic variables have sharp variables (to the extent required by desideratum d). Incidentally, it is fascinating, and perhaps fruitful, to consider the experimental search for very short-lived superpositions of radically differing states in mesoscopic systems.

h. *The modified dynamics should be capable of accounting for the occurrence of definite outcomes of measurements performed with actual apparatus, not just with idealized models of apparatus.* The Spontaneous Localization theory of Ghirardi, Rimini, and Weber (1986) has been criticized for not satisfying this desideratum. In the measuring apparatus that they consider, the macroscopic variable which is correlated with a variable of a microscopic object is the center of mass of a macroscopic system, and spontaneous localization ensures that within about ten nanoseconds this variable will be quite sharp. (At the 1990 PSA meeting I incorrectly stated that the macroscopic system had to be rigid in order to obtain such rapid localization, and my error was pointed out by Professor Ghirardi.) Albert and Vaidman (Albert 1990, 156-8) note that the typical reaction of a measuring apparatus in practice is a burst of fluorescent radiation, or a pulse of voltage or current, and these are hard to subsume under the scheme of measurement of the Spontaneous Localization theory.

3. The "Quantum Telegraph": A Promising Locus of Investigation

A great weakness in the investigations carried out so far in search of modifications of quantum dynamics is the absence of empirical heuristics. To be sure there is one grand body of empirical fact which motivates all the advocates of stochastic modifications of quantum dynamics and most of the advocates of non-linear modifications: that is, the occurrence of definite events, and in particular, the achievement of definite outcomes of measurement. But this body of fact is singularly unsuggestive of the details of a reasonable modification of quantum dynamics. What is needed is phenomena which are suggestive and even revelatory. No more promising phenomena for this purpose have been found than the intermittency of resonant fluorescence of a three-level atom.

H. Dehmelt (1975) proposed to study fluorescent radiation from a trapped atom (confined to a small region by techniques of which he was a pioneer) exposed to two laser beams, one labeled "strong" and one "weak." The first is tuned to the frequency of a transition from the ground state 0 to an excited state 1, and the second to the frequency of a transition from 0 to an excited state 2. The 1-0 transition is dipole-allowed, so that the state 1 has a lifetime of about 10^{-8} s, whereas the 2-0 transition is dipole-forbidden and the lifetime of 2 is about 1 s. Dehmelt anticipated that there would be fairly long periods (of the order of a second) in which the atom undergoes cycles of excitation and spontaneous emission about 10^{-8} s in duration. During such a period the radiation from the single trapped atom would be visible to the naked eye; Cook (1990, p. 367) says, "With a 10 x magnifying lens a point source of this strength would appear as bright as one of the stars in the Big Dipper"! Every few seconds, however, Dehmelt conjectured, the atom would absorb a photon from the weak laser beam and would be excited to state 2, where it would remain for a fairly long period, "shelved", in his descriptive term. Consequently, the fluorescent radiation from this three-level atom would be intermittent, with a pattern of alternating light and dark periods that has been described as the "quantum telegraph." (Of course, unlike the dots, dashes, and spaces of Morse telegraphy, the periods of light and dark would be of random durations.)

Dehmelt's reasoning seemed implicitly to accept the idea of quantum jumps from one state to another. It is reminiscent of the old Bohr theory of atomic transitions (1913), though to an advocate of a stochastic modification of quantum dynamics it could be construed as an intimation of the theory of the future. In any case, it was criticized for neglecting the superposition principle and the linearity of quantum dynamics, which seem to be inconsistent with "shelving" (Pegg, Loudon, and Knight 1986). But if the atom is always in a superposition of states 0, 1 , and 2 except when a photon is detected (at which point emission has occurred with certainty and the state is "reset" to 0), then it is straightforward to show that there is negligible probability of a dark period longer by an order of magnitude than the natural lifetime of state 1. It follows that the phenomenon of the quantum telegraph should not appear.

Dehmelt's intuition was confirmed by experiment (Bergquist et al. 1986, Nagourney et al. 1986, Sauter et al. 1986, Itano et al. 1987). These results are among the most dramatic in the history of optics. And they have given rise to a number of sophisticated analyses, attempting to show the consistency of the quantum telegraph with standard quantum mechanics (e.g., Cohen-Tannoudji and Dalilbard 1986, Porrati and Putterman 1989, Erber et al. 1989, Cook 1990). For the most part these analyses agree with each other, but there are some differences in emphasis and detail. I shall summarize the main ideas without examining the differences, since this procedure will suffice for the heuristic purposes of the present paper.

First, the system of interest is taken to be the atom together with the scattered part of the radiation field (which is discriminated sufficiently from the incident laser beams because of the precisely defined directions of these beams). The states of interest will be represented by superpositions of the form

$$|\Psi(t)\rangle = c_1(t)|1\rangle \otimes |0\rangle_F + c_2(t)|2\rangle \otimes |0\rangle_F + \sum c_{kp}(t)|0\rangle \otimes |kp\rangle_F, \quad (8)$$

where $|0\rangle$, $|1\rangle$, $|2\rangle$ respectively represent the ground state and the two relevant excited states of the atom, $|0\rangle_F$ represents the state of the scattered radiation field with no photons, and $|kp\rangle_F$ represents a state with a single photon of wave vector k (of variable direction but with magnitudes restricted by the energies of the 1-0 and 2-0 transitions) and polarization p. These states evolve from an initial state consisting only of the first two terms of Eq. (8), with coefficients $c_1(0)$ and $c_2(0)$.

Second, for simplicity it is assumed that a perfect photo-detector is in place, to respond to any photon in one of the permitted modes kp. A detection of a photon would "reset" the state to a superposition of the first two terms of Eq. (8).

In addition to such normal "positive" measurements, there is a recognition of "null measurements": i.e., the non-detection of a photon by the perfect photo-detector after a time interval which is long compared to the lifetime of the short-lived state 1. Non-detection has the effect of projecting the vector of Eq. (8) onto the two-dimensional space spanned by $|1\rangle \otimes |0\rangle_F$ and $|2\rangle \otimes |0\rangle_F$, so that the last term of Eq. (8) is projected out and the first two terms are preserved but with renormalization.

The questions of exactly when the projection occurs, and what the state looks like before the projection is fully accomplished, are evaded by making use of the enormous difference in order of magnitude of the lifetimes of states 1 and 2. The statistics of light and dark periods are therefore insensitive to answers to these two questions.

Once this projection is accomplished, the usual unitary evolution of the state will automatically account for a rapid diminution of the coefficient $c_1(t)$ relative to $c_2(t)$, thereby greatly extending the period of darkness to a length comparable to the natural lifetime of state 2.

Finally, an epistemic concept of probability is invoked. For example, Porrati and Putterman (1988, p. 3014) write, "In our picture the measurement of a period of time during which no photons are recorded changes our information about the system and thus the wave function. This null measurement increases the probability of successive periods of darkness." Cook (1990, p. 407) uses the locution "Bayesian transitions" to describe the consequences of null measurements, and he contrasts his point of view with Dehmelt's original suggestion as follows: "It is interesting that the quantum formalism attributes electron shelving to the lack of fluorescence, whereas the intuitive picture of the process attributed the lack of fluorescence to electron shelving."

A strenuous objection must be brought against the foregoing scheme of ideas, in spite of the elegance of the theoretical analysis based upon them and the agreement of this analysis with experiment. The scheme takes for granted that a photo-detector definitely has or has not registered the arrival of a photon in a certain interval of time. This assertion does not make a commitment to a definite instant beginning the interval and a definite instant ending it; the time-energy uncertainty relation and the operational uncertainties of the detector can be fully respected. The point is rather that a reduction of the wave packet has been assumed at the level of a macroscopic measuring apparatus, and the analogue of Schrödinger's cat — that is, a superposition of photon detected and photon not detected — has tacitly been excluded. This assumption underlies the Bayesian locutions about probabilities conditional upon the occurrence or non-occurrence of a certain event. Of course, working physicists regularly assume that at the level of macroscopic apparatus the superposition principle does not preclude definite outcomes. The opportunistic employment of the superposition principle in the early stages of a physical process and its suspension at the final stage is, in fact, a part of the ordinary practice of quantum mechanics, and as Bell forcefully reminded us (1990, p. 18), "ORDINARY QUANTUM MECHANICS (as far as I know) IS JUST FINE FOR ALL PRACTICAL PURPOSES." But the purpose of Section 1 of this paper was to review the argument that the opportunistic employment of the superposition principle is not understood from the standpoint of first principles.

My proposal is to avoid a merely "practical" explanation of the quantum telegraph in terms of ordinary quantum dynamics, but instead to let this remarkable phenomenon guide us heuristically to a modified dynamics. Two propositions seem to me to suggest themselves quite strongly. The first is that a stochastic modification of quantum dynamics is a natural way to accommodate the jumps from a period of darkness to a period of fluorescence. The second is that the natural locus of the jumps is the interaction of a physical system with the electromagnetic vacuum. Whether stochasticity is exhibited when the system in question is simple and microscopic, like a single atom, or only when it is macroscopic and complex, like the phosphor of a photo-detector, is not suggested preferentially by the quantum telegraph, for the simple reason that the single trapped atom and the photo-detector are both essential ingredients in the phenomenon. But whichever choice is made points to a stochastic modification of quantum dynamics that has little to do with spontaneous localization. There is hope, therefore, for a stochastic theory that will escape the criticisms leveled by Albert and Vaidman against the localization theories of Ghirardi, Rimini, and Weber, and of Pearle. I must admit, however, that the envisaged theory which I prefer to those of Professors Ghirardi et al. and of Pearle has one serious disadvantage relative to theirs — it does not exist, whereas theirs do!

4. Two Concluding Remarks

The search for a reasonable modification of quantum dynamics was motivated by a cluster of problems arising from the linearity of the standard time evolution operators. The implications of a modified dynamics, however, may reach far beyond the original motivation. In particular, a stochastic modification of quantum dynamics can hardly avoid introducing time-asymmetry. Consequently, it offers an explanation at the level of fundamental processes for the general phenomenon of irreversibility, instead of attempting to derive irreversibility from some aspect of complexity (which has the danger of confusing epistemological and ontological issues). Thus a stochastic modification of quantum dynamics is a promising way to satisfy the thesis of R. Penrose (1986) that the problem of the reduction of the wave packet is inseparable from the problem of irreversibility.

Finally, to the list of eight desiderata listed in Section 2 for a modification of quantum dynamics I want to add a ninth, highly personal one: that a satisfactory theory be found by some one during my lifetime.

Notes

[1]This work was partially supported by the National Science Foundation, Grant No. 8908264.

References

Albert, D.Z.(1990), "On the Collapse of the Wave Function", in *Sixty-Two Years of Uncertainty: Historical, Philosophical, and Physical Inquiries into the Foundations of Quantum Mechanics*, A.I. Miller (ed.). New York: Plenum Press, pp. 153-65.

Bell, J.S. (1987), *Speakable and Unspeakable in Quantum Mechanics*. Cambridge, U.K.: Cambridge University Press.

_ _ _ _ _. (1990), "Against 'Measurement'", in *Sixty-Two Years of Uncertainty: Historical, Philosophical, and Physical Inquiries into the Foundations of Quantum Mechanics*, A.I. Miller (ed.). New York: Plenum Press, pp. 17-31.

Bergquist, J.C., Hulet, R.G., Itano, W.M., and Wineland, J.J., (1986), "Observation of Quantum Jumps in a Single Atom", *Physical Review Letters* 57: 1699-1702.

Bohr, N. (1913), "On the Constitution of Atoms and Molecules", *Philosophical Magazine* 26: 1-25.

Cohen-Tannoudji, C. and Dalibard, J. (1986), "Single-Atom Laser Spectroscopy. Looking for Dark Periods in Fluorescence Light", *Europhysics Letters* 1: 441-8.

Cook, R.J. (1990), "Quantum Jumps", in *Progress in Optics XXVII*, E. Wolf (ed.). Amsterdam: Elsevier Science Publishers., pp. 361-416.

Dehmelt, H. (1975), "Proposed $10^{14}\Delta v$>v Laser Fluorescence Spectroscopy on Tl$^+$Mono-Ion Oscillator II", *Bulletin of the American Physical Society* 20: 60.

Diósi, L. (1988), "Quantum Stochastic Processes as Models for State Vector Reduction", *Journal of Physics* A 21: 2885-98.

_ _ _ _ _. (1989), "Models for Universal Reduction of Macroscopic Quantum Fluctuations", *Physical Review* A 40: 1165-74.

Erber, T., Hammerling, P., Hockney, G., Porrati, M., and Putterman, S. (1989), "Resonance Fluorescence and Quantum Jumps in Single Atoms: Testing the Randomness of Quantum Mechanics", *Annals of Physics* 190: 254-309.

Fleming, G. (1989), "Lorentz Invariant State Reduction, and Localization", in PSA 1988, A. Fine and J. Leplin (eds.). East Lansing, Michigan: *Philosophy of Science Association*.

Ghirardi, G.C., Rimini, A., and Weber, T. (1986), "Unified Dynamics of Microscopic and Macroscopic Systems", *Physical Review* D 34: 470-91.

Gisin, N. (1984), "Quantum Measurements and Stochastic Processes", *Physical Review Letters* 52: 1657-60.

_____. (1989), "Stochastic Quantum Dynamics and Relativity", *Helvetica Physica Acta* 62: 363-71.

Itano, W.M., Bergquist, J.C., Hulet, R.G., and Wineland, D.J. (1987), "Radiative Decay Rates in Hg^+ from Observations of Quantum Jumps in a Single Ion", *Physical Review Letters* 59: 2732-5.

Károlyházy, F., Frenkel, A., and Lukács, B. (1986), "On the Possible Role of Gravity in the Reduction of the Wave Function", in *Quantum Concepts in Space and Time*, R. Penrose and C. Isham (eds.). Oxford: Clarendon Press, pp. 109-28.

Nagourney, W., Sandberg, J., and Dehmelt, H. (1986), "Shelved Optical Electron Amplifier: Observation of Quantum Jumps", *Physical Review Letters* 56: 2797-9.

Neumann, J.v. (1955), *Mathematical Foundations of Quantum Mechanics*. Princeton: Princeton University Press.

Pearle, P. (1985), "On the Time it Takes a State Vector to Reduce", *Journal of Statistical Physics* 41: 719-27.

_____. (1989), "Combining Stochastic Dynamical State-Vector Reduction with Spontaneous Localization", *Physical Review* A 39: 227-39.

_____.. (1990), "Toward a Relativistic Theory of Statevector Reductions", in *Sixty-Two Years of Uncertainty: Historical, Philosophical, and Physical Inquiries into the Foundations of Quantum Mechanics*, A.I. Miller (ed.). New York: Plenum Press, pp. 193-214.

Pegg, D.T., Loudon, R., and Knight, P.L. (1986), "Correlations in Light Emitted by Three-Level Atoms", *Physical Review* A 33: 4085-91.

Penrose, R. (1986), "Gravity and State Vector Reduction", in *Quantum Concepts in Space and Time*, R. Penrose and C. Isham (eds.). Oxford: Clarendon Press, pp. 129-46.

Porrati, M. and Putterman, S. (1989), "Coherent Intermittency in the Resonant Fluorescence of a Multilevel Atom", *Physical Review* A: 3010-30.

Sauter, T., Neuhauser, D., Blatt, R., and Toschek, P.E. (1986), "Observation of Quantum Jumps", *Physical Review Letters* 57: 1696-8.

Shimony, A. (1989), "Search for a Worldview Which Can Accommodate Our Knowledge of Microphysics", in *Philosophical Consequences of Quantum Theory: Reflections on Bell's Theorem*, J.T. Cushing and E. McMullin (eds.). Notre Dame: University of Notre Dame Press, pp. 25-37.

Stein, H. (1984), "The Everett Interpretation of Quantum Mechanics: Many Worlds or None?", *Nous* 18: 635-52.

Wigner, E.P. (1971), "The Subject of Our Discussions", in *Foundations of Quantum Mechanics*, B. d'Espagnat (ed.). New York: Academic Press, pp. 1-19.

Part III

BIOLOGY: THE NON-PROPOSITIONAL SIDE

Are Pictures Really Necessary?
The Case of Sewell Wright's "Adaptive Landscapes"

Michael Ruse

University of Guelph

Biologists are remarkably visual people. Yet, the classics of logical empiricism never raised the general question of scientific illustration. Moreover, one suspects that the silence was, if anything, actively hostile. People did not talk about biological illustration, because they did not judge it to be part of "real science". This enterprise produces statements or propositions, ideally embedded in a formal system. It may be *about* the real world, but it is not in any sense *of* the real world, in being a copy or mirror image. Philosophers recognized that regretfully human weakness demanded the visual. But it was judged at best a prop. (See, for instance, Braithwaite 1953, Hempel 1965, and Bunge 1967; although see also Achinstein 1968. The best discussion of scientific illustration that I know is Rudwick 1976.)

As one who belongs to that growing school of philosophical naturalists, who think that one's philosophy must be informed and in accord with the methodological dictates of science, and that therefore one must be true to the real nature of science — not an idealized preconception — I am made most uncomfortable by this tension between the reality and the theory. At the very least, so major an item as biological illustration demands philosophical attention, whatever one's ultimate conclusion. Here, indeed, I shall look at but one example; although I hope that its great importance in the history of science will justify such selectivity. From among the many candidates, I chose the adaptive landscapes of the great population geneticist Sewell Wright. And the question I ask is: What was/is their status and role within evolutionary biology? (Although I differ from Provine 1986 in my assessment of the virtues of Wright's picture, my debt to him should be apparent on every page.)

1. Adaptive landscapes.

Sewell Wright's first job after leaving graduate school (Harvard) was with the US Department of Agriculture. In 1926 he was appointed to the faculty at Chicago, and it was about this time that he wrote his major paper in evolutionary theory (Wright 1931). Much of the text of this paper is given over to complex mathematics — at least by biological standards, especially by biological standards of the day. Wright concerns himself primarily with the fate of genes in populations, under given conditions of selection, mutation, and so forth, and he is interested in the consequences of population sizes being genuinely finite and thus subject to random factors in breeding (errors of sampling). He is able to show that if population numbers are large enough, and the forces are strong enough, then selection and like factors determine the fates of genes. For instance, a favoured gene or gene combination will establish itself in a population. However, what Wright is able to show also is that if population numbers are small (judged against the other factors), then genes will "drift" either to total elimination or total fixatio— despite counter-forces of selection and the like. Chance becomes a real phenomenon for change.

To illustrate the mathematical points, Wright gave graphs showing possible effects, and these together with the formal conclusions were used to launch Wright's own particular theory of evolutionary change: the "shifting balance" theory. [fig.1] Wright argued that very small populations would suffer from significant drift and rapidly go extinct. However, conversely, large populations under fairly uniform selective pressures would not truly be candidates for any significant change, good or bad — or at least that they could incorporate only *very* slow and stately change. For significant change, within realistic timespans, one needs a more dynamic mechanism. This is provided by the breaking of a species into sub-populations, of a size-order where drift could be effective — but not of a size so small that drift could be too effective! Every now and then such a sub-population would, by chance, come up with a highly adaptive gene complex, and then this combination could take over the species, either by direct selective elimination of rivals or by interbreeding.

Wright's theory transcends his formalisms. It is based on them but is not identical, being more inclusive (more falsifiable, in Popper's terminology). There is nothing in the formalisms about species subdividing, about new adaptive complexes being hit upon, about insufficient time for selection in large groups, and so on. This is added. Significantly, Wright and Fisher agreed on the mathematics, but because Fisher added different non-formal elements, he came up with a very different theory of change. (Most importantly, Fisher 1930 believed that selection in large groups *did* hold the key to evolution.)

Wright's paper, a long paper, appeared in the journal *Genetics*. The next year (1932) he had a wonderful opportunity to promote his theory, because he was asked (by E M East, his doctoral supervisor) to participate in a forum (with Fisher and with the third great theorist, J B S Haldane) at the Sixth International Congress of Genetics, at Cornell. Normally, Wright was as given to long mathematical demonstrations in lectures as he was in print, but here he was forced to keep his presentation very short — and urged to keep it simple. To do this, he dropped the mathematics entirely, presented his shifting balance theory in words (as he had done in his long paper) and backed up his thinking with a new metaphor, which he presented pictorially: the *adaptive landscape*.

Wright wrote, and illustrated, as follows:

If the entire field of possible gene combinations be graded with respect to adaptive value under a particular set of conditions, what would be its nature? Figure 1 [Fig. 2] shows the combinations in the cases of 2 to 5 paired allelomorphs. In the last case, each of the 32 homozygous combinations is at one remove from 5 others, at two removes from 10, etc. It would require 5 dimensions to represent these relations symmetrically; a sixth dimension is needed to represent level of adaptive value. The 32 combinations here compare with 10^{1000} in a species with 1000 loci each represented by 10 allelomorphs, and the 5 dimensions required for adequate representation compare with 9000. The two dimensions of figure 2 are a very inadequate representation of such a field. The contour lines are intended to represent the scale of adaptive value.

One possibility is that a particular combination gives maximum adaptation and that the adaptiveness of the other combinations falls off more or less regularly according to the number of removes. A species whose individuals are clustered about some combination other than the highest would move up the steepest gradient toward the peak, having reached which it would remain unchanged except for the rare occurrence of new favorable mutations.

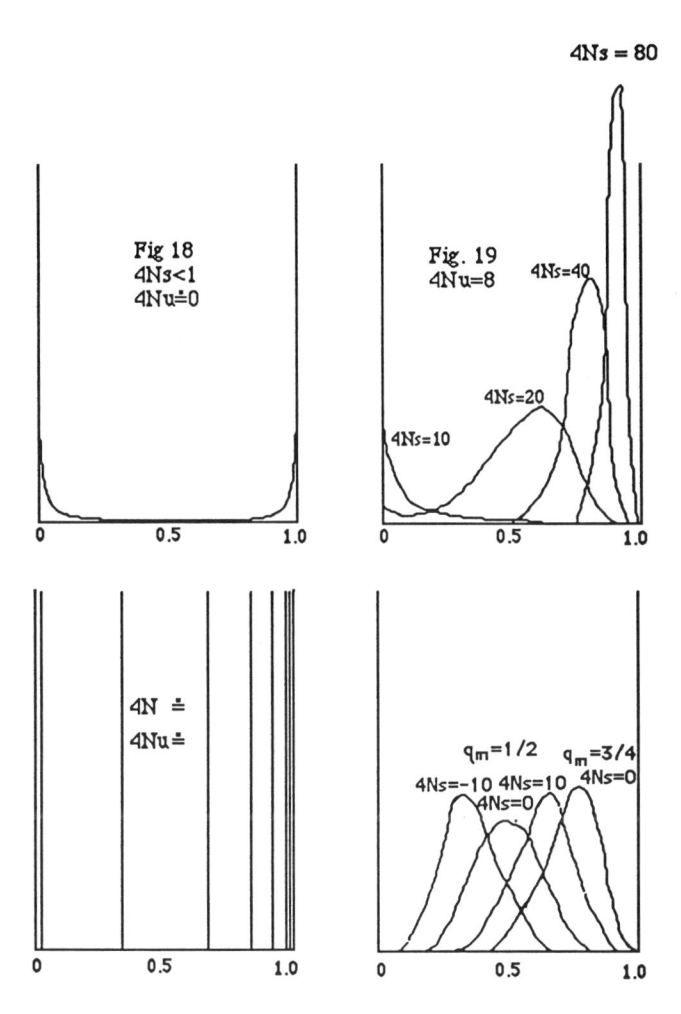

Figure 1. "Figures 18 to 21.–Distributions of gene frequencies in relation to size of population, selection, mutation and state of subdivision. Figure 18. Small population, random fixation or loss of genes ($y=Cq^{-1}(1-q)^{-1}$). Figure 19. Intermediate size of population, random variation of gene frequencies about modal values due to opposing mutation and selection ($y=Ce^{4Nsq}q^{-1}(1-q)^{4Nu-1}$). Figure 20. Large population, gene frequencies in equilibrium between mutation and selection ($q=1-u/s$, etc.). Figure 21. Subdivisions of large population, random variation of gene frequencies about modal values due to immigration and selection. ($y=Ce^{4Nsq}q^{4Nmq_m-1}(1-q)^{4Nm(1-q_m)-1}$)."

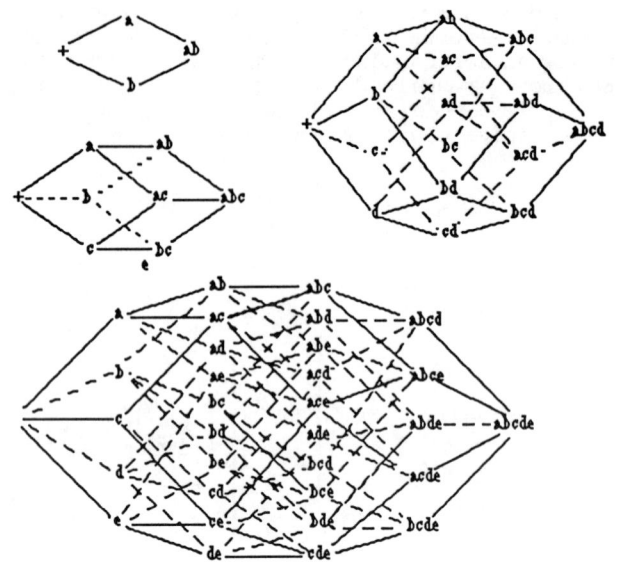

Figure 2. The combinations of from 2 to 5 paired allelomorphs.

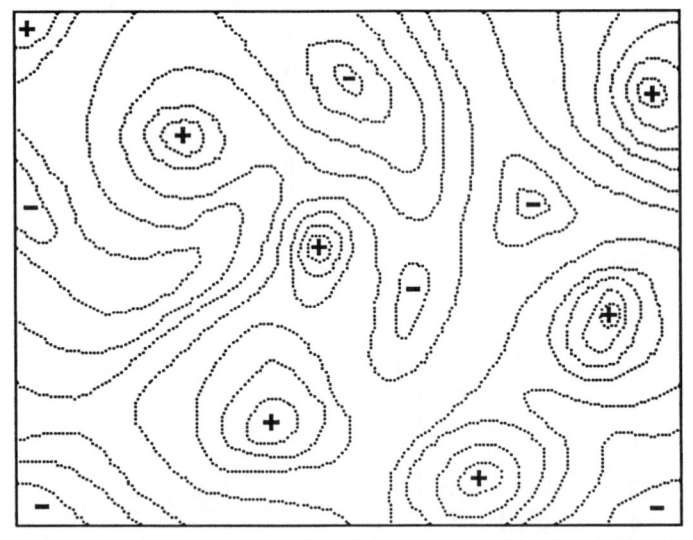

Figure 3. Diagrammatic representation of the field of gene combinations in two dimensions instead of many thousands. Dotted lines represent contours with respect to adaptiveness.

But even in the two factor case (figure 1) [fig 2] it is possible that there may be two peaks, and the chance that this may be the case greatly increases with each additional locus. With something like 10^{1000} possibilities (figure 2) [fig 3] it may be taken as certain that there will be an enormous number of widely separated harmonious combinations. The chance that a random combination is as adaptive as those characteristic of the species may be as low as 10^{-100} and still leave room for 10^{800} separate peaks, each surrounded by 10^{100} more or less similar combinations. In a rugged field of this character, selection will easily carry the species to the nearest peak, but there may be innumerable other peaks which are surrounded by "valleys." The problem of evolution as I see it is that of a mechanism by which the species may continually find its way from lower to higher peaks in such a field. In order that this may occur, there must be some trial and error mechanism on a grand scale by which the species may explore the region surrounding the small portion of the field which it occupies. To evolve, the species must not be under strict control of natural selection. Is there such a trial and error mechanism? (Wright 1932, 162-4)

Next Wright presented (without the mathematical backing) versions of the graphs of gene distribution that had been given in the large paper. [fig 4] He showed visually how drift and other phenomena can occur, given the right specified conditions. Then, using the landscape metaphor, Wright showed how the various options might or might not lead to change, and — as before — he opted for a position that involved a break into small groups, drift, and then reasonably rapid adaptive change in one direction. [fig 5]

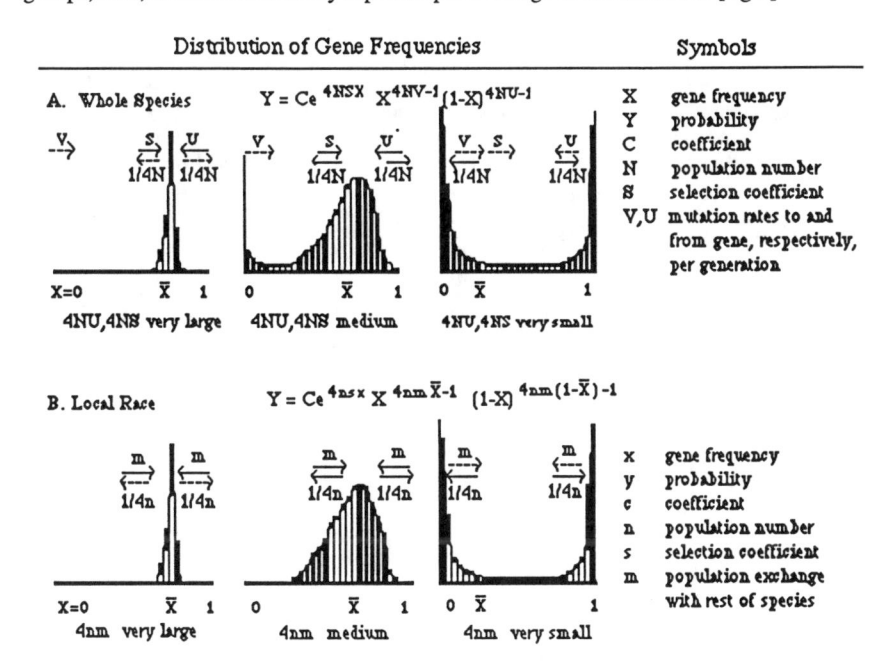

Figure 4: Random variability of a gene frequency under various specified conditions.

Finally (figure 4F) [fig 5], let us consider the case of a large species which is subdivided into many small local races, each breeding largely within itself but occasionally crossbreeding. The field of gene combinations occupied by each of these local races shifts continually in a nonadaptive fashion (except in so far as there are local differences in the conditions of selection). The rate of movement may be enormously greater than in the preceding case since the condition for such movement is that the reciprocal of the population number be of the order of the proportion of crossbreeding instead of the mutation rate. With many local races, each spreading over a considerable field and moving relatively rapidly in the more general field about the controlling peak, the chances are good that one at least will come under the influence of another peak. If a higher peak this race will expand in numbers and by crossbreeding with the others will pull the whole species toward the new position. The average adaptedness of the species thus advances under intergroup selection, an enormously more effective process than intragroup selection. The conclusion is that subdivision of a species into local races provides the most effective mechanism for trial and error in the field of gene combinations.(Wright 1932, 168)

A. Increased Mutation or reduced Selection
4NU,4NS very large

B. Increased Selection or reduced Mutation
4NU,4NS very large

C. Qualitative Change of Environment
4NU,4NS very large

D. Close Inbreeding
4NU,4NS very small

E. Slight Inbreeding
4NU,4NS medium

F. Division into local Races
4NU,4NS medium

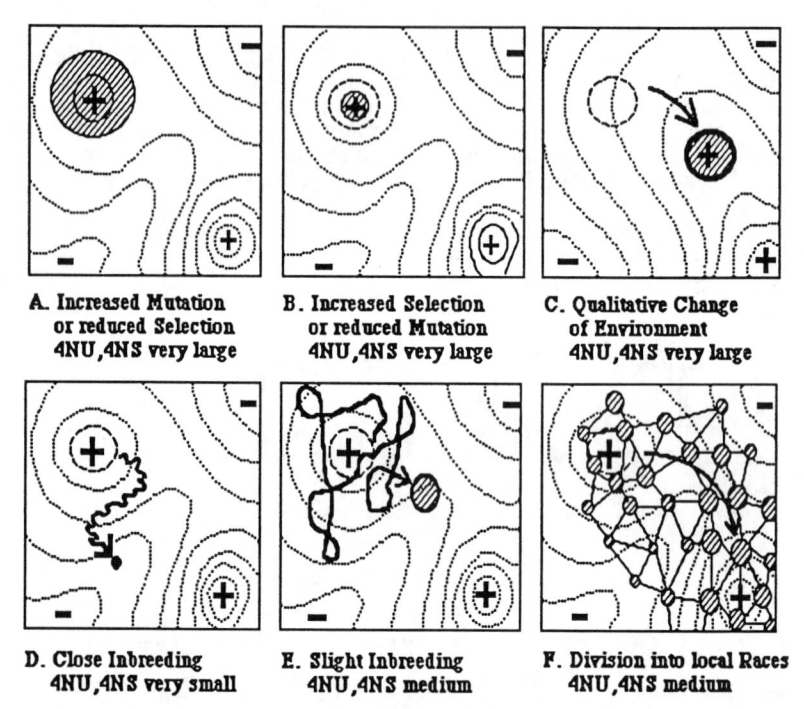

Figure 5. Field of gene combinations occupied by a population within the general field of possible combinations. Type of history under specified conditions indicated by relation to initial field (heavy broken contour) and arrow.

2. How important were the illustrations?

Let us start with the basic historical facts. Wright's talk was a great success. People grasped what he had to say and they responded warmly to his claims — at least, this seems to have been true of his American audience. Moreover, word seems to have got out, and Wright was flooded with reprint requests. Most important was the fact that among Wright's listeners at Cornell were active and ambitious young evolutionists, simply desperate for a good theory around which to structure their empirical research.

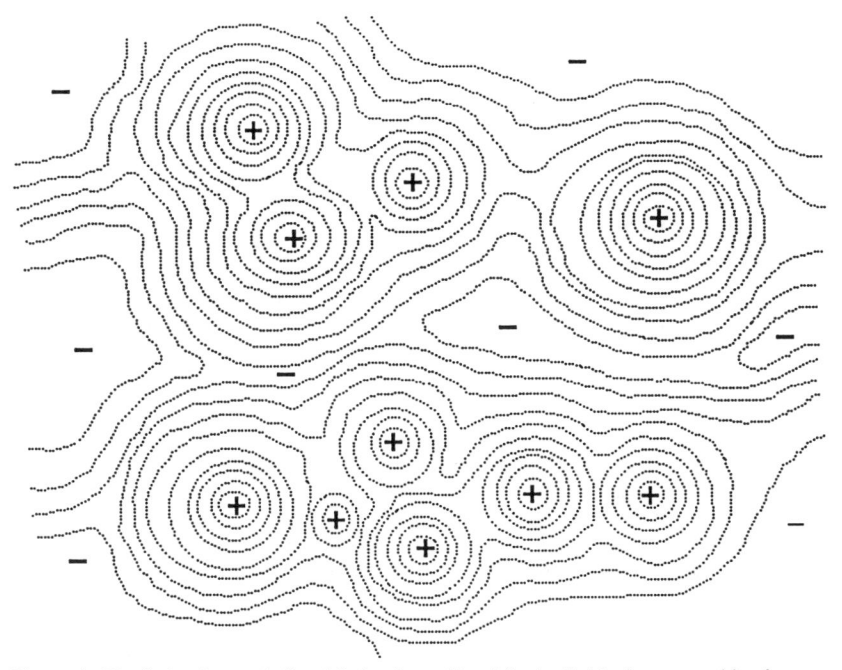

Figure 6. The "adaptive peaks" and "adaptive valleys" in the field of gene combinations. The contour lines symbolize the adaptive value (Darwinian fitness) of the genotypes. (After Wright.)

One of these people was the Russian-born Theodosius Dobzhansky, then working in Morgan's lab at Cal Tech. In his own words, "he simply fell in love with Wright", or at least with the ideas (Provine 1986, 328). Thus, when in 1936 Dobzhansky was invited to give the Jessup lectures at Columbia, Wright's shifting balance theory had pride of place, and in the published version next year — *Genetics and the Origin of Species* — Wrightian adaptive landscapes get (early) praise. It is not to much to say that the metaphor pervades the whole book.

> Every organism may be conceived as possessing a certain combination of organs or traits, and of genes which condition the development of these traits. Different organisms possess some genes in common with others and some genes which are different. The number of conceivable combination of genes present in different organisms is, of course, immense. The actually existing combinations amount to only an infinitesimal fraction of the potentially possible, or at least conceivable, ones. All these combinations may be thought of as forming a multi-dimensional space within which every existing or possible organism may be said to have its place.
>
> The existing and the possible combinations may now be graded with respect to their fitness to survive in the environments that exist in the world. Some of the conceivable combinations, indeed a vast majority of them, are discordant and unfit for survival in any environment. Others are suitable for occupation of certain habitats and ecological niches. Related gene combinations are, on the whole, similar in adaptive value. The field of gene combinations may, then, be visualized most simply in a form of a topo-

graphical map, in which the "contours" symbolize the adaptive values of various combinations (Fig. 1). [fig 6] Groups of related combinations of genes, which make the organisms that possess them able to occupy certain ecological niches, are then, represented by the "adaptive peaks" situated in different parts of the field (plus signs in Fig.1). The unfavorable combinations of genes which make their carriers unfit to live in any existing environment are represented by the "adaptive valleys" which lie between the peaks (minus signs in Fig. 1). (Dobzhansky 1951, 8-9)

Dobzhansky's book had immense influence. It has fair claim to having been the most important work in evolutionary theory since the *Origin*. And with the influence has gone the Wrightian landscape — reproduced again and again, in work after work (not the least of which were Wright's own writings, which were using the original illustrations right down to the 1980s). In America, all of the major evolutionists and most of the minor evolutionists used the notion of a landscape, and although the British were not so keen on Wright's actual theory, the metaphor itself found its way across the Atlantic.

Most interestingly, those evolutionists who could not use Wright's landscapes directly adapted them to their own ends. As a paleontologist, G G Simpson(1944) could not work at the genetic level, nor could he think in terms of individual populations of a species. So he hypothesized landscapes of phenetic or morphological difference, and he supposed taxa of higher categories working their ways across the landscapes, down valleys and up peaks. Wright, incidentally, approved of this extension. [figs 7,8 and 9, taken from the later Simpson 1953]

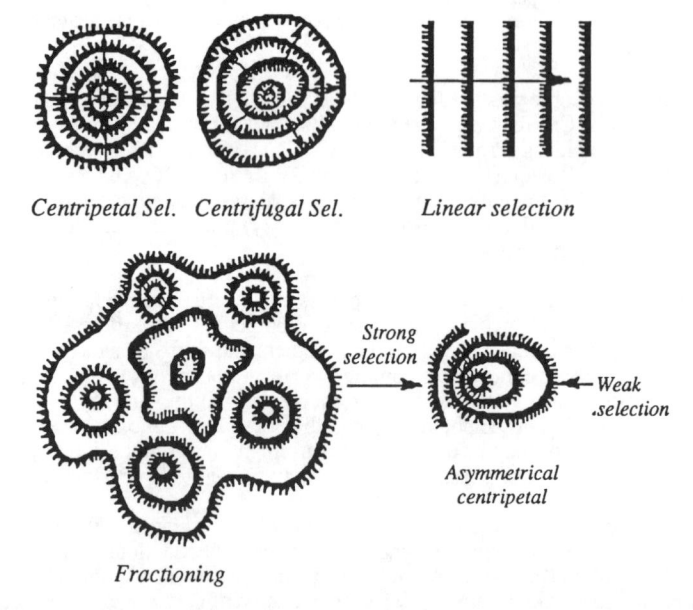

Centripetal Sel.   Centrifugal Sel.   Linear selection

Strong selection → ← *Weak selection*

Asymmetrical centripetal

Fractioning

Figure 7. Selection Landscapes. Contours analogous to those of topographic maps, with hachures placed on downhill side. Direction of selection is uphill, and intensity is proportional to slope.

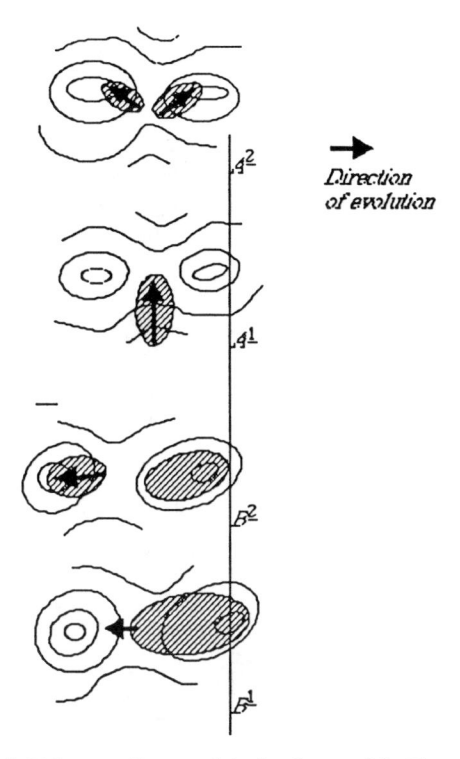

Figure 8. Two Patterns of Phyletic Dichotomy; Shown on Selection Contours Like Those of Figure 7. Shaded areas represent evolving populations. A, Dichotomy with population advancing and splitting to occupy two different adaptive peaks, both branches progressive; B, dichotomy with marginal variants of ancestral population moving away to occupy adjacent adaptive peak, ancestral group conservative, continuing on same peak, descendant branch progressive.

So much for history. Wright's idea of an adaptive landscape — where by "idea" I mean at the general level the metaphor, but at a specific level actual pictures, and usually the original pictures of Wright himself — became a commonplace in evolutionary thought. But speaking now at a philosophical level: Were the landscapes *really* part of evolutionary thought? Since Dobzhansky is generally taken as one of the founders of the "synthetic" theory of evolution, also known as "neo-Darwinism": Was Wright's metaphor in general, and his pictures in particular, *really* part of the synthetic theory of evolution, of neo-Darwinism?

The answer, of course, depends on what you mean by "*really* part of". The pictures were around in a big way, so they are clearly candidates for inclusion in a manner that for instance (to take an object entirely at random) the head of King Charles the First was not. The decision for inclusion must therefore depend on how one construes inclusion itself. Let us run through some possible senses.

At the most basic level, the pictures obviously are part of evolutionary thought. Evolutionists thought about them a great deal — and there is an end to the matter. I realize, of course, that many philosophers — all of those of the older cast of mind — will find this answer profoundly unsatisfying. They will claim that the question is not

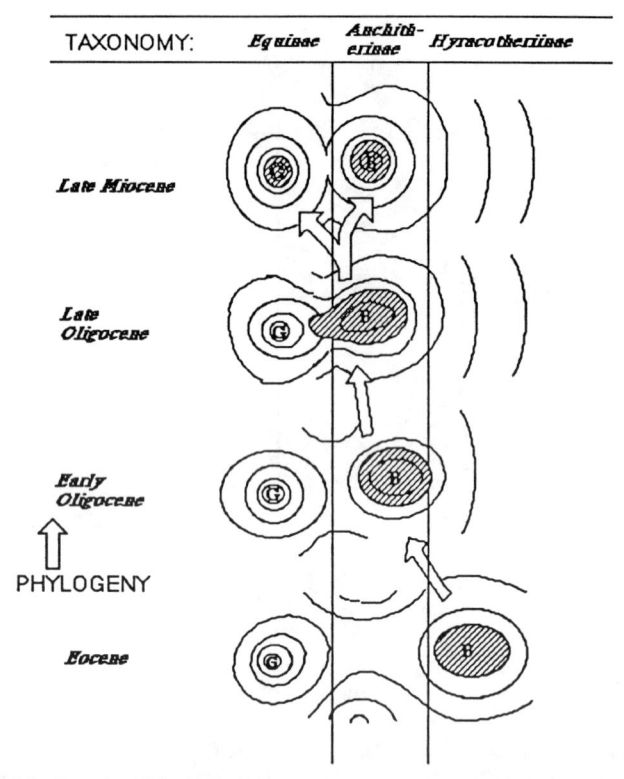

Figure 9: Major Features of Equid Phylogeny and Taxonomy Represented as the Movement of Populations on a Dynamic Selection Landscape.

whether people did think about them — we know that they did — but whether they *had to* think about them. Were the pictures an integrally necessary part of the science? Putting matters another way: The pictures were part of evolutionary thought. But, were they part of evolutionary *theory*?

In response, the argument for their necessity can easily be made a notch stronger. Not only were the pictures part of evolutionary thought, the scientists involved could *not* have done their work without the pictures. I speak now at the empirical level of psychological or intellectual ability. Wright's mathematics was simply too hard for the average evolutionist. It was certainly too hard for that very non-average evolutionist Theodosius Dobzhansky. He admitted again and again that he could not follow Wright's calculations. And he was not alone. G L Stebbins, another who heard Wright at Cornell, and later to provide the botanical arm to the synthetic theory, likewise was quite incapable of thinking mathematically.

But, they could understand the pictures! And so, as a matter of empirical fact, this was the level at which these men worked. They seized on the notion of an adaptive landscape and they experimented and theorized around it. Dobzhansky, for instance, studied natural populations of Drosophila, looking for evidence that they have drifted apart in a non-adaptive fashion (Lewontin *et al* 1981). As it happens, at first he did think he had evidence for his hypothesis. Then he found evidence against it. What is important is that, in both cases, in was at the picture level that he was thinking, because quite frankly he could do no other. In this sense, therefore, history supports the philosophical claim that the pictures were necessary. The science would not have been done without them.

"The science would not have been done without them"? Here the traditionalist will call a halt. The important point surely is whether the science *could* not have been done without the pictures. A philosophical analysis tries to strain out the fallibility of the individual and aim for the ideal. Moreover, the claim will probably be that the ideal, that which is in some sense preferable, would do away with the pictures. In a perfect world, the pictures could and would go.

Let me say simply that I find unconvincing the flat *a priori* dictum that the abilities of the scientists involved must necessarily (obviously?) be excluded from any adequate philosophical analysis. To the contrary, my feeling now is that the philosopher should start with the empirical necessity of the pictures and base his/her analysis on that. However, again for the sake of argument, let us grant the traditionalist the point. Still there are problems. At the least, one has to admit that the pictures were important, and may indeed now still be important, if not always in the future. And by "important" here I do not just mean "helpful". We have seen that the formalisms themselves did not express Wright's theory fully. The formalisms alone were shared by Fisher, who had an altogether different theory. The adaptive landscape idea went beyond the formalisms, expressing the notion that drift could generate variation in isolated populations, and that selection could then act to bring about rapid change. Moreover, let me point out that this, more than anything, was the *theory,* so the traditionalist cannot wriggle out of the claim that the adaptive landscape idea was (and may still be) part of Wright's basic science.

The response no doubt will be that although Wright's theory clearly did go beyond the formalism (because at that stage it was "immature"?!), the claim for the necessity of the pictures can be jettisoned. After all, in the main 1931 paper there were no pictures or even the metaphor. Everything that needed to be said, could be said and was indeed said, in words, literally.

In reply to this I will say three things. First, I simply do not know whether or not Wright had the landscape metaphor in mind when he first thought up his theory. We know that it predated publication of the 1931 paper, because it is used in an earlier letter to Fisher. Wright may have had it all along. I do know that the young Wright (and the old Wright, for that matter) was an Henri Bergson enthusiast, and something very much like the adaptive landscape metaphor occurs in *Creative Evolution* (published in 1912). It could well be that Wright was thinking seriously about landscapes even before he began his formalisms. The case for the necessity of the landscapes in the 1932 form of the theory does not depend on this, but I think the critic should tread warily before making sweeping claims about what *must* have been the case, historically.

Second, I would challenge the claim that the 1932 version of Wright's theory was simply the 1931 version, without the mathematics. The pictures do indeed add some factual claims — most importantly, that there are going to be some adaptive peaks for organisms to occupy, so long as one drifts far enough. The 1931 version really does not say much about why drift will eventually pay off. I have quoted the relevant passages and they are very vague. Indeed, Wright has already said that one small group drifting will probably go extinct. In the 1932 version, the pictures make it clear that there are all sorts of good opportunities waiting for drifters. Wright could have drawn a peak with a plain all around it, or with lots of (by definition) inhospitable sea or uncrossable rivers or chasms. But he does not, and it is certainly part of the plausibility of his theory that every peak seems to have other relatively accessible peaks in the vicinity.

Third, before it is immediately objected that one could have expressed all of Wright's new (post 1931) claims in words, let me point out that he did not. Moreover, let me point out also that (as people like Mary Hesse(1966) have pointed out generally about metaphorical thinking) there is a heuristic element to adaptive landscapes which escapes a simple list of factual claims that a scientist might make at a particular time (specifically Wright in 1932). Like all metaphors, they are "open-ended" in a way that the strictly literal is not.

In this context, consider Dobzhansky's 1937 rendering of the landscape. He has peaks clustering together in a way quite absent from Wright. Although, interestingly, he does not acknowledge the fact (that is he does not write it down in words), he is adding a distinctively new element to the theory — that adaptations are not random and that what works well in one way might have similar (although somewhat different) mechanisms also working well. The point is similar to someone noting the virtues of both gasoline and diesel motors, and noting also what a big gap there is between them and a steam engine or a jet engine.

There is therefore a forward-rolling aspect to Wright's picture. It stimulates you to push ahead with more claims. Just as in real life peaks tend to be clustered (the Alps, the Rockies), so Dobzhansky was stimulated to think of adaptive clustering. And it is certainly in this significant sense, centring on the heuristic value, that I would deny that Wright's adaptive landscape could, even in theory, be dropped without loss of content.

3. But is it good science?

We cannot conclude just yet. There is another line of argument which will tempt the traditional philosopher of science. It will be granted now that at least some science, at some level, incorporates pictures. But the complaint will now be that the *best* science does not. All science, even relatively good science, would be better were there no illustrations. Top quality science is just a formal system.

I confess that my general reaction to this line of inquiry is to query precisely whose criterion of value is being invoked here. Why is the best science non-pictorial? It seems to me that by just about any standard of excellence you might normally raise, the work of Wright and his successors like Dobzhansky rates highly. If anything, it defines the criteria rather than is measured by them. But since I have staked my position so firmly on one single case, perhaps the critic can come back on the basis of this case. Good though Wright's work may have been, there are reasons to think it might have been better without the adaptive landscape idea.

How might the critic argue? Most obviously, I suppose, by pointing out that the heuristics of the landscape are all very well, but if they lead one on false trails, their virtues are of dubious status. Take the question of other peaks surrounding any specified peak. Perhaps these exist. Perhaps they do not. One has no right to assume, as the metaphor forces on one, that they are always there. In fact, they are probably not.

In response, I would agree that perhaps Wright's picture does suggest false trails. But with respect: "So what?" No one wants to say that scientific hypotheses — exciting scientific hypotheses — always work or are always true (although sometimes philosophers have a yearning towards this last option). The point is that the theory is fertile, and with respect to something like available niches, can be tested and rejected or revised if necessary. In fact, as comments I have made already clearly imply, one can certainly redraw Wright's landscapes if one finds that niches are not readily available. And if no niches at all are available, then the whole theory must be rejected, not just the pictures.

I might add in this context that, although treatment of metaphor usually labels implications cleanly as good, bad or neutral heuristics, in real life (as our example shows) it is often not so easy to decide whether or not implications are such a very good or bad thing. Take the presumed stability of Wright's landscape. Although the possibility of change is certainly mentioned, generally — as with landscapes as opposed to water-beds — the terrain is supposed to be fairly solid. This suggests that organisms will scale ever-higher peaks, and that in short there will be progress. However, although many today — like George Williams(1966) and Stephen Jay Gould(1989) — would consider this the consequence of a negative heuristic, others are not so sure. I suspect that Wright himself endorsed progress. Certainly, the botanist G L Stebbins is a progressionist and has used Wright's ideas to make precisely such a case (Stebbins 1969). And active today someone like E O Wilson(1975) is an organic progressionist and would, no doubt, find any supporting implications of Wright's metaphor most comforting.

The critic might now argue in a slightly different way. Wright himself admits that in his diagrams he is collapsing down a huge amount of information into two dimensions (three if you consider the axis from eye to page). But is this legitimate? One is taking drift from many many dimensions and confining it to two dimensions. One of the things that Wright always prided himself on was the fact that he acknowledged the fact that genes in combination might well have very different effects from genes taken singly. What right therefore have we to assume that the many drifting genes will combine to behave like one drifting gene (or, rather, a line of such genes)?

There is an important point here — one which shows that although Wright may have been sensitive to gene interaction, critics like Ernst Mayr(1959) were not entirely off base when they accused the population geneticists of undue reductionistic thinking, in treating their subjects as beans in a bag. However, note that if there is a problem here — that the collapse of dimensions is too dramatic — it is one which affects all levels of theory and not just the illustrations. Again, therefore, I suggest that Wright's theory should simply be put to the test, and check made to see if genes do wander in the way that he suggested.

In fact, as I have intimated, a decade after Wright published, Dobzhansky and others found strong evidence that selection is far more powerful and effective than Wright and others had suspected. (I am not now referring to molecular genes which, by their very nature evolve at levels below the power of selection.) The shifting balance theory required modification. But I am not sure that such modification required/requires rejection of the very notion of an adaptive landscape. One can rework the landscape to show that factors other than drift are significant.

I conclude, therefore, that the criticisms of conservatively minded analysts are not well-taken. Wright's work was not perfect, in the sense of being absolutely true or totally without conceptual blemish. But this is a far cry from saying it was not first-rate science. Fortunately, scientific theories are like human beings — they are complex entities, with lives of their own, and the best are the best, not because they never do anything wrong, but because they do so many things right.

4. Conclusion.

What have I proved? I have certainly not proved that every scientific theory has to have pictures, or that every scientific picture is essential. By my own admission, I have been dealing with a picture or a special kind, namely one which expresses a metaphor. Nor am I claiming here that every scientific theory contains metaphors, although as a matter of fact this a claim I would be prepared to defend. I am not even claiming that every scientific metaphor gives rise, actually or potentially, to a picture. Indeed, this seems to me to be a false claim. Only in a very limited way do such important biological metaphors as natural selection or the struggle for existence give rise to pictures, and these are usually misleading.

Nevertheless, some scientific metaphors are pictorial — Wright's landscapes prove this. And those metaphors/pictures are in an important sense (any sense which is important) essential parts of the science — Wright's landscapes prove this. Moreover, the science containing these pictures can be good science — Wright's landscapes prove this also. These seem to me to be a good set of conclusions with which to end this somewhat preliminary foray into the philosophical significance of biological illustration.

I am indebted to David Hull and Ernst Mayr for typically thoughtful comments on an earlier version of this paper.

References

Achinstein, P. (1968), *Concepts of Science*, Baltimore: Johns Hopkins University Press.

Bergson, H. (1912), *Creative Evolution*, London: Macmillan.

Braithwaite, R. (1953), *Scientific Explanation*, Cambridge: Cambridge University Press.

Bunge, M. (1967), "Analogy in quantum theory: from insight to nonsense". *British Journal for the Philosophy of Science,* 18, 265-86.

Dobzhansky, T: (1937), (3rd ed. 1951) *Genetics and the Origin of Species,* New York: Columbia University Press.

Fisher, R. (1930), *The Genetical Theory of Natural Selection*, Oxford: Oxford University Press.

Gould, S.J. (1989), *Wonderful Life*, New York: Norton.

Hempel, C. (1965), *Aspects of Scientific Explanation*, New York: Macmillan.

Hesse, M. (1966), *Models and Analogies in Science*, Notre Dame, Ind: University of Notre Dame Press.

Lewontin, R., Moore, J., Provine, W., and B. Wallace eds. (1981), *Dobzhansky's Genetics of Natural Populations*, New York: Columbia University Press.

Mayr, E. (1959), "Where are we?" *Cold Spring Harbor Symposium on Quantitative Biology* 24,1-14.

Provine, W. (1986), *Sewell Wright and Evolutionary Biology*, Chicago: University of Chicago Press.

Rudwick, M. (1976), "The emergence of a visual language for geological science 1760-1840" *History of Science,* XIV, 149-95.

Simpson, G. (1944), *Tempo and Mode in Evolution*, New York: Columbia University Press.

_ _ _ _ _ _ _. (1953), *The Major Features of Evolution*, New York: Columbia University Press.

Stebbins, G. (1967), *The Basis of Progressive Evolution*, Chapel Hill, N.C.: University of North Carolina Press.

Williams, G. (1966), *Adaptation and Natural Selection*, Princeton: Princeton University Press.

Wilson, E. (1975), *Sociobiology: The New Synthesis*, Cambridge, Mass: Harvard University Press.

Wright, S. (1931), "Evolution in Mendelian populations", *Genetics*, 16, 97-159

_ _ _ _ _. (1932), "The roles of mutation, inbreeding, crossbreeding and selection in evolution". *Proceedings of the Sixth International Congress of Genetics,* 1, 356-66.

Material Models in Biology

James R. Griesemer

University of California, Davis

1. Introduction

Propositions are no more constitutive of science than they are of any activity: a body of knowledge is not all there is to the life of science. Thus I take the premise underlying the topic of this symposium to be uncontroversial, there *is* a "non-propositional" side of science and of biology in particular. From time to time, however, philosophers ask whether the "non-propositional" side of science is theoretically superfluous, or as Duhem put it, logically dispensable. What they mean to ask is whether science can be fully *analyzed* in propositional terms; must philosophy of science, in other words, consider the non-propositional side in order to adequately account for science?

Negative answers to the question often rest on the tacit view that the most (or only) *important* thing about science is scientific theories. Positive answers that are nevertheless framed in terms of theories usually take for granted the existence and function of "non-propositional" elements like non-literal uses of language (metaphors and analogies), diagrams, and pictures and try to show why they are irreducible to propositional content. What seems surprisingly uncommon is any attempt to search for systematic connections between the two as they function in science. There is of course a substantial literature on metaphor and analogy in science (e.g., Hesse 1966, Achinstein 1968, Leatherdale 1974), but this literature deals primarily with the role of figurative language in *contrast* to literal language rather than seeking systematic connections between the propositional and non-propositional contents of science.

In this paper I explore the boundary between the propositional and non-propositional sides of biological science, drawing on three cases with which I am familiar. It is important to stress that I focus only on the boundary between two perspectives on scientific *theory*, as exemplified by biological theories. I do not aim to address the larger questions of the non-propositional character of scientific practice *tout court*. I propose a picture of material model-building in biology in which manipulated systems of material objects function as theoretical models. Sometimes material models serve as direct theoretical models, as in non-mathematized fields of biology where structure may be abstracted directly from the material system without detour through a formal sentential apparatus. Sometimes material models serve as vicarious, or indirect, theoretical models. That is, work with material systems can serve as an important basis for manipulation in thought of abstract, formal objects for the purpose of articulating a theory.

It is irrelevant to my analysis whether theoretical modeling, in so far as it yields scientific models with empirical content, is essentially or ineliminably vicarious. I do think the analysis is generalizable beyond the cases I discuss. In fact, I think there is very little in science that is ineliminable *ceteris paribus*. Vicarious theoretical modeling is certainly more common in biology in particular and science in general than philosophers realize, and as such it is a phenomenon in need of explanation. My aim is to describe an interesting and important feature of biological practice, not to show that it is essential to biological science.

My story is thus in contrast to Ruse's (this volume), which concentrates on Sewall Wright's illustration of mathematical ideas, already formalized, in his adaptive landscape pictures. It is also in contrast to Taylor's (this volume), in its reconstructive use of diagrams as a tool with which scientists are invited to represent their research field during an interview. My topic accords best with those aspects of Wimsatt's discussion (this volume) of the generative role of diagrams in biology. (Ruse and Taylor do discuss the generative, and hence non-superfluous, role of diagrams, but their cases are different because they consider non-propositional elements of science which are consequent on (presumably) propositional forms.)

In each of the cases I will discuss, material systems play important generative as well as presentational roles. In the first case (Grinnell's remnant models of faunas), a material system stands in for propositions and functions as a material "theory" in lieu of a propositional system formulated in natural language or mathematical equations. In the second case (Wright's path analysis), diagrams serve as a way to express causal hypotheses for the sake of generating precise mathematical propositions which can be tested. In the third case (Weismannism), diagrams serve as abstractions of causal processes in a background theory for the sake of generalizing a theory in the foreground of biological concern (Darwinian evolutionary theory).

I wish to illustrate an important point about the non-propositional side of biology by means of each of these cases: (1) material models are able to serve certain sorts of theoretical functions *more* easily than abstract formal ones in virtue of their material link to the phenomena under scientific investigation. "Remnant models," i.e. material models made from parts of the objects of interest, are of this sort. They are robust to some changes of theoretical perspective because they are literally embodiments of phenomena. If these embodiments are preserved, they may be studied again and again under different perspectives. This stability in itself depends on the creation of certain scientific institutions for the sake of remnant preservation and these institutional forms may thus serve as indicators for the presence of the models in scientific research.

(2) Diagrams can function as statement-generators if a set of rules for "reading" them is adopted as part of a theoretical methodology. In such cases it can be argued that this possibility makes formalized sentences themselves subject to the argument of eliminability that a traditionalist would apply to diagrams, metaphors and other structures resistant to the axe of first order logic. Thus diagrams can serve to facilitate theory-construction, possibly even supplanting more traditional modes of mathematical modeling.

(3) Diagrams can also hinder theory-construction in virtue of their representational power: they can become so entrenched in our way of theorizing that they outlive their theoretical usefulness. In serving to represent causal theories which function heuristically in the background of a given theory of interest, diagrams may provide strong ontological constraints on the way the foreground theory is understood. This may be true even if theory has advanced beyond the framework specified in the diagram simply because diagrams are powerful modes of representation that may become entrenched. Ruse makes this point, in rare agreement with Provine, in his paper on adaptive landscapes: Wright's diagram was constructed as a vast simplification of a mathematical theory within one theoretical framework describing the evolution of gene combinations, and then shifted to another framework describing the evolution of sets of semi-isolated populations for which the adaptive landscape diagram appears inadequate in its details. But despite its many inadequacies, Wright's diagram lived on in all its particulars among a broad range of evolutionary biologists. While the landscape diagram was obviously heuristically fruitful in its early use, it may be an obstacle to further work.

2. Grinnell's Remnant Models of Faunas

My first case concerns specimen collecting for ecological study in the Museum of Vertebrate Zoology of the University of California, Berkeley (MVZ). Joseph Grinnell (1877-1939), the MVZ's first director, was a Darwinian naturalist interested in questions of the geography of speciation and the character of the ecological environment (see Griesemer 1990). He is noted primarily for two accomplishments: an early formulation of the niche concept and for founding the MVZ. Grinnell became convinced that California was an ideal place to study evolution in an ecological context because compared to the east coast, the Pacific states were relatively unspoiled.

Grinnell, along with his patron-collector Annie Alexander, anticipated that the settling of California and the conversion of its economy to large scale agriculture would cause dramatic change in the higher vertebrate fauna of the state. They founded the MVZ in part to make a record of vertebrate diversity as it was known at the turn of the century (Star and Griesemer 1989). This prediction also presented an opportunity for research: Grinnell thought that Darwin's evolutionary theory needed expansion to include a description of the evolution of the environment, because that was what drove the force of Darwinian natural selection. If the vertebrate fauna could be studied continuously from the founding of the MVZ in 1908 for one hundred years, say, then evolutionary change caused by habitat and micro-climatic alterations could be studied in real time: California could be used as an ecological-evolutionary laboratory. In a well-known 1924 paper, Grinnell summed up his perspective:

> Observation of species in the wild convinces me that the existence and persistence of species is vitally bound up with environments. The extent and persistence of a given kind of environment bear intimately upon the fate of the species we find occupying that environment. Environments are forever changing — slowly in units of recent time, perhaps. Yet with relative rapidity they circulate about over the surface of the earth, and the species occupying them are thrust or pushed about, herded as it were, hither and thither. If a given environment be changed suddenly its more specialized occupants disappear — species become extinct (Grinnell 1924, p. 153 in 1943 reprint).

The notion that ecological factors cause species to be herded into certain biological relations with other species and thus set conditions for Darwinian adaptation led Grinnell to classify environments in terms of ecological causes drawn from his experience as a biogeographer in the 19th century tradition of Wallace and Merriam. World realms, regions, life-zones, faunas, associations and niches comprised a hierarchy which could be used to classify the causes of the presence and absence of particular species or subspecies in a given location. Each level in the hierarchy was associated with a specific primary causal factor. Following Merriam, Grinnell thought of lifezones as defined in terms of physiological limits of temperature tolerance (e.g., Merriam 1894). Grinnell defined faunas primarily in terms of humidity, associations by local presence of plants that vertebrates might use for food, shelter or breeding sites, niches by the presence of constellations of other animal species to which the species in question was adapted to live.

Presence or absence of a taxon in a given place and time, along with information about the environment at that point thus constituted the basic data with which Grinnell would attempt to build a theory of the evolution of environment. Data gathered in particular field studies were typically summarized in maps indicating zones of homogeneity with respect to a given ecological factor; the most famous of these are the well-known "life-zone maps" (see, e.g., the life-zone map of California reprinted in Grinnell 1943). By studying shifting patterns in such data over time, Grinnell hoped to identify the significant causes and patterns of natural selection shaping species ranges and thus how their joint presence led to forces of organic adaptation. Fine-grained study of the distribution of subspecies would lead to an understanding of speciation (see Griesemer 1990 for a detailed account of Grinnell's argument).

We can conceptualize Grinnell's work on these subjects in terms of two notions: a remnant model of an ecological structure and an institutional specification for a vicarious program of theoretical model-building. The elemental structures that interested Grinnell were of the most mundane sort, the presence or absence of a given taxon in a location at a time. A list of taxon presences together with ecological information at a time can be taken to be a simple formal ecological model of that place and time. Alternatively, the specimens of those taxa with identifying tags linking them to a place, a taxon, and a set of environmental data, can be taken as the model of presences in a place at a time (i.e. an empirical substructure, see van Fraassen 1980). A theory of ecological change could be specified by assembling a collection of such models for different times and assigning ecological causes by classifying them according to their specific place in the ecological hierarchy.

The latter, material characterization of these simple ecological models has special significance. During the period Grinnell worked, there was great controversy in systematics over the proper characterization of species, in particular whether subspecies should be recognized as "real". Grinnell's models are what I called above remnant models. An environment is modeled by the organisms that occur in it and which are preserved as specimens along with information about ecological factors recorded in field notes by the collectors. Because the specimens are components of the model and ecological data recorded in notebooks can be keyed to specimens by labels, *models* can be preserved in a museum. This is significant because changes of theoretical perspective about the nature of species can be taken into account by pulling the specimens back out of their drawers or off the shelves and reanalyzing the model in terms of a different set of taxonomic designations. This is not possible in the isomorphic *formal* model because once the *information* is recorded that members of a particular taxon were present in a location, there is no recourse—through that information alone—to revise the assessment of specieshood that underlies it, should the theoretical perspective on the nature of species change. Only consultation with the original specimens will do. Even the collection of new specimens is insufficient because without comparison to the original ones there is no way to determine whether any differences are due to changes in those species' ranges, changes in the analysis of them as species, or whether evolution had perhaps changed them into different species. In short, these ecological remnant models are robust to changes in theoretical perspective on the nature of species.

In order to be robust, several conditions had to be met by the methods for constructing the models. Collecting, note-taking and preserving had to be standardized in order for these "snapshot" ecological models to be collected over time and compared; the materials would also have to be properly preserved for comparison to succeed. This is made all the more difficult in virtue of the goal of real time evolutionary analysis: Grinnell knew he would not live to see the significant results of his own efforts because evolution in nature happens on a longer time-scale than the careers and lives of scientists. Nevertheless, Grinnell was committed to the study of evolution "on the ground, as it happened" rather than by constructing short-term artificial experiments in an indoor laboratory (Grinnell 1910).

Evidence of Grinnell's zeal in standardization can be found in the extent to which he indoctrinated students into his methods. His course handouts beginning in 1913 on how to collect and take field notes cover everything down to the kind and size of paper to use, precisely where and how to enter information on each page, and the brand of ink that *must* be used (Higgins eternal). Although seemingly picayune, Grinnell's concern is quite significant: field notes must be curated with the same high standards as specimens if the ecological relations between organisms and environment are to be preserved. Grinnell was so successful in this regard that his student, the mammalogist E. Raymond Hall, published an elaborated version of Grinnell's handout as a manual showing not only Grinnell's method to the last detail, but also that Hall managed to preserve its every detail for 50 years (Hall 1962) (see Figures 1 and 2). Guidebooks describing Grinnell's method are still being turned out (Herman 1986).

Grinnell solved these methodological problems by what I have called vicarious theoretical modeling. To repeat, the theoretical models were themselves material models constructed from remnants of the particular assemblages of animals, plants and environmental factors that intersected in fleeting points in space and time. Grinnell worked doggedly to amass as much data as he personally could, publishing monographic articles running hundreds of pages each. But he worked at least as hard to build an organization, the MVZ, a network of professional and amateur collectors, and a set of institutional practices of collecting, note-taking, labeling, cataloging, preserving and storing. I think he did this as a vicarious means of modeling. Rather than construct all the models himself, which he could not do, he institutionalized his practices in such a way that others could complete what he had begun. If the institution was well-designed, it could persist on the same time-scale as the evolutionary change Grinnell hoped to measure. The vicariousness of Grinnell's specification of theoretical models is thus manifest in the extent to which his methods of museum work were institutionalized. The structure of the models can be studied vicariously as well, by scrutiny of the methodological and institutional writings that specify how models are to be made and of the implementation of those specifications in the organization of research work.

3. Wright's Path Diagrams

Sewall Wright developed path analysis as a method for quantitatively assessing the relative degree of influence of variables represented in a system of presumed causal interactions (Wright 1918, 1920, 1921, 1934; cf. Griesemer 1991, Irzik and Meyer 1987, Provine 1986 for review). The method consists of two components: the

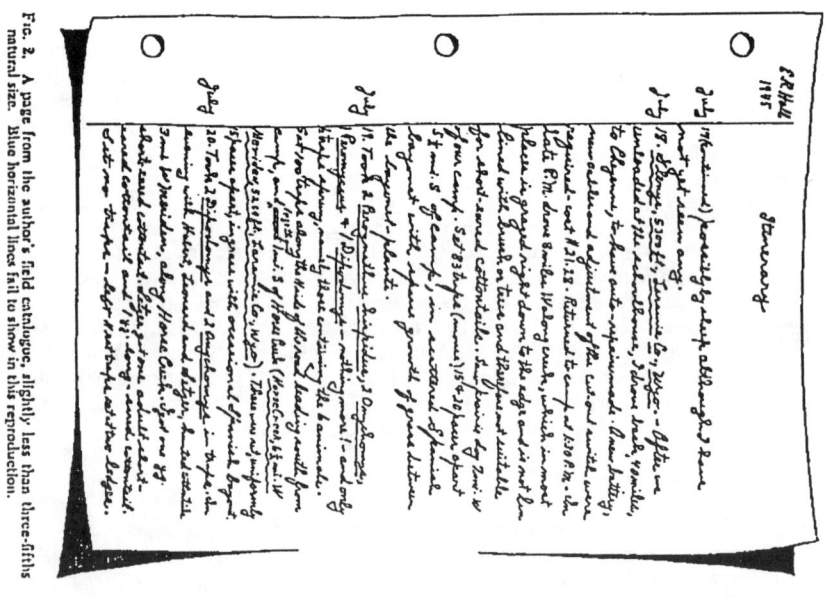

Figure 1. Illustration of how to take field notes, reproduced from Hall (1962, pp. 5-6, figs. 1,2).

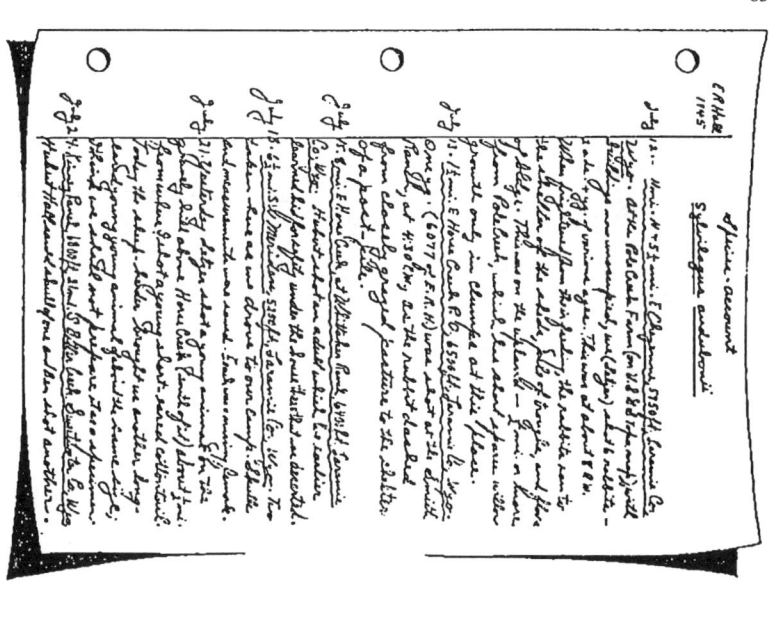

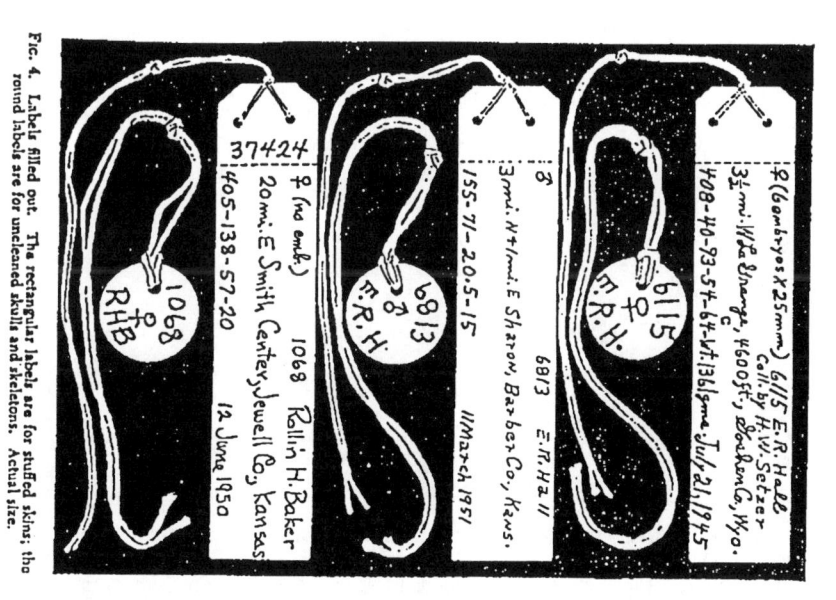

Figure 2. Illustration of how to take field notes and how to fill out specimen labels, after Hall (1962, pp. 7, 10, figs. 3,4).

formulation of a causal hypothesis by means of a path diagram and the quantitative analysis of relative degrees of influence by the method of path coefficients.

Wright developed the method as an improvement on Pearson's method of partial correlations, which indicated only the aggregate effect of each variable on the total correlation. Wright's method resolved the influence of each variable within a system into components for each specified causal pathway. A path diagram represents variables as letters and indicates causal influence of one variable on another by means of a straight, directed arrow pointing from one to the other. Unanalyzed correlations between variables thought to have influence on a given "effect" variable are represented by curved, double-headed arrows between pairs of causal variables (see Figure 3). Variables are expressed in standardized form, i.e. as deviations from the mean, divided by the standard deviation. Variables are thus treated in abstraction from their units of measurement. On Wright's interpretation, standardization plus the use of diagrams makes each analysis local to the system in which the data are collected and the causal hypothesis depicted.

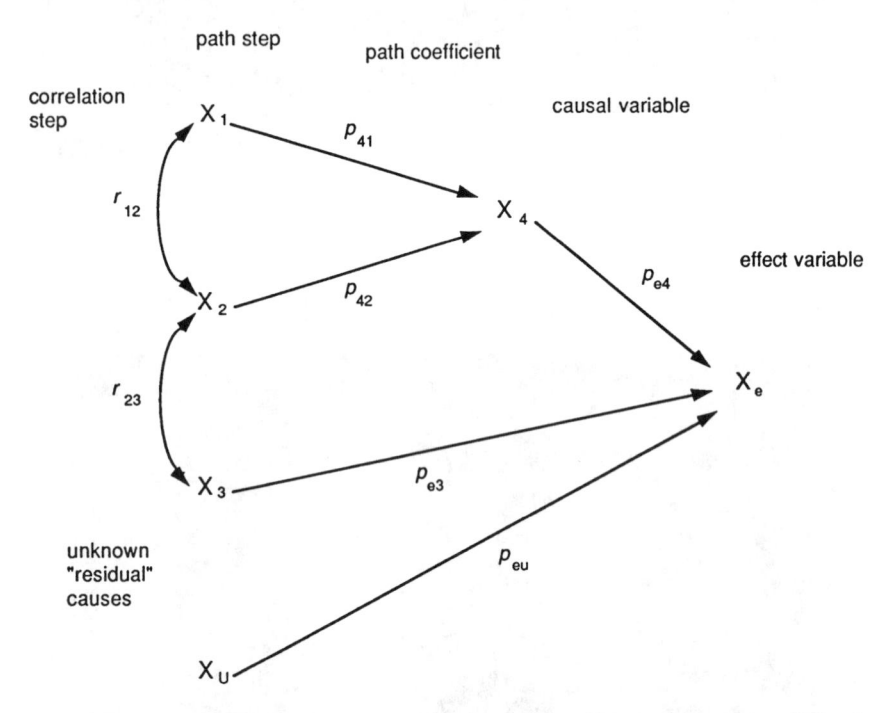

Figure 3. The elements of a path diagram. Note that correlation coefficients (r's) are path coefficients (p's) for correlation steps. The circuit from Xe to X4 to X2 to X3 to Xe is a valid path, while the circuit from Xe and back that includes both correlation steps is not (by rule 2, see text).

The method of path coefficients involves the construction and analysis of path equations, linear equations in the causal variables weighted by the path coefficients, describing relationships between the standardized variables in the path diagram. Path coefficients are regression coefficients on standardized variables, but path analysis is not merely regression analysis with diagrams thrown in for illustration. Indeed, Wright (1960) was adamant in distinguishing his method from so-called "causal" regression analysis, as advocated by Tukey (1954). The adamancy takes on significance in light of Wright's view that path equations could be "read off" the diagrams if certain rules were followed. Treating regression and path methods as alternatives, as Tukey did, blurred the distinction between the conceptual role of the path diagram as the causal hypothesis which *generates* path equations with the merely *post hoc* role of an illustration of equations.

Wright distinguished two uses of path analysis: "direct" and "inverse" (Wright 1921). The inverse use deduces path coefficients from measured correlations by solving a system of simultaneous linear equations read from an *a priori* path diagram. The direct use deduces correlations among variables from specified path coefficients (usually given by a previously accepted causal theory) and an *a priori* path diagram. Wright developed the inverse use first, while a graduate student, as a refined analysis of partial correlations in measurements on the bones of rabbits (see Provine 1986, pp. 133-134). The direct use has its most familiar application in Wright's development of a general theory of inbreeding, in which the correlation among relatives is calculated from path coefficients given by Mendel's laws and path diagrams given by pedigree diagrams of descent relations.

The crux of the difference between Wright's path analysis and standard regression analysis is in what sorts of inference are licensed by regression vs. path methods. In regression analysis, the data used to estimate regression coefficients are thought of as population samples and the goal is extrapolation of coefficients for the variables from the sample to the population as a whole. In path analysis, the estimates of path coefficients are relative not just to a sample, but to a particular causal hypothesis expressed in a path diagram. A change in causal hypothesis, i.e. a change of diagram, means that *no* extrapolation from previous data is directly licensed. In other words, concrete, i.e. unstandardized, regression analysis is suited to extrapolation across systems because regression coefficients are estimates for particular *variables* whose component influences in the various causal paths in which they participate is unknown. Path analysis interprets relative causal influence along component *paths* within a system of unitless, standardized variables. This renders variables with different units comparable within the context of a given causal model. Thus regression models are concrete with respect to variables and abstract with respect to systems, while path models are abstract with respect to variables and concrete with respect to systems (Griesemer 1991, p. 166).

For Wright, causal analysis required *a priori* stipulation of a causal hypothesis in a path diagram. Wright developed a set of rules for reading diagrams which allowed construction of correct linear path equations (see Griesemer 1991, pp. 176-180; Li 1975). A path is a valid traversal of a path diagram. Paths may be simple or compound, depending on whether one or more arrows are involved in the traversal. In traversing, one traces the arrows of a diagram and collects the path coefficients: the path is represented by the product of the path coefficients associated with each arrow of the path. Rules for traversal include the following: (1) once a traversal has been made forward along a path (i.e., in the direction of a single-headed arrow), backward traversal is not allowed, (2) traversals along multiple "path steps" (single-headed arrows linked through variables) may be included in a compound connecting path, but only one "correlation step" (double-headed arrow) may be, (3) a path may not traverse a given variable more than once.

For simple systems where the diagram would be simple enough to carry in the head, this rigor is unnecessary: one could write down a correct set of linear equations expressing the antecedent causal hypothesis directly. For more complex systems, it is unlikely that one could maintain a consistent causal hypothesis for the sake of experimentation and quantitative analysis this way. Moreover, at least for many of the sorts of applications Wright envisioned in biology, it is unlikely that a solution to the maximal set of equations would be desired. Indeed, writing just the right subset directly and correctly from a mental hypothesis seems impossible for the complex, irregular systems with which Wright routinely dealt.

There are thus two important reasons for viewing path diagrams as generative. In complex cases, correct statement of a system of statistical relationships that reflect causation can be generated by following diagrammatic construction rules. Moreover, the causal structure deemed implicit in the maximal set of consistent equations for a system is not only isomorphic to the causal structure explicit in the corresponding diagram, but this structure (and all substructures) are easily examined in terms of the latter. Indeed, some numerical questions about statistical relations can even be answered without recourse to equations in the usual sense at all. Calculation can be done simply by collecting term values as one traverses a path diagram visually without writing down equations: the formal apparatus of first formulating an equation symbolically and then setting about solving it is unnecessary. As Wright demonstrated in his development of a general theory of inbreeding, this diagrammatic approach can be generative not only for solutions to specific causal problems, but also for producing concepts applicable to whole classes of path diagrams (in this case systems of mating satisfying Mendel's laws).

4. Weismannism and the Generalization of Evolutionary Theory

My final example concerns the role of Weismannism in the interpretation of Darwinian evolutionary theory as a general theory. Darwinian theory is couched in terms of organisms, but modern evolutionists take the theory to apply to any entities with certain properties (e.g., heritable variance in fitness). While rich in its characterization of certain component processes such as selection, evolutionary theory provides no resources to identify and individuate the entities it is about: nothing in the theory tells us how to determine what count as organisms. Some of this ontological work is done by background theories which are therefore critical for understanding the structure of evolutionary theory.

The modern units of selection literature depends on one of two background theory assumptions in generalizing Darwinian theory: it either assumes a biological hierarchy of levels of organization or it assumes a causal relation between gene-like and organism-like entities expressed in the doctrine of Weismannism (Griesemer submitted). Weismannism is the view that the germinal material forms a continuous lineage while bodies, or somata, do not. The main role of Weismannism in evolutionary theory is to discount the possibility of the inheritance of acquired characteristics (Wilson 1896; cf. Hull 1988 ch. 11). But this is merely one implication of Weismann's theory and some of the others are carried along in its use as a device for generalization (Weismann 1892).

Weismann, it turns out, is no Weismannian. The way in which his theory departs from the standard account and from the modern molecular version of his view is significant for interpreting the character and success of the application of Weismannism to the generalization of evolutionary theory. The moral of the brief story to come is that Weismann's own theory has quite different implications for evolutionary theory than Weismannism. The latter view has become entrenched in the thinking of biologists and philosophers alike in part because it is presented in a simple and powerful but flawed diagram.

Weismannism expresses a causal asymmetry between germinal and somatic elements and this asymmetry is the basis for various distinctions between the evolutionary roles of "replicators" and "interactors" as generalizations of the concepts of "gene" and "organism" (Dawkins 1976, 1982; Hull 1980, 1981). The evolutionary function of replicators is to directly pass on their structure largely intact in a process of replication. The evolutionary function of interactors is to interact directly with their external environments in such a way that replication is differential (Hull 1980; cf. Griesemer submitted).

The doctrine of Weismannism is most prominently expressed by means of a widely-reprinted and copied diagram due to E. B. Wilson (1896) (see Figure 4; cf. discussion in Griesemer and Wimsatt 1989). The image is further simplified in its modern molecular rendition, made explicit by Maynard Smith (1965) though already implicit in the central dogma of molecular biology (see Figure 5). Wilson's diagram is ambiguous, however: it could reflect either or both of at least two distinct views: (1) the continuity of the germ-cells and discontinuity of the somatic cells, or (2) the continuity of the germ-*plasm* and the discontinuity of the somato-plasm. It is clear from the discussion in Wilson (1925) that he interpreted Weismann to mean the former, lumping Weismann's view with contemporaries who held to the former. But Weismann explicitly rejected this interpretation (Weismann 1892; cf. Griesemer and Wimsatt 1989). Maynard Smith's diagram, on the other hand, is unambiguously wrong of the view Weismann himself held even if we factor out the modern interpretation of the molecular constituents of germ-plasm (Griesemer submitted).

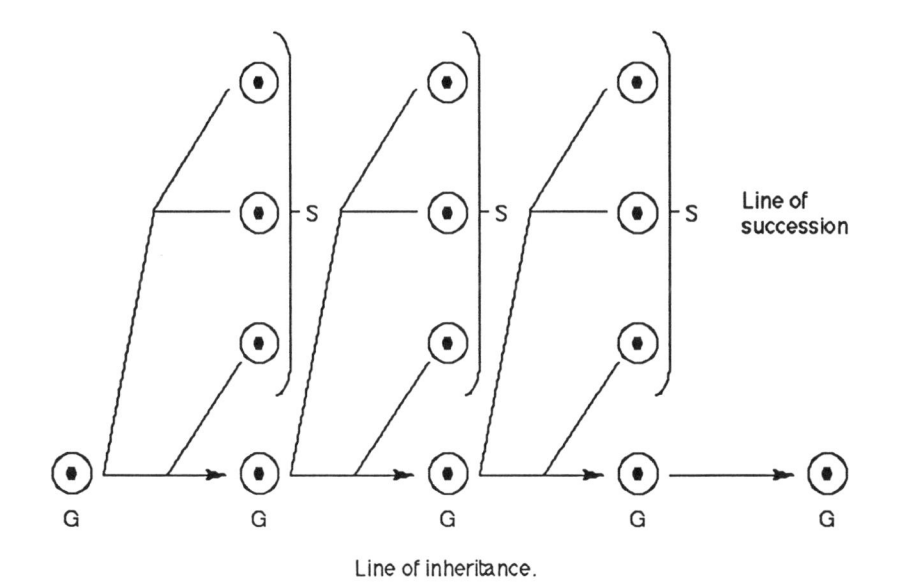

Figure 4. E.B. Wilson's representation of Weismannism, after Wilson (1896, p. 13, fig. 5).

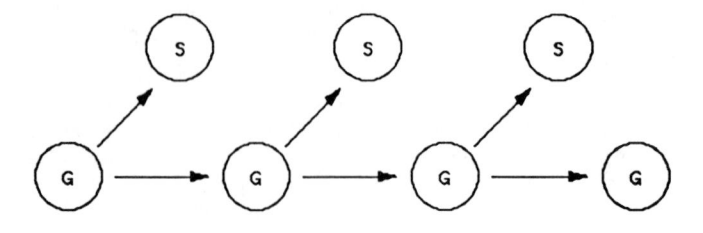

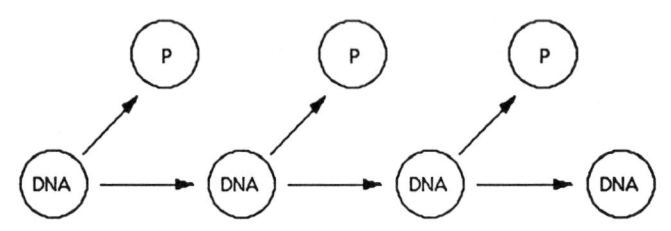

Figure 5. A simplification of Wilson's representation of Weismannism compared to a modern molecular representation of the central dogma of molecular biology, after Maynard Smith (1965, p. 67, fig. 8).

Moreover, it is clear from Weismann's own very different diagrammatic representation of his theory that he held the germ-plasm rather than the germ-cell view (see Figure 6). For Weismann, germ-cells were products of somatic differentiation, just like every other cell in the body. This is shown in his diagram in virtue of the fact that germ-cells first appear in the developing worm depicted in cell generation 9 (the cells are labeled "urKz", urKeimzellen). It follows that germ-*cells* do *not* form a continuous lineage, they are interrupted in development by a series of somatic cells, just as organismal bodies are interrupted in descent by their passage through gametes that must combine (in sexual species) to form zygotes. What is continuous is the molecular germinal protoplasm in the nucleus of the cells fated in development to become germ-cells. The implication of this fact is that the phenomena of heredity must ultimately be explained in terms of development. Thus, for Weismann, the problem of explaining hereditary transmission of germ-plasm is a problem of development, not an autonomous problem to be treated separately from development, as is represented in Wilson's diagram with a separate continuous line for germ-cells and a discontinuous "line of succession" for somata.

Returning to the problem of generalizing evolutionary theory, if the goal is to rely on a background theory to abstract causal roles that must be fulfilled for evolution to occur, then Weismannism is a bad choice for two sorts of reasons. First, it gets the facts about genes and organisms wrong (Griesemer submitted). Second, and more significantly, it gets the abstract structure of the causal relations between genetic material and somatic material wrong. If hereditary transmission is controlled by somatic development, then there *are* causal arrows from soma to soma (ibid.). This does not imply, however, that the flow of genetic *information* in its currently evolved state is

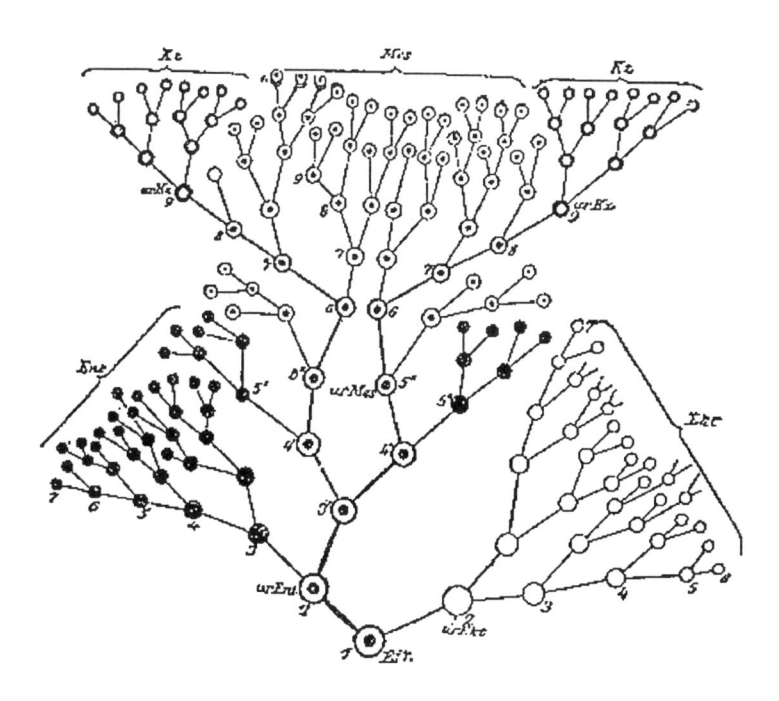

Figure 6. Weismann's representation of the continuity of the germ-plasm and discontinuity of the soma, reproduced from Weismann (1893, p. 196, fig. 16). The original figure legend title is: "Diagram of the Rhabditis nigrovenosa." The legend also describes the various elements of the picture.

otherwise than molecular biology tells us. It does imply that the specific mechanisms by which genetic information flows from gene to gene may be particular to that sort of material or that level of organization or that evolutionary state and hence undesirable as a basis for the generalization of evolutionary theory. Indeed, it should give some pause to those who would rely uncritically on the notion of information for the sake of interpreting the causal process of evolution by natural selection. We would be better off returning to Weismann's original, richly connected theory (and diagrams) of heredity-development as a basis for generalization by abstraction. It is indeed ironic that evolutionists trying to integrate development with evolutionary theory attack Weismann by arguing that Weismannism is contrary to the evolutionarily significant facts of development and to propose an alternative view that approximates Weismann's theory (e.g., Buss 1987). In so far as the Wilson diagram of Weismannism has *become* entrenched as the background theory, its inadequacies may block further progress in evolutionary theory.

References

Achinstein, P. (1968), *Concepts of Science, A Philosophical Analysis*. Baltimore: The Johns Hopkins University Press.

Buss, L. (1987), *The Evolution of Individuality*. Princeton: Princeton University Press.

Dawkins, R. (1976), *The Selfish Gene*. New York: Oxford University Press.

_____. (1982), *The Extended Phenotype*. New York: Oxford University Press.

Griesemer, J. (1990), "Modeling in the Museum: On the Role of Remnant Models in the Work of Joseph Grinnell", *Biology and Philosophy* 5: 3-36.

_____. (1991), "Must Scientific Diagrams Be Eliminable? The Case of Path Analysis", *Biology and Philosophy* 6: 155-180.

_____. (submitted), "The Informational Gene and the Substantial Body: On the Generalization of Evolutionary Theory by Abstraction".

Griesemer, J. and W. Wimsatt (1989), "Picturing Weismannism: A Case Study of Conceptual Evolution", in M. Ruse (ed.), *What the Philosophy of Biology Is, Essays for David Hull*. Dordrecht: Kluwer Academic Publishers, pp. 75-137.

Grinnell, J. (1910), "The Methods and Uses of a Research Museum", *The Popular Science Monthly* 77: 163-169.

_____. (1924), "Geography and Evolution", *Ecology* 5: 225-229.

_____. (1943), *Joseph Grinnell's Philosophy of Nature*. Berkeley: University of California Press.

Hall, E. (1962), *Collecting and Preparing Study Specimens of Vertebrates*. Lawrence: University of Kansas Press.

Herman, S. (1986), *The Naturalist's Field Journal, A Manual of Instruction Based on a System Established by Joseph Grinnell*. Vermillion, South Dakota: Buteo Books.

Hesse, M. (1966), *Models and Analogies in Science*. Notre Dame: University of Notre Dame Press.

Hull, D. (1980), "Individuality and Selection", *Annual Reviews of Ecology and Systematics*, 11: 311-332.

_____. (1981), "The Units of Evolution: A Metaphysical Essay", in U. Jensen and R. Harré (eds.), *The Philosophy of Evolution*. Brighton: The Harvester Press, pp. 23-44.

_____. (1988), *Science as a Process, An Evolutionary Account of the Social and Conceptual Development of Science*. Chicago: University of Chicago Press.

Leatherdale, W. (1974), *The Role of Analogy, Model and Metaphor in Science*. Amsterdam: North-Holland Publishing Co.

Li, C. C. (1975), *Path Analysis — A Primer*. Pacific Grove: Boxwood Press.

Maynard Smith, J. (1965), *The Theory of Evolution*. (2nd ed.), Middlesex: Penguin.

Merriam, C. H. (1894), "Laws of Temperature Control of the Geographic Distribution of Terrestrial Animals and Plants", *National Geographic Magazine* 6: 229-238.

Provine, W. (1986), *Sewall Wright and Evolutionary Biology*. Chicago: University of Chicago Press.

Star, S. and J. Griesemer (1989), "Institutional Ecology, "Translations," and Boundary Objects: Amateurs and Professionals in Berkeley's Museum of Vertebrate Zoology, 1907-1939", *Social Studies of Science* 19: 387-420.

Tukey, J. (1954), "Causation, Regression and Path Analysis", in O. Kempthorne, T. Bancroft, J. Gowen and J. Lush (eds.), *Statistics and Mathematics in Biology*. Ames: Iowa State College Press, pp. 35-66.

van Fraassen, B. (1980), *The Scientific Image*. Oxford: Clarendon Press.

Weismann, A. (1892), *Das Keimplasma, Eine theorie der Vererbung*. Jena: Gustav Fischer. English translation (1893) W. Parker and H. Ronnfeldt, *The Germ-Plasm, A theory of heredity*. New York: Charles Scribner's Sons.

Wilson, E. B. (1896), *The Cell in Development and Inheritance*. London: Macmillan Co., (2nd Ed. 1900, 3rd Ed. 1925).

Wright, S. (1918), "On the Nature of Size Factors", *Genetics* 3: 367-374.

_____. (1920), "The Relative Importance of Heredity and Environment in Determining the Piebald Pattern of Guinea Pigs", *Proceedings of the National Academy of Science* 6: 320-332.

_____. (1921), "Correlation and Causation", *Journal of Agricultural Research* 20: 557-585.

_____. (1934), "The Method of Path Coefficients", *Annals of Mathematical Statistics* 5, 161-215.

_____. (1960), "Path Coefficients and Path Regressions: Alternative or Complementary Concepts?", *Biometrics* 16: 189-202.

Mapping Ecologists' Ecologies of Knowledge

Peter J. Taylor

Cornell University

1. Introduction

Ecologists, particularly those who consider socially generated effects in the environment, grapple with complex, changing situations. Historians, sociologists and philosophers studying the construction of science likewise attempt to account for (or discount) a wide variety of influences, which make up what historian Charles Rosenberg has called "ecologies of knowledge" (Rosenberg 1988). This paper introduces a graphic methodology, mapping, designed to assist researchers at both lev-

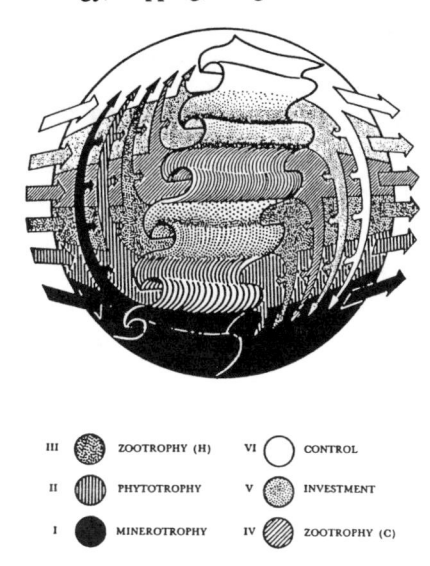

Figure 1. "A scheme showing a projection on the six tropic levels of the mainstream flow of energy (central part) [the waves], the supply of resources (left part), and the reinvestments (right part), as well as the import (left margin) and export (right margin)" (from Dansereau 1973, p. 48).

els—in science and in science studies—to work with the complexity of their material. By analyzing the implications and limitations of mapping, I aim to contribute to an ecological approach to the philosophy of science. Let me start with two diagrams to open up the territory I will be exploring.

Dansereau's diagram (Fig. 1) conveys a dynamic equilibrium of energy and resource import, export, flows up a trophic hierarchy, and down again to exert control over lower levels in an ecosystem (Dansereau 1973). The diagram is a pictorial metaphor, connoting the dualism in ecology of process (flows) and structure (levels and interdependencies), and doing so with a vividness difficult to achieve verbally. The interpretation of pictorial representation is not, however, my topic (see Taylor and Blum 1991a,b). Dansereau's diagram fills two other roles here. First, he reminds us that ecologists must address complexity and on-going process. In doing so they often partition the world into systems, clearly bounded and internally integrated, interacting simply with the context or "environment" of the system (Taylor 1992a). In fact, there is no explicit environment in Dansereau's diagram. Nor does he identify organisms, which exemplifies a popular way to simplify ecological complexity—reduce all organisms and activities to a common currency, e.g., energy or information, and thus obviate the need to address their heterogeneity or particularity (Taylor 1988).

Dansereau's diagram is a neat distillation of ecological complexity. Therein lies its second role—the neatness invites us to consider the opposite, the unruliness of complexity. My sense of ecology is that boundaries and categories are problematic; levels and scales are not clearly separable; structures are subject to restructuring; control, generalization and prediction are difficult. The exercise of mapping attempts to discipline this ecological unruliness, but without suppressing it. Figure 2 gives a preview

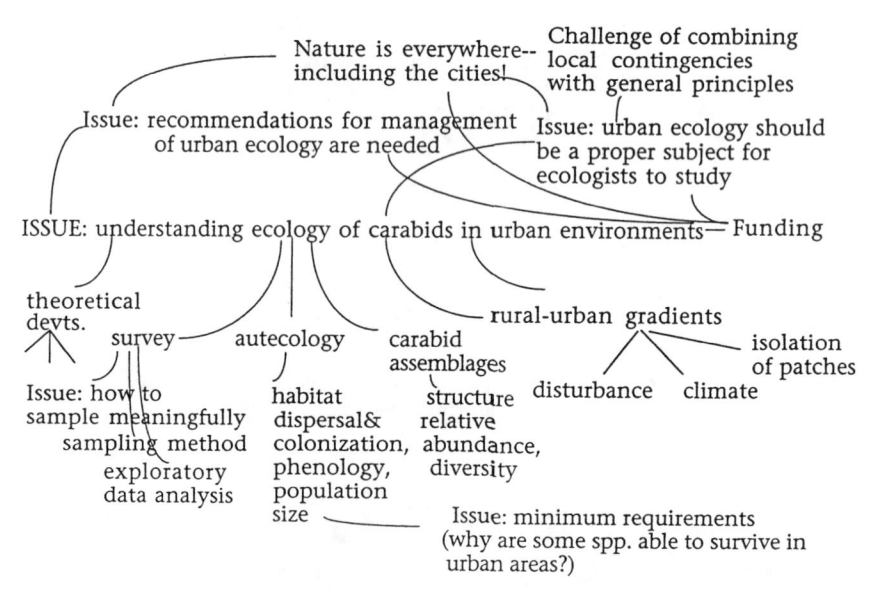

Figure 2. Outline of E's map concerning the urban ecology of carabid beetles in Helsinki.

of a map. This map, though already distilled from the original in order to prepare copy for publication, is clearly messy, detailed, and idiosyncratic. Moreover, it does not stand on its own but requires narration to give significance and weighting to the labels and arrows. Instead of narrating this map, however, let me step back and develop more systematically the context in which I intend such maps to be considered.

2. Ecologies of Heterogeneous Resources

Sociologists and historians of science have begun to advocate "ecological" analysis of scientific knowledge and activity, integrating diverse social and historical elements with conceptual and methodological developments (Rosenberg 1988, Star 1988). In this spirit, Law, Latour and other sociologists have described the "networks" of heterogeneous resources that scientists link together to support theories, resources such as equipment, experimental protocols, citations, colleagues, the reputation of laboratories, metaphors, rhetorical devices, funding, publicity, and so on (Law 1986, Latour 1987). Difficult questions immediately arise: What is entailed by the metaphor of ecology (or network), beyond the very general characterization that there are many factors interacting? What makes something a resource (or a node)? But these questions may be seen as jumping the gun, in that they presuppose an acceptance of the ecological/network image of science. So first we need to look at how philosophy of science is responding to sociologists' attempts to shift the dialogue between theory and reality into the wings and to usher onstage a rabble of unfamiliar things.

One response is to propose that the non-propositional aspects of an ecology of knowledge are theoretically superfluous. These aspects are unavoidable, perhaps even necessary (Hull 1988, Laudan 1990), but nonetheless the harnessing of resources such as citations, reputations etc. does not change what is justifiably held to be true about the natural world. This is a strong claim. Unless it is merely a claim of faith, it obliges us as philosophers to demonstrate that no changes in the resources employed by the scientist would have produced a significantly different theory. To soften the claim, we might hold that the original generation of a theory may be strongly driven by metaphor, funding, etc. which obscure aspects of the natural world or distort our understanding. But, by the time a scientific community reaches a strong consensus about a theory (or achieves a repeatable effect using a theoretical entity such as an electron), it is the natural world, and not the divergent, idiosyncratic situations of the members of the community, that accounts for the consensus (or repeatable effect) (Hacking 1983, Laudan 1990). Without wanting to doubt the effectiveness of theoretically informed interventions in laboratories, it's not clear to me how to determine whether, and for what aspects of a theory, we are in the temporary diversion period. How do we know when we have reached the reliable consensus? Instead, could we advocate a compromise position?—Admit equipment, reputations, etc. as ever-present contributors to theory generation and confirmation, but hold that the natural world, through the results of experiments and observation, carries a special weight in scientific deliberation. To develop such a hybrid theory we need to be able to assign relative weights to the real and artificial, nature and society, necessity and contingency. A difficult project; it is always tempting to allow nature "in the last analysis" to regain the causal throne.

A final philosophical response would be to embrace, rather than ward off, the conceptual challenges of social studies of science. Latour (1987, p.93) suggests a definition of reality that dissolves the boundary between philosophy and sociology: "Reality is that which has resisted all efforts at modification." The sociologist might emphasize how the organization of scientific work, schools, and institutions bolsters theories; the philosopher might emphasize the degree of confirmation achieved for theories. But for both it becomes an open question, under Latour's definition of reality, as to whether social or theoretical changes would be the most efficacious route to exposing problems of correspondence between theory and reality. Science then becomes simultaneously representation and intervention in natural and social realms. Although a convincing exposition of this last response is beyond the scope of this paper, I trust it is sufficiently plausible to be worth pursuing, and I take it as the point of departure for the mapping project.

Let us return to the questions of how to reconstruct the ecology of resources supporting theories and of what qualifies as a resource. Of the exponents of ecological or network accounts of science Latour, the French anthropologist of science who I have cited several times already, is the most explicit about methodology (see Latour 1987, pp. 258-9). He exhorts us to begin with no *a priori* assumptions about what science is, but to follow individual scientists "in action", or, more accurately, during controversies, and observe what they do in response to their opponents. Their actions define what constitute resources in the particular controversy—if they invoke the reputation of a laboratory or the reliability of an experimental control, then these are resources. This broad definition of resources allows Latour to produce rich descriptions of scientific controversies, but does not lead him to a clear specification of what it means for resources to be arranged into networks.

Many commentators label Latour as an actor-network theorist, but, in practice, Latour does not follow the approach of defining actors as nodes and connecting the nodes into networks according to their similarity of shared citations, words, and so on—the stuff of scientometrics. Instead, the image Latour paints is of agents who accumulate citations, invoking the support of colleagues or the track record of a piece of equipment, and so on, and link these "resources" around claims that a theory is justified or a practice is effective. These resources, in turn, have networks built around them. The resulting networks are very unruly, scarcely amenable to scientometric analysis. In practice, Latour disciplines his networks by concentrating on historical cases in which the controversy has been settled, not on cases still in the thick of action. With the benefit of hindsight he provides us streamlined accounts of the resources. Moreover, his networks collapse almost into piles of resources—it becomes the sheer numerical weight of citations, for example, that renders one scientist's claim harder to modify or discount than their opponent's (Latour 1987, pp. 92-3). The arrangement and mutual reinforcement of resources remains vague.

If we want to develop the network or ecology metaphor much further we need to remove four obstacles raised by a Latourian analysis of scientists harnessing heterogeneous resources. 1) Latour and others have attempted to expand the definition of agents to include non-human and non-living things. In order to keep the focus on what humans do to establish knowledge, let us not follow this tidy but unilluminating move (see Collins and Yearley 1991). 2) Instead of an image of scientists building networks simply in response to the "stimulus" of others building competing networks, we should view scientists as imaginative agents, who assess, in advance of acting, the practical constraints and facilitations of their possible actions. 3) Imaginative individuals are cognizant, at least to some extent, of the regularities or structuredness of constraints and facilitations of their actions. There is no need, therefore, for us to feign perpetual naivete about science and act as if we were starting afresh every time we began to follow a scientist in action. 4) Finally, the resources exposed during a controversy do not necessarily make up the full set used by a scientist. We need to expand our analysis to include resources taken-for-granted by opposing parties, and, moreover, resources used even when there is no apparent controversy.

Extending our analysis to non-contested resources suggests a generalized definition of a resource as something that makes a difference, that makes the claim more difficult to modify. If we accept this deceptively simple definition, then we face two challenging methodological problems: 1) how to derive ideas of what else could have been, even if there is no controversy to provide concrete evidence of alternatives; and 2) how to assess the theoretical and practical implications of pursuing the alternatives, even if no one was actually inclined to do so. In short, an ecological philosophy/sociology of knowledge must eventually be able to address complex causality and counterfactuality (Taylor 1992b).

Constructing such a conceptual apparatus is not, however, the purpose of this paper. Instead of defending propositions, I want to take an exploratory turn, to open up further questions that an ecological philosophy/sociology of science might address. My exploration starts from the recognition that every day science, not just controversial science, is built by linking resources. What emerges then if we try to get scientific researchers to reflect explicitly on their ecologies of knowledge? This is where mapping enters the picture.

3. Maps

Recall the spirit of ecological analysis—the integration of diverse social and historical elements with conceptual and methodological developments. We need, therefore, to lead researchers to consider how they address both the natural situations they are studying and the social situations in which they organize their research. The technique we call mapping requires researchers to focus on key issues and to trace the practical and theoretical "connections" that they see motivate, facilitate, or constrain their inquiry and action. Let me explain some of these terms. Issues include questions, disputes, and actions in which the researcher would like to know more or act more effectively. *Key* issues, because a researcher needs to be somewhat motivated to undertake the work of mapping. *Connections*, a more neutral term than resources, so that, at least initially, the map-maker doesn't evaluate the causal influence of every connection before including it. Connections might include theoretical themes, empirical regularities, methodological tactics, organisms, events, localities, agents, institutional facilities, disputes, debates, and so on. *Maps*, because a pictorial depiction, employing conventions of size, spatial arrangement, and perhaps color allows many connections to be viewed simultaneously. The metaphor of a map is not intended to connote a scaled-down representation of reality, but instead a map serves as a guide for further inquiry or action—to show the way.

The map-makers, to date, have been drawn from the fields of ecology and natural resources in two workshops of six or seven researchers: 1) ecologists at the University of Helsinki (where I collaborated with ecologist and philosopher, Yrjö Haila), and 2) resource ecologists/economists—let me call them "socio-ecologists"—at the Energy & Resources Group of the University of California at Berkeley. Almost all were advanced graduate students with several years of research experience, self-selected by their willingness to commit time to reflect on their current research and possible future directions. The workshops were not, unfortunately, recorded on audio or videotape so the only data surviving are the maps and my notes. Further details of the procedures adopted in these workshops are given in Taylor and Haila (1989).

What was the outcome of these "pilot projects", in which ecologists reconstructed their own ecologies of knowledge? Let me describe three maps. Fig. 2 by "E", a Finnish ecologist studying carabid beetles in the leaf litter under trees, is the most or-

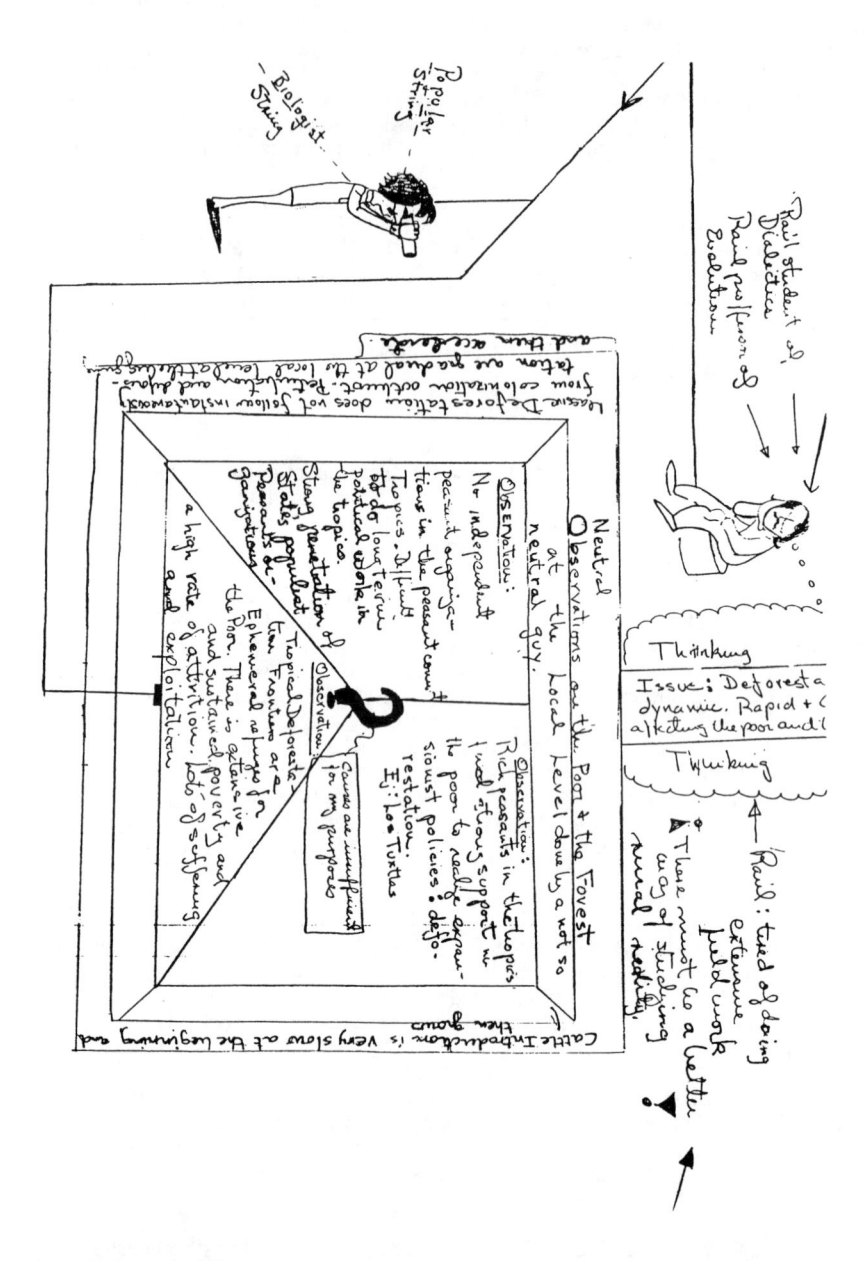

Figure 3. Extract from R's map concerning peasant economics and politics and tropical forest destruction in Mexico.

derly of the maps, having been redrawn on a computer for publication. His central issue was very broad: to understand the ecology of carabids in urban environments. Below this issue on E's map are many theoretical and methodological sub-problems, reflecting the conventional emphasis in science of refining one's issue into specialized questions amenable to investigation. Above the central issue are various background considerations, larger and less specific issues, situations, and assumptions that either motivate work on the central issue or are related to support for the research. E's research alone will not transform the urban public into recognizing that "nature is everywhere—including in the cities," but by combining the upward and downward connections, E reminded himself that work on the background issues, not just refining a working hypothesis, would be necessary to keep his research "do-able" (Fujimura 1987). This interpretation of E's ecology of knowledge is strengthened by mentioning some history not discernable from this map. In narrating his map, E mentioned that many of the ecologists with whom he collaborates were studying a forest area, but the group lost their funding when the Forestry Department asserted that forest ecology was their own domain (even though animals are barely mentioned in the ecology of forestry scientists). E and his collaborators self-consciously, but of necessity, turned their attention to the interconnected patches of forest that extend almost to the center of Helsinki, and explored novel sources of funding and publicity (including a TV documentary). The upward connections were thus a recurrent, if not, persistent, influence on E as he defined his specific research questions.

Historical background depicted in a narrative format is more evident in a large map by "R", which depicted how he had come to specialize on the economic and agronomic dynamics which lead to impoverishment of peasants in Mexico, their migration into forest areas, and subsequent clearing of those forests. Fig. 3 is just one section of that map. Although radically different from E's redrawn map, R also highlighted simultaneous issues of building the disciplinary and collaborative context in which to pursue his many concerns: as a biologist with rainforest destruction, as a political activist with rural poverty, and as a graduate student with framing technical questions that he could answer.

In Fig. 4, however, "M", an American studying land degradation and impoverishment among nomadic pastoralists in West Africa, depicted a more conventional conception of research: questions form the bulk of the map, separated from methods—the strip along the bottom. M omitted the movements, arrangements, alliances, and negotiations he built so that he could monitor milk production, elicit from the herders rules governing herd movement, assess herd ownership, measure the effect of grazing on pasture growth, complete surveys to "ground truth" satellite images, and so on. M's map also located him in his remote field area, and omitted the audiences in the U.S.A.—sponsors and critics alike—for his current and future research. In short, resisting the guidelines given to mappers, M included his research object and left himself out.

4. Problems and prospects for mapping

None of these maps is very remarkable, and mapping might be viewed merely as a stimulating tool for working with advanced students (Taylor and Haila 1989). Furthermore, no generalizations can be drawn from a sample of three maps. Remember, however, the exercise was intended to be exploratory, not to be conclusive. The two mapping workshops raise or highlight issues needing to be pursued, both in developing the mapping approach to expose scientists' ecologies of knowledge, and, more broadly, in exploring ecological approaches to philosophy/sociology of science.

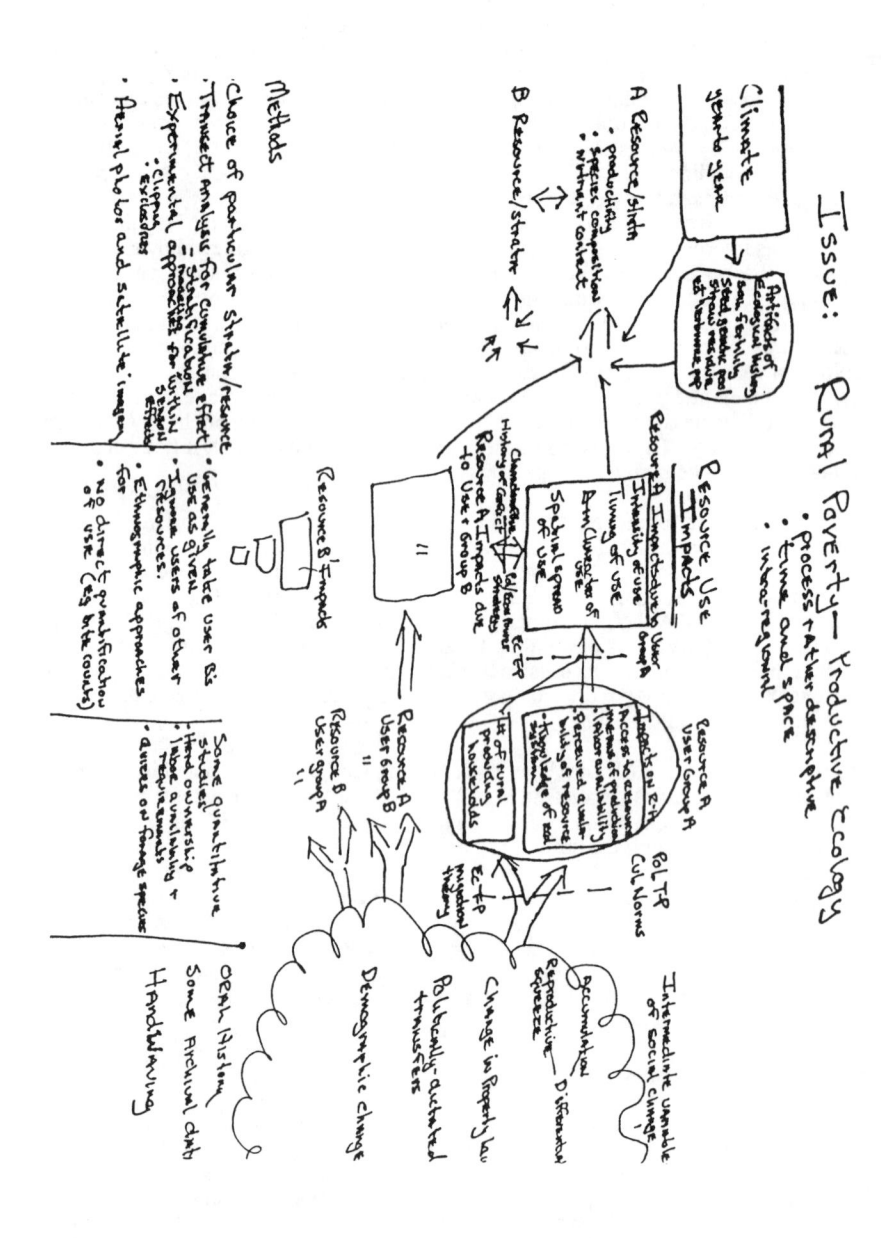

Figure 4. M's map of his research into ecological degradation and impoverishment among nomadic pastoralists in West Africa.

1) Maps vs. mapping: Maps, as I mentioned earlier, do not stand on their own but are accompanied by the mapper's narration. For example, E's map must appear quite bland to anyone not present at the Helsinki workshop or otherwise familiar with his research. However, the snippet of additional history about funding of forest research indicates the additional detail that emerged in the mapping process. If the workshops had been taped this detail would be retrievable. Of course, including this detail on the one map would produce a crowded result, difficult to take in visually. Overlays or blow-ups could be used to separate the connections into a first-level "skeleton" and second-level "tissue."

2) Idiosyncracy of maps: To the extent that this is seen as a problem, creating an obstacle to systemization or generalization, workshop leaders could become more directive, urging a standard format for maps or injecting into the process selected interpretive frameworks from the literature and model taxonomies of connections derived from previous workshops. However, idiosyncracy could also be viewed as salient data—this may be what you get when scientists reflect on their ecologies. In the second workshop, in fact, despite shared vocabulary and common problems, the various socio-ecologists drew widely divergent kinds of maps. Could diversity-underlying-a-shared-core be the norm, even in sharply defined research areas? Entertaining such a possibility is consistent with the "naturalistic turn" that Callebaut (1991) has observed among philosophers of biology.

3) Self-reporting: Skilled researchers are by no means skilled self-observers. The workshop format was adopted so that participants' thinking would be exposed to questioning by other participants. The questioning was intended to result in clarification and reorganization, forcing the maps to be revised (but see 7 below). In this way, we might elicit M's knowledge of his own action and context and encourage him to include it in his map.

As an alternative to self-mapping, a third party could construct the maps of the researchers, drawing on interviews, archives, published material, ethnographic observations, and so on, and then perhaps use these maps to elicit further detail and revision from the researchers.

4) Workshops vs. other contexts of reporting: A workshop is, of course, not a neutral forum for scientists to present their maps. Sociologists of science have analyzed the diversity of versions of scientific knowledge and method that scientists produce in different contexts (Gilbert and Mulkay 1984). Trevor Pinch (pers. comm.) has suggested to me that the blandness of the maps may indicate more than a visual pruning of detail evoked in the workshops (see 1 above), but be a result of scientists talking in front of their colleagues. In-depth, one-on-one interviews elicit juicier detail about the constraints and opportunities to which scientists have been responding. Furthermore, what people do does not necessarily correspond to what they say (Briggs 1986). From a sociologist's perspective, therefore, mapping should be, at most, one tactic among many (see 3 above) for exposing ecologies of knowledge.

5) Connections vs. resources: The instruction to the researchers to identify connections sidestepped the question of what constitute resources. Certainly, some assessment of which connections are important is reflected in the choice of which connections to report and which to defend under "cross-examination" by the other workshop participants. In effect, such choices indicate that the connection makes a difference to the researcher, that modifying or eliminating it would significantly alter their ecology of knowledge.

In order to stimulate researchers to weight the connections, the workshop leaders could prescribe the drawing of "shadow" maps in which alternative connections were specified and the researcher imagined the consequences for their actions or inquiry. A skillful choice of alternative connections could expose what connections make a significant difference to the researcher. In addition, a third party's map (see 3 above) could enhance this process (but see 7 and 8 below).

6) Unrepresentative mappers: Most of the map-makers were graduate students; more established scientists did not choose to join the workshops. Furthermore, the participants were to some extent attracted by the "transformative agenda" of the workshops. For example, for the first workshop the pre-publicity stated that one question motivating the workshop was "how to steer ecological science so it is not overspecialized, but instead remains responsive and relevant to environmental concerns." The researchers responding to such an agenda would not be representative of all ecologists. The problem of representativeness would not necessarily vanish if mapping were done by a third party, because the sample of researchers who provide access through interviews and opening up their records is not guaranteed to be representative.

7) Resistance to revision: Maps take many hours to prepare by hand, and not surprisingly the mappers resisted revising or redrawing their maps. For example, M never redrew his map to include his own context. To help overcome that resistance computer software that combines diagramming with a textual data base is being developed that will facilitate redrawing of maps. In addition, the time period of the workshops (five or six afternoons over a two week period) was insufficient for the researchers to digest the insights derived from the workshop. Continuing meetings at three or four week intervals after the initial intensive two week workshop might be necessary to elicit deeper reflection on and reconceptualization of the researchers' ecologies of knowledge. Ultimately, however, we can expect there to be a residual resistance to modifying the connections, insofar as the connections are perceived as resources for the researcher's actions.

8) Individual-centeredness: Maps are drawn by individual researchers who have little experience in identifying institutional or structural regularities that may run through the ecologies of many researchers. The focus on individual scientists in action almost ensures a limited horizon that doesn't encompass social structure. In an attempt to help bring such regularities into the maps, the shadow mapping exercise (see 5 above) could help expose higher level resources that had been taken for granted.

For example, E's central issue became more likely to command research support when the need for recommendations for management of urban ecology was perceived. Research support might also become more available if urban ecology became a proper subject for ecologists to study. The idea that Nature is everywhere—including in the cities—would certainly promote or reinforce these changes. A little reflection on this last idea reveals that a major obstacle to these changes is the conventional wisdom that Nature worth preserving is Nature that affords us an escape from culture, undisturbed and far from the cities (Williams 1973, 1980) This conventional wisdom is a regularity of E's social context, a taken-for-granted resource for most ecological research, but a constraint for E's research.

9) Lack of a temporal dimension: The spatial arrangement in a map allows many connections to be considered simultaneously. On the other hand, it privileges a static idea of resources, whereas, in fact, many resources are only mobilized or made visible over the course of events. R clearly recognized this in adopting a narrative superstructure for his maps. Certainly, it would be fascinating to see how researchers drew successive

maps over time, but even then each map would still have static connotations. Further exploration of graphic convections for signifying temporality and change is needed.

5. Problems and prospects for philosophy of science

The problems and possible solutions sketched in the previous section indicate that mapping is at a very preliminary stage of development. More experience is needed before mapping can shed well-focused light on researchers' ecologies of knowledge. Nevertheless, the initial exploration has already raised deeper issues for an ecological philosophy/sociology of science to address whether or not mapping is pursued.

1) Incommensurability as the rule: Nowadays most model-centered philosophies of science admit that the terms of models achieve meaning not just in relation to the object represented, but from the terms' embeddedness in larger fabrics of propositions and in language in general. Marked change in the context therefore raises the question of incommensurability—can we really claim that a model, embedded in a new paradigm or advocated by a changed linguistic community, is a better representation than an earlier model once we conclude there is no way to compare the models *ceteris paribus*? (Kuhn 1990). The idiosyncracy of maps suggests a different perspective, one which undermines the problem of incommensurability by seeing it as the rule. Meaning or shared understanding should be seen as always a negotiated state, a temporary elevation against a background of, at best, partially overlapping "maps." The indeterminacy of this perspective is similar to an interpretation of metaphors that locates their power in their indeterminacy, being open at three different levels: the associations of one field which illuminate the other are never fully explicated, the associations vary among users or readers, and the associations vary among contexts or over time (Taylor and Blum 1991b).

2) Representing or representing-intervening?: Although I have discussed maps as representations of researchers' ecologies of knowledge, to be consistent with the fusion of philosophy and sociology, mapping should have been presented in terms of the actions or "interventions" facilitated by the researchers reflecting explicitly on the construction of their research. Did this form of reflection help them evaluate the effect of modifying some of the connections, and, eventually, to reconstruct their work?

Several participants at the Helsinki workshop, in particular, claimed that the mapping workshop had expanded the range of influences, both theoretical and methodological, that they would bring into planning their future work. Nevertheless, although the workshops provided the opportunity to link up with others around revealed affinities, no new coalitions have emerged; the researchers are making their way more or less as before. I am not surprised by this outcome—if science is simultaneously representation and intervention then enhanced representations (the maps) of our science would be unlikely, by themselves, to modify the resources and constraints shaping that science. On the other hand, I reject the claims made most forcefully by the literary critic, Stanley Fish, that becoming "more self-consciously situated [does not enable us to] inhabit our situatedness in a more effective way" (Fish 1989, p. 347) Reflection upon the construction of one's research is, admittedly, no guarantee of being able to change the available resources, but it remains an open empirical question how much such reflection can stimulate scientists to modify or even reconstruct their ecologies of knowledge. Nevertheless, at this early stage in conducting mapping workshops I make no grand claims to be facilitating significant new representation-interventions among scientists. In fact, it may be noted that in this paper I have myself stayed with a textual rather than a graphic mode of presentation, using the conventional linear structure and not some spatial arrangement.

The astute reader may also detect in these last two sentences an attempt to insulate myself from any deeper questions about my own agenda as a representor of/intervener in several areas—ecology, socio-ecology, philosophy and sociology of science (see Taylor 1992b). I must concede that if I had explored mapping with just one of these areas in mind I would have been obliged to address more directly the concerns of the practitioners in that area. Instead my translation of mapping into this text can, at best, be expected to invite, rather than convince others to develop an ecological perspective in science and science studies.

3) Resources and counterfactuals: As I remarked earlier, the judgement of what connections makes a difference depends on a sense of what else could be or could have been, in other words, on an assessment of counterfactuals. This issue is of more general relevance than the mapping project. While political activists always face the problem of creating something that is, at first, hypothetical, scholars usually feel more comfortable if their interpretations rest on a solid basis, either of logic or of empirical evidence about what actually happened. (For a trenchant argument by a sociologist of science against entertaining counterfactuals see Collins and Yearley 1991.) Yet even the most neutrally descriptive historians select and juxtapose in composing their narratives, thus giving weight to different factors and making implications about causality, that is, about what made a difference. Analysts who employ the comparative method make implications of causality more explicit. Two (or more) situations are chosen, similar in many aspects and different in a few identified factors. Different outcomes are then attributed to the combined effect of those few factors. In the more general case, no single comparison exists that can expose the relevant causal factors and a composite of comparisons must be employed (Taylor 1992b). In this spirit, the challenge for an ecological philosophy/sociology of science is to make sense of the mutual reinforcement of different resources and not to place too great a weight of causality on any isolated factor.

In the case of contemporary science we (or the researchers making maps) could conceivably decide to "test" our assessment of the resources supporting some given science, and deliberately intervene to see what gives and what holds. Mapping, even in a workshop that produces vigorous questioning, is gentler than intervention, generating only "accusations" that certain connections are resources. Yet the boundary between interventions and accusations is not distinct. In fact, we might view descriptive history and conscious activism as two ends of a spectrum of accounting for what is by reflecting on what could be or could have been—even if the account actually attempts a suppression of alternatives.

4) The shape of ecologies: Although I accused Latour of vagueness concerning the arrangement of resources into his networks, neither the maps included in this paper nor the others produced in the workshops demonstrate anything systematic about the shape or structure of ecologies of knowledge. The maps have, however, put a little flesh on the image of unruly complexity I associated with ecology. The *boundaries* of maps call out for negotiation—how far away from the individual researcher should the "horizon" of the map be drawn? And, in what directions should we attempt to stretch the conventional representations of science? Should we follow the ethnomethodological emphasis on the definition and redefinition of specific research questions and outcomes (Lynch 1991), or tease out the structure of indirect background considerations, such as E's issue of ideologies of Nature? In fact, if we expand outwards from the researcher's specific question, should the researcher remain at the center of the ecology? Instead of an ecology of resources supporting a scientific claim, why not an ecology of scientific activity, in which, for example, the rise and

fall of autonomous peasants' movements in Mexico or of international concern for pastoralists in West Africa become central conditions in the generation of R's and M's scientific activity. Once we entertain such shifts in focus the appropriate *categories* for philosophy/sociology of science hardly remain obvious. Yet, even if we maintain the traditional philosophical focus on scientific claims, to stabilize that focus we have to separate research questions from research work and social support. Unfortunately scientists' actions traverse such *levels*, even though they (and we) can perform from the "science *versus* society" repertoire. If we acknowledge the existence of resources in the work situation or in the wider social context and recognize regularities or structure in those resources, should we attempt to *generalize* about ecologies? Yet, if, by generalizing we discount or filter-out the contingency and idiosyncracy of scientists' actions, do we inject a degree of determinism which is not apparent in the individual situations? Finally, as we decide, perhaps by default, on our answers to these questions of representation, what *interventions* are we privileging? In a seamless philosophy/sociology of science we can no longer rationalize our choices as flowing from the goal of representing the activity of scientists as faithfully as possible (Taylor 1992b).

Too many questions; not enough argument or evidence? Maybe; in every part of my sketch fine brush work is clearly called for. Noticing such gaps, however, on a much enlarged canvas seems to be what is entailed once philosophy turns towards the "non-propositional side."

Note

[1] Greg Tewksbury suggested mapping to me; Yrjö Haila collaborated on the first application in the Helsinki workshop; and the ecologists and socio-ecologists who attended the Helsinki and Berkeley workshops participated in subverting our pre-conceptions. Anonymous reviewers of a related manuscript, members of the audience at PSA '90, and Trevor Pinch provided searching criticisms, many of which remain unanswered. Lori Hamilton's last minute secretarial assistance enabled the manuscript to reach the editors on time. To all these people I am grateful.

References

Briggs, C.L. (1986), *Learning how to ask: A sociolinguistic appraisal of the role of the interview in social science research.* New York.: Cambridge University Press.

Callebaut, W. (1991), *How to take the naturalistic turn: Exchanges on the new theory of science.* Chicago: University of Chicago Press.

Collins, H.M. and Yearley, S. (1991), "Epistemological chicken", in *Science as practice and culture,* A. Pickering (ed.). Chicago: University of Chicago Press.

Dansereau, P. (1973), *Inscape and Landscape.* Toronto: Canadian Broadcasting Corporation.

Fish, S. (1989), "Anti-foundationalism, theory hope, and the teaching of composition", in *Doing what comes naturally.* Durham: Duke University Press, pp. 343-355.

Fujimura, J. (1987), "Constructing doable problems in cancer research: Articulating alignment", *Social studies of science* 17:257-293.

Gilbert, G.N. and Mulkay, M. (1984), *Opening Pandora's box: A sociological analysis of scientists' discourse*. Cambridge: Cambridge University Press.

Hacking, I. (1983), *Representing and intervening*. Cambridge: Cambridge University Press.

Hull, D. (1988), *Science as a process: An evolutionary account of the social and conceptual development of science*. Chicago: University of Chicago Press.

Kuhn, T. (1990), Presidential address, Philosophy of Science Association, Minneapolis, Minnesota, October 20, 1990.

Latour, B. (1987), *Science in action*. Cambridge, MA: Harvard University Press.

Laudan, L. (1990), *Science and relativism: Some key controversies in philosophy of science*. Chicago: University of Chicago Press.

Law, J. (1986), "On the methods of long-distance control: Vessels, navigation and the Portuguese route to India", in *Power, action, belief*, J. Law (ed.). London: Routledge & Kegan Paul, pp. 231-260.

Lynch, M. (1991), "Science in the age of mechanical reproduction: Moral and epistemic relations between diagrams and photographs", *Biology and Philosophy* 6: 205-226.

Rosenberg, C.E. (1988), "Woods or trees: Ideas and actors in the history of science", *Isis* 79:565-570.

Star, S.L. (1988), "Introduction: The sociology of science and technology", *Social Problems* 335:197-205.

Taylor, P.J. (1988), "Technocratic optimism, H.T. Odum, and the partial transformation of ecological metaphor after World War II", *Journal of the History of Biology* 21:213-244.

_____. (1992a), "Community", in *Keywords in evolutionary biology*, E.F. Keller & E. Lloyd (eds.). Cambridge, MA: Harvard University Press.

_____. (1992b), "Re/constructing socio-ecologies: System dynamic modeling of nomadic pastoralists in sub-Saharan Africa", in *The Right Tool for the Job: At Work in Twentieth Century Life Sources*, A. Clarke and J. Fujimura (eds.). Princeton, NJ: Princeton University Press.

Taylor, P.J. and Blum, A.S. (1991a), "Pictorial representation in biology", *Biology & Philosophy* (In press).

_____. (1991b), "Ecosystems as circuits: Diagrams and the limits of physical analogies", *Biology & Philosophy* 6: 275-294.

Taylor, P. and Haila, Y. (1989), "Mapping workshops for teaching ecology", *Bulletin of the Ecological Society of America* 70:123-125.

Williams, R. (1973), *The country and the city*. N.Y.: Oxford Univ. Press

Williams, R. (1980), "Ideas of Nature", in *Problems in Materialism and Culture,* R. Williams (ed.). London: Verso Press, pp. 67-85.

Taming the Dimensions–Visualizations in Science[1]

William C. Wimsatt

The University of Chicago

The role of pictures and visual modes of presentation of data in science is a topic of increasing interest to workers in artificial intelligence, the psychology of problem solving, and increasing numbers of scientists in all fields who must deal with problems of how to represent large quantities of complex multidimensional data in an intelligible fashion. The use of pictures is marvelously illustrated by but not limited to the biological sciences, so I will use examples from elsewhere as appropriate. With the development of our visual technology—television, videotape, and the computer, the uses (and misuses) of visualization has properly become a matter not only of theoretical but also of practical concern. I will start with a practical story because it has multiple morals, both practical and theoretical—most of which lead elsewhere than I wish to go here.

A colleague of mine, J. Z. Smith, got into a discussion about pictures with his freshman humanities class. Whereas we academics trust and pay close attention to the written word, and find visualization mysterious and more than a little suspicious, he was quite astounded to find out that *their* characteristic reaction was quite the reverse. The evening "Eyewitness News", because pictured—often "live" by portable videocam—was to them irrefutable and trustworthy, whereas (partly through our efforts) they were quite aware of the means of manipulating the written word, and thus much more suspicious of it. Smith's message was that we academics spend all of our time telling our students how to read critically and suspiciously, but ignore the information channel from which they get most information—which to them is also the most real and reliable. (Similarly Bruno Latour (1987) comments that data presented in graphic format seems "more real" than the numbers from which the graph was drawn.) This should all seem quite surprizing and anomalous to most of us—it did to me.

Smith was able to make a deal with an audio-visual teacher at a local college (we don't have such things at my university, of course!) to tape a meeting of his class and edit it down to one minute, as if it were going on the nightly news. The A-V teacher was delighted —the taping and editing became a project for *his* class. They did an excellent job. But when Smith's class saw the finished result, they were outraged. Class incidents were collapsed, taken out of context, and even permuted (so that a student's question was "answered" by another's remark given earlier in response to another question). The students got a new respect for the foibles of the media.

But this is not the issue of my paper. Nor is this a good example to start an analysis with—it is too real, messy, and uncontrolled. It makes nicely the point of the ubiquity, immediacy, and supposed reliability of visualizations, but it doesn't suggest why. The promising issue of biases induced by visualizations in the presentation of data (an important topic—see Tufte (1983, 1990)) are in this case confounded with too many biological (sensory), psychological (cognitive and affective), and social (economic, cultural, and ideological) variables. It is a good case to keep in mind however, in response to questions like: "Why work on pictures? Aren't words (or equations) all that really count?" As Mike Ruse points out in his contribution, a surprisingly large proportion of scientists' work is devoted to taking, making, designing, and evaluating pictures. (See also Lynch and Woolgar 1990). There are things which are at least virtually impossible (to scientists or engineers, I would say impossible, but philosophers would misunderstand that) to do without visualization, and—whether possible or not—no-one in their right mind would try to do so. These are usually also things that visualization is best suited for. For pragmatic reasons, I will only sample these in a biased fashion: I have not the space, nor the facilities (color, high resolution graphics, motion pictures, or stereoscopic presentation) to do more. I will instead discuss (and illustrate) some of my favorite issues, focussing on robustness and multidimensional complexity.

Part of my message with these examples is that in using pictures, we are for many problems giving the most natural, economical, and inferentially fruitful representations of data. To the extent that theories aim to organize and simplify (in their role as templates for organizing data for analysis), pictures can have a theory-like status. (It would be tempting to say that they are in that respect propositional—but that biases the case just as strongly as if I said that all propositions were picture-like). Even if something is *in principle* representable in another way (conceivable, or even plausible if we place no constraints on complexity) we should be interested in finding the most natural and simplest ways of representing the information and structure of a problem—usually thus aiding both its robust understanding and its solution. This solution is often visual.

If something is easier, we will do it that way by choice, or—in a race of different methods against time—do it that way first. Getting there first is important in evolution, and indeed in all selection processes. Biologists are often asked, if it is so easy to make life, why isn't it happening all of the time? The standard answer is that it probably is, but it is getting eaten immediately by those who got there first. (See Simon (1981), ch. 7 for a more serious argument along the same lines.) Another variant of the same point is the observation that, once invented, it is easier to modify a method or an adaptation originally designed for different purposes than to construct a new one from scratch. (Schank and Wimsatt 1988)). In this light, it is sobering to realize that visualization is a far older adaptive mode than language. Visual thinking should turn out to be far more pervasive than those of us raised on "linguistic philosophy" have given it credit for.

1. Some simple examples (without pictures)

Can you intersect a cube with a plane in such a way as to make a regular hexagon? Most people say they don't know, or if pressed say that they are pretty sure you can't. (This is impressive because just *asking* the question in this way suggests that the answer is surely "yes" and they are supposed to figure out how.) This problem is rapidly solved through visualization, but almost impossible without it. One also can see (with the picture, but not before) that this construction *requires* a cube—a rectangular parallelpiped will not do—and that the solution has unanticipated symmetries. A very strong hint is that the solution will involve connecting the midpoints of edges of the cube (and remember—you need six of them!), but why this is a good hint is not obvious until one draws the visual solution. (I don't even know of a non-visual solution, though you could undoubtedly laboriously construct one by working backwards from the solution with an analytical (though still geometric) proof. I won't show you this picture—you should discover it yourself to feel the force of the visual solution. Draw the cube, and then try the hint.

Problem-solving often requires visualization, but so also does pattern recognition and discrimination. Indeed, this is probably the most basic skill for which the more complex visual systems were designed. (Orientation serving directed locomotion surely came earlier in simpler systems which could only tell light from dark, and thus could not discriminate patterns.) Edward Tufte (1983) opens his landmark book, *The Visual Display of Quantitative Information,* with an example due to statistician F. J. Anscombe. Anscombe presents 4 data sets of 11 pairs of points, each of which fit the same linear model equally (and tolerably) well. One cannot tell by looking at the tabulated data sets that one is a typical regression line of moderate slope with good Gaussian scatter, one is an absolutely straight regression line with no scatter save for one severe outlier, one is a beautiful but assymetric part of a parabola, and one is a degenerate case of a vertical line (no variation in the independent variable) with an outlier. Which case applies is of course very important both for causal inference and for judging the quality of the data.

Anscombe's point was to show that only a fool depends on statistical analysis alone without ever *literally* looking at the data, but the case is a set-up—a trap—in a revealing way. It is normal (at least for experimental data, where values are usually produced by systematically changing the values of the independent variable) to present the data ordered by monotonic increases in the independent (x) variable. If Anscombe's data-sets are rearranged in this way, one can immediately "see" the different character of the four sets of data, and "read off" the graphs from the tables of numbers. It is tempting to say that Anscombe's (abnormal) procedure shows that we don't really need to look at the data if we present it right. But this is a double-edged sword. It just as strongly supports the conclusion that we have internalized visual constraints in our normal manner of presenting the data! *Note that the statistics do not discriminate among different orders for presenting he data, so there would be no analytical help in telling us that we should order our data in this way.* Indeed order-independence would surely be held up as a necessary condition for any useful statistical measures. Things get very embarrassing for data analysts if data have to be presented in a certain order—statistics must be "aggregates" (Wimsatt 1985)). Storage and analysis of multi-component quantitative data in terms of systematic (monotonic) change of important variables is a characteristic —and characteristically visual—way of representing spatial order.[2] This is true whether the data is presented in analytical or tabular form or in visual 2- or 3-dimensional space.

2. Picturing multidimensional complexities using heuristics of the visual system

Representation of multi-component (more commonly, "multi-dimensional") data is now a "hot" topic. 2 and 3-spaces sound very limiting in these times of multiple regressions with hundreds of variables. We can't go to several hundred, but we can do a lot better than 2 or 3 by utilizing other discriminatory powers of the visual system. It is very easy with color, variation in size or intensity of symbols, and temporal "slicing", to represent 4 to 6 components or dimensions in a 2-D representation of a 3-

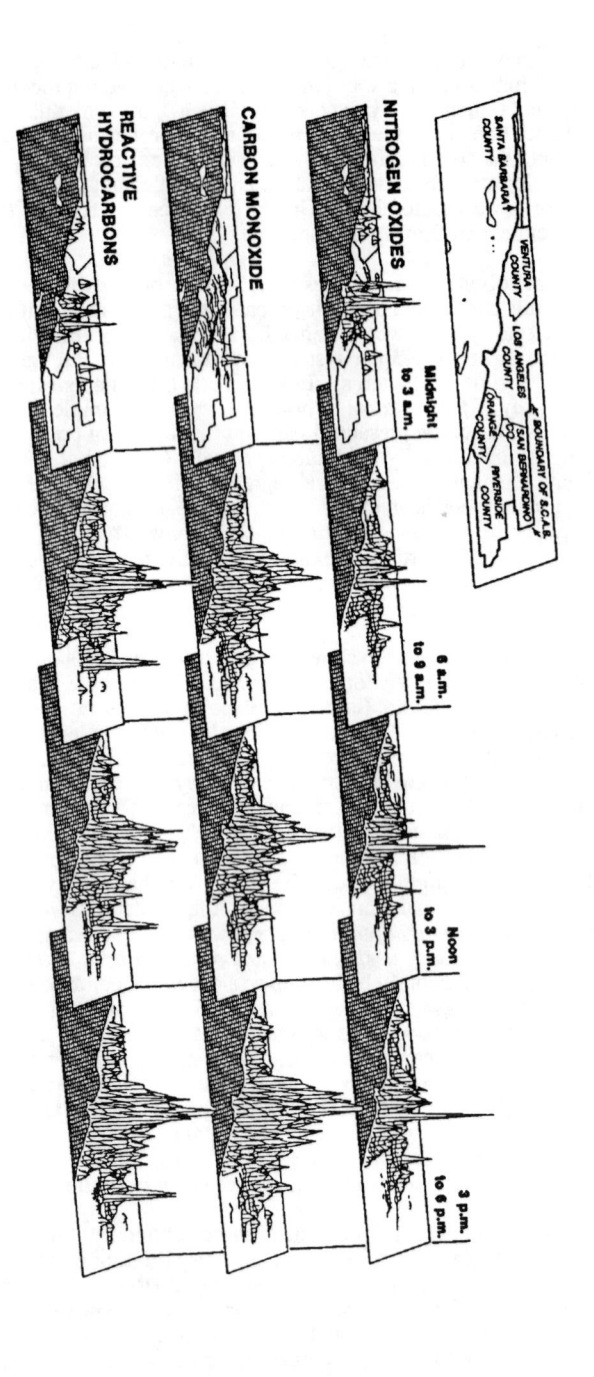

Figure 1: The use of "small multiples" (with a geographical key) permits compact representation of multidimensional data in a format which is not confusing. Here changing concentrations of three components of airborne pollution in Los Angeles are graphed over space and time—4 dimensions of data. *Los Angeles Times*, January 22, 1979, based on work of Gregory J. MacRae, California Institute of Technology, reprinted from Tufte 1983, p. 42.

Marey (1890-91), Pole Vaulter (From Marey, 1984, #44)

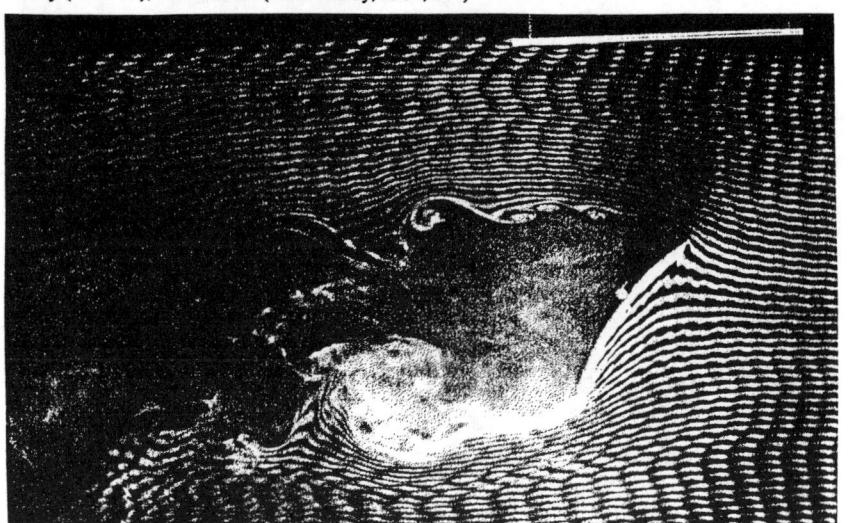

Marey (1900), Air flow in a smoke tunnel around a curved plate at a high angle of inclination. (From Marey, 1984, #57.)

space, as almost any good microcomputer statistics program readily demonstrates. The use of "small multiples" (Tufte 1983, pp. 42, 48, 50, and below) in which small graphs are ordered in a row or rows according to the values of one or two additional parameters can easily add one or two more dimensions, and various schemes have been proposed for the visual representation of data with higher numbers of dimensions—easily up into the range of 10-20 variables.

One of the most interesting of these devices is "Chernoff faces", which exploit our natural biological curiosity about and ability to discriminate faces accurately in terms of their characteristics. Faces have lots of quantifiable features, so they can represent data with many dimensions. A critical threshold is passed in moving up from "small multiples" to "Chernoff faces" and other like schemes, in which there is no natural ordinal mapping from values of the represented to values of the representing variable. (Thus, which end of the smile-frown continuum should be mapped to the higher values?) With small multiples (thus with up to about 6 variables) it is possible to look at a

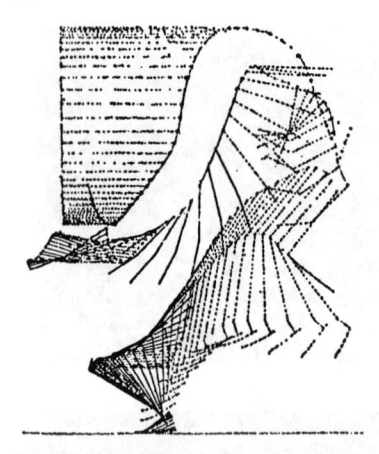

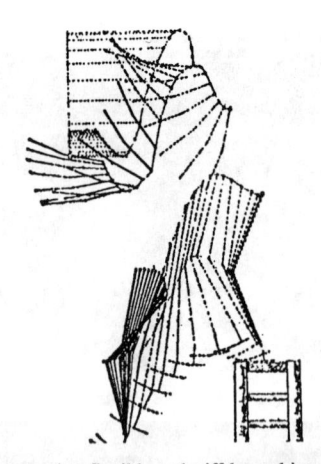

Figure 4: Jumping stick figures derived from photographs, comparing flexible and stiff-legged jumps. Note differ- ent trajectories of various parts of the body, and ease of measuring velocities and accelerations. Marey deliberately used stick figures (and had subjects wear black suits with white stripes paralleling all of their major anatomical members) to reduce the complexity of the visual images to manageable proportions. Marey 1895, pp. 142, 143.

point and without further help tell immediately "where it is"—i.e., what its values are in its 6-space. This is given up with Chernoff faces and similar schemes (though decoding is always possible later, and more familiarity may give it immediately), but one retains the ability to make fine multidimensional discriminations of points without knowing exactly "where" they are. Tufte (1983, p.142) has an excellent example of the recognition of a multi-dimensional outlier as a "stranger"—somepoint who "looks different", and one can also see "boundaries" and evaluate "connectivity" between regions in different subsets of the dimensions without knowing any more.

Perhaps even more important is the processing of visual information about motion. There are at least three important components here: (1) motion detection, (2) the integration of similar, temporally ordered, and spatially contiguous stimuli into ordered motion of identifiable though possibly changing objects, and building on these reifications, (3) "hard wired" tendencies to hypothesize causal relationships—though not always correctly. (The lack of general validity of these inferences is not problematic if we can either be warned when errors are likely, or restrict their application, through careful analysis and design, to situations where they work. See the discussion of *heuristics* in Wimsatt 1981, Griesemer and Wimsatt 1988.) In the modelling simulation used to generate the chaos pictures below, motion detection is useful in detecting equilibrium, approach to equilibrium, divergence from unstable equilibrium, bifurcations, and differential sensitivity of variables to parameter changes in the same and different parts of its range. (Motion can't be shown here because the PICT files from which they are generated are "frozen" in the ink of figures 9 and 10, but the desire to have effective ways of visualizing and measuring motion led biophysicist Jules-Etienne Marey in the late 19th century to the invention of highly effective photographic methods and static representations derived from them (figures 2 and 4), the smoke tunnel (for visualizing and analyzing airflow around aerodynamic surfaces, figure 3), and most of the technology for motion pictures. Marey did not have a computer to make the static PICT files printed here come alive. (They are redrawn by the computer in the same order as they were produced, so text-figures re-create the simulation live without the program used to generate them!) On a Macintosh, one can even paste many of these (color) PICTS into the "scrapbook", and scroll through them as a movie —doing Tufte's "small multiples" as successive runs of an experiment for different values played back rapidly in time.[3]) Seen as static objects, these figures often show a confusion of points. But presented in temporal order, the trajectories of these same points are integrated as ordered transitions—demonstrating the second point above. Finally, all of the simulations with this program are families of curves in which the same system is run many times for systematically changing values of a key parameter. The visual impression of a causal relation between changes in the parameter value and in the behavior of the system is both overwhelming and usually justified.

I will now move from phenomena I cannot show to other cases—many related to these—which can be appreciated with static images. A major theme of these cases will be that visual presentation is particularly appropriate when one needs to see or to track (where they are systematically changing) similarities and differences among complex entities which are changing in space, time, or—generalizing from these—in some subset of properties comprising the base components of an n-dimensional property space. These tasks make visual or spatial representation particularly appropriate:

(1) analysis of highly multidimensional data. (In effect, this means two or more dimensions.)

(2) The need to see, or to find, in such data (a) similarities in diverse cases in spite of their differences, or (b) differences in spite of their similarities, or (c) to factor perceived global dissimilarities into localizeable similarities and differences.

(3) The need to look for relationships among patterns on several different size scales or levels of organization (Wimsatt 1976, Tufte 1983).

(4) The need to combine or use jointly information from different perspectives on (i.e., different ways of accessing or measuring) the same object (Wimsatt 1974), or

(5) to determine where there are boundaries between objects, or to identify whether there is a common object, property, or cause, behind apparently correlated patterns (Wimsatt 1981).

The visual system has been selected for the efficient, reliable, and rapid solution of tasks like those of 2-5 above in a complex variegated natural environment. The discrimination, identification, and reidentification of objects, the identification of camoflaged predators, and the tracking of prey—all in the rapid service of feedback controlled locomotion, (McClamrock 1992) require an on-board computer that could well make our specialized language capacity seem like little more than a late add-on for the input and generation of relatively specialized kinds of information. (This needn't be true, but it may be, and it is only our language-using ethnocentrism which makes the possibility seem incoherent.) Even the term "perspective" here, in (4), betrays the visual origin of this metaphor for the multi-dimensional integration of diverse kinds of information about the same object. The connections among realism, objectification, and multiple independent means of access to a common object—beginning with our three spatial senses, vision, hearing, and touch—go very deep, (Wimsatt 1974, 1976, 1981), and almost certainly represent an interleaved family of heuristics with a long evolutionary history. The assumed salience of vision as an indicator of the real by Smith's students is not accidental.

Fig. 161. Pelvis of *Archaeopteryx*.

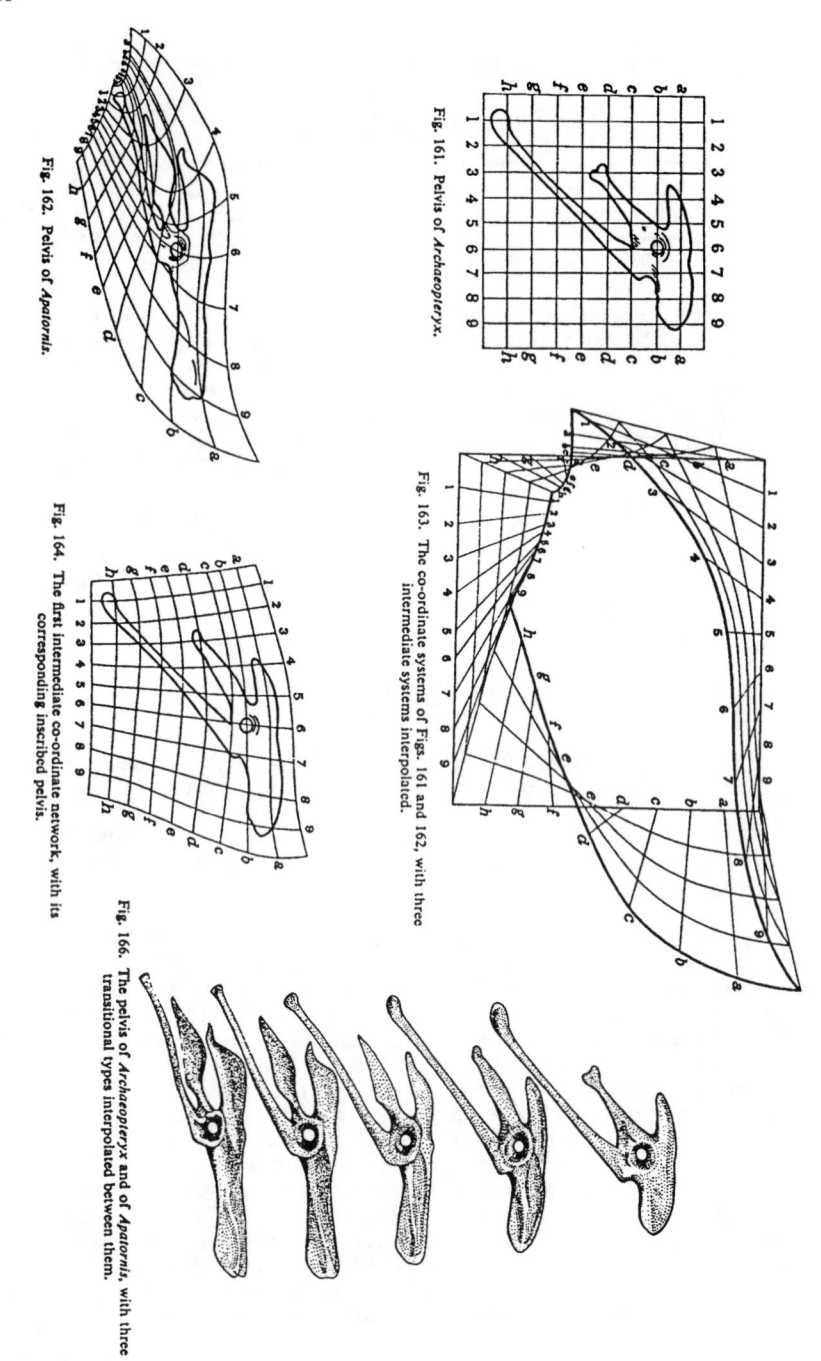

Fig. 162. Pelvis of *Apatornis*.

Fig. 163. The co-ordinate systems of Figs. 161 and 162, with three intermediate systems interpolated.

Fig. 164. The first intermediate co-ordinate network, with its corresponding inscribed pelvis.

Fig. 166. The pelvis of *Archaeopteryx* and of *Apatornis*, with three transitional types interpolated between them.

Since in much of what follows, I will be interested in the joint use of a number of images, some further comments are in order. Most such applications require the comparative analysis of images, (1) looking for similarities and differences, (2) for transformations between images, or (3) for a common reference frame to use when integrating the information from different images. As noted by Tufte (1983), these cannot be done effectively unless the images are juxtaposed so that they can be seen together. This is plausibly due to two factors: (A) we have the highest sensitivity for comparative differences and similarities in local analysis. This is supported by what is known about neurophysiological mechanisms (e.g., the local influence scale of lateral inhibition networks so useful in edge detection, and the foeveal concentration of cones in a small region in the retina, and reflexes through which peripheral motion trigger fixation bringing this small region to bear).[4] Wilson 1990 gives a recent review. (B) More global comparisons require some way of transporting local information, but we aren't very good at doing this simultaneously for multiple dimensions of that information.[5] So multidimensional comparisons have to be done locally, or one dimension at a time, using or supplemented by symbolic and analytical techniques. Thus, for example, a graph grid allows local comparison with grid elements (providing a common reference frame) to give more global comparisons of the location of features, involving comparison of only one spatial component at a time. We can change from local to global comparisons via a scale change, preserving multidimensional information *by making what was global local*, but do so under pain of losing fine detail.

3. Continuous multi-dimensional transformations in time and space

Continuous transformations (or discrete approximations to them) make use of our powerful system for processing information concerning motion and object-reidentification. Marey's pole jumper and jumping stick figures (figures 2 and 4 above) make use of this. We naturally re-identify correctly key moving points of the figure, which is essential to seeing what the jumpers are doing (and calculating trajectories, velocities, accelerations, and jerks[6] for the parts) In nature (when there may be many other moving things in the visual field) analysis of broader correlations of local motion is essential to picking out figure from ground—detecting the object, and conjointly determining what sort of object it is. The need to be able to deal with continuous deformations carries over to cases where motion is present but not directly represented (because there is only a single static image), or even where no motion is implied. These two kinds of cases are nicely exemplified by Marey's smoke tunnel image (Figure 3) and by morphological transformations from D'Arcy Thompson's (1917,1961) classic *On Growth and Form*, (Figure 5a-e).

Marey built perhaps the earliest modern smoke tunnel for visualizing and studying airflow around various streamlined and lifting bodies. Slow speed airflow may be treated as incompressible, and there was already a well developed mathematical theory[7] for treating such flow around regular bodies as long as it remained laminar. (Turbulent flow is another matter, and is still unsolved today, though it is yielding in part to analysis in terms of chaotic dynamics.) If such flow is laminar and incompressible, then the width of ideal lamina is inversely proportional to its velocity, and therefore (from Bernoulli's law) directly proportional to the pressure exerted perpendicular to the direction of flow. If these lamina can be visualized as they flow around an airfoil, it would be possible to calculate the pressure above and below it, and thus its lift. Marey accomplished this by drawing thin streams of smoke into the airflow at regular vertical separations, allowing them to be treated as streamlines. Also visible in the smoke streams is a 10 Hz oscillation (generated with a tuning fork[8]) allowing direct calculation of local velocities. Thus the photograph gives direct measurement of spatially distributed relevant aerodynamic parameters, and—to modern eyes—also a nice demonstration of chaotic mixing in the turbulence downstream.

D'Arcy Thompson's work founded the science of allometry (the dependence of morphology on size) and he gave us many insights on and techniques for studying morphological relationships. Perhaps best known is his "theory of transformations" which he uses to compare corresponding parts of diverse organisms. His methods have since intrigued evolutionists as promising ways through which changes could be analyzed and (see Figure 5—his figures 161, 2, 3, 4, 6) unknown evolutionary intermediates could be interpolated. Thompson's method involved drawing the pelvis of the then earliest known bird, *Archaeopteryx*, in rectangular coordinates, and constructing a new transformed coordinate system around the pelvis of one of ite presumed descendants, *Apatornis*, constructed so that major morphological features had the same coordinates. These coordinate frames were placed over one another and used to linearly interpolate three intermediate coordinate systems. The hypothetical intermediate pelvises were then drawn (with corresponding coordinates) in the intermediate coordinate systems. This problem would be simply impossible without visual representation. His critics have complained that Thompson's grids did not faithfully represent all of the changes in the two structures, but he clearly preferred relatively simple transformations that captured the main deformations in a coordinated manner (p. 275). Finer scale transformations which were left out (e.g. virtually all of the finer variations in the top of the pelvic bone of *Apatornis*, from the "notch" at a-9 to the reverse curvature of the forward point in the region bounded by a, b and 3, 4) indicate differences not captured by the global transformations, and are possible local targets of selection. The linearly interpolated hypothetical intermediates have a similar status. These transformations assume that the natural transformational scale is linear, and that the various features evolved at the same time and at the same rates. On the other hand, this linear model can act as a "null hypothesis" against which any residual differences stand out. As Thompson says, "the deformation of a complex figure may be a phenomenon easy of comprehension, even though the figure itself has to be left unanalyzed and undefined." (p.271) If applied properly both to his transformations and to the residual differences, his method may help not only to accomplish what he set out to do, but also to generate natural coordinates for the relevant morphogenetic processes, and also identify (as deviations) plausible foci which were subject to locally specific selection.

4. Multiple independent views of the same object—the product and process of linkage mapping

I have argued earlier (1974, 1976, 1981) that our theories often give us 1-dimensional or severely reduced descriptions of our objects, and that practitioners (both theoreticians and experimentalists) take for granted and use multiple characterizations or means of access to their objects of study which are often stripped away as redundant in insensitive definitions or axiomatizations. Almost by definition an object is something that has a multiplicity of properties, and as such, it also has a multiplicity of independent means for detecting it and delineating its boundaries. This characteristic of multiple detectability, derivability, or measureability (baptized *robustness* after Richard Levins' use of a similar notion) is the most central and fundamental criterion for realism and trustworthiness, and is readily extended to theorems and properties as well, as is readily demonstrated by examples from across the sciences (Wimsatt 1981). Crucial to this account of robustness and realism is the fact that by what are in effect multiple means of bootstrapping (Glymour 1980), we successfully use differences as well as similarities in the results of our determinations to simultaneously learn more about the object of study (it is after all that thing which can be simultaneously accessed by *these*

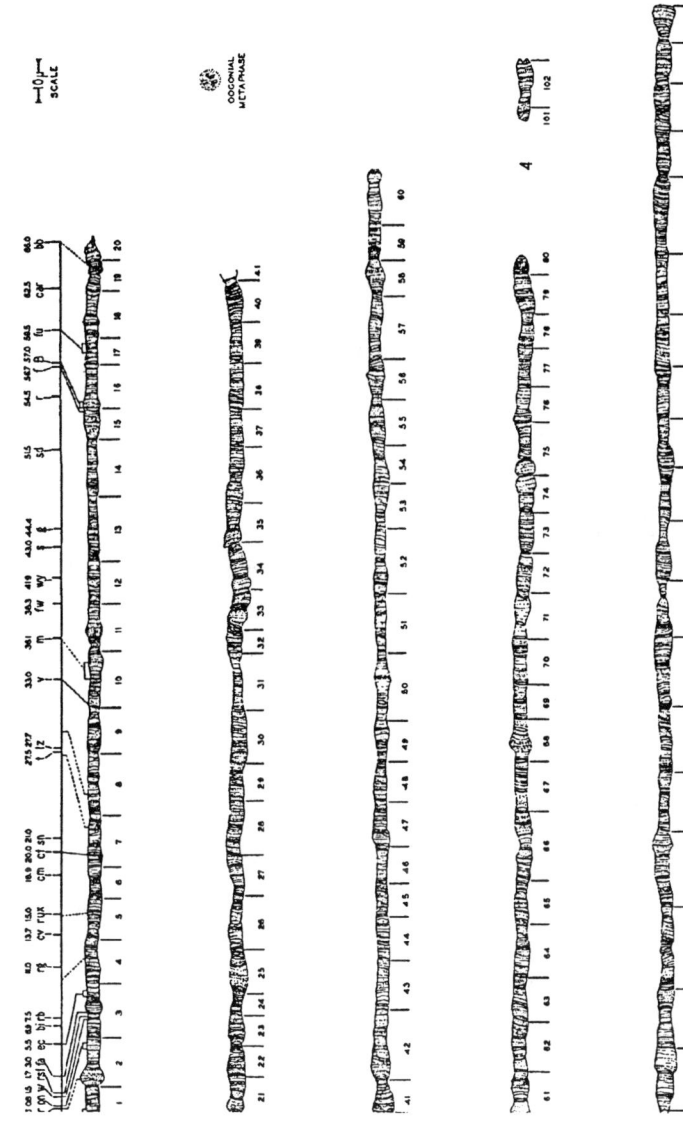

18.—The salivary gland chromosomes of Drosophila melanogaster. The upper line shows the linkage map of the X chromosome, with corresponding points on the salivary gland chromosome. The numbered sections below each chromosome are arbitrarily chosen, for convenience in referring to specific regions. The chromocenter, in which the base of each chromosome is embedded, is not shown. The oogonial metaphase figure (above, to right) is drawn to the same scale. (After Bridges, Journal of Heredity.)

different means) and to calibrate the means of study (it after all produces *these* disparities in these respective detectors). Thus Richard Levins' (1966) comment that "Our truth is the intersection of independent lies." Cartwright (1983), Hacking (1983), and Galison (1988) have since offered related accounts of realism.

I want to show how our view of the linkage map and its chromosomal basis are enriched by having multiple pictures of the same entity—a sort of "picture essay" on robustness. While doing a study of the early years of linkage mapping, I discovered several "new" kinds of representations—two of which are particularly interesting because they give information on the practice or procedures of map construction as well

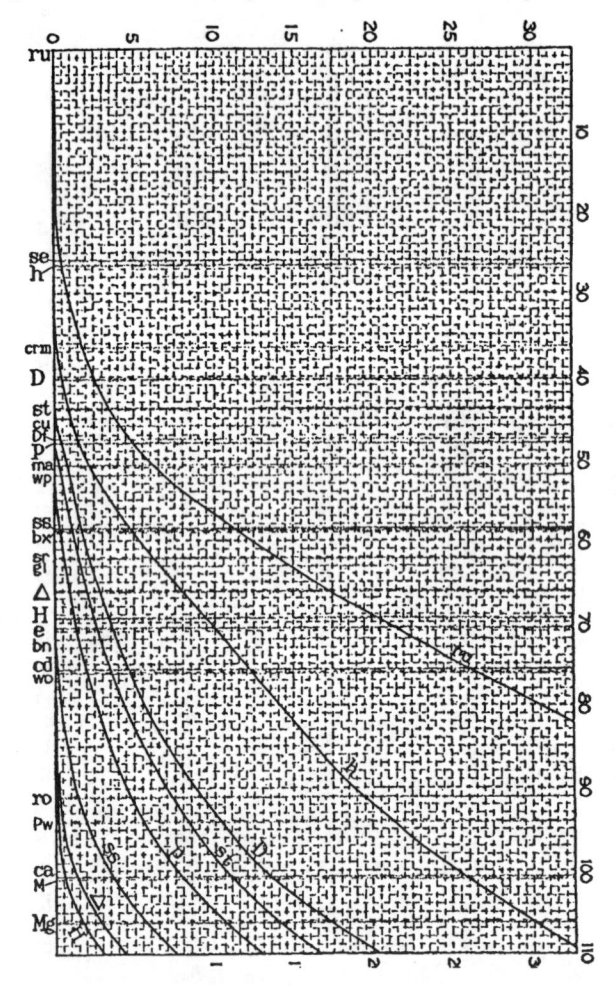

Figure 7: Linkage (or "line-map") of the third chromosome of *Drosophila melanogaster* with correction curves for differences between map distance and observed recombination frequency, showing curves of different forms from the different genes located at various points along the chromosome. (From Bridges and Morgan, 1923, p. 12.) Used by courtesy of the Carnegie Institute of Washington.

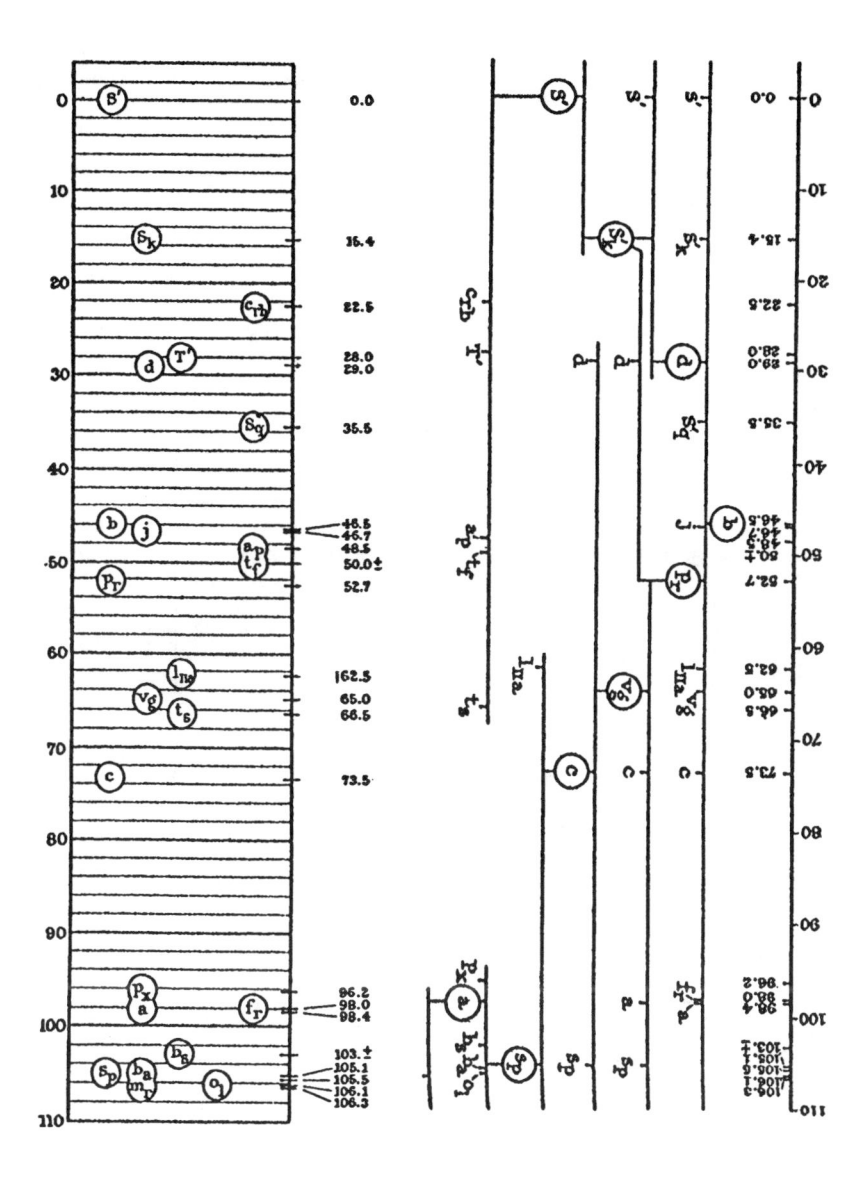

Figure 8: Figures 88 and 87 from Bridges and Morgan, 1919, pp. 303, 302.
"Figure 88: Working and valuation map for the second chromosome. The loci mapped at the left margin represent the most valuable mutants, those further to the right, progressively less useful. Those next to the right margin are mutants no longer extant. Figure 87: Constructional map of second chromosome, giving bases of reference and indicating various cross-over values used in calculating mean position for each locus." Used by courtesy of the Carnegie Institute of Washington.

as depicting the product. These older representations never make it into today's textbooks in part because of a revolution in linkage mapping that took place in the early 1930's with the rediscovery of the giant salivary gland chromosomes in *Drosophila*. These gave microscopically observable banding patterns which rendered immediately visible judgements on many questions (e.g., the detection and understanding of inversions or translocations) which had earlier taken many years to resolve. In the interests of space and time, I will violate historical order, and present this latest map first since it helps to explain the others. Many of these points can be found in my (1987), but more are topics of current research, (e.g.Wimsatt 1992).

I will here tell a painfully abbreviated (and occasionally permuted) history of linkage mapping—so named because some characters appeared to be "linked"—to appear together in offspring more frequently than the random expectation predicted by Mendel's law of independent assortment. These characters formed linked groups in terms of their frequency relations. Each pair of characters (or the factors or genes whose presence caused the characters) has characteristic frequencies of association with all other factors in the same "linkage group". In 1911, Morgan hypothesized that this was due to the factors being linearly arranged along the chromosomes, homologues of which, according to a hypothesis of Janssens in 1909, twisted around one another and exchanged corresponding segments. Morgan also guessed that the frequency of separation of pairs of factors should be a measure of the distance between them along the chromosome. In 1913 Sturtevant shows how cross-over frequencies can be used to generate the relative order of the factors or genes on the chromosome in what came to be called a "linkage map". He had a very sophisticated conception of this theoretical structure, and the complex physical interactions through which it is tied to the material chromosome, pointing out that relative distances in the map do not imply relative distances along the chromosome because different parts of the chromosome may differ in their breakage resistance. At the time this was conjecture—it could not have been tested.

It becomes immediately visible however in figure 6, (from Sturtevant and Beadle 1939, p.130) in which the laboriously constructed linkage map of the X-chromosome is itself mapped onto X-chromosome locations in the (much more visible) giant chromosomes of the salivary gland of *Drosophila*. (This was the revolutionary technique of the 1930's mentioned above—revolutionary because the banding patterns were characteristic of chromosomes, the bands could be mapped to genetic loci, and the linkage map was thus immediately visualized.) Correlations between linkage map and bands in the salivary gland chromosomes were accomplished thru "deletion mapping" —looking for chromosomes which had missing bands, and seeing what character changes resulted. With this, some macroscopic chromosomal mutations (deletions, duplications, translocations and inversions) became visible under the microscope even without breeding experiments. These phenomena had taken many years (from 1913 thru 1926, in the case of inversions) and hundreds of experiments to understand, but all of these finely wrought theoretical constructs were now visible at a glance! This visual hindsight naturally suggests a massive underestimate of the tasks and the sophistication of classical geneticists—a mistake characteristic of certain molecular biologists today. Here, subsequent ease of visualization contributes a distorted view of the history.

The so-called "line-map", and its association with the salivary gland chromosomes, provide a reference frame for understanding other maps of the chromosome, and integrating the information that their different perspectives provide. Sturtevant noted in 1913 that factors which are relatively close together should be used in constructing the map, for they and they only will have additive or nearly additive recombination frequencies or map distances. Factors which are further apart may have two or more crossovers occurring between them. These double crossovers will look like no crossovers—they will not be detected, since the two factors being tracked will end up on the same side, and in the same chromosome, as if nothing at all had happened. Double and higher-order crossovers will therefore give a biased measure—a "deviation" or an underestimate of the distance between the factors being mapped. Finally, he also noted that one crossover appears to make a second cross-over less likely nearby than had that first crossover not occurred—a process that Muller later names "interference". The judgement that this inhibition occurs presupposes a model or assumption for how frequent double crossovers should be if there is no mutual inhibition or "interference", This last basically gives the model behind what we call the Haldane Mapping Function—one of 5 models proposed in a classic paper by J. B. S. Haldane in 1919. For reasons I will not discuss here (Wimsatt 1992), the Drosophila group rejected Haldane's pretty modelling attempts, and argued that the chromosomes were too variable (primarily in their strength of interference) both between and within chromosomes for any single mapping function to be adequate. The empirical basis of their opposition is found in figure 7, from Bridges and Morgan's 1923 monograph on the mutants of the third chromosome. Here the deviation (the amount by which observed recombination frequency is less than theoretically constructed map distance) is graphed as a function of map distance. Such a curve constitutes a "mapping function", of the sort Haldane proposed. But in this illustration, there are different curves starting from different factors located at different points along the map. And one can see—in effect making a visual multidimensional comparison—that the curves are not the same functions of increasing distance. *They are not superimposeable, and this demonatrates their case* (against Haldane) *that no single mapping function will do.*

But they do not stop with this proof. They comment (here and elsewhere) that the changes in mapping functions (basically with less interference in the middle regions of the chromosome, symmetrically higher interference at the ends, and other features not considered here) correlate with and are probably caused by the central location of the spindle attachment point, thus relating features that emerge in this view to features that appear in the chromosome map (the spindle attachment point is the dividing point between left and right arms in the second and third chromosomes) but not in the corresponding line maps. The line maps thus serve as a reference frame for relating these other two maps of the chromosome.

We have two more kinds of maps of the chromosome which also relate to each other and the other types of maps through line maps. These are the evaluation and constructional maps. (There is but one example of the constructional map. These two must be of the same chromosome, so they are of the second, rather than of the third chromosome). These two kinds of maps are particularly revealing because they together show important details about the construction of linkage maps which neither would alone, and in part because they are not representations produced after the fact, but are intended as tools for use in the construction of the more common "line maps" and in planning experiments. If "line maps" show the results, these show parts of the process of getting there. In the evaluation map, (their figure 88) an additional dimension is added to the standard line map, with the location along the x-axis indicating how useful the character is for the experiment being considered. Mutants closest to the left are the best, growing gradually worse with displacement to the right (by failing on one or more of 5 desiderata for mutants given elsewhere in the text), ending up with those mutants which have been lost and thus cannot be of further use. The evaluation map thus simplifies a multi-dimensional decision problem by collapsing 5 dimensions to one, and using that with another crucial dimension—the location of the gene in the line map—to plan experiments. This kind of map would have to be much more frequently modified than most maps or graphs. Locations would presumably need to be updated as more information becomes available. Even more frequently, the evaluation rankings are not absolute, but relational, and can change depending on the purposes of the particular experiment or what other mutants are going to be used, so it is imperative that the map be easily modifiable. Indeed, it appears that this map was a rendition of their working version posted in the lab which "...can be kept subject to continuous changes in the valuations or locations of the different mutants by drawing the map scale on a soft board and mounting the symbols for each mutant on the head of a thumb-tack." (Bridges and Morgan 1919, p. 303). This was in effect a scientific ancestor of the air operations maps of the second world war, which were used to plan and document the progress of individual missions.

Corresponding to the evaluation map of figure 88, is the constructional map of figure 87 (Bridges and Morgan 1919, p. 302.), which shows the order in which the major loci are located. This despiction of the serial dependency of map construction gives an immediate idea of the significance of different loci, and which loci might have to be relocated if the site of a key locus is revised. (Clearly, the choice of "pivot" loci has been carefully made using the evaluation map, to minimize just this possibility!) A comparison of the two maps shows that of the "pivot" loci, only S_k, d and v_g are not of the first rank. They are however "nearly first rank" (their language!), and they are used only because no first rank loci are to be found in their regions of the map. All of the "pivot" loci are located relative to each other using multiple reliable experiments, weighting most strongly those with lots of flies in the relevant recombinant classes and distances that are sufficiently close to rule out double crossovers. The close relation between the two maps is shown by the fact that, given just the evaluation map and the rule, "construct the map using first the leftmost loci, and going to the right only as necessary to avoid separations that permit double crossovers", the pivot loci of figure 87 are the first to be chosen from figure 88.

In sum then, we know much more about the chromosome, and the processes of its mensuration, by having multiple pictures of it, all coordinated through that elegant theoretical construction of classical genetics, the linkage map. These pictures range from the highly theoretical (the "deviation map") to the almost directly observational (pictures of the giant salivary gland chromosomes) all of which *become* maps and comparable to each other when related to the standard line or linkage map.

5. Analytical complexity and multiple dependent views of the same object—picturing chaos

Robustness in the sense of Wimsatt 1981 presupposes multiple independent means of getting to the same result or object. But in the study of very complex phenomena, we may learn a great deal by presenting the same data in a variety of different ways. Even though they may be identical or analytically related, our visual system picks out different orders in these different presentations— orders which may be there in other views, but which are not visually accessible. This is nowhere more obvious than in the study of deterministic chaos, where simple deterministic systems exhibit enormous complexity and pragmatic unpredictability (Wimsatt, 1980), and at other levels of analysis—surprizing general features. This too is a "hot" topic, and there are many good books on it. My own favorite is by mathematician Ian Stewart (Stewart 1989). Deterministic chaos is one of the favorite topics of the new "empirical" mathematics, explorations of analytical results using the computer, and—inevitably—visualization. The example presented here is one of many generated using software we designed for the teaching of model-building[3], and one aim of this software has been to encourage students to explore the data by visualizing it in a variety of different ways.

This particular example was discovered using this software and has never been described in the literature. It has a number of features which should interest philosophers of science. In particular, it shows a system which is (1) deterministic at the micro-level, (2) probabilistic at the macro-level, but (3) has its macro-behavior describeable in terms of what I would call "sloppy, gappy generalizations" which (4) are riddled with exceptions, (5) whose nature can be probabilistically characterized. (6) These generalizations support subjunctive conditionals of a real and useable sort, (7) whose exceptions are explicable by recourse to the micro-level, but (8) whose interesting behavior is to be found in the regularities at the macro-level. They are, I think, the simplest known, and perhaps the simplest possible, examples of these kinds of generalizations which are endemic in the study of complex systems throughout the biological, social, and (dare I say the more interesting?) parts of the physical sciences. This example is interesting here because there are patterns in it which (a) cannot be readily understood without visualization, and even more, (b) cannot be correctly understood without the simultaneous use of *multiple* visualizations.

In this example, which is an analysis of behavior in a given domain of what physicists call "the logistic map" (and biologists call "the finite difference equation for logistic growth") I will use three of the four modes of data visualization supported by the program. (A fuller presentation would use all four, but would require 8 rather than 4 pictures, so will be left for another occasion.) Also, I will not explain these representations in any more detail than minimally necessary. The first mode of presentation (the bifurcation plot or Feigenbaum tree) is probably one of the two best known diagrams of chaotic behavior. I use it, (figures 9a and 9b) at two different scales (a) to visually and qualitatively locate the parameter domain of interest and (b) to demonstrate the phenomenon of self-similarity (in which similar dynamical patterns appear recursively on different size scales in the state space—a characteristic feature of "fractal" organization).

The logistic growth equation is an equation in which the change in population size from one generation to the next is a quadratic function of population size. What is important here is that this change has a maximum for an intermediate value of population size, and is actually negative if the population size is large enough. This nonlinearity has the effect that as one increases a parameter (r, the intrinsic growth rate) which acts as linear multiplier for the rate of change, the system becomes increasingly unstable, going first through various oscillatory modes in which it exhibits all periods which are integral powers of 2. It then passes into a regime of so-called chaotic behavior, in which it behaves in an apparently random fashion, and more importantly, in which nearly everywhere within its domain, trajectories which start arbitrarily close together will diverge until their behavior is uncorrelated. The bifurcation plot or Feigenbaum tree is produced by picking a starting population size, and then for each value of r, running the trajectory for a predetermined large number, g, of generations (200 to 1000 is common), and then plotting the next p generations (100 to 300 is common). If r is plotted along the x-axis, and population size along the y-axis, the effect is that at each x-value, p points are plotted, indicating the population size in generations g thru $g + p$. The idea is to make g large enough to have the system settle down into a stable behavior pattern if there is one (i.e., if there are no transients lasting longer than g), and then see from the points plotted for that r-value what that pattern is. (Pseudo-random behavior is indicated by an apparently random scatter of points for that r-value.)

In the first plot (figure 9a) we see a characteristic bifurcation plot—a single line which divides into two branches, each of which divides into two, each of which divides into two, and so on. (This plot actually starts right at the 1—>2 branch transition in order to show more structure in the "canopy".) These indicate the regions in which the stable behavior of the system is a single equilibrium point (1-point), and stable 2-point,

128

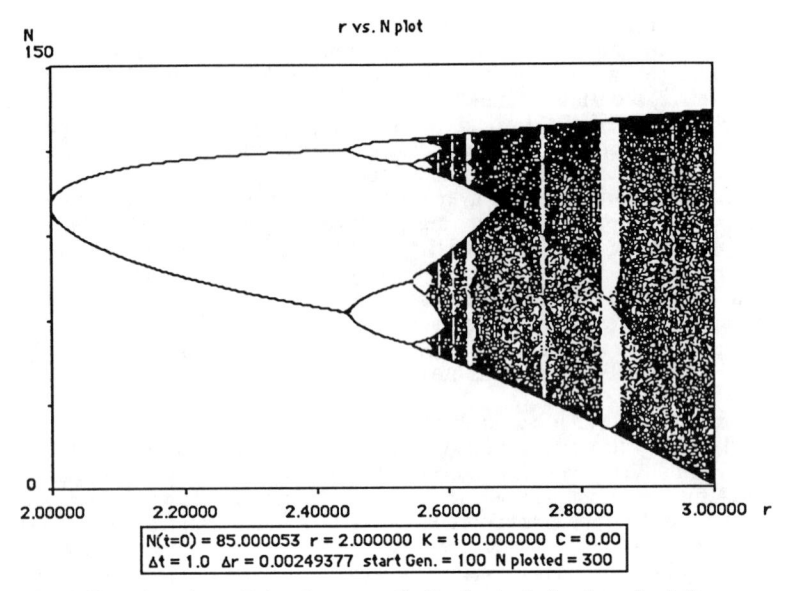

r vs. N plot

N
150

0

2.00000 2.20000 2.40000 2.60000 2.80000 3.00000 r

| N(t=0) = 85.000053 r = 2.000000 K = 100.000000 C = 0.00 |
| Δt = 1.0 Δr = 0.00249377 start Gen. = 100 N plotted = 300 |

Figure 9a: Bifurcation plot or Feigenbaum tree, indicating loci of trajectories in between generations 100 and 400. The blowup of figure 9b is of the middle branch in the period 3 window between 2.82 and 2.85, and is a magnification of 25x in the horizontal and 6x in the vertical directions.

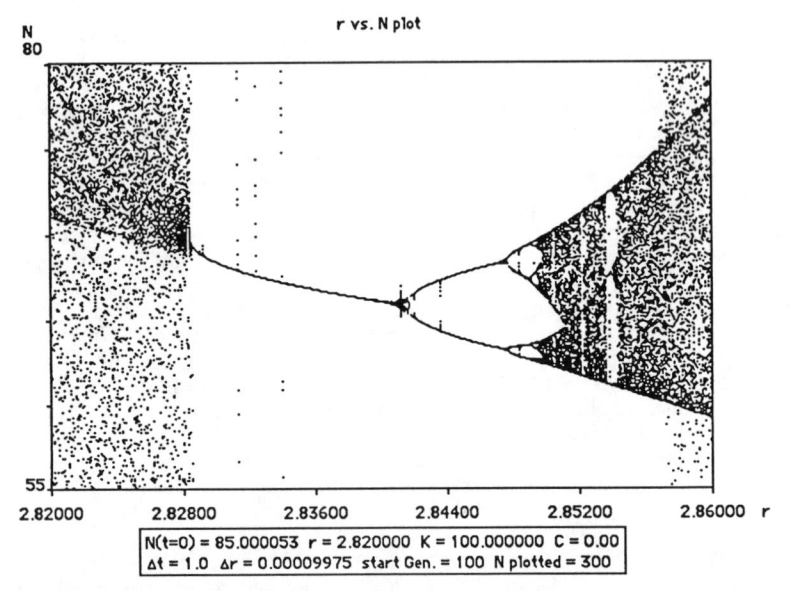

r vs. N plot

N
80

55

2.82000 2.82800 2.83600 2.84400 2.85200 2.86000 r

| N(t=0) = 85.000053 r = 2.820000 K = 100.000000 C = 0.00 |
| Δt = 1.0 Δr = 0.00009975 start Gen. = 100 N plotted = 300 |

Figure 9b: Blowup of figure 9a, showing similarity of middle branch with the whole tree, and the locus of attention for figures 10a and 10b, at r = 2.856820. Note that some trajectories are still showing transient behavior at generation 100.

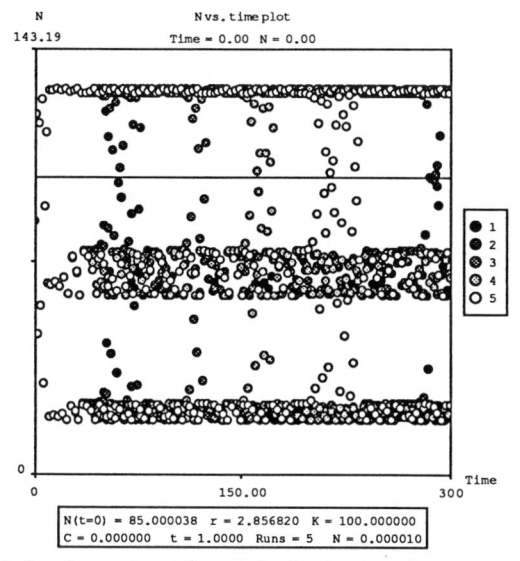

Figure 10a: Population size vs. time (N vs. t) plot for 5 trajectories starting very close together and then diverging, showing the infinite sensitivity to initial conditions characteristic of chaotic dynamical systems. Note that in this case, the trajectories spend most of their time circulating regularly among three regions, with occasional excursions, exhibiting a *"sloppy, gappy generalization"*.

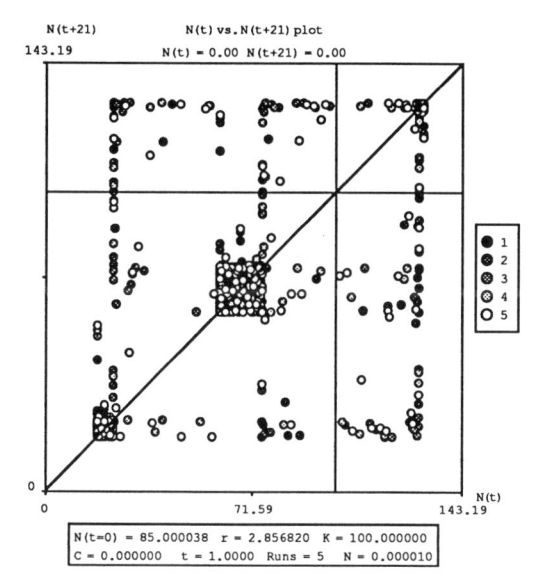

Figure 10b: Phase plot of the same trajectories with a time lag of 21 generations. The regular patterns of this phase plot are very revealing of the statistical dynamics of the system. See the text.

4-point, 8-point, etc. cycles. The ratio of x-distances (or values of r) for which successive bifurcations or "period-doublings" happen converges rapidly to a number, δ, (Ferigenbaum's number, approximately 4.67), and at the limit, there is a transition to the chaotic regime, marked by an apparently random scatter of points within an expanding envelope. The population size range covered by recurrent chaotic behavior expands with increasing r, until the lower limb of the tree intersects the x-axis (at an r-value of 3.0, or 4.0 for the somewhat different parameterization used by physicists.) Before this happens, there are several visible "bands", in which behavior settles down into fairly short periodic cycles, the widest of which occurs for period 3 cycles between about 2.82^8 and 2.85^7. Figure 9b is a blowup of the middle strand in theis band, demonstrating a "flipped" sub-tree, in which this single strand itself (with the other two off-screen) goes through a pattern of period-doubling bifurcations like that of the original whole tree, producing periods of 6, 12, 24, ..., before going chaotic in its subdomain. (And in its chaotic regime, there is another period three band, containing periods of 9, bifurcating to periods of 18, 36, 72, ..., and so on.) Notice that not everything has quite settled down in the second figure (transients last longer for higher period cycles), and that there is a lower relative density of points in this because it is a blowup. This second figure serves to locate the domain of interest analyzed in the next two figures. This domain is at the transition where behavior has gone chaotic within the three sub-regions, and the sub-regions have expanded to the point where points are just beginning to boil out of the restricted sub-domains into the larger state-space, indicated by the vertical point cloud about 2/3rds of the way between 2.852 and 2.860 (This is thus right at what appears to be the "back wall" of the period-3 band in the first figure—for a value of $r = 2.856820$, the r-value for all trajectories in the last two figures).

We now turn to figure 10a, in which population size is plotted vs. time starting in generation 0, for the next 300 generations for 5 different starting values. These values are very close together, starting at $85.00000 + k(.00001)$ for $k = 0,1,2,3,4$. If you look closely at the start, you see that the last trajectory (with the white circles), starting at 85.000038 (there is a cumulative roundoff error of .000002 here) covers over the preceding 4 trajectories for about the first 25 generations, after which there is a rapid divergence, and the trajectories get totally desynchronized. This divergence ("almost everywhere") is a defining property of chaotic dynamics. These trajectories are also captured in the 3-region cycle by the 10th generation, and after that time spend most of their time there (with only some 118 out of 1455 points on excursions), but continue to diverge nonetheless. They go on their excursions independently (the dark—>light order is an accident), follow different trajectories out, and stay out for different lengths of time, though all tend to be recaptured fairly quickly. Their pattern of behavior follows a *"sloppy, gappy* generalization", in that they follow a "three region" rather than a "three point" cycle, within which they *"slop"* around randomly on successive returns. (Their slop follows a "U-shaped distribution" rather than a normal distribution however!) Occasionally, they sneak out through a *"gap"* in the fence of the attractor, and bounce around for a while before returning.

There are suggestions of statistical regularity in the N vs. t plot, but these are glaringly obvious in our last graph, figure 10b, which is a phase plot. Plots like this are used for time-series analysis in the time-domain, and have also been used to analyze and explain the origins of chaotic behavior through successive "stretching and folding" (see Stewart, 1989), though it is usually used by chaos analysts for time lags of no more than a few generations. I here deliberately picked a time lag of 21, a longish integral multiple of 3, so that there was no visible correlation in locations within the chaotic sub-regions. (For another demonstration of fractal self-similarity, compare phase plots of 1, 2, and 3 for the behavior of the whole region at $r = 3.00$ with phase plots at 3, 6, and 9 at $r = 2.856820$ for behavior within the sub-regions.) In this phase plot, all pairs of points (within, but not across trajectories) 21 generations apart are graphed, with the x-coordinate given by the trajectory's N value at t and the y-coordinate given by the trajectory's N value at $t + 21$. (Since there *is* no point 21 generations after the last 21 points, we have 279 points for each trajectory rather than 300.)

This graph looks like a scotch plaid with spots. It turns out to be the most dynamically informative phase plot I have ever seen. There are 5 x 279 or just short of 1400 points in this graph, so obviously the vast majority of them are in the squares symmetrically located on the diagonal line. The diagonal line is the locus of points which have the same value at $t + 21$ as they have at t, so this just says (1) that *most trajectories spend most of their time in the 3-region cycle*. That they are random within the three subregions is indicated by the uniform square random coverings within those regions, indicating (2) no correlation *within the sub-regions* between where they are at times t and $t + 21$. But why the horizontal and vertical bands intersecting the square sub-regions? A vertical going through a sub-region indicates points which were in that subregion at t, but which are randomly spread around at $t + 21$. (3) These are the points that had escaped 21 generations later! Similarly, (4) the horizontal bands indicate captures after 21 generations. (5) The white squares indicate that almost nothing stays out longer than 21 generations (and you can try different time lags to see how many stay out how long). (6) The off-diagonal intersections of the horizontal and vertical bands represent escapes and recaptures with phase shifts of ± 1 or 2 generations from the three subregions. Finally, there is information to be captured from this plot which is not only not obvious in the other plot, but which shows just the opposite of what that plot appears to suggest. If you look back at white-dot trajectories in the middle region of the N vs. t plot, the trajectories appear to start near the top and accelerate downwards, suggesting to the gullible eye that the escapes occur at the bottoms of the middle and bottom regions or at the tops of the top region. (Try it for this and other shades.) This is an illusion, as is clearly shown in the phase plot. In this plot, (7) *the vertical bands intersect the bottom and middle regions at their tops and the top region at its bottom, demonstrating clearly that escapes occur there.* Finally (8), inspection of the horizontal bands shows no corresponding biases for captures, either *from* differ- ent *outside* regions, or *to* different *inside* regions.

Finally, a caution—an illusion on the other side: the character of the phase plot exaggerates for the eye the number of events causing the trends in question. Thus, a trajectory that goes on an excursion for 15 generations will add 15 points to the vertical bars (for the 21-generation ancestors of its 15 points spent out), and 15 points to the horizontal bars (for the 21-generation descendants of its 15 point excursion). The only cure for this is to look back at the N vs. t plot, and recognize that *all of the points in the off-diagonal horzontal and vertical bars are contributed by 7 macroscopically statistically independent events*—the starting 9-generation transient before first capture of the trajectories, and the 6 independent excursions of the 5 trajectories. Any inferences suggested by *this* statistical scotch plaid should be further tested by more trajectories with different starting points and/or by tracking these trajectories for longer periods of time. Some additional surprises remain, but I'll leave that for you to find out.

6. Two closing remarks

The focus of this paper has been on visualization as the source of natural modes of representation for certian kinds of data. But we shouldn't stop there. First of all, vision is not the only sense which can contribute significantly in the analysis of pattern. In astronomer Fred Hoyle's science fiction novel, *The Black Cloud*, a living being (with a somewhat different size and time scale and form of life than we are used to) regarded our music as more informative than any of our other creations. This might seem somewhat far-fetched, but a phenomenon discovered almost by accident by biologist Bob Gross of Dartmouth (personal conversation, Fall, 1988) suggests the contrary. Gross had written a program to display DNA sequences so that one could look for similar patterns in different regions of the sequence. On a whim, he also had the computer "play" the sequence, with 4 different notes for the 4 bases. To his surprize his ear proved far better at picking up similarities (he heard the later ones as variations on the earlier theme!) than his eye. To be sure, it was not a rigorous test: the use of letters (A, T, G, C) may invoke additional symbolic processing—it would be better to use a sequence of color patches, or rectangles of different heights. But, as Eric Nelson has suggested to me, the possibility that the ear—selected for speech recognition across different speakers and situations—is simply better than the eye for such tasks deserves serious consideration. Would a theory of taste give us a different taste for theory? We have nothing but riches to gain from making better use of the strengths of our specialized natural integrated transducers and data analyzers.

Finally, I present one last exhibit for those who believe that I have communicated to them only linguistically after all:

. How much do practices in the presentation of text make use of visual modes of organization? (Anyone who thinks they don't should compare an `ASCII file with no formatting` with the more imaginative productions of desktop and **MAGAZINE** publishing.) Different (1) fonts,   (2) font sizes, (3) emphasis (**bold** and *italics*), (4) column formats, (5) boxed text and quotes, (6) bullet and number markers, and (7)   spacings   are   used   to   do   a variety of things from *seduction* and maintaining interest to segmenting and organizing text. As misused here, such things can impede comprehension, but properly used, markers like these facilitate scanning the text at different levels of detail and   rapid transitions from one part of the text to another—something further   elaborated in so-called "outliner" programs. With multiple paths from one point in   the text to another, we break away from strictly sequential modes of access.   This is often helpful—though not always!
All of these are characteristically offered, (e.g., by Tufte), *as properties of visual modes of presentation*, but as with most things, *we never appreciate how important visual organization is until it breaks down.*

Notes

[1] I would like to thank Jim Griesemer, Ron McClamrock, Mike Ruse, Jeff Schank, Edward Tufte, and Hugh Wilson for thought provoking conversations, Tufte, Wilson, and Dan Margoliash for bibliography, and for Marey's photographs and knowledge of his work, I owe major debts to Joel Snyder, and to Marta Braun, respectively. From each of them I have gotten more than I gave. I hope that I can repay the favor some time. Much of this work would have been impossible without support through National Science Foundation Grant, SES-8807869 from the History and Philo- sophy of Science Panel.

[2] It could also be regarded as derivative from various practices of causal manipulation where it is common to increase a variable monotonically until the desired linear or non-linear (e.g., threshold) response is achieved. Given the absolute centrality of vision, touch, and hearing—all spatial modalities—in causal judgement, it is not clear how such claims could be easily settled. It seems more important at this stage to point to connections than to propose probably premature answers.

[3]These simulations, and others by Jeffrey Schank and me are intended to teach skills in the modeling of some of the simplest dynamical systems important to evolutionary genetics and ecology. They will be available in 1992 from Addison-Wesley as part of the BioQUEST package of strategic simulations. We expect also to use them in a book on model-building.

[4]What work has been done comparing vision with audition is very suggestive here. Thus Schuller and Pollak (1979) report the functional analogue to a foevea in some (FM) bats, whose frequency discrimination is much more accurate in a narrow frequency range. When tracking prey, they are constantly modulating the frequency of their emitted echoes so that the doppler shifted frequencies produced by changing relative motion of the prey are shifted into their range of maximal sensitivity. (It's too bad that Tom Nagel didn't know a little better what it is like to be a bat!). And von Bekesy (1967) reports widespread use of lateral inhibition networks in the processing of auditory and tactual signals.

[5]See R. N. Shepard (1964), for experiments showing how bad we are at estimating even linear weightings of multidimensional data in perceptual tasks.

[6]The "jerk" derivative is the time derivative of acceleration (the third derivative of displacement). Contrary to intuition, for real (not perfectly elastic) materials it is in most cases high values of the jerk derivative (and not of acceleration) that causes failures. See any text on cam design.

[7]This theory was developed in the late 19th century by a variety of hydrodynamicists and mathe- maticians. A particularly elegant example of morphological transformations was the use of the theory of conformal mapping (an application of the theory of complex functions to 2-dimensional potential flow) in which a specific class of coordinate transformations took an eccentrically located circle into an airfoil-like shape (called a Joukowski profile). More importantly, this also trans- formed the streamline flow around the circle (which was comparatively easy to solve for) into the correct streamline flow around the airfoil shape, which they had no way to generate directly. Here again, although the transformation was generated and applied mathematically, one needed to see the particular Joukowsky profile produced to see that it had the right properties. See Prandtl and Tietjens (1934), and especially figures 106-108, p.176, for more details.

[8]Marta Braun,whose fascinating book on Marey will be published by the University of Chicago Press in 1992 is the source for the tuning fork story. With a time marker that could have been traced thru the flow, it also should have been possible to judge momentum loss (and thus drag) in the wake of the airfoil, but turbulence would have rendered this impossible.

References

Bridges, C.B., and Morgan, T.H. (1919), *The Second Chromosome Group of Mutant Characters*, Carnegie Institute Of Washington, Publication No. 278, Part II, pp. 123-304.

_____. (1923), *The Third Chromosome Group of Mutant Characters in Drosophila melanogaster*, Carnegie Institute Of Washington, Publication No. 327.

Cartwright, N. (1983), *How the Laws of Physics Lie*, Oxford: Oxford University Press.

Galison, P. (1987), *How Experiments End*, Chicago: University of Chicago Press.

Glymour, C. (1980), *Theory and Evidence*, Princeton: Princeton University Press.

Griesemer, JR., and Wimsatt, W. C. (1989), Picturing Weismannism: A Case Study in Conceptual Evolution, in M. Ruse (ed.), *What Philosophy of Biology Is*, Dordrecht: Martinus-Nijhoff, pp. 75-137.

Hacking, I. (1983), *Representing and Intervening*, Cambridge: Cambridge University Press.

Latour, B. (1987), *Science in Action*, Cambridge: Harvard University press.

Levins, R. (1966), The strategy of model building in population biology, *American Scientist*, 54:421-431.

Lynch, M., andWoolgar, S. (1990), *Representation in Scientific Practice*, Cambridge: MIT Press.

Marey, E. (1895), *Movement*, (International Scientific Series, v. 53), New York: Appleton.

_ _ _ _ _. (1984), *Etienne-Jules Marey*, (introduction by Michel Frizot), Paris: Centre National de la Photographie.

McClamrock, R. (1992), *Existential Cognition: Computational Minds in the World*, Chicago: University of Chicago Press.

Prandtl, L., and Tietjens, O. (1934), *Fundamentals of Hydro- and Aeromechanics*, Engineering Society Monographs, reprinted by Dover Books, 1957.

Schank, J. C. and Wimsatt, W. C. (1988), "Generative Entrenchment and Evolution", in *PSA-1986*, Volume 2, A. Fine and P. K. Machamer (eds.). East Lansing: The Philosophy of Science Association, pp. 33-60.

Schuller, G. and Pollak, G. D. (197), Disproportional frequency representation in the inferior collicus of horseshoe bats: evidence for an accoustic foevea, *Journal of Comparative Physiology*, 132: 47-54.

Simon, H. A. (1981), *The Sciences of the Artificial*, Cambridge: MIT Press.

Shepard, R. N. (1964), "On subjectively optimum selections among multi-attribute alternatives", in, *Human Judgements and Optimality*, M. W. Shelley and G. L. Bryan. New York: Wiley, pp.257-81.

Stewart, I. (1989), *Does God Play Dice?*, London: B. H. Blackwell.

Sturtevant, A.H. and Beadle, G.W. (1939), *An Introduction to Genetics*, Philadelphia: W. B. Saunders. (facsimile reprint by Dover Books, 1962.)

Thompson, D.W. (1961), *On Growth and Form* (abridged edition), Cambridge: Cambridge University Press.

Tufte, E. (1983), *The Visual Display of Quantitative Information*, Cheshire, CT: Graphics Press.

_____. (1990), *Envisioning Information*, Cheschire, CT: Graphics Press. von Bekesy, G., (1967), *Sensory Inhibition*, Princeton: Princeton University Press.

Wilson, H.R., Levi, D., Maffei, L., Rovamo, J. and DeValois, R. (1990), "The perception of form: retina to striate cortex", in *Visual Perception: the Neurophysiological Foundations*, L. Spihlmann and J. Werner (eds.). New York: Academic Press., pp. 231-272.

Wimsatt, W.C. (1974), "Complexity and Organization", in *PSA-1972*,(Boston Studies in the Philosophy of Science, volume 20), K.F. Schaffner and R.S. Cohen (eds.). Dordrecht: Reidel, pp. 67-86.

_____. (1976), "Reductionism, Levels of Organization and the Mind-Body Problem", in *Consciousness and the Brain*, G. Globus, I. Savodnik, and G. Maxwell (eds.). New York: Plenum, pp. 199-267.

_____. (1980), "Randomness and perceived-randomness in evolutionary biology." *Synthese* 43: 287-329.

_____. (1981), "Robustness. reliability and overdetermination" in *Scientific Inquiry and the Social Sciences,* M. Brewer and B.Collins (eds.). San Francisco: Jossey-Bass. pp. 124-163.

_____. (1985), "Forms of aggregativity" in *Human Nature and Natural Knowledge,* A. Donagan. N. Perovich. and M. Wedin (eds.). Dordrecht: Reidel. pp. 259-293.

_____. (1987), "False Models as means to Truer Theories", in *Neutral Models in Biology*, M. Nitecki, and A. Hoffman (eds.). London: Oxford University Press, pp. 23-55.

_____. (1992), "Golden generalities vs. co-opted anomalies: alternative approaches to the theory and practice of linkage mapping", in a forthcoming volume edited by S. Sarkar, on Fisher, Haldane, Muller and Wright and the origins of the mathematical theory of evolution. Dordrecht: Martinus-Nijhoff.

_____. and J.C. Schank (1988), Two Constraints on the Evolution of Complex Adaptations and the Means for their Avoidance, in *Progress in Evolution*, M. Nitecki (ed.). Chicago: The University of Chicago Press, pp.231-273.

Part IV

HISTORY OF PHILOSOPHY OF SCIENCE

Conventionalism and the Origins of the Inertial Frame Concept

Robert DiSalle

University of Western Ontario

The obvious metaphysical differences between Newton and Leibniz concerning space, time, and motion reflect less obvious differences concerning the relation between geometry and physics, expressed in the questions: what are the invariant quantities of classical mechanics, and what sort of geometrical frame of reference is required to represent those quantities? Leibniz thought that the fundamental physical quantity was "living force" (mv^2), of which every body was supposed to have a definite amount; this notion violates the classical principle of relativity, since it makes a physical distinction between uniform velocity and absolute rest. But Leibniz did not try to represent this physical quantity in a spatio-temporal reference frame, assuming, instead, that all such frames are equivalent so long as they agree on the relative motions (changes in the mutual Euclidean distances) among bodies. Newton, in contrast, explicitly incorporated the relativity principle into his dynamics: he characterized the invariant quantity of force by acceleration rather than velocity, and recognized that the quantity of a body's moving force depends on the frame of reference in which it is described. Yet he tried to represent this dynamical conception in absolute space, which entails precisely the distinction between motion and rest, and hence the violation of classical relativity, that Leibniz's dynamics postulates. In the subsequent history of dynamics, serious methodological issues have been raised by the problem of expressing what is physically invariant in an appropriate geometrical structure.

As far as Newtonian mechanics is concerned, this problem was essentially solved by James Thomson in 1884, when he introduced the term "reference frame" and defined what we now call the inertial frame. His insight was that the laws of motion themselves postulate certain spatial and temporal structures, or, in his words, a certain kind of reference-frame and "dial-traveler" (time scale). He therefore proposed a "Law of Inertia," which I paraphrase: for any system of interacting bodies, it is possible to construct a reference-frame and a time scale with respect to which all accelerations are proportional to, and in the direction of, impressed forces. (Thomson 1884, p. 387.) To assert this as a law of nature is to assert, in modern language, that space-time is a flat affine space with a projection on time, and that deviations of a body's motion from the geodesics of the affine structure correspond to forces acting on the moving body. Of course we must add to this statement the demand that all of these actions are matched by equal and opposite reactions, so that a true impressed force can be distinguished from a pseudo-force, but Thomson's basic approach remains satistfactory; the main reasons to doubt it are the reasons to doubt Newton's laws generally.

Yet the end of the nineteenth century saw, not a recognition of the clarification that had been achieved, but an often bewildering philosophical debate over Newton's laws and the reference frame that they require, a debate that continued even after special relativity might have made the issues seem outdated. This is because to many of the participants in that debate, the (now seemingly straightforward) problem that Thomson addressed inevitably raised philosophical issues that affected the foundations of mechanics. Thus Ludwig Lange, for example, independently recapitulated Thomson's work (more or less) in his definition of "inertial system" (Lange 1885), but, while Thomson modestly thought he had clarified a point in Newtonian mechanics, Lange thought that he had set Newton's laws on an entirely new footing — a footing that purportedly replaced absolute notions with relative, real with ideal, and, especially, factual with conventional. Even though Lange's definition and his general point of view were given much attention by Ernst Mach and other physicists and philosophers, they seem to have had little influence on physicists working on the electrodynamical questions that led to special relativity, and so Lange's historical impact is difficult to assess. Yet something like Lange's general understanding of the role of reference-frames in mechanics became quite widespread in the philosophy of physics of the late nineteenth and early twentieth centuries, and it continues to make its presence known. It seems worthwhile, therefore, to look at the origins of this view and to see how a problem essentially concerning the invariant structure of Newtonian mechanics became a problem concerning a conventional choice of reference frames.

Lange developed his concept of inertial system from precedents in the German literature on the foundations of physics, especially Carl Neumann (1870), Heinrich Streintz (1883), and Ernst Mach (1883). Taking it as a basic principle that all motion is relative, these authors held that the classical law of inertia raised the fundamental "problem of the reference-system": relative to what system of reference is the motion of a free body rectilinear, and relative to what time-scale is it uniform? More generally, relative to what do the laws of motion hold? Neumann's infamous "Alpha body" was his answer to this question: he postulated that "at an unknown place in the cosmos," beyond our observation, sits a rigid and unchanging body, and that relative to this "Alpha-body" the motion of a free body is rectilinear (1870, p. 15). Streintz, attempting to improve on Neumann, proposed that the law of inertia holds with respect to *any* body found to be free of external forces, and named the appropriate reference-bodies "fundamental bodies." Both approaches had serious difficulties (see DiSalle 1988, especially chapter III), but the questions they were designed to answer seemed unavoidable once Newton's theory of absolute space was recognized to be unsatisfactory.

As Thomson's work shows, however, these questions are not really appropriate. The laws of motion do not hold "relative to inertial systems," or relative to anything. Again, by claiming that accelerations occur only in the presence of forces, the laws of motion assert the possibility of constructing systems in which every acceleration is traced to an interaction — in other words, they assert that inertial systems *exist* — and so one can't sensibly say that the laws are *true* only in such frames. True or false, the laws make *frame-independent* claims about the distinction between free motion and motion influenced by interactions. Ernst Mach recognized this when he pointed out that we could state the laws of motion without specifying any frame of reference at all, since the laws enable us to find a suitable frame (Mach 1883, p. 269); his chief objection to the laws was not that they are essentially relative, but that all of the *evidence* for them comes from motions relative to the fixed stars, and so there is no empirical reason to regard them as anything more than an inductive generalization about *those* relative motions. Even general relativity, while it suggests that one of the postulated Newtonian distinctions (between free motion and gravitational free fall) cannot be justified empirically, does not claim that such distinctions in general are relative to frames of reference. Yet by the end of the nineteenth century, a number of physicists and philosophers had come to regard the "relativity of the laws of motion" to a restricted class of reference frames as a fundamental principle, and by 1916 Einstein regarded this as the fundamental "epistemological defect" of the classical laws. It may be possible to gain some insight into the historical passage from Newton's physics to Einstein's from the nineteenth-century discussions of relativity and frame-dependence and the methodological issues that they raised.

Ludwig Lange arrived at the idea of "inertial-system" from Neumann's (1870) definition of equal times, which Lange called the "inertial time-scale": equal times are those in which a free particle travels equal distances. As Neumann pointed out (1870, p 18; see also Thomson and Tait, 1867, sections 247-48), this is an arbitrary stipulation, since one can always find a time-scale relative to which *any* particle motion is uniform. Given the stipulation, however, it is a factual claim that any *second* free particle will travel equal distances in equal times relative to the first. Since the comparison of distances and times for the two particles presupposes absolute simultaneity, and the two particles provide a dynamical definition of "equable flow" of time, Neumann's proposal is just an explicit dynamical version of Newton's absolute time. Lange's definition of the spatial "inertial system," and the corresponding law, are modelled on Neumann's formulation:

> **Definition I.** "Inertial-system" means any coordinate system with the characteristic that, with reference to it, the concurrent paths of three [material] points, simultaneously projected from the same point of space and then left to themselves (but which do not lie in a straight line) are all *rectilinear.*
>
> **Theorem I.** With reference to an Inertial-system, the path of *any fourth* point left to itself is also *rectilinear.* (Lange 1885a, p. 544-45.)

The requirement of *three* moving points follows from an argument analogous to Neumann's, presented more or less rigorously in Lange (1885a, 1885b). For three or fewer point motions, we can almost always construct a coordinate system in which these motions are rectilinear, and so it is a matter of convention whether three "points left to themselves" move in straight lines; we can first make a factual claim (Lange's "theorem") about the rectilinear motion of some *fourth* point. On these lines Lange believed he had solved the "problem of the reference-system": the laws of motion do not describe motion in absolute space, or even motion relative to the fixed stars, but motion relative to inertial systems.

The most important philosophical consequence of his work, according to Lange, was that one could capture the empirical content of what Newton called "absolute" in a relativistic treatment of motion. Since "Neumann's convention" had provided a "substitute" for absolute time, he mistakenly asserted, the concept of absolute time "has almost disappeared from present-day dynamics." (Lange 1885b, p. 336.) Since he viewed the inertial system as an analogous substitute for absolute space, he was convinced that his version of the law of inertia had accomplished the "avoidance of any *absolute* concept" (1885, p. 278). He therefore proposed some new terminology:

1. A point in uniform motion relative to an inertial system can also—with reference to another given inertial system—be treated as *inertially at rest.*

2. A point with curvilinear motion relative to an inertial system cannot be called inertially at rest relative to any other, nor can one so designate a point whose motion relative to one inertial system is rectilinear but not uniform. (1885, p. 279)

Points that are not "inertially at rest" are "inertially accelerated" or "inertially rotating." But since the latter quantities are independent of the choice of an inertial system and so correspond exactly with what Thomson called absolute rotation and acceleration — even Lange conceded that "all the same, there remains in the concept of absolute motion a valuable core consisting of that which it has in common with the concept of inertial rotation" —it should be obvious that Lange is proposing only a verbal change. Furthermore, in one respect in which his proposed changes are *not* entirely verbal, they are seriously misleading. They suggest since the laws of motion themselves make sense only with respect to a certain kind of coordinate system, acceleration, rotation, and uniform motion are meaningful only in such coordinate systems. Obviously, however, the possibility of inertial frames assumes the possibility of dynamically distinguishing these states of motion; what is "relative to a coordinate system" is just what cannot be so distinguished, namely position and velocity. Thus, Lange's "relativistic" way of thinking about inertial systems obscures a crucially important aspect of the transition from absolute space to inertial frames: the abandonment of the search for a background against which dynamical laws are supposed to be valid, and a corresponding focus on the spatio-temporal structure intrinsic to Newtonian dynamics.

The methodological implications of his work are also difficult to gather from Lange's own statements. For one thing, his talk of a *procedure,* projecting material points from a given spatial point, seems to suggest that he is trying to base the law of inertia on an operational definition. (For this interpretation see, for example, Barbour 1989.) Yet Lange's stated philosophical aim was not to describe operational procedures, but to find the most abstract formulation possible of the law; for example, he disagreed with Mach's expression of the law by reference to the fixed stars, precisely because "abstract mechanics" demands a form of the law that "does not rest on any given object of physical astronomy, but rather consists of *purely dynamical* concepts." (1885, p. 269.) Moreover, he repeatedly emphasized that the inertial system was supposed to be an "ideal construction" which "could never find immediate application," but from which all practical methods could be derived (e.g., 1885a, p. 544). He described the "material point left to itself" on which the construction is based as a "mathematical abstraction" and a "requirement that is never completely fulfilled"; on that account, the content of the law of inertia is "never factually given, but rather *assumed*" in order to comprehend mechanical phenomena. Lange's project, therefore, was only a conceptual "completion" of the law, a "purification of the hypothesis of superfluous elements" (1885, p. 270), and was not *intended* to be operationalistic.

We can get a clearer idea of Lange's methodology, and of the consequences it had for his broader philosophy of space and time, from his understanding of the conventional stipulation that he placed at the core of the law of inertia. First, he saw Neumann's convention for equal times as an illustration of a profound and general methodological prescription, which he called the "principle of particular determination": a scientific "theorem" should be expressed

> so that the theorem is conventionally valid for the smallest possible part of its objects; the contents of the theorem relate, then, insofar as they are more than mere convention, insofar as they are new empirical results, only to all of the rest [of the objects to which the theorem refers].(1885, p. 278.)

In other words, given a hypothesis that generalizes about a class of objects, we ought to determine how much arbitrariness there is in the generalization, and to express the non-arbitrary, empirical content of the hypothesis with the minimal number of objects. Lange's determination that the law of inertia is a convention for three or fewer particles, and an empirical claim for more than three, was intended to be a direct application of this principle. Indeed, he saw his recognition of this "partial convention" as the crucial difference between his formulation of the law inertia and James Thomson's (Lange 1885b, p. 351).

Considering his place in the history of the philosophy of space and time, Lange elevated his "partial conventionalism" into a metaphysical challenge to the Newtonian belief in the existence of absolute space. To Newton's argument that centrifugal effects reveal the absolute rotation of a sphere in absolute space, Lange responded that

> For the satisfaction of our epistemological need it suffices completely to introduce the inertial system as an *ideal convention* and to refer the motion of the sphere to this. We are not in the least advanced by the assumption of a *really-existing* [emphasis added] immaterial coordinate system.

Like some contemporary commentators, Lange blamed Newton's metaphysical realism concerning space and time on his "religious conception of nature," according to which space and time are "creations of the Eternal and Omnipresent." The modern scientist, however, more accustomed to the "systematic separation between belief and knowledge," does not need to claim real, "transcendent" existence for the mental constructions that are created in order to give coherence to the phenomena. Thus Lange viewed absolute space as Newton's realist answer to the "problem of the reference system," and considered that he himself had solved the problem in a conventionalist way, showing that the law of inertia is essentially founded in a coordinate system chosen by convention.

To understand clearly the philosophical significance of Lange's work, however, his own interpretation notwithstanding, we have to examine the precise role played by convention. His mathematical arguments make it a matter of convention whether three particles are moving in straight lines, but what does this imply about the dynamical case of free particles in rectilinear and uniform motion? One such particle can always be represented as moving uniformly in a straight line by suitable adjustments of the coordinates and the time-scale, but two or three particles whose mutual distances increased and decreased could not be represented as moving each in a straight line in the same sense. What we can say about the requirement of three free particles is more restricted than the "conventionalist" result about arbitrarily moving points: a reference-system in which one, two, or three free particles move uniformly may not be one in which all free particles move uniformly. Lange's illustration of this supposes, first, that all free particles move uniformly relative to some reference-system (which we can call the inertial frame). Obviously there will be systems relative to which some one of these particles moves uniformly, but which rotate (for example, about the path of the particle) relative to the inertial frame; we can even find systems in which two of the particles move uniformly, but which rotate (for example, about the line joining those two particles) relative to the inertial frame — and, evidently, relative to any system in uniform motion relative to the inertial frame. After the third particle, however, the freedom to adapt the frame comes to an end: as long as the particles do not travel collinearly or in parallel paths, any movement of the frame that preserves the uniform rectilinear motion of these particles must do the same for all the free particles in the inertial frame. So the *dynamical* significance of Lange's minimal "partial convention," as he pointed out (1885b, pp. 344-46), is that if there is at least one coordinate system and time scale in which arbitrarily many free particles move uniformly, any three free particles satistfying Lange's conditions define another system and time scale in uniform motion relative to the first.

All of this indicates that Lange did not really provide a conventionalist answer to the "problem of the reference system." In fact he answered a different sort of question, one more nearly related to Thomson's approach than one might have thought: *assuming* that inertial frames exist, in what circumstances can we say that any given frame is inertial? If such a class of frames exists, it is not a matter of conventional choice whether a given frame belongs to the class; what Lange has shown is that a factual determination can indeed be made, but only with at least three free particles. Far from supporting the relativity of motion, then, Lange's spatial "particular determination" (that three particles are required) postulates an objective distinction between the state of free motion and other possible states: for particles moving freely, frames are always possible in which aritrarily many move uniformly in straight lines, while for particles moving anyhow, such frames are generally possible for any three particles. Lange recognized this point in 1902, when he published a reconsideration of his 1885 work in light of critical reactions to it (1902, pp. 9, 37). Now, instead of presenting the inertial system as that relative to which the law of inertia holds, he claimed (analogously to James Thomson) that "the pronouncement of the law...is simply reduced to the assumption of the *phoronomical possibility* of a system in which *arbitrarily many* ($n > 3$) points left to themselves" move uniformly; the three particles projected from a point, formerly part of the definition underlying the law, Lange now recognized as "the simplest possible prototype of all practical real methods" of constructing an inertial system (1902, pp. 38-39).

Thus in the course of Lange's philosophical development, his "ideal construction" lost its foundational role in the expression of the law of inertia and became a way of considering the possible application of the law. The law itself, meanwhile, became something that Lange had criticized Newton for expressing, namely a kind of "existence-hypothesis," insofar as it asserts the existence of reference systems in which free particles move uniformly. Lange himself did not fully appreciate this point, and never abandoned the view that he had established some form of relativism combined with conventionalism, and it is not difficult to understand why (especially in light of the fact that even contemporary physics texts occasionally say that the laws of motion hold relative to inertial systems). When we speak of the "kinematical possibility" of constructing a certain kind of reference system, namely one in which every acceleration is proportional to an impressed force, we seem to be naming just one of (obviously) many possibilities for the assignment of coordinates; this leaves an opening for the popular but vague statement that inertial frames are merely the simplest possible ones — a statement that evidently contains a certain element of truth, but that ignores the grounding of inertial frames in the lawlike dynamical structure of the Newtonian universe. Therefore the assertion that they are possible seems scarcely comparable, in its "metaphysical" significance, to a claim about the affine structure of Newtonian space-time. For similar reasons it is frequently said that Minkowski's space-time formulation of special relativity has essentially different metaphysical implications from Einstein's original formulation. In the time before the invariant four-dimensional version of Newtonian mechanics was developed, then, it was (at first glance) comparatively easy to suppose that the abandonment of absolute space for an equivalence class of possible frames was really the abandonment of a realistic picture of spatiotemporal structure for a conventionalist one.

In order to clear this matter up, and to understand from a methodological point of view what Thomson and Lange really accomplished, we need to recognize exactly where convention really enters into the construction of an inertial frame. That an inertial frame is possible is evidently a strong dynamical claim: it means that every component of every acceleration within a dynamical system is objectively traceable, by virtue of the equality of action and reaction, to some source within the system. This implies, for example, in the case of our solar system, that of all the possible resting points we could choose in order to "frame" the system, those are distinguished *by law* — not necessarily by simplicity or any other kind of "convenience" — in which all accelerations are determined by Newtonian interactions. (The dynamical analysis of all of these interactions warrants Newton's assertion that, assuming the laws of motion and astronomical phenomena, we can "demonstrate the frame of the system of the world." 1729, p. 323.) Thus, as was noted above, Thomson's and Lange's claim is precisely equivalent to the claim, in the four-dimensional picture, that of all the geometrically possible worldlines, those are distinguished which represent physically possible trajectories of free particles. However it is formulated, this structure provides a definition of matter in its passive inertial state, and so it is the necessary foundation for the Newtonian account of active forces as causes of accelerations. Whatever conventionalist challenge affects the proposed geometrical structure, then, affects at the same time the entire classical picture of "fundamental forces of nature."

The opening for conventionalism occurs precisely with the principle through which the commitment to the Newtonian program (and its accompanying spatio-temporal structure) is first made, a principle which Lange, for all his stated conventionalism, was content to take for granted. This is just the principle that there is an objective distinction between free motion and motion under the influence of a force. Lange spoke of the free particle as the "element" of dynamical theory, in the sense that constructions (like his construction of the inertial system) are "derived from" this element just as constructions in pure geometry are derived from the geometrical point (1885, p. 350). But he did not see that, while this ideal physical thing, the free particle, is indeed the element of a physical theory, its spatio-temporal *path* is the element of a spatio-temporal affine geometry in *precisely* the sense in which spatial length is the element of spatial metrical geometry; the worldlines of freely moving particles thus provide a "coordinative definition" of the affine structure of space-time just as the rigid body provides a coordinative definition of Euclidean length. We can give empirical arguments for or against the suitability of these definitions: for example, Newton's Scholium on space, time, and motion argues empirically that the definition of true motion through force and acceleration makes sense and has a clear application, while Einstein's argument about the equivalence principle suggests that the traditional definition cannot be unambiguously applied, and therefore (in effect) that the affine structure of space-time is best coordinated to the paths of freely falling particles. In either case, the justification for the definition extends only as far as the successful application to the phenomena that the theory is supposed to address.

Of course it was impossible to say all of this explicitly in the language available to Lange, before the concept of space-time was developed, but, in a review of Lange's work written in 1891, Gottlob Frege made essentially the same point. He criticized Lange for exaggerating the difference between his "inertial" concepts and those that Newton called absolute, and suggested that the question whether motion is "real" was only a verbal dispute: the important question, he pointed out, was whether there is a real distinction between accelerated and unaccelerated motion. The distinction is real, he asserted, "in the same sense in which the constancy of a length is real": "in both cases we have arbitrary stipulations, which however are so closely linked to the lawfulness of nature that they are thereby distinguished from all other stipulations which are mathematically and logically equally possible." (Frege 1891, p. 157.) Both stipulations connect a physical process (respectively, measuring forces by accelerations and measuring lengths by rigid measuring-rods) with an aspect of geometrical structure, and the success of the stipulation suggests its connection with some regularity of nature. This sort of argument follows a pattern familiar from discussions of the measure of time: it resembles the claim that a pendulum clock is a better choice to measure time than my heartbeat because the rate of the clock stands in a simpler relation to natural laws than the rate of my heart. We therefore make no special ontological claim when we call the quantities thus defined "absolute, true, and mathematical," as long as we mean, "as opposed to relative, apparent, or common," insofar as only the former are imbedded in a coherent system of laws.

The difficulty Lange's contemporaries had in seeing Frege's point brings out an interesting difference between the inertial frame postulate and its four-dimensional counterpart. In the latter case, the connection between the elementary physical process and its geometrical representation (between the path of a free particle and the affine geodesics of spacetime) has (*modulo* the essential difficulty of having four-dimensional intuitions in the first place) the same intuitive obviousness as the connection between the congruence of rigid rods and Euclidean length: the physical invariant "looks like" the geometrical invariant. In the former case, however, we can only picture the projection of a geodesic in a given inertial frame (a rectilinear motion in a three-dimensional coordinate system), and we know that this is just one of an equivalence class of allowable projections. Before there was an invariant representation of what is common to members of the equivalence class, it was perhaps a natural mistake to look here for some conventional element in the theory of motion. And so it seems to me even more remarkable that Thomson should have recognized that the intrinsic spatio-temporal structure of Newton' laws could be expressed in nineteenth-century language.

More recently, misunderstanding of the fundamental coordination underlying space-time structure has created difficulties for us, who are accustomed to the four-dimensional picture. In the modern version of the metaphysical debate between Newton and Leibniz, spatial relations are assumed to be epistemologically unproblematic, because the geometry of these relations is easy to think of as the structure of possible distances between objects — as a set of rules governing the motion and comparison of rigid bodies. Space-time, by contrast, represents states of bodies *through time*, and bodies are therefore thought to have states of motion "relative to" it; inertial forces arising from different states of motion are thought to be *caused* by it. Therefore space-time seems to take on the aspect of a thing, where space alone could be thought of as an order; the *existence* of space-time therefore seems to require some special metaphysical argument. But space-time is no more the cause of differences in states of motion than Euclidean space is the cause of differences in length. Both are "structures of possibilities" coordinated to elementary physical processes; space embodies a set of laws governing possible momentary distances among bodies, space-time a set of laws governing possible dynamical evolutions. Neither one can be coherently reduced to a set of relations, since both structures *govern* possible relations. Yet such a structure is not really a thing either, at least not like the things to which its structure is coordinated. The history of the theory of space-time structure, beginning with its origins in the three-plus-one dimensional account of inertial frames, helps us to see that the structure is an aspect of our laws of motion, and that the important philosophical questions about it are not ontological, but methodological after all.

References

Barbour, J. (1989), *Absolute or Relative Motion?* v. 1. Cambridge: Cambridge University Press.

DiSalle, R. (1988), *Space, Time, and Inertia in the Foundations of Newtonian Physics, 1870-1905*. Ph.D. dissertation, University of Chicago.

Frege, G. (1891), "Uber das Tr gheitsgesetz." *Zeitschrift fHr Philosophie und philosophische Kritik* 98: 145-61.

Lange, L. (1885), "Ueber die wissenschaftliche Fassung des Galilei'schen Beharrungsgesetzes." *Wundt's Philosophische Studien* 2: 266-97.

_ _ _ _ _. (1885a), "Nochmals ueber das Beharrungsgesetz." *Wundt's Philosophische Studien* 2: 539-45.

_ _ _ _ _. (1885b), "Uber das Beharrungsgesetz." *Berichte der Koniglichen Sachsischen Gesellschaft der Wissenschaften zu Leipzig, Mathematisch-physische Classe* 37: 333-51.

_ _ _ _ _. (1886), *Die geschichtliche Entwickelung des Bewegungsbegriffes und ihr voraussichtliches Endergebniss. Ein Beitrag zur historischen Kritik der mechanischen Principien.* Leipzig: Wilhelm Engelmann.

_ _ _ _ _. (1902), "Das Inertialsystem vor dem Forum der Naturforschung. Kritisches und Antikritisches." *Wundt's Philosophische Studien* 20, v. 2: 1-71.

Mach, E. (1883), *Die Mechanik in ihrer Entwickelung, historisch-kritisch dargestellt.* Leipzig: F.A. Brockhaus. 9th edition, 1933.

Neumann, C. (1870), *Ueber die Principien der Galilei-Newton'schen Theorie.* Leipzig: B.G. Teubner.

Newton, I. (1729), *The Mathematical Principles of Natural Philosophy.* Translated by Andrew Motte. London. Reprint: New York: The Philosophical Library, 1964.

Streintz, H. (1883), *Die physikalischen Grundlagen der Mechanik.* Leipzig: B.G. Teubner.

Thomson, J. (1884), "On the Law of Inertia; the Principle of Chronometry; and the Principle of Absolute Clinural Rest, and of Absolute Rotation." *Proceedings of the Royal Society of Edinburgh* 12: 568-78.

Thomson, W. and P.G. Tait. (1867), *Treatise on Natural Philosophy.* Cambridge: Cambridge University Press.

Designation and Convention: A Chapter of Early Logical Empiricism[1]

Thomas A. Ryckman

Northwestern University

We have yet to fully understand the manner or the measure to which logical empiricism emerged as a conventionalist response to both traditional Kantian and empiricist epistemology and to the apparent triumphs of "conventionalist stratagems" (in Popper's aspersive locution) in the foundations of science. By "conventionalism", however, is here understood a broader sense than customary, an extrapolation of views on the foundations of geometry and physics (associated in the first instance with Poincaré) to an encompassing epistemological consideration of the development and validity of scientific concepts generally. In this new construal, the concepts of science are neither derivable from sense experience, nor are they transcendentally valid *a priori* conditions of its possibility. Rather they are "free creations of the human mind" whose provenance is "logically arbitrary", as Poincaré and (subsequently) Einstein put it.

In the initial phases of the period which was to produce the "linguistic turn" in philosophy, Schlick and, some ten years later, Carnap, provide seemingly quite different reconstructive accounts directed towards clarifying and systematizing the role of conventional elements in the formation of scientific concepts. For each, the conventional — reincarnate as the purely formal —paradoxically affords the requisite means to distinguish and privilege scientific knowledge. For Schlick, the precision and, for Carnap, the objectivity, of scientific knowledge are to be accounted for by reconstructive projects aimed at showing how it is possible for the concepts of the sciences to be purely formally determined. In so doing, the intuitive, ineluctably private and incommunicable content or meaning of these concepts is declared irrelevant or epistemologically inert. Of course, fashioning a rigorous separation between the formal and the intuitive, contentual elements of experience proves in each case to be an ideological conceit. But the failure is an instructive one, in particular, in view of the subsequent trajectory of logical empiricism.

In his (1918) Schlick provides the new conventionalism with its characteristic separation of form and content, arguing for a doctrine of concepts as purely formal, implicitly defined **signs** which are coordinated (*zugeordnet*) to objects through the paradigmatically conventional act of designation. In this very influential book, Schlick attempts to provide a general account of scientific knowledge in terms of a "merely semiotic" doctrine of implicitly defined concepts and their designata, while coupling lengthy critical scrutinies of Machian positivism and Kantian *a priorism*, with a "hypothetical realism" owing much to Helmholtz.

Employing a judo-like maneuver, Carnap in the *Aufbau* holds that logic itself (in the sense of the second edition of *Principia Mathematica*) is conventional; nonetheless, logical structure is to be the reconstructive medium of scientific objectivity, i.e., of the intersubjectivity, of scientific knowledge. Following Poincaré in holding that science treats of, and so describes only relations between sensations, not the private and incommunicable content of the sensations themselves, Carnap's goal is to sketch a method for redefining all the concepts of science using only the resources of pure logic and a sole non-logical relation obtaining among autopsychological experiences. The procedure is targeted at a reconstruction of the sign relation, together with its inherent arbitrariness, the lynchpin of Schlick's semiotic conventionalism. Within the *Aufbau*'s *Konstitutionstheorie*, the designative concept- and relation-signs of science are to be systematically recast *à la* Russell as *Kennzeichen*, indicator-signs ("structural definite descriptions") that unambiguously identify (and thereby "constitute") the objects of the various sciences. In this enterprise at the end of the day, the only remaining non-logical signs are signs for a basic relation holding among autopsychological experiences; ultimately, even these too are to be formally conjured away. After a brief preliminary look at the doctrine of concepts in Schlick's early classic, I will suggest that this 'semiotic' reading of the *Aufbau*, emphasizing its shared problematic with Schlick, effects something of a reconciliation between two heretofore widely varying interpretations of this work current in the literature (cf. Friedman, 1987 and Richardson, 1990).

1. A "Merely Semiotic" Account of Scientific Knowledge

In the *Allgemeine Erkenntnislehre*, Schlick argues that in science (particularly, theoretical natural science) what we find is that, as in ordinary life, knowledge (*Erkenntnis*) is not acquaintance (*Kennen*) but a matter of recognition (*Wiedererkennen*) or rediscovery (*Wiedererfinden*), the finding of one thing in another. As opposed to everyday knowledge, the process of knowing in science has two distinctive features: it is reductive and it seeks to employ a minimum of explanatory principles. But knowledge *qua* recognition assumes a comparison and, in an exact science such as physics, what is compared must be fixed with "absolute constancy" and "determinateness". Only concepts, not ideas, impressions, or intuitive images satisfy this requirement; they alone can be clearly individuated and identified with complete assurance. When concepts in science are clearly defined, this is not due to whatever intuitive meaning or content might attach to them. Rather it is to consider them as mere signs (*Zeichen*) for all the objects whose properties include the defining characteristics (*Merkmale*) of the concept. This is neither to say that concepts are not invariably or often accompanied by intuitive images (*anschauliche Vorstellungen*) nor that concepts are real, in the sense of mental or Platonic entities. A concept is a mere fiction (*blosse Fiktion*), and strictly speaking, only the conceptual function — the designation of an object by a concept, considered solely as a coordination (*Zuordnung*) of concept to object — is real (§5).

Schlick goes on (§7) to urge that Hilbert has accordingly indicated "a path of the greatest significance for epistemology" in showing that concepts can be defined merely by the fact that they satisfy chosen axioms. Thus the method of implicit definition — which, Schlick argues, is profitably extendable outside the realm of pure mathematics to the sciences generally — shows how it is possible to completely determine the content (*Inhalt*) of concepts in a manner which does not fall back on the intuitive (*im Anschaulichen*), and that it is possible to speak of the meaning (*Bedeutung*) of concepts without reduction to intuitive images. With implicit definition, an instrument has been found that enables complete determination of concepts and the attainment of "strict precision" in thinking. Satisfying only the requirement of consistency, the method of implicit definition provides purely formal specification of both the sense (*Sinn*) and the designata (*Bedeutungen*) of concepts in the mathematical sciences; the sense of mathematical concepts is acquired solely in virtue of the particular axiom system, while the *Bedeutung* of such concepts is nothing real, but consists in "a determinate constellation of a number of the remaining concepts". Not mincing implications, this means that mathematics "has only the significance of a game with symbols"; whereas, more generally, in the empirical sciences of nature:

> A system of truths created with the aid of implicit definitions does not at any point rest on the ground of reality. On the contrary, it floats freely, so to speak, and like the solar system bears within itself the guarantee of its own stability (p.35).

In so many words, "the bridges are down" between "concept and intuition, thought and reality".[2]

Due to a "correlation" between judgements and concepts, the result of implicit definition is a network holism of conceptual meaning. On the one hand, "implicit definitions determine concepts by virtue of the fact that certain axioms — which are themselves judgements — hold with regard to these concepts; thus such definitions make concepts depend on judgements". Concepts must occur in several different judgements if they are to have any sense and meaning (*Bedeutung*) at all. On the other hand, "judgements are linked to one another by means of concepts: one and the same concept appears in a number of judgements and thus sets up a relation between them". The mutual dependence of concepts and judgements anticipates the favored metaphor of a later, and more well-known, holist:

> "every concept constitutes, as it were, a point at which a series of judgements meet (namely, all those in which the concept occurs); it is a link that holds them together. The systems of our sciences form a net (*ein Netz*) in which concepts represent the knots (*die Knoten*) and judgements the threads that connect them (p.43).

But it is only in two further sections, on the nature of judgement (§8) and truth (§10), that we begin to see how a semantically self-contained system created through implicit definition is to be tempered with what Carnap will term, in an appropriately vivid phrase, "the blood of empirical reality". Restoration of links to the empirical world is established through acts of designation or coordination (*Zuordnung*). For judgements are also merely signs designating not only relations between objects but **that** a certain relation obtains between the designated objects. Judgements therefore designate facts or sets of facts (*Tatbestände*) which are presupposed (p.40). However, designation by judgements is doubly holistic: it is only the judgement as a whole that is coordinated to a fact as a whole, whereas **which** fact a given judgement (proposition) designates is determined solely by the position occupied by the proposition (*Satz*) in our system of judgements (p.62). Moreover, as against the purely formal requirement of consistency to ensure the truth of mathematical theories, an empirical judgement is to be considered true if and only if it "unambiguously designates" (*eindeutig bezeichnet*) or is "coordinated" to (*zugeordnet*) a set of facts. Weaving together the formal mechanism of implicit definition with the conventionalism of designation yields a structural, though holistic, account of truth as *eindeutige Zuordnung*:

> ...it is the structural connectedness of the system of our judgements which produces the unambiguous coordination and conditions its truth;...(p.62).

Elsewhere, Schlick observes that in these sections of his (1918) scouted above, the case is made

> that such a mapping (*Abbildung*) of the lawfulness of the real (*des Wirklichen*) with the help of a sign-system comprises in general the essence of all knowledge...(1921 p.156, n.15).

The radically formalist method of determining concepts through implicit definition, together with his thesis of "the merely designative (or semiotic) character of thinking and cognition," that thinking involves — as does any act of coordination — "somewhat arbitrary conventions" (1922, p.105), decisively settles, in Schlick's eyes, the whole question against Kantian philosophy (1918, p.306). Yet Schlick's account cannot be fully satisfactory, requiring us to accept as unproblematic the considerable weight placed on the notion of *Zuordnung*,[3] by all appearances a mysterious and seemingly synthetic act which is difficult, if not impossible, to reconcile with empiricist precepts against synthetic *a priori* principles as well as with strict (non-holistic) verificationist doctrines of meaning. In his attack on intuition and alleged intuitive means of acquiring knowledge, Schlick had been content to leave the characterization of *Zuordnung* as only "a fundamental act of consciousness...a simple ultimate...to which every epistemologist must, in the end, advance"; in other words, an unexplained explainer, a fundamental, not-further-reducible act of thought.[4]

2. The *Aufbau*: Reconstructing Designation Itself

Although he is far more explicit than is Schlick that he is giving a "rational reconstruction" of actual concept formation (cf.§§ 100, 143) Carnap in the *Aufbau* also maintains the purely symbolic (or designative) view of scientific knowledge to confront "metaphysical" accounts of the acquisition of knowledge through intuition. With Carnap no less than Schlick, to exclude the intuitive, the subjective and the private, entails upholding the possibility of a purely formal determination of scientific concepts as requisite to the precision and intersubjectivity of scientific knowledge. For Carnap, the system of conceptual knowledge that is science (§§ 180,182) is itself symbolic; hence, an object must be designated by a sign —at least in principle —if it is to be an object of conceptual knowledge (§ 19). Venturing some rare irony, Carnap even states his agreement with Bergson on the latter's definition of metaphysics as "That science which wants to get by without symbols", glossing this position as advocating

> metaphysics would seek to grasp its object not through the detour (*Umweg*) of concepts, which are symbols, but immediately through intuition (§ 182).

While in agreement with Schlick on the necessity of a purely formal determination of scientific concepts, Carnap will offer in the *Aufbau* a completely different method for its attainment.

Now science is distinguished, preconstructionally as it were, from metaphysics by the fact that it pursues not problems of essence (*Wesenprobleme*) but only problems of coordination (*Zuordnungsprobleme*), i.e., of determining which objects or object-types are associated with each relation. This has been especially emphasized by Mach, and in several places Carnap repeats Mach's strictures that science seeks only fixed functional dependences (§§ 21-2, 169). The reconstructive method of *Konstitutionstheorie* (hereafter 'KT') is itself part of science, since in order to be able to make statements about objects at all, these objects must first be constituted, else their names have no meaning (*Sinn*); for this reason indeed, "**the formation of the constitution system is the first task of science**" (§ 179). As part of science, KT is similarly bounded by the injunction to pursue only *Zuordnungsprobleme*, whereas the particular relation of concern to KT is the sign relation (*Zeichenbeziehung*) obtaining between a concept-sign, or a relation symbol, and its designation or *Beudeutung*, i.e. its extension.[5] KT is given the task of rationally reconstructing the sign relation,[6] showing that merely structural statements (*blosse Structurangaben*) can yield a univocal, i.e., unambiguous, sign-indication (*eindeutige Kennzeichnung*) for the domain of objects of science: if no difference in formal sign-indication can be given for two objects, these objects are not distinguishable by scientific methods (§§15-16). Hence identity in science is identity of "relation number" i.e., structural isomorphism (§§11-12). Recalling Reichenbach's puzzlement (expressed in his 1920 monograph on relativity[7]) over the "remarkable fact that, in cognition we carry out a coordination (*eine Zuordnung*) of two sets, the elements of one of which are first defined through this coordination", Carnap maintains that the rational reconstruction of the sign-relation as purely structural indicator signs gives an explanation (*Erklärung*) of this fact. For thereby it is demonstrated how it is possible through a sign-coordination (*Zeichenzuordnung*) to first carry out the individuation of single objects, thus fixing an intersubjective domain of discourse in science which can then be subjected to conceptual treatment (§15).

Picking out individual objects through univocal coordination of signs is precisely what Schlick's otherwise laudable, because purely formal, method of implicit definitions does not allow. Implicit definition Carnap notes here, and argues at some length in an important, but little-known, 1927 paper, succeeds in defining only classes of objects; hence empirical concepts that are implicitly defined are "improper concepts". KT, on the other hand, uses the method of explicit definition which, taken in a "widened sense", comprises both real (eliminative) definitions and Russellian definitions-in-use. For it is only through real definition that "the blood of empirical reality enters and...flows into the most remote veins of the thus-far empty schema" (1927, p.373) of theory created by implicit definition; thereafter, definitions-in-use single out and identify the objects (to be called "quasi-objects" in the *Aufbau*) of the ostensibly different domains of the special sciences (e.g., 'cow' in zoology and in economics, (1927, p.357)). However, these are ultimately eliminable through further logical decomposition in favor of real definitions and symbols of logic alone.[8] As against the unavoidable meaning holism stemming from conceptual determination through implicit definition and the implacable arbitrariness of restoring the linkage to the empirical through mere *Zuordnung* of concept and object, KT's approach begins with the structural relations of the empirical at the ground level and then builds upward. In consequence, it becomes possible to institute a **non-network-holist** verifiability theory of meaning for the individual statements of a science. Strict verificationism — as a semantic doctrine demarcating the legitimate concepts of science — is to be vindicated.

It is important to see that the derogation of objectivity to purely formal structure and thus to logic — which "**consists only in conventional posits concerning the use of signs**" (§107, original emphasis) — goes hand in hand with the rational reconstruction of the sign-relation. The signs of logic and mathematics do not designate at all; there are no real logical objects. However, empirical concept-signs of the special sciences do purport to designate objects. Within KT, an explicit translation can be formulated which will show that in fact, each of these concept-signs designates only a "quasi object", a logical construction from the sole non-logical basic relation(s) located at one or another level of stratification (corresponding to the various domains of the special sciences). The thesis of KT maintains that any legitimate scientific statement can thus be step-by-step recast as a statement 'about' "quasi-objects" on successively lower levels, and ultimately can be transformed into a statement containing only logical symbols (which do not designate) and signs for the basic relation(s) which "indicate" (§106) or "express a definite (formal and extensional) state of affairs relative to the basic relation".[9] Since it is possible to recognize whether the basic relation obtains or not between elementary experiences, and because in principle there are only finitely many elements of the basic relation extension (i.e., combinations of the elementary experiences), the truth or falsity of every legitimate scientific statement can in principle be ascertained in a finite number of steps.

The *Aufbau* might thus seem to be sounding a Russellian theme about how impoverished is the class of "logically proper names", holding that the only genuinely denoting (designating) terms are signs for the basic relation, i.e, relations of immediate acquaintance. Strictly speaking, however, these sole remaining empirical concepts are held to not really designate objects or to have *Bedeutungen* at all; indeed, to suppose they do is to mistakenly inquire into the "metaphysical essence" of the object (§161). Rather the only linkage remaining between concept-sign and object lies not in designation but in a **sententially** holistic "indication of the truth criteria" — the epistemological bedrock of experiential content — "for those sentences in which the sign of this object can occur" (§§160-1). In this evaporation of the relation of designation through rational reconstruction, the arbitrariness inherent in designation is systematically replaced by constitution based upon explicit conventions governing logical signs themselves. The *Aufbau*-project of the constitution of objects — providing unambiguous sign-indications of them — is the first aim of science, and it is attained through convention (*eine Festsetzung*); the second aim of science is empirical, and lies in determining the various properties and relations of these objects within the special sciences. In the conception of KT, the only two components of knowledge are the conventional (*konventionelle*) and the empirical, separate but equally important (§179).

It is because KT is thus crafted as a **semiotic** doctrine reconstructing concept and relation **signs** that Carnap can maintain that the program of the *Aufbau* abides metaphilosophical neutrality: KT is a means of cleaning up the problem of designation, independently of the realism-idealism dispute. Moreover, in "explaining" the relation obtaining between sign and designatum through reductive and eliminative explicit definition via constructional chains, ultimately from logical symbols and the basic relation alone, KT demonstrates as a lemma, as it were, the thesis of unity of science, i.e., that all and only objects (concepts) of science are constructible — and hence there can be intersubjective agreement upon **what** is under discussion — within the domain of a single *Konstitutionstheorie*. If this reading is correct, then we can perhaps locate the *Aufbau*'s most illustrious philosophical precedent in Leibniz's universal characteristic (as did Carnap himself (§3) as well as Schlick and Heinrich Scholz in their contemporary reviews of the *Aufbau*), similarly imbued with the spirit of metaphilosophical neutrality in proposing a purely symbolic means of resolving otherwise intractable philosophical disagreements ("*Calculemus!*"). Looking forward from the cusp of the "linguistic turn" a bit into the 1930s, we can identify a continuity with Carnap's method of syntactical designation via "the formal mode of speech" as a corrective for "pseudo-object" sentences, as well as with the later metaphilosophy of logical tolerance, and of linguistic methods for solving problems internal to a chosen framework.

On this interpretation, the *Aufbau* is attempting to simultaneously balance two ostensibly incompatible positions: first, a Russellian and Machian verificationist doctrine of meaning which requires that rational reconstruction of empirical concept formation reflect the epistemic primacy of autopsychological experience and, secondly, a view that the objectivity of scientific knowledge is accounted for by a shared language whose intersubjectivity rests on the possibility of formulating purely structural statements. The equilibrium solution advanced is targeted at the sign relation, whose semantic arbitrariness is systematically reconstituted — conventionally, but not arbitrarily — by pure logic alone. A strict verificationist account, providing for the meaningfulness of all and only scientific statements and couched in the autopsychological terms of (near-) positivist orthodoxy, can be advanced while still securing intersubjectivity through a method that constitutes away the relation of designation, systematically installing purely formal *Kennzeichen*. In this enterprise, logic is a *Hilfsmittel*, but it is more than that: it is the sole locus of structure which alone is communicable. The step-by-step dismantling of the designation relation — down to the sole empirical signs for the basic relation and perhaps even beyond —and its systematic replacement by a purely formal criterion of identity, structural isomorphism, is intended to reconcile the competing demands of verificationism and intersubjectivity. Thus verificationism and formalization go hand in hand; to the extent we maintain the former, we need the latter, so to speak, as compensation. We should not, however, slight Carnap's interest in providing only a **rational reconstruction**: just as "science wants to speak about what is objective" (§16), so too the issue of complete formalization (i.e, including the basic relation(s)) arises only in the abstract, as the limiting point of this reconstruction of scientific intersubjectivity.[10] Of course, the *Aufbau*-project of explicitly defining scientific concepts in the primitive idiom of sense experience flounders long before this point is reached (see note 8).

In this paper, I have argued that despite opposing assessments of the designation relation — an end point for Schlick, a starting point for Carnap — a central link is forged between their major initial works in that each offers a general epistemological program for eliminating all intuitive, imagistic, private and hence incommunicable 'material' or content from scientific concepts. Within a few years, Tarskian semantics would provide an entirely novel framework for discussion of semantic terms such as 'designation' and 'truth'. But we should not lose sight of the fact that much of the ground upon which the "linguistic turn" transpired was already in place, prepared by recognition and accommodation of conventionalism within empiricist doctrines of scientific concept formation.

Notes

[1] I wish to thank Michael Friedman and Thomas Uebel for comments and criticisms of previous drafts, and Alan Richardson for preliminary discussions on these matters.

[2] Of course, the bridges are not really "down"; see the remarks linking "concrete" (ostensive) to "implicit" definitions in §7.

[3] Cf. Coffa (forthcoming, chapter 11).

[4] Schlick 1918, p.326; on the significance of the notion of *Zuordnung* in the epistemology of science in this period, see Ryckman (1991).

[5] As Proust (1989, p.176) notes, the Fregean distinction between concept and object is rendered "a mere matter of words" by Carnap's adherence to a thesis of extensionality.

[6] Rolf George's translation of *Zeichenbeziehung* as "designation relation" thus obscures the cardinal point of the interpretation ventured here, that the *Aufbau's* reconstructive project is precisely to 'syntacticize' this semantic relation.

[7] Reichenbach (1920, p.38); cited in §15.

[8] As has repeatedly been observed, Carnap steps beyond the strict limits of explicit definition in the constitution of physical objects in assigning qualities to space-time points, where this is done via postulates, i.e., implicitly. See Kraft (1950, p.100), Quine (1951, p.40) and Carnap (1967, p.viii).

[9] §180; cf. the formulations in §§ 16, 112, 119-21, 144 and 160.

[10] Does Carnap's understandable inability to establish "foundedness" as a purely logical concept (§§ 154-5) thus undermine the entire *Aufbau* program (cf. Friedman 1987, pp. 532-3)? If we view his concern with objectivity merely as counterpart to a corresponding concern with strict verificationism, then the failure here plausibly indicates only the *explicative* (to anachronistically retrofit this term) limits of the method of KT.

References

Carnap, R. (1927), "*Eigentliche und Uneigentliche Begriffe*", *Symposion* I, pp. 355-74.

_____. (1928), *Der Logische Aufbau der Welt*. Berlin-Schlactensee,Weltkreis-Verlag.

_____. (1967), *The Logical Structure of the World*, translated by Rolf A. George, with a Preface by Carnap. Berkeley and Los Angeles, University of California Press.

Coffa, J.A. (forthcoming), *To the Vienna Station; Semantics, Epistemology and the a priori from Kant to Carnap*, L. Wessels and G. Steinhoff (eds.). New York and London, Cambridge University Press.

Friedman, M. (1987), "Carnap's *Aufbau* Reconsidered", *Noûs* 21, pp. 521-45.

Kraft, V. (1950), *Der Wiener Kreis; der Ursprung des Neopositivismus*. Wien, Springer-Verlag.

Proust, J. (1989), *Questions of Form; Logic and the Analytic Proposition from Kant to Carnap*, translated by A. Brenner. Minneapolis, University of Minnesota Press.

Quine, W.V.O. (1951), "Two Dogmas of Empiricism", reprinted in *From a Logical Point of View*, 2nd. rev. edition. New York, Harper and Row, 1961, pp. 20-46.

Reichenbach, H. (1920), *Relativitätstheorie und Erkenntnis A Priori*. Berlin, J. Springer.

Richardson, A. (1990), "How not to Russell Carnap's *Aufbau*", *PSA 1990*, vol. 1, pp. 3-14.

Ryckman, T.A. (1991), *"Conditio sine qua non? Zuordnung* in the Early Epistemology of Cassirer and Schlick", *Synthese*, 88.

Schlick, M. (1918), *Allgemeine Erkenntnislehre*. Berlin, J. Springer.

_____. (1921), *Hermann v. Helmholtz Schriften zur Erkenntnistheorie*, P. Hertz and M. Schlick (hg.). Berlin, J. Springer.

_____. (1922), *Raum und Zeit in der gegenwärtigen Physik*. Vierte Aufl. Berlin, J. Springer.

_____. (1929), *Rezension von Carnap (1928), Die Naturwissenschaften*, 27, pp. 550-1.

Scholz, H. (1930), *Rezension von* Carnap (1928), *Deutsche Literatur Zeitung*, Heft 13, Sp. 586-92.

Hidden Agendas: Knowledge and Verification

Joia Lewis

University of San Diego

1. Introduction

Schlick has been accused of a number of philosophical sins over the years, most notably his rather casual, and frequent, traversing of the borders between language, experience, and reality. While we allow our scientists the freedom to roam creatively throughout the peripheral regions of Epistemology and Metaphysics, we are not so tolerant of our philosophers. We know that Schlick gave up the physics laboratory for the philosopher's armchair, and we expect him to stick to a particular position.

Schlick's colleagues in the Vienna Circle were not totally blameless in this regard, either, and it may be that part of the charm of their work is precisely their inability to stick to the positions that solidified over 2500 years of western philosophy. Empiricism and rationalism appear to coalesce in their writings, idealism and realism cohabitate the same theories, the Senses follow the Intellect right out of Plato's Cave and into the light of certainty. We become seasick in Neurath's boat, wondering whether his rebuilders can keep it afloat, or if it will eventually hit the empirical ground after all.

In Schlick's case, we wonder whether he really did 'convert' from realism to something other than realism, and what exactly Wittgenstein had to do with it. We wonder what his realism entailed: in the inimitable words of Alberto Coffa,

> In Schlick's hands, realism had been turned from the boring, trivial common sense view that it was before Kant, into an exciting, bold and utterly unbelievable conjecture. ...The world of common sense has been torn to pieces: its time is subjective and transcendentally ideal and so is its space (Coffa, 188).

We also wonder how Schlick could lodge meaning safely within logical relationships in one paragraph and then go on in the next to talk about meaning and empirical content. Michael Friedman has noted (see Friedman, 1983) that Schlick vacillated throughout his philosophical career between a holistic and formalistic account of knowledge and meaning, and an atomistic and foundationalist account, apparently without ever committing himself completely to either one. Tom Ryckman has written that Schlick's original linking of thought with reality by means of the notion of *Zuordnung*, or coordination, constituted a very minimalized epistemology:

> It is therefore all the more surprising to find him actively engaged, within a decade or so, in a truly traditional dispute over the "foundations of knowledge" (Ryckman, 56).

This latter dispute over the foundations of knowledge is perhaps what we wonder about most of all: why was the exchange between Schlick and his colleagues in the famous protocol debates of the 1930's so bizarre? Carnap, Neurath and Hempel allegedly found a "detestable metaphysics" in Schlick's comparison of a statement in his Baedecker with the number of spires on a certain cathedral; Schlick accused them of being relativistic and rationalistic, as well as irrationalistic, if they could not allow him to determine the truth of the matter by simply checking his guidebook while standing in front of the cathedral.

What was at issue was Schlick's theory of *Konstatierungen*. These were, according to Schlick, our momentary affirmations of reality, which functioned as the foundation of empirical knowledge. However, their foundational authority was short-lived: as soon as they illicited our assent or dissent of a particular claim, they metamorphosed into hypotheses and joined the ranks of all other claims about empirical reality. Schlick's colleagues had dispensed with the idea that we could ever have an absolutely certain foundation of empirical knowledge, and therefore rejected Schlick's theory on the basis of its metaphysical claim about the possibility of comparing statements to facts. Tom Oberdan has pointed out quite correctly that there was a great deal of misunderstanding going on in the debates: "[Schlick] seems to have assumed—incorrectly—that his commitment to conventionalism was apparent to anyone who knew his work." (Oberdan, 18) If Schlick was indeed as committed to a conventionalist account of language and meaning as his colleagues, what then were they arguing about?

There is a growing literature on this topic, which includes many dissections and interpretations of the discussion between Schlick, Carnap, Neurath, and Hempel, on the nature of basic empirical statements and their role in verification (see Haller, Gadol, *Synthese 64* (1985)). What I would like to focus on specifically is the dual nature of Schlick's major philosophical theses. In the case of Schlick's verificationist theory of meaning, why did he require an account that would be both intersubjectively accessible as well as privately incorrigible? In the case of his 'affirmations of reality,' why did Schlick need to invent, or more modestly, to discover, something that would function as a foundation of knowledge but that would at the same time not clog our systems of conceptual claims with any debris from our perceptual apparatus? I would find this a very frustrating assignment, if asked to create a model fulfilling these terms.

I believe the answer to these questions can be found by exposing the assumptions behind what we can consider to be Schlick's two separate agendas, one formalistic and conventional, the other foundational and empiricist. The agendas are present and fully formed in his earliest philosophical writings. Given that the hypothesis of split personality is not warranted in Schlick's case, the question to ask is, why would he deliberately develop inconsistent lines of thought? What makes the situation interesting is not only that the two agendas are not compatible, which is evident in his early work, but also that Schlick consciously attempted to *make* them work together in his later accounts of meaning and verification. Schlick's conviction that the assumptions in each agenda were equally correct can explain a good deal about the development of his mature theses and is, I believe, responsible for much of the criticism that his work has received, both before and after his death in 1936. Making the two agendas explicit can also provide us with a better framework within which to judge Schlick's work. Finally, the story provides an instructive example of how one's firmest beliefs constrain and dictate one's choice of solutions and perhaps define the problems themselves.

2. Agenda 1

The first agenda had to do with the uncompromising distinction Schlick made between conceptual knowledge and intuitive experience. Athough he subscribed to the general empiricist claim that all knowledge stems ultimately from sense experience, Schlick was always careful to point out the vast difference between precisely formulated and enduring concepts as opposed to imprecise and ephemeral intuitive experiences. Schlick sought to clarify this distinction by means of Hilbert's notion of *implicit definition*, whereby the basic or primitive concepts of a system of truths are defined by virtue of the fact that they satisfy the axioms. He contrasted implicit definitons to *concrete definitions*, in which "the defining terminates when the ultimate indefinable concepts are in some way exhibited in intuition...[they involve] pointing to something real, something which has individual existence." (*General Theory of Knowledge*, abbrev. *GTK*, 37) He described the relationship between concrete and implicit definitions as follows:

> it is through concrete definitions that we set up the connection between concepts and reality. Concrete definitions exhibit in intuitive or experienced reality that which henceforth is to be designated by a concept. On the other hand, implicit definitions have no association or connection with reality at all.... A system of truths created with the aid of implicit definitions does not at any point rest on the ground or reality. On the contrary, it floats freely, so to speak, and like the solar system bears within itself the guarantee of its own stability. None of the concepts that occur in the theory designate anything real; rather, they designate one another in such fashion that the meaning of one concept consists in a particular constellation of a number of the remaining concepts. (*GTK*, 37)

We begin, then, by assigning concepts to designate specific experiences. Once we have the concept, we can find an implicit definition for it among concepts already incorporated into our system of knowledge. Thus the relationship between the concepts utilized in concrete definitions and sense-experiences is that of names to the objects they name. Our implicitly defined concepts, however, bear no systematic relationship at all to sense-experience.

Besides the influence of Hilbert, Schlick was also following Planck in his separation of conceptual knowledge from intuitive experience. Schlick accepted Planck's mandate that science should move away from descriptions of reality that originate with our sensations and toward a purely quantitative picture of the world. It is for this reason that Schlick also rejected Mach's restriction of science to descriptions of phenomena within the realm of our sensations. Further, following Planck, Schlick agreed that only abstract quantitative formulations could assure a *unified* picture of the world; such a goal could not be reached by focusing exclusively on sense-dependent descriptions. Schlick identified Planck's distinction between the subjective and qualitative as opposed to the objective and quantitative with his own distinction between concepts and intuition.

Consonant with this first agenda was Schlick's abhorrence of any philosophy which elevated the role of non-conceptualized experience or intuition above that of systematically determined concepts, either with respect to questions of truth or to existence. Schlick considered these positions to be direct attacks on scientific thought, and on any philosophy designed to reflect the scientific spirit. Passages like the following appear repeatedly throughout Schlick's early works:

> the arrogant structures of idealist thought, which created the bitterness and brought the philosophic spirit into discredit, have long since crumbled...these doubts about philosophy rest merely upon hasty judgment, upon deliberate neglect of the ultimate problems of science, in short, upon lack of clarity. (*Philosophical Papers, Vol. I*, abbrev. *Vol. I*, 104)

Clarity, for Schlick, as we have seen, required that knowledge be wholly pruned of its roots in intuitive experience.

3. Agenda 2

Schlick's second agenda had to do with the epistemological foundation of empirical knowledge. Under this agenda are his doctrines of *unique coordination* and *verification*.

Schlick considered the notion of *unique coordination* to be the defining characteristic of *truth*; the method to check uniqueness was then *verification*. In his words,

> truth is defined by a single, extremely simple characteristic: the uniqueness of the correlation of judgments with facts. (*GTK*, 162)

In "The Nature of Truth in Modern Logic" (1910), Schlick likened the notion of judgment to sensation and ideation, in that all three notions involved designation of some sort. He was expanding on Helmholtz's theory that our sensations are signs of things-in-themselves, not in the sense of images of reality, but by giving us the formal features of the otherwise unknowable noumenal realm. According to Schlick, sensations and ideas are "merely signs for the *content* of experience, the world given to us, without regard to the form in which this content makes its appearance" (*Vol I*, 91) But we are aware of order among our experiences, of the fact that "elements of experience are connected in relations to one another...[that] they possess *order* and *form.*" (*Vol. I*,91) Another type of sign is therefore required, not one that is given to us but one which we actively employ to designate the relations among our experiences. Judgments fulfill this function. They are "designations of facts, of forms of experience, of the ordering and association of elements of the given." For a judgment to be true, it must isolate a single fact unambiguously, or, in Schlick's words, "*A judgment is true if it univocally designates a specific state-of-affairs.*" (*Vol. I*, 94) The relationship is not one of similarity or picturing; a judgment cannot be "more than a sign in relationship to a set of facts.... A judgment pictures the nature of what is judged as little as a musical note pictures a tone, or the name of a man pictures his personality." (*GTK*, 60) It is the rather the *interconnection* between judgments that allows an adequate designation of reality:

> By virtue of the interconnection of judgments a new truth receives a specific place in the circle of truths; the fact corresponding to this new truth is thereby assigned to the place that, by virtue of the interconnection of facts, it occupies in the domain of reality. ...Hence it is the structural connectedness of our system of judgments that produces the unique coordination and conditions its truth. (*GTK*, 67)

Recall that concepts, for Schlick, were originally connected to experience through concrete definitions. Introducing a concept by means of a concrete definition was a conventional act, "a quite arbitrary stipulation, and consists in introducing a particular name for an object that has been singled out in one fashion or another." (*GTK*, 69) These were contrasted to implicit definitions of concepts within the system of interlocking judgements, which "will then be connected to one another by a system of judgments coinciding fully with the network of judgements that on the basis of experience had been uniquely coordinated to the system of facts." (*GTK*, 70)

Schlick's conviction that this was the case could not have been stronger:

> Obviously, to suppose that the world is intelligible is to assume the existence of a system of implicit definitions that corresponds exactly to the system of empirical judgments. (*GTK*, 70)

So far this fits in with Agenda 1, in that we have only a conventional link between knowledge and experience, through our concrete definitions of concepts. But Schlick's second agenda also had a new nonconventional element.

The system of empirical judgments was uniquely coordinated to the system of facts on the basis of experience. Schlick called the system of empirical judgments *historical* or *descriptive* judgments, and distinguished them from *definitions* and *conventions*. Historical judgments that hold for facts not being immediately observed were *hypotheses*. Schlick noted that this distinction cannot strictly be maintained, however, since "the class of historical judgments dwindles to zero if we consider that strictly speaking it can embrace *only such facts as are immediately experienced in the present moment*." (*GTK*, 73; my emphasis) Schlick thus introduced the idea of *fundamental*, or *perceptual* judgments, to distinguish the judgments embracing "immediately experienced facts" from other descriptive or historical judgments: these were the 'building blocks' of the system, the "propositions...by virtue of which the system rests directly on real facts.... If the whole edifice is correctly built, then a set of real facts corresponds not only to each of the starting-points—the fundamental judgments—but also to each member of the system generated deductively." (*GTK*, 78)

4. Schlick's Early Account of Verification

In Schlick's early account, verification is the method to check for *uniqueness of coordination* between judgments and facts, which, as we have seen, is the defining characteristic of truth. What is involved in verification, according to Schlick, is the identity of two judgments, one of which must be a fundamental or perceptual judgment, as just described.

Schlick first illustrated his account of verification with an example from the history of science, also in his (1910) article, "The Nature of Truth in Modern Logic." He explained that the verification of the existence of the new planet Neptune consisted in the identity that obtained between the prediction that a certain planet would be found and the perceptual judgment made at the time that it was first observed. The prediction was deduced by Leverrier from Newtonion law and observed facts about the perturbed orbit of Uranus. When Galle looked through the telescope and perceived the presence of a planet exactly where it was predicted to be, the perceptual judgment that he formed upon making this observation was identical to the prediction-judgment originally formulated by Leverrier.

Later in the *General Theory of Knowledge*, Schlick presented a more detailed analysis of what happens when a judgment is verified. The procedure to verify an arbitrary assertion about reality, J, is as follows, where J', J", ...are auxiliary assumptions:

(A) Derive J_1 from J and J', where the truth of J' is considered established, for one of the following reasons:

    (i) J' is an assertion about reality
    (ii) J' is a definition
    (iii) J' is a purely conceptual proposition

(B) Derive J_2 from J_1 and J"... until finally J_n is derived, which is a prediction in the following form:

"At such and such a time and at such and such a place under such and such circumstances such and such will be observed or experienced." (*GTK*, 163)

(C) At the appointed time and place, we make a perceptual judgment, P. If P and J_n are identical then J_n is verified, and so is the original judgment J.

Since both P and Jn designate the same fact, even though we arrived at them through "two entirely different paths," we have, according to Schlick, established a unique correlation. The judgment is therefore true. (*GTK*, 163) Thus, when J_n and P designate the same fact, we may consider the original J to be true. But in order to do this, we must first accept P to be true. Schlick considered the truth of P noncontroversially established as the expression of the fact. In other words, for J_n to be correlated with the fact, it simply had to be identical with the perceptual judgment P that designated this fact. But, if verification is the method by which we establish the unique coordination between a judgment and a fact, thereby establishing the truth of the judgment, how can we then simply accept that the 'perceptual judgments', or judgments of experience are themselves true? The perceptual judgment P, which embraces "immediately experienced facts" is not in need of verification itself, according to Schlick.

Schlick also stated that not only is the original claim J verified, but also the whole chain J_1, J_2, ..., leading up to the prediction-judgment J_n:

since the last member of the chain of judgments led to a unique correlation, we take this as a sign that the other members, hence the starting-point and the endpoint J, also fulfill the truth condition, and we count the entire process as a verification of judgment J. (*GTK*, 163)

An identity of two judgments was also the basis of the verification of purely conceptual or analytical judgments, for Schlick (*GTK*, 166). What is involved in our recognition of an identity in either case is some kind of intuitive process, a mental picturing of the sense of the proposition. But even though empirical and purely conceptual statements share an identity experience at the end of verification, "a vast difference separates these two classes of judgments, an abyss that no logic or epistemology can bridge." (*GTK*, 168) Verification of empirical judgements can only be probable rather than certain. Schlick gave two accounts of the source of the uncertainty of empirical judgments. One considered the inductive problem of knowing the future: this is the problem of relying on the law-like regularities of nature in our assertions about reality. Since a true proposition should be confirmed "always and without exception...what we can infer, strictly speaking, from a limited number of verifications is not absolute truth but only probability (*GTK*, 168)."

The second source of uncertainty cited by Schlick, in a later passage in the *General Theory of Knowledge*, is that associated with the imprecise nature of intuitive experience:

> Due to the fleeting character of experiences, this act of comparing and finding the same is always subject to an uncertainty that, although harmless and of no significance for the practical conduct of science and everyday affairs, is always present theoretically and stands in the way of absolute infallibility. (*GTK*, 342)

Schlick needed a nonconventional link between his perceptual judgments and the facts of experience to make his account of verification work. But verification was supposed to *produce* this assurance that our judgments are true of reality. Schlick knew that he could rely on the certainty produced through manipulations of his *implicitly* defined system of conceptual judgments, and that his originally concretely defined concepts did not contaminate the system with extra-linguistic uncertainty. But his account of *truth* required the confrontation of a perceptual judgement with an extra-linguistic state-of-affairs. The experience linking one's judgments to reality had to provide the basis of one's designation by Schlick's foundational Agenda 2. But, experience with respect to Schlick's formalistic Agenda 1 was not to be trusted; specifically, it was not to be considered knowledge at all. Coffa wrote:

> It would be hard to exaggerate the significance of this difficulty....the link between knowledge and reality depends entirely upon the link between basic statements and reality, and, as Schlick's example illustrates, it was widely assumed that the key to this link was experience. (Coffa, 356)

And we note a profound ambivalence in Schlick's attitude toward experience.

5. Meaning and Affirmation

We can see Schlick's two agendas still working in his later account of verification, which he presented in articles written from the late 1920's to the middle 1930's. What is new, following Schlick's acquaintance with Wittgenstein's work, is a focus on *meaning* as well as on *truth*. Schlick distinguished now between verifiability and verification: *verifiability* refered to a criterion of the *meaningfulness* of a statement; it was a purely philosophical notion having to do with the logical possibility of a corresponding state-of-affairs. *Verification* referred to a procedure for establishing the *truth or falsity* of a statement; it was a scientific activity having to do with the empirical possibility of confirming a particularly occuring state-of-affairs. (see *Vol. II*, (1936), and *The Problems of Philosophy in Their Interconnection*)

Schlick's discussion of meaning in terms of verifiability for the most part follows the holistic assumptions built into his Agenda #1; his discussion of truth in terms of verification follows those of his foundationalist Agenda #2. But in each there is a struggle that is not evident in his earlier work. Schlick raises questions in each account that show that he is trying to accomodate his commitment to a conventionalist account of knowledge along with his conviction that perceptual judgments can be considered true records of "immediately experienced facts."

Schlick wanted to show that what is meaningful is not based on private psychological experiences but on publicly accessible information. He wrote that "it would be nonsense to say 'We can mean nothing but the immediately given'." (*Vol. II*, 462) What is required when we give the meaning of a sentence is a knowledge of the rules that tell us how the sentence is to be used: these include both *ordinary definitions*, in terms other than the term itself, and *ostensive definitions*, which Schlick formerly called 'concrete definitions'. Schlick stated that "there is no way of understanding any meaning without ultimate reference to ostensive definitions, and this means, in an obvious sense, reference to 'experience' or 'possibility of verification'." (*Vol. II*, 458) He admitted that he had insisted both on an "empirical-meaning requirement" and on the fact that "meaning and verifiability do not depend on any empirical conditions whatever, but are determined by purely logical possibilities." (*Vol. II*, 467) He dealt with this apparent contradiction himself by explaining that the problem stems from the ambiguous use of the term 'experience.' It is used in one sense to refer to 'immediate data', and in another sense, the sense in which Hume and Kant used the word, to refer to the process by which we gather information inductively (*Vol. II*, 468). Schlick's 'ostensive definitions' are the link between his two senses of experience: "through them verifiability is linked to experience in the first sense of the word. No rule of expression presupposes any law or regularity in the world...but it does presuppose data and situations, to which names can be attached." (*Vol. II*, 468) So verifiability is not 'independent of experience' with respect to immediate data, but only with respect to experience as inductive learning. In other words, there is no Kantian synthetic *a priori* knowledge about the laws of nature, but an account of meaning must presuppose the immedate data of sense-perception.

In Schlick's ultimate discussions of verification as a criterion of truth, we find assumptions associated with his second foundational agenda. What is new here is his reference to our moments of perceptual certainty as *Konstatierungen*, or 'affirmations' of reality. We know that for Schlick "all verifications terminate in perception" (see above); it is *Konstantierungen* which function as these endpoints, and which fulfill the same role that his perceptual or fundamental judgements did in his early account. They have, according to Schlick, the "positive value of absolute certainty and the negative value of being useless as an enduring foundation." (*Vol. II*, 386) It would be absurd to question the validity of observations made by oneself, in the present, according to Schlick: we may therefore consider our current and subjective experiences as absolute indicators of the truth or falsity of empirical claims. Schlick's colleagues read his 'affirmations' as basic *statements*, subject to all of the criticisms levelled against Carnap's original construal of incorrigible protocol sentences. Neurath had pressured Carnap into a modified version of protocols as statements which were as fallible as any other empirical statements; they were simply *chosen* as basic statements. A number of writers have noted that had Schlick presented his 'affirmations' unambiguously as perceptions rather than as a type of perceptual *statement*, the debates would have taken a different turn, or perhaps not have occurred at all. (see Oberdan, Chisolm and Hilpinen in Haller) Schlick's position is at least consistent with his Agenda #1 if affirmations are merely the perceptual moments that illicit linguistic responses, rather than statements themselves. They remain outside of science for the reason that they remain outside of language in general:

> Science does not rest on them, but leads to them, and they show that it has led aright. They are really the absolutely fixed points; we are glad to reach them, even if we cannot rest there. (*Vol. II*, 383)

The phrase 'absolutely certain foundation' was also guaranteed to ruffle his colleagues' philosophical feathers, and they viewed Schlick as stubbornly hanging on to an outdated correspondence theory of truth. But was this criticism warranted? In what sense do Schlick's affirmations provide an absolutely certain foundation of knowledge?

> The question behind the problem of the absolutely certain foundation of knowledge is, so to speak, that of the legitimacy of the satisfaction which verification fills us with. Are our predictions actually realized? In every single case of verification or falsification an 'affirmation' answers unambiguously with yes or no, with joy of fulfillment or disillusion. The affirmations are final. (*Vol. II*, 383)

Schlick's finality has to do with a personal sense of fulfillment or disillusion, not with an incorrigible base of knowledge. While it appeared to others that Schlick retained a Cartesian desire for certainty about the empirical world, Schlick had settled for momentary flashes of perceptual certainty rather than the constant but distant illumination of a self-evident principle.

6. Concluding Remarks

If Schlick's colleagues misunderstood his commitment to a formalistic account of knowledge and meaning, they also misread his account of verification by affirmation as producing certain rather than probable truth. As we have seen, Schlick had given a detailed analylsis of the verification of judgments of reality in his *General Theory of Knowledge* in terms of probability rather than certainty and very likely assumed that his remarks about an empirical foundation of knowledge were read in this light. Schlick considered his role in the debates "nothing but a gentle warning of a true empiricist against certain tendencies towards...a rather dogmatic irrationalistic formulation of positivistic principles." (*Vol. II*, 400) We may note that Schlick preferred the label 'consistent empiricist' to 'logical positivist' (see *Vol. II*, 283). A consistent empiricist would tend to be interested in what we actually do when we say that we know something, and how we got that information. *Konstatierungen* are, after all, not a bad *description* of what happens when we decide to affirm or deny a particular claim. Schlick's ideas do not even sound that far out of line with current views on perception and cognition. With respect to his conviction that the experiential foundation of knowledge could not be considered part of knowledge itself, since it lacked the precision of linguistic concepts, compare the following passage by Patricia Churchland in her book, *Neurophilosophy*::

> Although some cognitive activity probably is understandable as the manipulation of sentential representations according to logical rules, many cognitive processes likely are not. Indeed, the processes underlying sentential representation are surely themselves nonsentential in nature. (Churchland, 452)

With respect to Schlick's apparent ambivalence about the linguistic status of his perceptual judgments, or affirmations, compare this statement by Gerald Edelman in *Neural Darwinism*:

> Perception...is close to the interface between physiology and psychology...[it] involves categorization, a process by which an individual may treat nonidentical objects or events as equivalent. (Edelman, 26)

We have become accustomed to distinguishing between those who deal in empirically equivalent theories and those who have faith in our ability to 'get it right', in our potential someday to fashion a true description of reality. Schlick understood well that knowledge could encompass competing theories (see *Vol. I*, (1915)), but was also convinced that the role of perception in verification was to guide us in theory selection, ultimately to better theories. We tend to read the second sentiment as an override on the first, as if wanting to 'get it right' precludes acknowledging the hypothetical nature of our knowledge. Schlick watched his colleagues suffer from the unique hubris of the logical empiricist, which made it impossible to avoid murdering one's empiricism and marrying one's logic. Perhaps Schlick's hubris was of a different sort. Perhaps what fueled his conviction that both his logical and empirical agendas were correct was his very strong desire for clarity and precision at all costs. Verification, for Schlick, was the only road to clarity and precision on the empirical front; keeping conceptual knowledge systems floating above the empirical ground guaranteed clarity and precision on the logical front. The price for each of these assumptions, however, was to give up absolute certainty about the empirical world. This, as we have seen, Schlick was willing to do.

References

Churchland, P. S. (1986), *Neurophilosophy, Toward a Unified science of the Mind/Brain*. Cambridge: MIT Press.

Coffa, J. A. (forthcoming) *The Semantic Tradition from Kant to Carnap: To the Vienna Station*.

Edelman, G. (1987), *Neural Darwinism, The Theory of Neuronal Group Selection*. New York: Basic Books, Inc.

Friedman, M. (1983), "Moritz Schlick, Philosophical Papers", *Philosophy of Science* 50: 498-514.

Gadol, E. (ed.) (1982) *Rationality and Science, A Memorial Volume for Moritz Schlick in Celebration of the Centennial of His birth*. Wien: Springer-Verlag.

Haller, R. (ed.) (1982), *Schlick und Neurath—ein Symposion, Grazer philosophische Studien* 16/17. Rodopi, Amsterdam.

Oberdan, T., (1989), "Conventionalism in the Protocol Sentence Controversy: The Case for Schlick". (manuscript)

Ryckman, T., (1990), "Conditio Sine Qua Non? The Concept of *Zuordnung* in the Early Epistemologies of Cassirer and Schlick". (manuscript)

Schlick, M. (1985), *General Theory of Knowledge*. La Salle, Illinois: Open Court Publishing Co. Translation by A.E. Blumberg of *Allgemeine Erkenntnislehre*, Second German edition (1925).

_ _ _ _ _ _. (1978), *Philosophical Papers,* Volume I (1901-1922), ed. by H.L. Mulder and B.van de Velde-Schlick, Dordrecht: Reidel.

_ _ _ _ _ _. (1978), *Philosophical Papers,* Volume II (1925-1936), ed. by H.L. Mulder and B.van de Velde-Schlick, Dordrecht: Reidel.

_ _ _ _ _ _. (1987), *The Problems of Philosophy in their Interconnection*, Winter Semester Lectures, (1933-34), ed. by H. L. Mulder, A. J. Kox, and R. Hegselmann, transl. by Peter Heath (Dordrecht: Reidel.

Philosophical Interpretations of Relativity Theory: 1910-1930[1]

Klaus Hentschel

University of Göttingen

1 Introduction

This paper is a summary of my doctoral dissertation on philosophical interpretations of Einstein's special and general theories of relativity, submitted to the Dept. for History of Science, Univ. of Hamburg, in 1989, which was recently published in the Series *Science Networks* at Birkäuser.[2] After a brief overview of its content I will focus on a discussion of the method employed to analyse philosophical interpretations of a physical theory.

My analysis is based
- firstly on about 2500 contemporary published texts about the theories of relativity written both by scientists and philosophers. These texts have not been of particular interest to historians or philosophers of science up to now; this is understandable from the fact that many of them contain gross oversimplifications, misinterpretations and incorrect statements about the theories of relativity. But I claim that it is just these misinterpretations that can serve as a clue to a better understanding of the general process by which philosophical interpretations are formed (see next section).

- Secondly, collections of the many unpublished documents preserved in the estates of physicists of that time were another very important source, most notably, the duplicate files of the 'Einstein Archive' at the *Hebrew University of Jerusalem* located at the *Collected Paper of Albert Einstein* in Boston, which contain hundreds of letters by philosophers asking Einstein to explain features of his theories. These include Bergson, Bridgman, Cassirer, Metz, Meyerson, Petzoldt, Reichenbach, Schlick among others.

- Thirdly, other unpublished materials of interest were found in the estates of philosophers, e.g., of Reichenbach and Carnap (Pittsburgh), of Schlick (Amsterdam), of Bavink (Bielefeld), of Petzoldt (TU Berlin), or of Friedrich Adler (Vienna). All these letters are very telling about the different strategies used by each philosopher in attempting to understand a particular aspect in the theory to be interpreted. This struggle for a closer understanding of scientific matters outside the traditional realm of philosophy was largely omitted or hidden in the published papers.

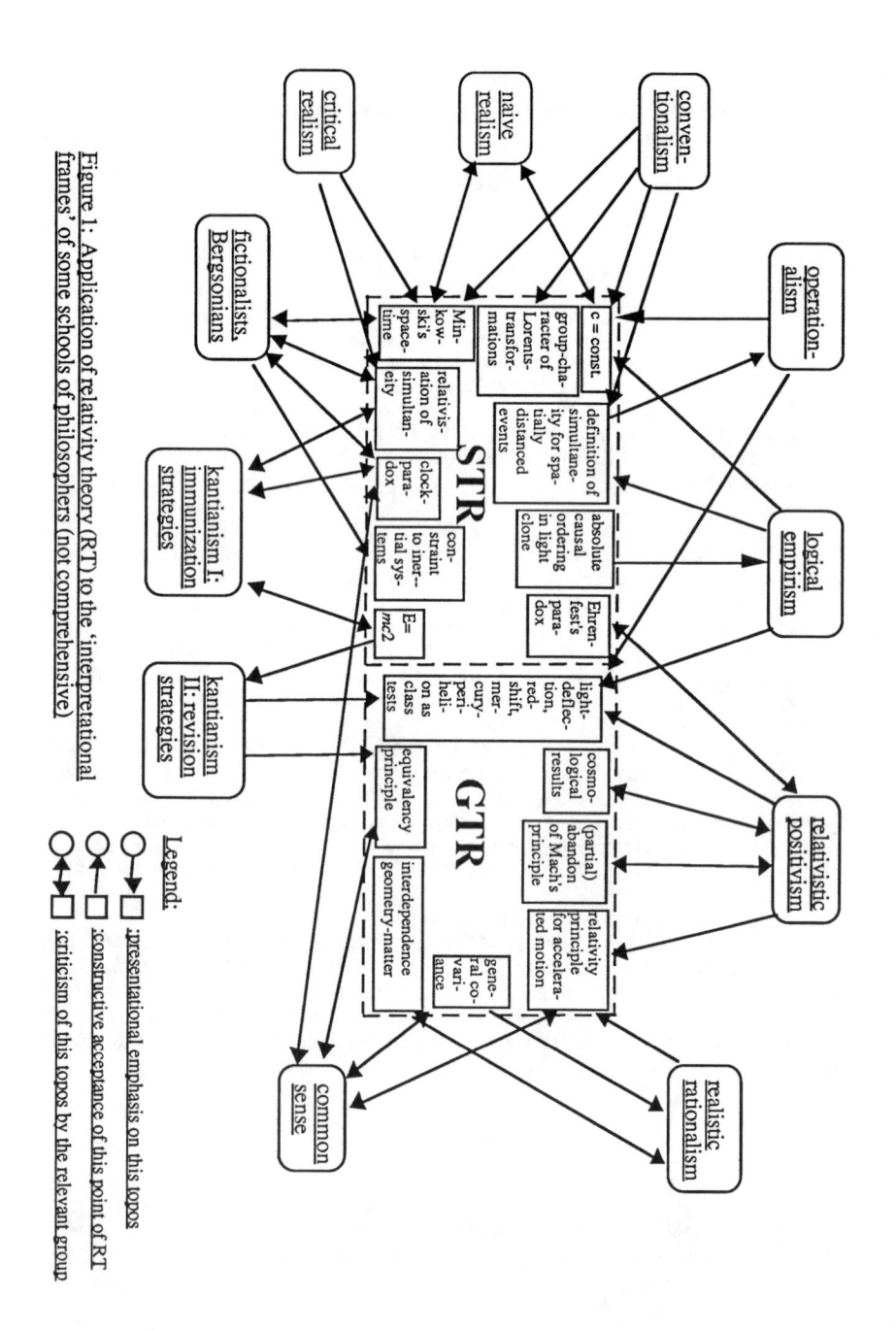

Figure 1: Application of relativity theory (RT) to the 'interpretational frames' of some schools of philosophers (not comprehensive)

2. Some General Traits of Philosophical Interpretations

In figure 1 (= figure 6.1. of my thesis (footnote 1).), some (but by no means all) contemporary philosophical schools [ovals] are depicted as they relate [arrows] to selected issues in the theories of relativity [boxes in the center]. Although I could not, of course, achieve completeness with this two-dimensional sketch, it illustrates some general features of philosophical interpretations:
- The various schools discussed the special (left) and general (right) theories of relativity electively, picking out and emphasizing drastically different parts.
- While several groups only dealt with a very few isolated topics from both theories but showing no interest in their context, some schools were ambitious enough to try to incorporate substantial parts of the theories into their interpretations.
- Representatives of some groups of philosophers tried only to understand and to highlight parts of Einstein's theories, while others did not confine themselves to this restricted role, but openly refuted parts of both theories as being in conflict with their general opinions (these cases are marked with double arrows in fig. 1). Although you can often find psychological reasons stemming from the individual personalities of the interpreters behind these different reactions to a compelling scientific issue (an arrogant Oskar Kraus versus a modest Ernst Cassirer), there are also different models about the relationship between physics and philosophy which ultimately account for these astonishing variations. For example, a *critique* of scientific results was forbidden for the logical empiricists; they could only legitimately analyze them, put them into a broader context and, perhaps, make them thereby more comprehensible to the public. On the contrary, a realm of fundamental issues existed for the neo-Kantians, that could only be approached using a transcendental form of philosophy, which sought to deduce the a priori patterns of the world excluding any possible interference with empiricial issues. Since space and time were among them, a conflict with the special theory of relativity (STR) was to be expected.

3. The Guiding Idea of My Analysis and Presentation: The Dependence of Philosophical Interpretations on a Framework of Underlying Assumptions

Now let me briefly discuss how the various philosophical interpretations of relativity theory came about. The issues involved in the special and general relativity theories, which were essentially spelled out since 1905 and 1915, resp., intrinsically involved what could be called an **interpretational framework** consisting of general epistemological and methodological convictions which ultimately determined the resulting philosophical interpretations of the theories. The vast differences in these often implicit background assumptions explain the formation of completely different, in a sense, incommensurable interpretations of one and the same physical theory.

Some scholars (e.g., John Stachel, Arthur Fine, and at the meeting in Minneapolis, David Stump) have remarked that it is misleading to speak of 'one' theory. Of course, I do not contest that there were a lot of theoretical alternatives to Einstein's general theory of relativity (GTR), construed by Abraham, Nordström, Weyl, Mie, etc., all differing on specific points from the standard general theory, and I also agree that certain points were not resolved for a long time (e.g., value of the cosmological constant, preference of cosmological solutions, singularities). These clearly defined identifiable theoretical alternatives to Einstein's GTR might also deserve a philosophical interpretation in their own right, but their existence does not preclude us from speaking of a standard GTR as physicists do, e.g., in their textbooks on the special or the general theory of relativity. Despite what might be called the not yet settled boundary of the theory ('Forschungsfront', 'research frontier'), where contemporary physicists were in disagreement on many points, what formed the 'hard core' of the theory was undis-

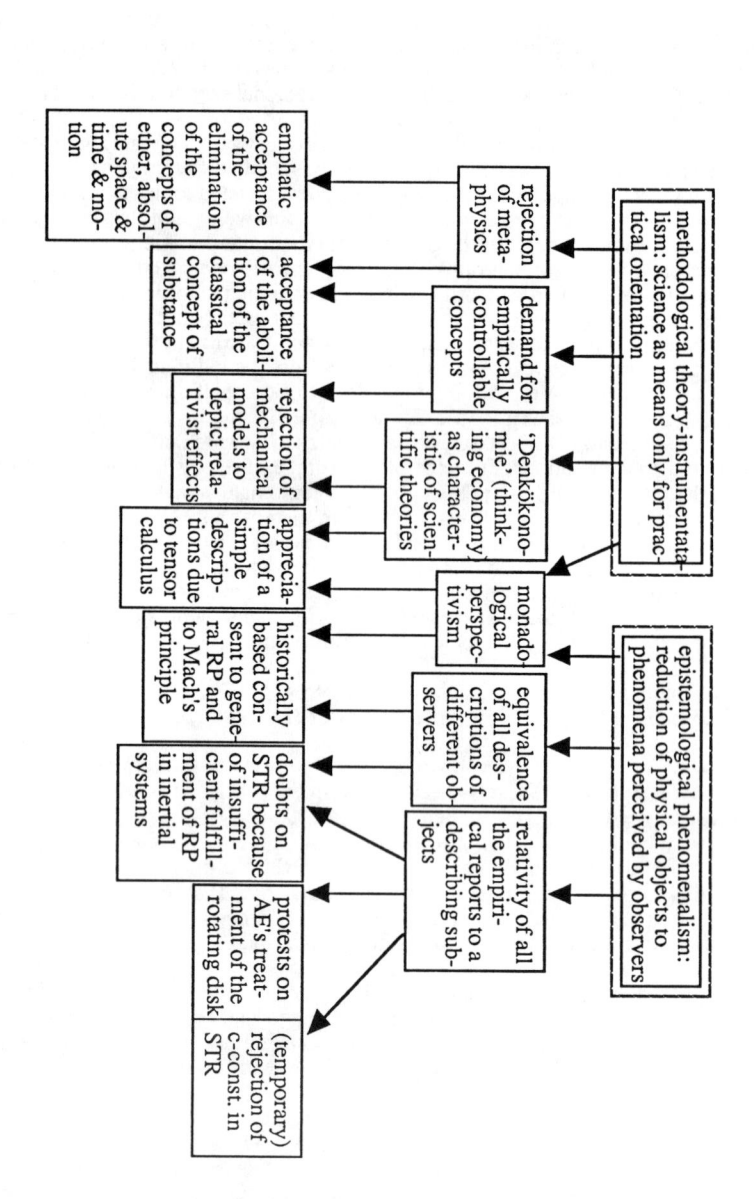

Figure 2. Overview of phenomenalism (Mach school)

puted and was not altered by further developments at the research frontier.[3] So for example, the two axioms of the constancy of light in vacuo and the principle of relativity form the 'hard core' of the SRT. In this sense, my model assumption of one theory being interpreted in different ways is justified at least pragmatically. At the end of my paper another argument for this model assumption will be given, based on considerations of what would result if one would *not* make it.

To get a clear picture of the complicated relationship between general but often implicit background assumptions and specific interpretative statements, between internal and surface structures, to employ the terms of linguists, I modelled the **arrival upon an interpretation as a multilevelled process of specialization that I** will discuss now within the Machianism example[4] (Cf. fig. 2).

In the beginning of this multigrade process of interpretation we find very few fundamental premises (for the Machians, those of epistemological phenomenalism, coined by Berkeley as 'esse est percipi', and methodological theory instrumentalism). Those basic convictions are obligatory and not questioned by members of this group of philosophers: They form a core (analogous to Lakatos's hard core in scientific theories).

General statements are derived from these core premises in a first specialization step, e.g., about the aim of scientific theories, the status of their results, etc. [see the arrows from the top of figure 2 to the second level]. These general opinions form the level of understanding, so to speak, on which the scientific theory to be interpreted is projected; they define what I earlier referred to as the 'interpretational framework'. In a further step of specialization, the concrete interpretational statements about relativity theory are derived from this interpretational framework [arrows from level 2 to level 3 of figure 2]. The latter is usually fixed *before* the interpreters started their study of relativity theory. Described in another way (as seen from what will be discussed ahead), those topoi in the theories that offered some connection to their prefixed interpretational framework were selected for further interpretation; others, that did not fit were either left out or reformulated so that they would fit within the procrustean bed. This mechanism already explains the astonishing similarities in the selections of independent interpreters from similar backgrounds.

Let us regard our example. Again and again the Machians emphasized, that for them the physical principle of relativity, one of the axioms of special relativity, was nothing but a trivial consequence of a much broader 'epistemological relativity' between a describing subject and the resulting description. For Petzoldt, whose own philosophical system was consequently called "relativistic positivism", this 'relativity' of motion, of size, of time (and of virtually everything else) was the central feature of both theories of relativity—he largely ignored its complementary aspect, namely the prescript of finding observer-independent invariants.

And it is in this area of obliterated parts of the theory where the defects of his interpretations show up. Not only Petzoldt, but all Machians had serious trouble with an adequate understanding of the postulate of the constancy of light c in vacuum, the second axiom of special relativity. You can often find their claim in letters to Einstein and other physicists, but also in their publications, that a properly understood theory of relativity should not contain absoluta independent of observers. They demanded the complete relativization of all physical quantities, thereby completely misunderstanding the aim of Einstein's relativity theory,[5] and they openly rejected an absolute magnitude like c.[6] The Machians operated for a couple of years with this obviously philosophically motivated pseudo-argument against the special theory of relativity, until Petzoldt and some of his allies (e.g., Lämmel) found a way to make their peace with the constancy axiom of special relativity via their methodological demand of simplicity. This step required a sort of reordering of their hierarchy of norms (simplicity of the axiomatically formulated theory now ranked above proper relativization, formerly it was the other way round), but this example demonstrates once again, that inspite of considerable mental work it was possible to adapt the interpretative framework to the challenges of a new scientific theory, if and only if one was only willing to revise this framework instead of artificially remolding the theory to their own guidelines. So much for the Machian example.

Many examples and case studies in my thesis show the following **bias in philosophical interpretations of physical theories.** On the one hand, most of the interpreters of relativity theory highlighted and clarified at least one of its facettes, thereby helping in the understanding of its meaning and of its context, or brought to light historical roots of some of the steps made in it. Let me give you some examples: Bridgman's operationalism focussed on Einstein's operational definition of simultaneity for spatially distant events; Cassirer's form of neo-Kantianism emphasized the conceptual change in the constitution of physical objects—away from the classical (material) substances, towards formal (mathematical) invariants of tensor calculus. Poincare and the later conventionalists helped to realize the existence and to understand the significance of free conventions, e.g., in the description of the geometry of space; logical empiricists further clarified the complicated relationship between theoretical terms and observational statements, rules of correspondence and various layers of theories. Several groups also proposed reasons why Einstein's relativity theories were preferable to their historical alternatives.

On the other hand, most of the competing contemporary interpreters of relativity theory fell into the all too human mistake of overstressing their point, e.g., Bridgman's operationalism fit well for an understanding of Einstein's redefinition of simultaneity, but did not help in understanding the meaning of Christoffel symbols of the second kind or the Ricci tensor. It was a mistaken view of the meaning of relativity theory that led Heisenberg to expose the view (in a discussion with Einstein), that it was the moral of relativity theory to use only operationally definable concepts.[7] In fact, many of the contemporary philosophers unfortunately did not even reach an adequate level of understanding of the factual content of statements derived from the theory; some of them even ended up in open conflict with it by contradicting some relativistic results caused by their disharmony with philosophically inspired assumptions, for example, the Bergsonians and their insistence upon the uniqueness of time as ultima ratio of life and consciousness, or Dingler and his unshakable belief in the euclidity of space, that he regarded as proven once and for all by his method of exhaustion. Very often in cases of conflict, the philosophical presuppositions were not reconsidered or questioned, but rather the scientific theory, irrespective of all scientific motivations for it. Either the theory was declared to be incompetent for dealing with matters reserved to philosophers (as the conservative neo-Kantians said with respect to pure space and time as opposed to their empirical measurement which they held to be the task of physics), or the value and truth of the theory was radically questioned.[8]

4. Further Application of this Method: The Formation of a Protective Belt of Einstein Defenders and Resulting Incommensurability in Philosophical Debates on the Theories of Relativity

The method of setting apparently independent interpretative statements within an interpretative framework that explains how they came about, how they were motivated, is not only useful for a better understanding of the merits and limits of individual interpretations, but is also helpful in bringing to light interesting consequences that appeared in the discussions of competing groups of philosophers about their resp. interpretations.

One such consequence was the formation of what I have called a 'protective belt' of defenders of both theories of relativity—see figure 3 (= fig. 3.1. of my thesis). By 1920, Einstein had become tired of repeatedly countering the antirelativists with the same arguments. So, philosophers with strong pro-relativistic opinions like Reichenbach and Petzoldt took over the defense of his theories, by inevitably evoking norms and arguments stemming from their own interpretational frameworks. Now the Einstein opponents responded to these defenses by confusing relativistic statements with the corresponding philosophical interpretations by Einstein's defenders. For in-

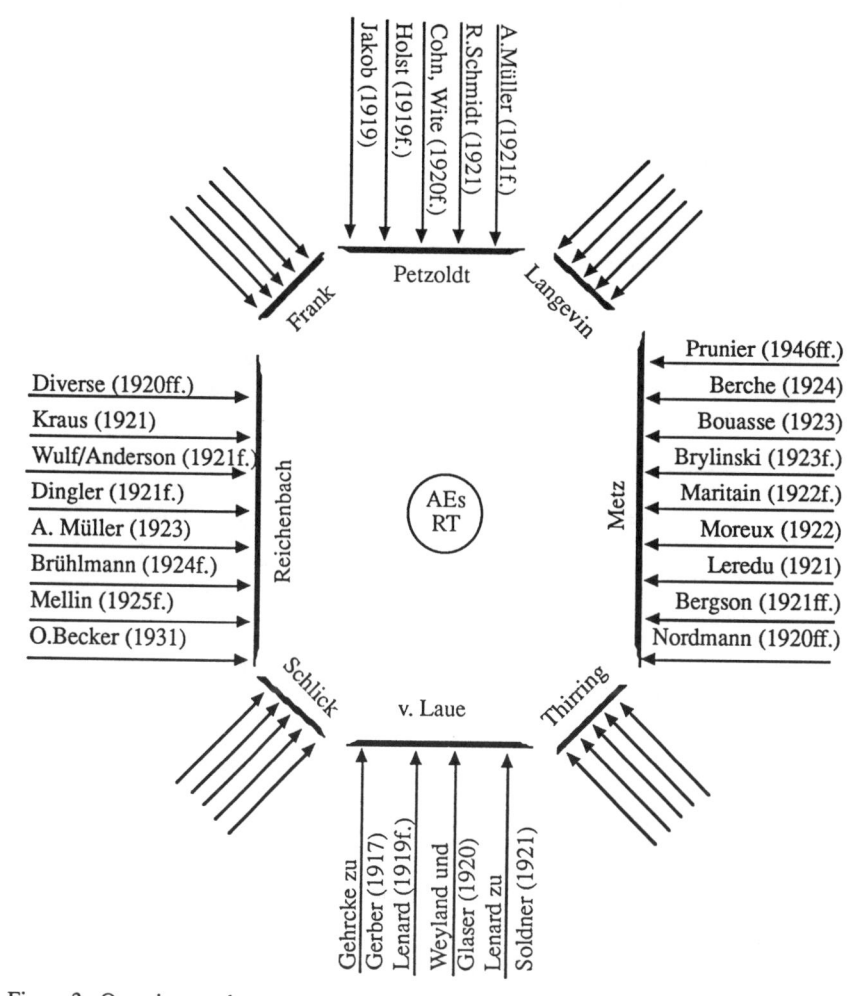

Figure 3. Overview on the contemporary disputes about relativity theory and the most pertinent defenders of Einstein

stance, Petzoldt used his monadological perspectivism to defend the special theory of relativity against charges of missing absolutes and thereby provoked a further and more serious misunderstanding of this theory as subjectivism and relativism, which Einstein clearly had not meant.

An even more radical form of misunderstanding occurred in most contemporary discussions between members of different groups. Here different interpretational frameworks clashed with each other. Both parties usually thought they had found compelling arguments to defeat their opponents but felt misunderstood by the other side, whose arguments simply seemed to go pointlessly astray. These heated discussions, full of metaphors of strife and battle,[9] were subsequently forgotten but they are a highly interesting area for incommensurability studies.

5. Preconditions and Criteria for a Successful Interpretation

As mentioned before, I did not simply want to discuss all the historic cases of successful or unsuccessful interpretations, but also to find out what could be learnt from the mistakes made there—Which of these mistakes could have been avoided and how? In this context, only those cases where something went wrong and misinterpretations resulted were important. The abundance of contemporary interpretations of Einstein's theories of relativity between 1910 and ca. 1930 helped me to formulate the following **preconditions** for sound interpretations that avoid sidetracks, such as the distortion of the idea or statement at hand:

1. Interpreters should show **modesty and open-mindedness**—it is they who are walking on foreign territory, so they have to learn about the procedure, the rules of argumentation, and the scientific facts and laws to be obeyed. Although it sounds quite trivial, this precondition was by no means fulfilled in the case of Einstein's interpreters.[10]

2. Philosophical interpreters should aim at a **precise understanding** of the technical, mathematical language, in which the theory is formulated. Only then can they really understand its structure; only then can they get a feeling for the strength of intratheoretical derivations and for the harmony (or as scientists often say: the beauty) of it as a whole. Otherwise, a partial, fragmented understanding is the unavoidable result, in which some axioms, theorems or statements are overemphasized while others are illegitimately ignored. The selectivity of most contemporary interpretations is a clear indication of the inobservance of this precondition. Also, the overwhelming majority of the most accurate interpreters were not accidently educated physicists or mathematicians, i.e., Schlick, Reichenbach, Frank, Eddington, Weyl, Metz, Bavink.

3. If conflicts arise between the interpretational framework and some results of the scientific theory being interpreted, there should be a **willingness to revise these underlying philosophical assumptions**, that is, to accomodate the traditional philosophical outlook to the new challenge. Although philosophers seem to be very conservative in the revision of traditional concepts and assumptions, the examples of Cassirer, Elsbach, Winternitz and of the early Reichenbach show, that all of them were willing to revise the Kantian interpretative framework about how to apply the transcendental method and what to count as an *a priori* condition for human experience. Consider their far-reaching concessions in contrast to the majority of Kantians who insisted on the rigid adherence to their basic premises, immunizing themselves against any possible refutation by scientific results. (Natorp, Hönigswald, Sellien, etc.).[11]

4. Philosophical interpretations are **not** the right place for a **critique** of the methods, goals or results of scientific theories, because only scientific criteria decide about their adequacy in scientific discourse. Philosophers do not have to judge about the accuracy of theories. Philosophers do not have to judge about the accuracy of theories, but they can help to make them more accessible to nonscientists by analysing their content and their methods and putting them into broader context ("Wissenschaftsanalyse" in the sense of Reichenbach).

5. In order to a avoid unbalanced presentations and to avoid the danger of only noticing what is in accordance with one's own prefixed interpretational framework,[12] it is very helpful to **study competing interpretations** made from a very different perspective, even if one doesn't agree with them. It is by no means an accident, that the richest interpretations were made by Reichenbach and Metz, both philosophers who very actively discussed all kinds of aspects of relativity theory with proponents of virtually all the other contemporary schools.[13] All these disputes helped them to overcome the natural tendency toward a one-sided interpretation.

6. When the previously mentioned preconditions are fulfilled, the crudest forms of misinterpretation can be avoided. But still, there will remain a large spectrum of competing interpretations of one and the same theory. It is useful to have some **criteria for a comparison of these interpretations**. I propose the following ones:

Breadth: The total amount of scientific material incorporated into the interpretation. Compare the difference in quality between Reichenbach's *Philosophie der Raum-Zeit* or Meyerson/Metz's accounts of *la Relativité* against the lean traces of relativity theory in the writings of Natorp, Driesch, Brunschvicg and many others.

Depth: Only a few authors tried to include more complicated themata in their interpretations, for example, Reichenbach's discussion of the repetition of the Michelson-Morley experiments, Metz's clarification of the meaning of relativistic length contractions, Eddington's idiosyncratic but at least stimulating treatment of the field equations of general relativity in his *Space, Time, and Gravitation*.

Currentness: Only a few interpreters really understood those fields that were still under debate among scientists of the time, for example, the issue of singularities in general relativity, the debates about different models of cosmology, and also Russell's vision of a coordinate-free form of general relativity.

Historical context: E.g., Cassirer's well-founded claim that Einstein's relativity theory constitutes a further stage in the gradual substitution of material substances with mathematical forms, Meyerson's original, but debatable thesis of Hegelian traces in Einstein's *Déduction Relativiste*).

Naturalness of interpretation: That is, avoidance of conceptual gymnastics to accommodate the factual content (Reichenbach, Schlick, Cassirer), and finally

Uniformity and harmony in presentation (Cassirer, Schlick, Meyerson, Reichenbach in his axiomatics).

Among the contemporary interpretations of the theories of relativity, Hans Reichenbach fulfilled all preconditions and most of the criteria in the sixth point above. Schlick's interpretation, preferred by Einstein between 1915 and 1925, was too cursory—his early violent death prevented a more detailed account from his pen. Meyerson's interpretation, favored by the later Einstein, and Metz's later contributions very much in the spirit of Meyerson, worked as a sort of complementary interpretation focussing on points underestimated by the logical empiricists. But in toto it was certainly much less exhaustive and broad than Reichenbach's œuvre on relativity—mainly on special relativity. The general theory of relativity doesn't seem to have received an equally balanced treatment by contemporaries—its complexity was only fairly appreciated much later, e.g. in writings of A. Grünbaum, M. Friedman or C. Ray (to name just a few).

It is only by giving this set of criteria to compare the quality of competing interpretations, that **interpretational relativism can be avoided.** In my opinion, these main criteria (perhaps among others, less trivial and not yet formulated ones) allow us to treat competing interpretations of scientific theories in the same objective, rational way in which philosophers of science have tried to deal with competing scientific theories for many decades. I am well aware of the fact that the preconditions and criteria listed above will be as controversial as the analogous proposals for criteria of theory evaluation have been. But at least they might trigger the beginning of a discussion about comparative evaluations of philosophical interpretations of scientific theories. This should, in the light of the ongoing discussions about the philosophical interpretation of quantum mechanics, quantum field theories, etc., be an important topic on the agenda of philosophers of science.

Notes

[1] First of all, many thanks for the invitation to the organizers of the PSA meeting and to Prof. Don Howard (Univ. of Kentucky, Lexington), who had initiated the Colloquium on "Recent Work in the History of the Philosophy of Science". My dissertation was supervised by Prof. Dr. Andreas Kleinert (Institut für Geschichte der Naturwissenschaften, Hamburg) and Prof. Dr. Lothar Schäfer (Philosophisches Seminar, Univ. Hamburg). The editors at the Collected Papers of Albert Einstein in Boston and many archivists and librarians elsewhere have supported my research. Miss Ann M. Lehar helped me a lot in improving my English for the written version of this paper.

[2] *Interpretationen und Fehlinterpretationen der speziellen und der allgemeinen Relativitätstheorie durch Zeitgenossen Albert Einsteins*, Birkhauser, Basel, 1990; ca. 800 pp., more than 3000 bibliographic entries, many figures and tables.

[3] See also Chr. Ray: *The Evolution of Relativity*, Hilger, Bristol, 1987, for a similar view of relativity theory.

[4] See also section 4.8. of my thesis (fn. 2).

[5] Einstein later reflected about alternative names, such as 'theory of absoluta' or 'theory of invariants' to distinguish his theory from what was commonly refered to as 'relativism' - see sect. 2.4. of my thesis (fn.2).

[6] When Einstein made c a variable dependent on the gravitational potential in his Prague theory of 1911, several Machians (and I would also expect Mach himself) regarded this (erroneously) as a fulfillment of their demand of rigorous relativization.

[7] A recipe that Heisenberg later tried to implement in quantum mechanics.

[8] Vaihinger's pupils declared it as merely a convenient fiction; Dingler's certicism repeatedly declared relativity theory as stillborn and portrayed himself in the role of a high priest ringing its death knell; see sections 4.4. and 4.5.4. of my thesis. (fn.2).

[9] In my thesis, I treated the discussion between Kraus, Urbach and Frank in more detail (sect. 5.3.) and I also studied one person (Reichenbach) in the context of a multiple front war (sect. 3.4.3.). But other figures such as Dingler and Reichenbach are worth studying in more detail.

[10] For example, the neo-Kantian journalist Drill wrote angrily in reply to Max Born: "I'm not willing to talk about Einstein and I'm not competent to do so. [...]. But nobody has to accept it, when a scientific theory tries to cheat common sense, whoever may have formulated it." And many professional philosophers did not react much better.

[11] The Kantian Drill emphatically wrote against the reformist wing: "A philosopher should know that physics can't lead the way for philosophy. It is the latter's task to find out the necessary conditions for all knowledge, including all sciences."

[12] Somehow according to the principle: one sees what one expects to see. Note that theory-ladenness of observation was emphasized by Duhem and Quine!

[13] See sect. 5.1. and 5.2 of my thesis.

Part V

DEDUCTION FROM THE PHENOMENA

Newton's Classic Deductions from Phenomena[1]

William Harper

University of Western Ontario

I take Newton's arguments to inverse square centripetal forces from Kepler's harmonic and areal laws to be classic deductions from phenomena. I argue that the theorems backing up these inferences establish systematic dependencies that make the phenomena carry the objective information that the propositions inferred from them hold. A review of the data supporting Kepler's laws indicates that these phenomena are Whewellian colligations—generalizations corresponding to the selection of a best fitting curve for an open-ended body of data. I argue that the information theoretic features of Newton's corrections of the Keplerian phenomena to account for perturbations introduced by universal gravitation show that these corrections do not undercut the inferences from the Keplerian phenomena. Finally, I suggest that all of Newton's impressive applications of Universal gravitation to account for motion phenomena show an attempt to deliver explanations that share these salient features of his classic deductions from phenomena.

Newton's deductions from phenomena have not always been approved by philosophers of science. Twenty years ago in his "Classic Empiricism" (1970 p.160-164) Paul Feyerabend suggested that Newton's phenomena were no more than vivid illustrations of his theory made possible by ad hoc assumptions and that Newton's inferences from them were logically vacuous, even if they were effective pieces of rhetoric. The classic examples I shall consider certainly were effective pieces of rhetoric. They led Huygens and Leibniz, the two most accomplished advocates of the rival vortex theory, to accept that there are inverse square centripetal forces centered on the sun and the planets.[2] As to the status and nature of the phenomena and the inferences from them, that shall be our topic here.

1. The Classic Arguments

a) Three Forms of Argument

Propositions 1 and 2 of *Principia* book 3 are the classic examples of Newton's "deductions" of propositions about forces from phenomena of motion. They are the opening moves of his argument for universal gravitation. Proposition 1 applies to the moons of Jupiter and, in the second and third editions, to the moons of Saturn as well.

Proposition 2 applies to the primary planets. In both propositions it is asserted that the forces holding the satellites in their orbits are

> directed to the center of the primary
>
> and
>
> are inversely as the squares of their distances from that center

Here is the argument for Proposition 2.

> The first part of the proposition is evident from phen.5 and from prop.2 of book 1, and the latter part from phen.4 and from prop.4 of the same book. But this latter part of the proposition is demonstrated (demonstratur) with the greatest exactness from the fact that the aphelia are at rest. For the slightest departure from the doubled ratio would (by bk.1, pro.45, corol.1) necessarily result in a noticeable motion of the apsides for a single revolution and an immense such motion in many revolutions (Cohen and Whitman pp. 496-487).[3]

There are three separate forms of argument here. The argument for the first part, which refers to phenomenon 5 and proposition 2 of book 1, is an argument from the phenomenon that orbital motion satisfies Kepler's areal law

> K2 The orbiting body traverses areas proportional to the times of description by radii to its primary

to the proposition that the force deflecting the body into its orbit is centripetal. Newton used this same form of argument to infer the first part of proposition 1 from the phenomenon that Jupiter's moons satisfy the areal law. The first argument for the second part of proposition 2, which refers to phenomenon 4 and proposition 4 of book 1, is an argument from the phenomenon that a system of orbits satisfies Kepler's harmonic law

> K3 The periodic times of the satellites are as the 3/2 power of their mean distances from their primary.

to the proposition that the forces deflecting these satellites into their orbits are inversely as the squares of their distances from the center of the primary. This form of argument was, also, used in the inference to the second part of proposition 1 from the phenomenon that the orbits of Jupiter's moons satisfy Kepler's harmonic law. The second argument for the inverse square variation exhibits a third form of argument. This is an inference to inverse square variation of a centripetal orbital force from the stability of the apsides of the orbit. It is based on Newton's precession theorem, proposition 45 of Book 1.

b) Beyond Bootstrap Confirmation

Ten years ago Clark Glymour (1980, pp.204-214) established that these classic examples of Newton's deductions from phenomena satisfied the conditions on bootstrap confirmation. Glymour's bootstrap confirmation is a form of *deductive* inference upward from data to a theoretical claim based on theoretical background assumptions.[4] The key constraint on a bootstrap confirmation is that the background assumptions be compatible with alternatives to the data that would (together with those assumptions) have contradicted the theoretical claim inferred from the actual data.

I shall argue that these Newtonian deductions from phenomena satisfy constraints stronger than bootstrap confirmations. First, in Newton's inferences the background assumptions yield the converse conditional from the proposition to the phenomenon as well as the conditional from the phenomenon to the proposition. Secondly, such equivalences hold for a whole range of alternatives to the phenomenon and a whole range of corresponding alternatives to the proposition so that the dependencies between the phenomenon and the proposition inferred from it are systematic. Finally, Newton's phenomena are not just any sets of data, but are generalizations obtained by constructing best fitting curves for open-ended bodies of data, as we shall see in part two of this paper.

c) The Converse Conditional

Consider the first form of argument. The theorem referred to is proposition 2 of book 1. It asserts that for orbits in a plane the areal law is sufficient for the proposition that the deflecting force is centripetal.[5] Given the background assumptions from which this theorem follows we have a mathematical deduction of the proposition from the phenomenon. In theorem 1 of book 1 Newton also established, from his assumptions, the converse conditional that if the deflecting force is centripetal then the areal law holds. These background assumptions are compatible with increasing and decreasing areal rates. Therefore, theorem 1 establishes that Glymour's bootstrap condition is met by implying that *some* alternative to the phenomenon, e.g. having the areal rate be increasing, is incompatible with the centripetal direction of the deflecting force. Having this converse conditional, however, is stronger than the bootstrap requirements, because it implies that *any* alternative to the phenomenon is incompatible with the proposition.

Such a converse conditional is important for Bayesian methodology because it concentrates the likelihoods (the prior conditional probabilities of the evidence given the theory) on the phenomenon. This will maximize the support of the theory as more and more data supporting the phenomenon comes in. Roger Rosenkrantz (1981, section b.2) has used such considerations to make much of the point that a unified theory which concentrates the likelihoods on the evidence will receive bigger boosts of Bayesian confirmation as the evidence comes in than a rival which fails to so concentrate the likelihoods.

d) Systematic Equivalence Theorems

Theorems 1 and 2, together, establish an equivalence between the proposition and the phenomenon from which it is inferred. The phenomenon is the condition that the rate at which areas are swept out is constant—that is that the second derivative of the areas with respect to time is zero.[6] Alternatives to this phenomenon are alternative values of this magnitude. Positive values specify increasing areal rates, while negative values specify decreasing rates. The direction of the force deflecting the body can be specified by angles in the plane of motion so that zero is the direction toward the given center and positive angles are off center in the direction of tangential motion. Corollary 1 of Proposition 2 Book 1 establishes that increasing areal rates correspond to positive angles and decreasing rates correspond to negative angles. This establishes a systematic connection between these magnitudes which can be extended so that the second derivitive of the areas being swept out by radii to a center *measures* the angle of the deflecting force with respect to the radius from the planet to that center.

The harmonic law argument is also backed up by theorems establishing such systematic dependencies between the phenomenal magnitude and the corresponding theoretical magnitude. Corollary 6 of proposition 4 book 1 is an equivalence between having the periods be as the 3/2 power of the radii and having the centripetal forces be inversely as the squares of the radii. Corollary 7 generalizes this to having the periodic times be as any power n of the radii and having the centripetal forces be as the power 1-2n of the radii. The phenomenal magnitude specifies a power of the radii to which the periods are proportional, while the theoretical magnitude specifies a power of the radii to which the centripetal forces are proportional.

Corollary 1 of proposition 45 book 1 backs up the inference to inverse square variations from stable apsides. The theorem establishes systematic equivalences between orbital precession and the law of the centripetal force. The phenomenal magnitude is the total angular motion n in degrees of the planet in the course of returning to the same apside. Having n be greater than 360 is to have precession forward, while having n less than 360 is to have precession backward. Newton's theorem establishes that the force is as the power $(n/360)^2$-3 of the distance. Having n be 360 is to have a stable orbit and is equivalent to having the force be as -2 of the distance or inverse square. Here the theoretical magnitude specifies that the centripetal force on the body is proportional to a given fixed power of the distance from the focus which is the center of the force.

e) Measurement and Objective Information

In each case the systematic equivalences between alternative values of the phenomenal magnitude and corresponding alternative values of the theoretical magnitude make the phenomenon measure the value of the theoretical magnitude specified in the proposition inferred from it.

These systematic equivalences establish that, relative to the background assumptions, the phenomenon carries the objective information that the proposition inferred from it holds. A number of accounts of objective information based on counterfactual or causal dependencies have been proposed by writers such as Fred Dretske (1980), David Lewis (1980) and Robert Stalnaker (1984). Such accounts have been used to specify conditions under which one can know some proposition on the basis of something else (Dretske, 1980). Given the background assumptions, the dependencies expressed by the equivalence theorems make it such that one can know the proposition inferred from it on the basis of the phenomenon. I shall explore the extent to which this continues to hold as the idealizations and approximations in the assumptions and the phenomenon are taken into account.

2. Phenomena

a)The Areal Law

Here is Newton's formulation of the areal law phenomenon for the primary planets, together with the evidence he marshalls for it (Cohen and Whitman pp. 494-495).

> Phenomenon 5. The primary planets, by radii drawn to the earth, describe areas that are in no way proportional to the times, but, by radii drawn to the sun, traverse areas proportional to the times.
>
> For with respect to the earth they sometimes have a progressive motion, they sometimes are stationary, and sometimes they even have a retrograde motion; but with respect to the sun they move always forward, and they do so with a motion that is almost uniform—but, nevertheless, this motion is a little swifter in their perihelia and slower in their aphelia, in such a way that the description of areas is uniform. This is a proposition very well known to astronomers and is especially demonstrated (demonstratur) in the case of Jupiter by the eclipses of its satellites;

by means of these eclipses we have said that the heliocentric longitudes of this planet and its distances from the sun are determined.[7]

The dynamical significance of the areal law revealed by Newton's equivalence theorems makes it appropriate to search out centers toward which radii from a body sweep out constant areas in order to locate a center toward which deflecting forces are directed. The first part of Newton's discussion illustrates how to select for such centers from among candidates under consideration. The rough condition—a little slower in the aphelian, a little faster in the perihelian so as to be compatible with the areal law—may be all that is required to select between the candidate bodies. The significance of locating a center of force *in a body* may suggest that this rough agreement is all that is required by way of support from the data. Consider the following remark Newton makes about phenomenon 6:

> Actually, the motion of the moon is somewhat perturbed by the force of the sun, but in these phenomena I pay no attention to minute errors that are imperceptible (Cohen and Whitman pp. 495).

This would seem to support the suggestion and extend it to allow an idealization known to be false to count as long as it is compatible with the data.[8]

Newton goes on to suggest a somewhat more robust support for the areal law from the data. It is said to be very well known to astronomers and demonstrated in the case of Jupiter by the eclipses of its moons. When Jupiter eclipses a moon it is on the line from the sun to that moon. The direction of this line with respect to the fixed stars can be computed from data about the orbits of these moons made available by the telescope. Such data were not available to Kepler, but a review of the data he reasoned from in the investigation which led to his discovery of the areal law and elliptical orbit may help to see how the heliocentric longitudes and distances can be fixed. More importantly, this review will serve to illuminate, in a general way, the relationship between phenomena and their data.

b) From Data to Phenomenon

Kepler's data are locations of a planet against the fixed stars as seen from the earth. The path of the sun among the fixed stars over a year is called the ecliptic. It defines the plane of the earth's orbit. Familiar constellations, the signs of the Zodiac, mark off the ecliptic circle into twelve 30 segments. Celestial longitudes are measured (counter clockwise, looking down from the north) from the point on the ecliptic where the sun crosses the equator at the vernal equinox.[9] Celestial latitudes are measured north or south of the ecliptic circle. Here is an example. On 18 November 1580 at 1 hr 31 minutes Uraniborg time Mars was at 66 28' 35" longitude and at 1° 40' north celestial latitude.

This is an opposition of mars to the sun. Mars was on the exact opposite side of the earth from the sun or, equivalently, the earth was on the line from the sun through Mars to the fixed stars. At opposition the celestial longitude of Mars observable from the earth is the same as its celestial longitude with respect to the sun. Oppositions, thus, allow rather direct determinations of heliocentric longitudes of superior planets. Usually, they have the added advantage that the planet is high in the sky so that refraction and parallax are minimized. Tycho Brahe's oppositions could, perhaps, be trusted to within one or two minutes of arc.

On 17 November at 9 hours 40 minutes Uraniborg time, Brahe observed Mars at 66° 50' 10" and at 1° 40' north celestial latitude when it was in opposition to the mean sun (Caspar, vol.3, pp.110-111). The mean sun is what corresponds in Brahe's geocentric system to what would be the center of the earth's eccentric circular orbit in an equivalent heliocentric model. Our example datum is Kepler's calculated correction of Brahe's observation to give an opposition to the true sun rather than to the mean sun (Caspar, vol.3, pp.142-143).[10] The first six chapters of Kepler's *Astronomia Nova* argue Kepler's strong case for taking heliocentric longitudes to the true sun, rather than to the mean sun as Copernicus and Brahe had been doing.

Heliocentric periods for superior planets can be estimated approximately from the time differences between oppositions together with their angular separations.[11] Kepler had available 12 successive oppositions starting with this first one of Brahe's in 1580 and continuing after Brahe's death in 1600 to ones he made himself in 1602 and 1604. Ancient observations of angular separations of oppositions from known fixed stars can be used to refine estimates of periods.[12]

Comparing time differences with angular separations reveals that arcs of heliocentric longitude on one side of the ecliptic take longer to traverse than equal arcs on the other side. An apside line splits the orbit into symmetrical sides and connects the aphelion, which is the point of slowest angular rate, with the perihelion, which is point of fastest angular rate. Kepler (AN chap.16) was able to use four oppositions scattered about the ecliptic to set the apside line and other parameters of an eccentric circular orbit model. This is the model he came to call his *vicarious* theory. In it the apside line passed through the true sun. It, also, had an equant—a point about which equal angles are swept out in equal times—on the apside line on the other side of the center from the sun. This made the planet's speed along the orbital path greatest at perihelion, when the planet is closest to the sun, and least at aphelion, when the planet is furthest from the sun. This speed relation was eventually transformed by Kepler into the law of areas. The vicarious theory predicted heliocentric longitudes to within about 2' of arc when checked against the other eight oppositions.

Kepler (AN chap. 19) was able to use latitudes of oppositions near aphelion and perihelion to compute mars-sun distances by triangulation. These distances disagreed with the eccentricity of the sun that had been fixed from the four oppositions the vicarious theory had been constructed to fit. Kepler refuted the theory by establishing that if the sun-center eccentricity were set according to the triangulated aphelion and perihelion distances then, no matter where the equant point was placed, the predicted heliocentric longitudes would be out by at least eight minutes at some parts of the orbit. He, nevertheless, was able to continue to use the original version of the vicarious hypothesis to calculate heliocentric longitudes.

Even without a method of determining heliocentric longitudes for the times they are made observations other than oppositions can be used to triangulate distances. Observations of mars at intervals corresponding to the period of the earth can be used to triangulate mars-sun distances at several different positions in its orbit from the same position of the earth with respect to the sun. Kepler (Chapt. 44) used several sets of such observations to help argue that the correct orbit for mars could not be any circle.

Similarly, several observations of mars at intervals of the period of mars' orbit can be used to triangulate earth-sun distances at several points on its orbit with respect to the same sun-mars distance. Kepler (Chapt. 24) used such observations to correct Tycho's solar theory and to improve his own model of the orbit of the earth.

Once one has a relatively accurate model of the orbit of the earth and a relatively accurate theory of another planet's heliocentric longitudes than any observation of that planet can be used to triangulate its position with respect to the sun. Kepler eventually arrived at the elliptical orbit after a rather tortuous struggle in which the law of areas functioned as a premise. The elliptical orbit with the area law gives more accurate heliocentric longitudes than the vicarious theory and it gives accurate distances as well.[13]

c) Phenomena Colligate Their Data

We can argue that the elliptical orbit with the areal law is a Whewellian colligation of this open ended body of data. According to Whewell, colligation involves three steps which he calls selection of the idea, construction of the conception, and the determination of the magnitudes (NOR, p.187). He also tells us that

> these three steps correspond to the determination of the independent variable, the formula, and the coefficients in mathematical investigations (Aphorism 35, NOR, p.186).

The independent variable is the position of the earth with respect to the sun. The dependent variable is the position of mars triangulated from its observed position against the fixed stars, its heliocentric longitude at the time of observation, and the position of the earth at the time of observation. Each observation of the planet fixes a corresponding position, at the time of observation, of the planet with respect to the sun by triangulation. The construction of the conception is the determination of a specific elliptical orbit satisfying the law of areas as the best fitting curve for this open-ended body of triangulated positions at times. The determination of coefficients is the fixing of the parameters of this orbit, the period, the mean distance (which is the major semi-axis of the ellipse) and the eccentricity from the data. As more and more observations come in the confidence intervals on estimates of these parameters will become tighter and tighter if the fit of the orbit to the data is good.

Curtis Wilson (1974, p.250) has called attention to the fact that Kepler pointed out that his data allowed for errors in distance of about 100 to 200 parts where 100,000 parts represents an *Astronomical Unit* (the mean earth-sun distance). Wilson (p.249, 250) suggests that this allows for alternative oval orbits since the maximum discrepancy of the ellipse from a circle was only 650 of those parts. Owen Gingerich (1989, p.68) argues that Kepler was mistaken in the calculation which led him to purport to find an error of 5 1/2' in longitude by which to reject the *via Buccosa*, one of the competing ovals he considered. It appears that the differences are actually under a minute. (Whiteside, 1974, p. 14)[14]

According to Whewell a colligation is not just a summary of the data it was constructed to fit but a generalization which makes additional predictions. These predictions provide the basis for a test. The elliptical orbit with the areal law predicts positions at future times, past times, and times in between the times at which the given observations were taken. Kepler used these orbits to construct tables (*The Rudolphine Tables*) for reaching such predictions. According to Gingerich (p.77) these predictions were generally about 30 times better than those of earlier or of competing tables.

The use of the telescope gave much better data to feed into the colligation of the orbits. One of the early triumphs was Gassendi's successful observation of a transit of Mercury in 1631. Gingerich (p.77) points out that Kepler's prediction erred by only 10' compared to 5 for tables based on Copernicus and others. The use of eclipses of Jupiter's moons to fix its heliocentric longitudes that so impressed Newton is a good example of the improved data that became available. The goodness of the fit of the elliptical orbit with areal law shows up in more and more closely fixed estimates of its parameters as the quality and extent of the body of colligated data grows. For the inner planets, earth and mars, the fit is very good indeed. There are, however, long term perturbations of Jupiter and Saturn that could begin to show up on the data available in Newton's day.

d) The Harmonic Law as a Colligation

The harmonic law is a higher order phenomenon where the data are the parameters of the orbits colligated from the observations of the different planets. The independent variable can be taken to be the squares of the periodic times, while the dependent variable will be the cubes of the mean distances. For each of the planets there are as many distance estimates as there are observations to triangulate distances. The cube of any such estimate can count as a data point. So above the squared value for the periodic time for each planet there will be a host of cubed distances values. The construction of the conception here is the selection of a straight line $R^3/T^2 = K$ as the best fitting curve for these data points. If we take the period of the earths orbit—one sidereal year as a unit of time and an astronomical unit (the mean earth-sun distance) as our unit of distance then we get the straight line of 45° slope $R^3/T^2 = 1$ as the best fitting curve. As more and more data come in one can improve our estimate of the R^3/T^2 ratio by more and more closely fixing our estimates of it.

Newton's review of the evidence supporting the Harmonic Law accurately reflects the fact that the estimates of the periods are better than the estimates of the mean distances. He tells us that all astronomers are agreed on the periods and cites them to 4 places for Mars, earth and the inner planets and to 3 places for Jupiter and Saturn, where the unit of time is the decimal day. This corresponds to about 7 places and 6 places in decimal years. Newton gives distance estimates from Kepler's *Rudolphine Tables* and from Bulliau's tables. These distances show agreement to only about 1 or 2 decimal places in astronomical units, the distance unit corresponding to decimal years.

One mark of the empirical fit of the harmonic law was that Streete, following Horrocks, was able to improve significantly on Kepler's tables for predicting locations at times by taking the harmonic law as exactly true and using it to compute the distances from the periods. Wilson (1989b p.247) argues as follows:

> A computation of the aphelian and perihelian distances of Venus and Mars during the 1960s as determined by fourth-order Everett interpolation from the *radii vectores* listed at ten-day intervals in *Planetary Coordinates for the years 1960-1980* leads to the conclusion that the mean solar distances of these varied from Streete's values by no more than 0.00001.

We now know that the harmonic law estimates are far more accurate than the triangulation on which tables of Kepler and Bulliau were founded. For Mars and the inner planets they are not very much less accurate than the estimates of the periods themselves. We shall see, however, that the harmonic law estimate for Jupiter requires a significant correction.

3. Newton's Glorious Project

a) Assumptions and Ideal Centripetal Forces

The theorems backing up these inferences from phenomena are examples of the mathematical principles of natural philosophy Newton referred to in the title of his book. The general background assumptions on which these principles rest include the Laws of Motion Euclidian geometry, the version of the calculus Newton used in *Principia*, and other, less controversial, mathematical methods.

I want to call attention to corollary 6 of the Laws of Motion. It was appealed to to extend the results of theorems 1 and 2 to centers in non-uniform motion.

> If bodies are moving in any way whatsoever with respect to one another and are urged by equal accelerative forces along parallel lines, they will continue to move with respect to one another in the same way as they would if they were not acted upon by those forces (Cohen and Whitman p. 35).

This is important because it shows that one need not assume that a center is even approximately unaccelerated in order to apply the theorems to it. All that is required is an assumption that no external forces or impediments are producing significantly differential accelerations on the bodies in the system.

There are a number of specific assumptions about the systems for which the theorems are proved. The bodies are treated as point masses moving in non-resisting spaces. Corollaries 6 and 7 of proposition 4 are proved for concentric circular orbits. What I have room to discuss here, however, is what I regard as the central idealization. All the theorems are about one-body systems. Even in the harmonic law theorems, which apply to relations among several orbits, the only forces considered are the separate actions of the centripetal force on the arbitrary bodies. There are no forces of interaction among these bodies, nor any action by any of them on the central body. This suggests that Newton's theorems reveal the dynamical significance of Kepler's laws to be the laws of orbital motion for test bodies orbiting under an inverse square centripetal force field.

b) Universal Force of Interaction

The transformation from these inverse square centripetal force fields to the inverse square universal force of gravitational interaction of proposition 7 introduces interactions that require corrections to the Kepler law phenomena. Newton expressed his reaction to this implication of the idea that gravity is a universal force of interaction in the following remark from a draft of *De Motu* (tentatively dated in December 1684, Wilson 1989, p.253).

> By reason of the deviation of the Sun from the center of gravity, the centripetal force does not always tend to that immobile center, and hence the planets neither move exactly in ellipses nor revolve twice in the same orbit. There are as many orbits of a planet as it has revolutions, as in the motion of the Moon, and the orbit of any one planet depends on the combined motion of all planets, not to mention the action of all these on each other. But to consider simultaneously all these causes of motion and to define these motions by exact laws admitting of easy calculation exceeds, if I am not mistaken, the force of any human mind.

George Smith (1990) has pointed out the interesting irony that, if the dating is correct and he thinks it is, then within about six months or so of writing this apparently pessimistic passage Newton was hotly engaged upon his glorious and distinctively original project of attempting to use universal gravitation to account in detail for all the observable motions of the bodies in the solar system, including the perturbations.

c) Perturbations and Corrections

One of the largest perturbations is produced by the interaction of Jupiter and the Sun. Proposition 60 of book 1 is a transformation of Kepler's harmonic law distances, to corresponding distances for a two-body interaction where the planet and the primary orbit their common center or gravity.[15] Where R' is the corrected distance, R is the harmonic law distance, s is the mass of the sun and p is the mass of the planet proposition 60 tells us

$$R'/R = (s + p) / ((s + p)^2 s)^{1/3}$$

Newton estimated the ratio of the mass of the sun to the mass of Jupiter to be as 1067 to 1, by comparing the R^3/T^2 constants for their systems of satellites. Using his numbers we get

$$R'/R = (1068) / ((1068)^2 (1067))^{1/3} = 1.0003123$$

The Harmonic Law distance Newton computed was 5.2 0096 Astronomical Units. The corrected distance

$$R' = (1.00031)(5.20076) = 5.20257 \text{ AU}$$

This is a slight improvement over the harmonic law estimate with respect to the data available to Newton, but it is a very significant improvement relative to the data we now have. Danby (1962) cites 5.2027 AU as the present estimate of Jupiter's mean distance.

The Harmonic Law estimate is wrong! Newton knew it. So, how could he appeal to the Harmonic Law argument? Can we say more than that he knew the harmonic law estimate to be an approximation to the corrected mean distance? The systematic dependency expressed in Prop.60 *explains* the perturbation—the ratio R'/R, —as the correction factor appropriate to the result of transforming the one-body harmonic law system, which takes into account only the acceleration field of the sun, to the two-body interaction where the acceleration field of Jupiter makes its proper contribution. The new distance estimate is defined as the result of applying this perturbation to correct the harmonic law estimate. If this new estimate is the correct way to colligate the data then the harmonic law estimate is colligated, from that same data, as the mean-distance corresponding to the Keplerian component of the combined motion. Such a correction cannot undercut the inference to the inverse square law for the centripetal acceleration field of the sun from the harmonic law for the component of the total motion produced by it.

d) Explanations

The same dependencies that *explain* the perturbation as the result of taking into account the contribution of the acceleration field of Jupiter also make the perturbation carry information fixing the mass of Jupiter relative to that of the sun. An algebraic transformation of the equation of Prop.60 yields

$$p/s = (R'/R)^3 - 1.$$

This makes the perturbation *measure* the cause which explains it. The information theoretic situation is like that of the classic deductions from phenomena, in which the phenomenon measures the corresponding value of the theoretical magnitude expressed in the proposition inferred from it.

This two-body interaction is especially simple. Some three-body interactions such as the lunar precession required approximation techniques that were developed into perturbation theory, because no one could find analytic equations to express the required systematic dependencies. These perturbation techniques do, however, characterize such dependencies, as is demonstrated by the use of perturbations caused by them to estimate the masses of planets such as Venus which have no moons to measure their acceleration fields.

Newton's classic deductions from phenomena are not just special applications he used to motivate his theory. They exhibited a new ideal for explanation of phenomena by causes.

All Newton's applications of universal gravitation to explain phenomena of motion are attempts to generate explanations that share the salient features of the classic deductions from phenomena we have been discussing. The phenomenon to be explained, whether the motion of a comet, a lunar perturbation, a tide ratio, or the precession of the equinoxes is characterizable as a generalization corresponding to the specification of a best fitting curve colligating an open ended body of data. The explanation is an attempt by Newton to generate appropriate systematic dependencies from the theory and specific assumptions about the system under consideration. These dependencies are to be such that the phenomenon *measures* the cause which explains it. Thus, it was that he even attempted to measure the relative masses of the sun and the moon from the tide ratio.

These impressive attempts to realize this new ideal were the major support for the theory. They deserve a much more careful look than they have been given so far in the literature. I suggest that these characteristics they share with Newton's classic deductions from phenomena can illuminate a more detailed investigation into what Newton achieved by them.

Notes

[1] I want to thank Bryce Bennett, Ram Valluri, Rob DiSalle, Kathleen Okruhlik, John Nicholas and, especially, Curtis Wilson for help, encouragement and advice.

[2] Koyré 1965 provides translations of relevant passages from Huygens, pp. 121-122 and Leibniz pp. 132-133. Howard Stein 1967, p. 178 and Eric Aiton 1972 also discuss this aspect of the reaction of Huygens and Leibniz to *Principia*.

[3] I have translated "demonstratur" as "demonstrated", rather than as "proved" as Cohen and Whitman do. Howard Stein has suggested that Newton usually uses "demonstratur", "probare" which he translates as "prove" and "deducere" which he translates "deduction" in sharply distinguished fashion.

The first of these is Newton's characteristic term for *purely mathematical reasoning*. The second–"deduction"–is used by him in a quite wide sense, for reasoning *competent to establish a conclusion as warranted* (in general, on the basis of available evidence.) As for "proof", Newton typically means by it *the*

>   *subjection of a proposition to test by experiment or observation* (with a successful outcome). Stein 1990, p. 18.

Newton's use of "demonstratur" here suggests that he intends the inference from a phenomenon to the proposition to have something closer to the force of a mathematical demonstration than some of the weaker sorts of warrant that his use of "probar" or "deducere" would suggest on Stein's reading of his usage.

In the translations to follow I shall generally adhere to those of Cohen and Whitman 1987 except that I shall systematically translate "demonstratur", "deducere", and "probare", in accordance with Stein's reading. This will allow us to see if we can *probare* Stein's reading by these passages.

[4] Here we use "deductive" as philosophers do today to contrast with "inductive" and to indicate something like having the force of a mathematical demonstration or as Newton would use "demonstrature" on Stein's reading.

[5] The areal law holds for any point with respect to which a body is in uniform motion and, as Leibniz pointed out, with respect to a center of repulsive force that the body is being deflected directly away from. The theorems connecting the areal law with centripetal forces assume that the motion of the body is concave to the center in question, as is the case in any orbit.

[6] An analytic argument, using polar coordinates, establishes that having the areal velocity constant is equivalent to having all the acceleration be radial, from purely kinematical assumptions (Goldstein 1981, p. 73).

[7] Here we have "demonstratur" applied to the fixing of the areal law for Jupiter's orbit from the eclipses of its moons. Newton also uses "demonstratur" to indicate the force of the argument to the claim that Venus and Mercury orbit the sun from the fact that they exhibit phases, in his discussion of Phenomena 3 (Koyré and Cohen, p. 561). Apparently, these are examples where Newton does not adhere to what Stein suggests is his characteristic restriction of "demonstratur" to purely mathematical reasoning.

[8] Ron Laymon (1983) has suggested that it was Newton's regular practice to take as phenomena even idealizations incompatible with the data.

[9] The equinoxes precess clockwise at about 1°23 every century. this requires augmenting measured counterclockwise from the vernal equinox by around 50" per year.

[10] Such corrections and interpolations are often used to make up for the fact that observations at exactly the time of opposition (or other significant configurations) are not available.

[11] Here is an example of such a calculation. At the opposition of 1595 on October 31 at 0:39 Uraniburg time Mars was at 47.528° heliocentric longitude. This was just 18.94° short of the longitude of the opposition of 1580. The opposition of 1595 was the seventh after that of 1580, so Mars had traversed 7.947 revolutions in the 5,459.644 days between 01:31 on 18 November 1580 and 0:39 on 31 October 1595. This gives an estimate of about 687 days for the period.

[12] Curtis Wilson (personal communication) helped extract the following example of Kepler's use of Ptolemaic data to fine tune his estimate of Mars' period from chapter 69 of *Astronomia Nova*. After some corrections introduced to transform Ptolemy's

opposition from mean sun to the true sun and to account for precession of equinoxes and certain other difficulties, Kepler estimated from Ptolemy's data that the difference in heliocentric longitude between Mars and the star Cor Leonis was 128° 48' 30" at 18:00 on 26 May in 139 AD. From Tycho's data he found this difference to be 216° 31' 45" at 18:00 on 27 May 1599. The difference between these longitudes is 87° 43' 15". The time interval is 1460 Julian years or 1460 x 365.24 = 533265 days. With the approximate value for the period of 687 days one can find that mars completed 776 whole cycles plus the difference of 87.72° for a total of 776.24 revolutions. This gives a period of 686.98 days.

[13] In this sketch I ignored the important role played by Kepler's appeal to causal arguments in the course of his attempt to establish these laws as giving the true motions of the planets. Even without these considerations the role of the vicarious theory in giving heliocentric longitude, and the role of a theory of the earth's orbit show that Kepler did not triangulate positions from the data alone.

[14] Curtis Wilson helped Bryce Bennett and me check this. He was, also, unable to get differences of as much as 1' of an arc when the calculation is done using the same mean anomaly for both theories. A plausible, but incorrect, calculation comparing them a the same 45° eccentric anomaly gives a difference of about 6'.

[15] I.B. Cohen p. 224 has identified proposition 60 as Newton's transformation of Kepler's harmonic law from the one body idealization of proposition 4 Book 1 to a more realistic two body system. He sees Newton's practice of first proving results for radically idealized systems and then successively transforming these systems to get corresponding results that hold for more complex models that better approximate real physical systems as a salient feature of what he calls Newton's "mathematical style" (e.g. 1980 pp. 52-64).

References

Aiton, E.J. (1972) *The Vortex Theory of Planetary Motions*. London: Macdonald: New York: American Elsevier.

Butts, R.E. and David, J.W. (eds) (1970) *The Methodological Heritage of Newton*. Toronto.

Caspar (ed.) *Johannes Kepler Gesammelte Werke* (Munich, 1938—).

Cohen, I. B. (1980), *The Newtonian Revolution*. Cambridge University Press:.

Cohen, I. B. and Whitman, A. (1987). (Translators) Isaac Newton: *Mathematical Principles of Natural Philosophy*. Cambridge, Mass. Manuscript to be published by Harvard University Press and Cambridge University Press.

Danby, J.M.A. (1962), *Fundamentals of Celestial Mechanics*. Macmillan Press.

Dretske, F. (1980) *Knowledge and the Flow of Information*. M.I.T. Press

Earman, J. (ed.) (1983) *Testing Scientific Theories*. Minneapolis: University of Minnesota Press.

Feyerabend, P.K. (1970) "Classic Empiricism". In Butts, R.E. and David, J.W. (eds.)

Gingerich, O. (1989) *Johannes Kepler*, Chapt.5 in Taton and Wilson (eds.) 1989.

Glymour, C. (1980) *Theory and Evidence.* Princeton: Princeton University Press.

Goldstein, H. *Classical Mechanics.* Addison Wesley Publishing Company, 2nd edition, 1981.

Kepler, J. (A.N.) *Astronomia Nova* Vol.3, in Caspar (ed.)

Koyré, A. (1965), *Newtonian Studies.* University of Chicago Press.

Koyré, A. and Cohen, I.B. (eds.) (1972) *Isaac Newton's Philosophiae Naturalis Principia Mathematica.* Cambridge, Mass: Harvard Press.

Layman, R. (1983) "Newton's Demonstration of Universal Gravitation and Philosophical Theories". In Earman (ed.)

Lewis, D.K. (1980) "Veridical Hallucination and Prosthetic Vision". *Australasian Journal of Philosophy*, 58: 239-49.

Rosenkrantz, R. (1981) *Foundations and Applications of Inductive Probability*, Ridgeview, Publishing Company, Atascadero, Calif.

Stalnaker, R. (1984) *Inquiry.* M.I.T. Press.

Stein, H. (1990) "From the Phenomena of Motions to the Forces of Nature: Hypothesis or Deduction?". Manuscript (this volume?)

Whewell, W. (1858) *(NOR) Novum Organon Renovatum*, 3rd edition with additions London: John W. Parker and Son, facsimile reprint by University Microfilms, Ann Arbor, Michigan (1971)

Whiteside, D.T. (1974) "Keplerian Planetary Eggs Laid and Unlaid, 1600-1605", in *Journal for the History of Astronomy*, Vol. 1974, pg.1-21.

Wilson, C.A. (1974) "Newton and Some Philosophers on Kepler's Laws". *Journal of the History of Ideas*, Vol.35, pp.231-158.

_____. (1989) "The Newtonian Achievement in Astronomy" Chapt.13 in Taton and Wilson (eds.)

_____. "Horrocks, Harmonies, and the Exactitude of Kepler's Third Law" in Wilson, *Astronomy from Kepler to Newton*, Chap.VI.

Reasoning from Phenomena: Lessons from Newton

Jon Dorling

University of Amsterdam

1. Introduction

On the model of Newton's Principia, the great majority of successful new theories in physics have been introduced by deduction from the phenomena arguments. In such arguments an explanatory theory is deduced from one or more of the empirical facts, or lower level empirical generalizations, which it is designed to explain, by the device of adjoining suitable higher-level theoretical constraints on the form of the required theory. Those theoretical constraints leave certain parameters, the precise form of certain functions, and so on, in the new theory, undetermined, except with the help of the lower level empirical premises.

Although Newton's own complete argument to his inverse square law did contain at least one additional inductive step, it is not difficult to show that deduction from the phenomena arguments can be rigorously deductively valid within modern formal logic, (i.e. all the inductive steps can often be confined to the justification of the premises of such arguments) and that nearly all theoretical advances in physics since Newton have depended partly or wholly on the use of arguments of this general form. Sometimes the high-level theoretical constraints invoked are claimed partly or wholly to follow from a priori justifiable principles, but more usually they are either merely claimed to be plausible inductive generalizations from all experience (as Newton claimed for his three laws of motion which functioned as theoretical constraints in the deduction of his gravitational force law), or, as in most later examples, they are merely claimed to be derived by inductive extrapolation from the successful parts of previous theories.

Now most philosophers would probably be inclined to suppose that it is a truism of probability theory that the conclusion of such a deductively valid argument, namely the new theory thus deduced, must, if validly deduced, end up with at least the initial probability/plausibility of the conjunction of its premises. Hence, since in virtually all examples the higher-level premises are deliberately chosen so as already to appear (possibly with the help of arguments the innovatory theorist has himself just adduced) reasonably, or highly, probable/plausible, and since the lower-level premises have the status of relatively uncontroversial empirical facts and low-level laws, it would then seem that any new theories introduced in this way must themselves automatically be granted a reasonable, or high, degree of probability/plausibility by scientists. P is probable, P implies Q, therefore Q is probable.

Unfortunately the situation is by no means so simple. For not only have nearly all successful innovations in physics been introduced by arguments of this general form, but so have nearly all unsuccessful innovations in physics, and of course historically there have been far more of the latter. And in most of the latter cases the corresponding deductive justifications from the phenomena, in spite of their logical validity, never did cut much ice with other scientists, even before any direct evidence emerged for the falsity of their conclusions. For those other scientists took these arguments' conclusions as discrediting their theoretical premises: Q is improbable, P implies Q, therefore P is improbable.

To clarify this situation we need to appreciate that such deductive discoveries inevitably lead to revisions of the subjective probabilities initially assigned both to their conclusions and to their premises.

Thus if a seemingly highly unlikely conclusion (for example a surprisingly complicated equation relative to the apparent simplicity of the data requiring explanation) is deduced from premises which had previously seemed very likely to be true, then of course scientists are far more likely as a result of such a deductive discovery to lose confidence in, and to begin to question, one or more of the premises of the deductive argument, than to give their assent to its conclusion. In fact what often happens is that scientists conclude that one or more of the apparently plausible theoretical premises must after all probably be false, without their being able to say which that is. They lose confidence in the truth of the conjunction of premises without necessarily concluding of any particular conjunct that it has become less probable than not.

The opposite kind of example is the following: if theoretical premises which seemed at first, to most other theorists, theoretically unlikely, nevertheless lead to the deduction of an unexpectedly simple new explanatory theory, then as a consequence of that deductive discovery both the new theory and the premises which led to it can end up being assigned much higher degrees of assent than that originally granted to the initial conjunction of premises.

It is possible to give numerous historical examples of both these types of situation. The general rule seems to be that if the theory deduced proves to be more complicated/implausible-looking, than theorists reasonably expected it would have to be, in order to explain the relevant empirical data, then the deductive discovery does more to discredit its premises than to raise confidence in its conclusion; while if the theory deduced proves to be simpler than theorists had anticipated that it would be, then the deductive discovery strengthens scientists' confidence both in its premises and its conclusion. Let us call examples of this second kind impressive deduction from the phenomena arguments. (Of course there are many intermediate examples where the deductive discovery is noted as an interesting technical result, but doesn't induce any marked change in scientists' prior opinions.)

Newton's deduction from the phenomena argument to his inverse square law was, of course, of the impressive kind just mentioned. The same is true of many examples in Einstein's papers of arguments of a similar logical form. Though Einstein himself seems never to have noticed how essentially Newtonian in structure his principal derivations of his own succesful new theories were, or to have commented anywhere (as nearly all great previous theoretical physicists had done) on the apparent superiority of such a Newtonian justificational strategy over naive hypothetico-deductivism.

(For further discussion of Einstein's use of deductions from the phenomena see Dorling 1991.)

What I want to consider in the present paper is how much logical weight such impressive deduction from the phenomena arguments (which certainly have a persuasive effect on their scientific audiences) really carry. Do they, when deductively valid, in fact legitimate the assignment of high rational probabilities to their conclusions? Newton plainly thought they did. I shall argue that Newton was wrong, but that he was wrong in a surprising way, which he could hardly have anticipated, and which would probably in fact have delighted him.

For I shall argue that what is wrong with 'impressive' deduction-from-the-phenomena arguments is not that the theories which they lead to turn out to be too simple to be exactly true, thus not that they have to be succeeded in the long run by more complicated theories, but rather the reverse: such theories, in spite of superficial appearances of simplicity, have never turned out in the long run to be simple enough: the theories which later replace them, far from being, as they superficially appear to be, more complicated replacements, constitute in fact simpler explanations of the original data—though this often only becomes apparent later, when a sufficiently deep level of mathematical and logical analysis becomes available.

In a later section I shall illustrate this by considering the line of successors to Newton's theory of gravity. But before doing this it will be useful if I digress a little so as to bring readers up-to-date with our current very satisfactory understanding of inductive logic.

2. Digression: modern simplicity-based inductive logic

The centuries-old problem of setting up a completely general, powerful, plausible, and mathematically coherent system of inductive logic, was finally solved by the work of R. Solomonoff in the early 1960's, with some technical improvements by L. Levin in the early 1970's. The Solomonoff-Levin solution treats theories as computer programs for regenerating all the data as output, that is to say as encodings of the data, and assigns prior probabilities to theories according to the number of bits they require as prefix-free programs in the theorist's internal programming language. Such prior probabilities fall off by a factor of two for each extra bit required in the statement of a theory.

The mathematical background to this solution is well-known as Complexity-theory, or as the theory of Kolmogorov Complexity, though it would have been historically more accurate (since Solomonoff's publications anticipated Kolmogorov's), and philosophically more illuminating, to call it Simplicity-Theory or the theory of Solomonoff-Simplicity. One of the fundamental theorems in this theory shows that for any reasonably non-trivial theories their relative priors so-assigned are negligibly dependent on the choice of original programming language.

A recent review of this approach can be found in Li and Vitányi 1991. These authors, apparently following Solomonoff himself, seem to regard these simplicity-based rational priors as a sort of elegant mathematical substitute for unknown philosophically correct priors. I, however, maintain that these simplicity-based priors really are the unique philosophically correct priors.

The attempt to link rational probabilities to simplicity has of course a long history. One can consider Newton's rules of reasoning (with their missing ceteris paribus clauses added) as simplicity constraints on inductive steps in theory-construction. However those simplicity constraints did not quite work, for Newton's successors such as Boscovich pointed out that Newton's rules of reasoning applied indiscrimately could lead to inductive generalizations inconsistent with one another. In the modern approach it is the overall simplicity of the complete theory which determines its rational probability, not the requirement that each of the theory's epistemically distinguishable ingredients should be the simplest encoding of some particular fraction of the data. But we could not forge the necessary precise link between simplicity and rational probability prior to Solomonoff's and Kolmogorov's precise formal explications of simplicity.

So we now have an adequate theory of rational probabilities, ones which are language-independent for all practical purposes, and this theory yields an inductive logic satisfying the aspirations of all probabilistic inductivists from Laplace through Carnap to the Finns, an inductive logic which is in no way restricted to observational-language theories, since theoretical terms and parameters will be preferred whenever they can be introduced so as to shorten our theoretical encodings of the data.

This does not mean that different theorists now have to assign essentially the same priors to new theories. On the contrary, the number of bits required to add a new theory to a given epistemic system will depend not only indirectly on all the other evidence that that system has been developed to account for, but also in addition on just what definitions and constructs have already been introduced into the system in the interests of overall reduction of total epistemic program-length. The relative simplicity assigned by a theorist to a new theory, and hence the relative prior probability he should assign to it, is given by the number of bits he has to add to his total epistemic system to incorporate it, but this number will depend not only on his whole intellectual and experiential background, but also on the logical and mathematical skill with which he has constructed his epistemic system to date, for example on what notations he has already introduced so as to effect earlier encoding economies. For any new abbreviations he introduces in order to shorten a theory will themselves count as part of the cost in bits of that theory. Thus a differential geometer may give General Relativity theory a higher rational prior than an experimental physicist would, simply because the former's total epistemic system has already economically encoded much of the necessary technical apparatus. The rational prior a theorist should give to a theory depends on the actual economy of its encoding in his own epistemic system (or perhaps, if the approach is extended to a meta-level, on the economy of encoding he believes he could give to it in that epistemic system) and not on some hypothetical optimal encoding which he has not yet discovered. Thus different encodings are treated in the first instance as different theories for the individual in question, and an improved encoding as an improved theory. (Our more usual concept of theory would only emerge at a later level by adding the rational priors of different programs provably delivering the same output.)

We also have to be careful to distinguish the apparent number of raw data bits from the number of bits required for encoding that data. For example it is not the case that in deduction from the phenomena arguments, we can simply add the raw data bits for the experimental premises to the bits originally needed for the theoretical premises of the deduction, in order to bound from above the bits needed for the newly deduced theory, and hence to bound from below its rational prior. For that theory may have a very low simplicity ranking among theories consistent with those theoretical premises on their own. This means that the raw data bits for the experimental premises may grossly underestimate the number of bits we need to add to the epistemic system as a whole (to retain its coherence), in order to accommodate that data. Thus an unexpected null result of an experiment may require adding lots of bits to the total epistemic system, even though it is (when we ignore the theoretical background) formally the simplest possible result and thus involves fewest raw data bits. In fact until we've carried out theoretical investigations, such as the construction of deduction from the phenomena arguments, we won't know how many bits each particular experimental result adds to our epistemic system. The Solomonoff-Levin approach thus only allows us to derive relative rational probabilities from computable relative simplicities when the latter computations are carried out on the theorist's total epistemic system. Formal simplicities of data, or of whole theories, considered in isolation from a total epistemic system, may thus often be a very poor guide to Solomonoff-simplicity based rational priors. Thus in practical applications the new inductive logic behaves much more like subjective-Bayesianism, than like what one might first expect of a formal-simplicity-based inductive logic.

However while Solomonoff-Levin inductive reasoning seems satisfactory, indeed correct, as our underlying theory of epistemic rationality, there are tricky issues concerning its effective implementation in any finite physical system such as ourselves. The problems are twofold. First the number of alternative programs, of alternative epistemic systems or theories, which theoretically might need to be taken into consideration at any point, is, though finite, unacceptably large, indeed of astronomical dimensions even for a negligibly small amount of empirical input data. Hence, in practice, all alternative theories, whose theoretical priors are large enough to be relevant to a given theoretical choice, cannot possibly even be explicitly listed. Secondly, we are confronted with a problem of noncomputability. There is no effective decision procedure as to whether even a quite short program will yield any data predictions as output at all, let alone the data we are trying to regenerate.

It might at first seem that the latter problem could be solved by discarding only those theories actually computed to be inconsistent with the data so far, and re-normalizing the absolute Solomonoff-Levin priors assigned to the remainder to unity. However in practice this strategy won't do, because in practice inductive reasoning at a meta-level will make it, for most of the theories which would be left in by such a procedure, wildly unlikely that they would actually be consistent with the data so far. (For example most of the high-theoretical-prior theories left in by such an elimination procedure would be crazy theories such as the theory that all the input data bits so far received agreed precisely with the binary expansion of π from the billion billion billionth place onwards. Even a Popperian would surely not wish to assign a high position in his epistemic rank-ordering of surviving theories, to not-yet-actually falsified theories for which there was not yet the slightest reason to suppose that they were consistent with the existing data.)

Moreover, as we shall see, in practive there will often be actually simpler theories which yield all the data and which have been overlooked by the theorists. This means that in practice we only get relative rational probabilities from the Solomonof-Levin approach and not absolute rational probabilities. Whenever we have overlooked yet simpler theories, the absolute rational probabilities can be much lower for the known theories than most of us suppose. Only the relative rational probabilities for the known theories are computable and accessible to us.

Finally I should emphasize the obvious fact that this new inductive logic makes no pretence at being a logic of discovery. How far one can go in writing computer-implementable discovery algorithms for the discovery of shorter encodings of realistic data remains an open question. (I am personally optimistic: presumably we should be able to write discovery algorithms for discovering at least all the theories which humans could discover, and this may well include all theories which are actually true in the empirical world.)

3. The replacement of Newton's theory of gravity by successively simpler theories

Consider the series of successors of Newton's theory of gravity.

(i) The exponent 2, in the inverse square law, functioned as an unexplained constant in Newton's own theory: its value could only be derived, and then only approximately, by arguing backwards from the astronomical data of Kepler and others. The unsatisfactoriness of this was emphasized by Newton's eighteenth and nineteenth century successors. They therefore, without actually changing its predictive content, replaced Newton's action-at-a-distance theory of gravitation by a field theory in which the exponent 2 emerged as a consequence of the 3-dimensionality of space. This was a simpler theory, even on Popperian criteria: for it require fewer empirically-derived parameters.

While this first change did not lead directly to new predictions, it did lead nevertheless to a change in scientific strategy with respect to the treatment of recalcitrant data. From Newton's point of view there was nothing to rule out the possibility of additional terms in the gravitational force formula depending on different powers of the distance. Such additional terms were indeed proposed from time to time from the eighteenth century onwards. But believers in the new, more geometrically-explanatory, formulation of the theory were prohibited from taking such additional terms seriously, and had to seek elsewhere for the explanation of any predictive anomalies. From Newton's point of view there was also nothing to rule out an exponent in the force formula close to 2, but not exactly 2. Nineteenth century data (in particular data on the perihelion of Mercury) in fact led Newcomb and Hall to 'deduce from the more-accurately known phenomena' that this exponent was not exactly 2, but rather 2.0000001573. But no believer in the moregeometrized version of the theory, could entertain this change as a very serious theoretical possibility, since it sacrificed encoding gains which had already been made and added many additional bits to the theory, so he had to consider other possible explanations of the recalcitrant behaviour of Mercury. (One might at first think that such a small deviation from the inverse square law could be accommodated within a geometrical theory by postulating a slightly curved spatial geometry for the universe: but it was already known that the introduction of a curved geometry would not yield a correction of this kind.)

(ii) Newton had, through the kinematics he adopted as part of the mathematical framework of his theory, committed himself to a peculiarly degenerate geometry of temporal intervals. Among other disadvantages, Newton's choice here had the consequence that although his own theory required forces to explain accelerations, accelerated motion was not distinguishable within his own kinematics from non-accelerated motion: unlike the Euclidean case where a curved line differs in its intrinsic metrical properties from a straight line (only the latter is an extremal relative to the metric), Newton's geometry of temporal intervals failed to distinguish a curved line in space-time from a straight one, i.e. failed to distinguish accelerated from uniform rectilinear motion.

Only after the work of Minkowski did it become clear that Newton had not in fact chosen the simplest mathematical possibility for the geometry of space-time. Newton's implicit transformation group, which simply added the Galilean transformations to the transformations of the Euclidean group, was mathematically less simple, more artificial, and less unified, than one which picked instead the combination of the Lorentz and Euclidean groups. And later work, from a more synthetic-geometrical point of view, showed that more of the structure and simple theorems of Euclidean geometry were retained in Minkowski's proposed space-time geometry, than could be retained in Newton's space-time geometry. Thus Newton's geometry proved not to be the simplest generalization of Euclidean spatial geometry to include temporal intervals as well as spatial ones, i.e. to extend spatial geometry into a space-time geometry, but a more complicated alternative requiring more independent geometrical axioms.

This inductive error—a failure to assign the higher probability to the mathematically simpler theory consistent with the data—was what had forced Newton into the additional epistemically improbable conclusion that his dynamics required forces to produce sometimes epistemically unascertainable effects. But this latter defect simply disappeared once the preferred space-time geometry was adopted: absolute accelerations became identifiable with independently-measurable metrical curvature: thus kinematics and dynamics were no longer at odds with one another, and the conjunction of their respective axioms no longer reduced their joint probability and simplicity.

(iii) However this could only take us to a special relativistic theory of gravitation. (Einstein seems at one time to have thought that such a theory could not be consistent with all the data then known observationally, e.g. with that from the experiments of Eötvos. But Nordström and others soon proved Einstein technically mistaken in this belief.) There remained nevertheless unnecessary data bits in any such special relativistic theory of gravitation, ones which had already been mysteriously introduced by Newton; namely the unexplained proportionality of gravitational and inertial masses in Newton's gravitational theory. In fact Newton had quite bizarrely introduced a force field when all that the data evidently required was not a force field but an acceleration field. Only through a seemingly accidental cancellation of unnecessarily-introduced terms arbitrarily set equal to one another, did Newton's force field reduce to the equivalent acceleration field.

One way to solve this difficulty was simply to introduce a special-relativistic acceleration field for gravity. But Einstein found an even simpler alternative. For any such theory, when fully formalized, would contain two axioms where one would do: namely one axiom requiring that the space-time geometry was everywhere flat (i.e. requiring that the full Riemann curvature tensor vanished everywhere) and another axiom specifying gravitational departures from geodesic trajectories as a function of the gravitational source distribution in space-time. But given that gravitation was already at an observational level an acceleration field and not a force-field, these two axioms could, with a saving in bits, simply be combined into one axiom specifying the curvature in a curved space-time geometry as a function of the gravitational source distribution, i.e. one would no longer prescribe that the Riemann curvature tensor was everywhere zero, but make its value a function of the gravitational source distribution. Einstein's general theory of relativity introduced precisely this formal simplification, by combining two axioms into one with a small resultant saving in bits.

(iv) Einstein's theory was not the only theory which would do this or, it first seemed, the simplest. For Nordström's scalar theory of gravitation, in which the fully contracted curvature tensor, the curvature scalar, is simply taken proportional to the rest-mass density, seemed to have a much simpler field equation. However Einstein observed that if one began with a Lagrangian, rather than with field equations, his own theory was formally the simpler, and that it also unified the gravitational behaviour of matter and light in a way which no scalar theory could do, since light has zero rest mass. Both these considerations argued that Einstein's theory was simpler within the context of the rest of physics, since the similarities between matter and light had been a source of encoding gains in physical theory from Newton to the twentieth century, and it had been clear since the eighteenth century (if not earlier: Fermat) that action principles were one of the formally simplest way of formulating dynamical laws.

(v) Newton had already established the wave properties of light, and had conjectured connections between light and electrical attractions and repulsions, though it took more than another century before a consistent theory embodying these phenomena was available, namely the electromagnetic field theory of Maxwell. From the point of view of the Newtonian space-time geometry that theory's equations seemed quite complicated, but in the simpler space-time geometry of Minkowski, it became clear that all that was involved was the replacement of Faraday's geometrical lines-of-force explanation of electro-statics, with point-charges in space as sources, by planes-of-force emanating from the world-lines of charges in space-time as sources. Entirely analogous geometrical constraints then ensure that where the world line of the source becomes curved there is necessarily a propagated change in its associated planes of force, with all the properties of an electromagnetic wave. So taking into account the simplest space-time geometry, and the need to relate inverse power laws to geometry, if unnecessary data bits were not to appear in the theory, Maxwellian field theory became the simplest explanation of electrostatics.

However, a field theory of radiation coupled to a particle theory of matter was not really internally coherent. It led to infinities at the location of point-particles (or at the boundaries of extended particles), and to further infinities (the Rayleigh-Jeans catastrophe) when energy exchanges (later also momentum and angular-momentum exchanges) between the particles and the field were considered. Attempts to resolve the latter difficulty by returning to a particle theory of radiation coupled to a particle theory of matter never really got off the ground theoretically, and the only viable theoretical alternative was then to set up a wave-theory (i.e. a field theory) of matter coupled to the existing wave-theory (i.e. field theory) of radiation. The simplest wave-theory of matter consistent with the simplest space-time geometry was then discovered, by the combined efforts of Schrödinger and Dirac, to be modern relativistic wave-mechanics.

Due to an unfortunate quirk of history, the mathematical techniques for properly understanding this new theory not yet being available, physicists for the next seventy years (1926-1996) did not realise that this theory already without more ado also predicted and explained the particle properties of matter and radiation, and they thought that a further mysterious complication known as second-quantization was necessary, and not being able to understand this, pretended that the fundamental entities were not waves, but 'wave-particles'. In fact mass, charge, and energy quantization already comes out of the exact equations of the ordinary wave theory as a mathematical consequence of the self-coupling of the matter field via the electromagnetic field (and therefore does not need to be added independently) but this self-coupling term was ignored because it made the equations too difficult to calculate with prior to the nineteen eighties. And the so-called 'wave-packet-collapse' is really a pseudo-phenomenon due to the fact that physicists had neglected half the solutions (the advanced solutions) of their coupled time-symmetric equations, and ordinary wave-interference between these solutions and the others already predicts and explains at a classical level all the supposed wave-packet collapse phenomena. Thus the move that had been made in 1929 and 1930 to a more complicated and epistemically wildly improbable theory was, in hindsight, unnecessary. The simpler earlier theory was really actually correct. (For more extended discussion of my unorthodox contentions here, Dorling 1987.)

(vi) However there remained a gross improbability in fundamental physical theory due to the now wildly disparate treatment of gravitational and other forces. The former force was built into the geometry while the latter were still treated as essentially classical forces (i.e. as non-geometrized potentials in quantum-mechanical Lagrangians). This disunity was removed in the later '60's and '70's, when it was realised that once one treats the internal non-spatio-temporal degrees of freedom of the elementary particle fields as determining an internal geometry analogous to space-time geometry, one can perform the same trick as Einstein performed and replace all the other forces by geometrical curvatures, in each case combining two equations into one, with a resulting small saving in bits. This change in the direction of greater simplicity constituted the recent gauge-theoretical revolution in physics. Within this broader geometrical framework, Newton's second law of motion in effect reduces to Newton's first law of motion: all matter now moves uniformly along the straightest possible lines in the surrounding (generalized) curved geometry. Accelerated, non-geodesic, motions, no longer really exist in this fuller geometry.

(vii) As a by-product of this inductive simplification it became clear that Einstein's theory of gravitation had itself not gone quite far enough. For mathematically, although Einstein had made the curvature a function of the gravitational sources, there was an analogous tensor, the torsion tensor, which was still required to vanish everywhere and played the role of a flat background geometry. And this feature of space-time geometry was related in precisely the same way to the six-parameter sub-group of rotations and boosts in the ten-parameter Poincaré group underlying special relativity as the ordinary curvature was related to the four-parameter Abelian sub-group corresponding to translations in space and time. Einstein's theory was the gauge theory of the latter subgroup, not of the whole group. Einstein's argument from cause-effect reciprocity for space-time curvature depending on the distribution of matter was equally applicable to space-time torsion, and taking this consideration seriously generates a further twenty-four equations determining the torsion in addition to Einstein's sixteen field-equations determining the curvature. The result is the so-called U4 theory, or Einstein-Cartan-Sciama-Kibble theory, of gravitation. It is less mathematically arbitrary than Einstein's original theory. It also allows elementary particles such as fermions to function as sources of gravitational fields, which was not really possible in Einstein's original theory since the the natural energy-momentum tensor for fermions is anti-symmetric rather than symmetric as would be required by Einstein's original unmodified field equations. However this new less arbitrary theory still yields the same predictions as Einstein's original theory as far as ordinary macroscopic gravitational effects are concerned.

(viii) However even with these improvements physical theory is still not formally as simple as one might reasonably expect. For it still contains arbitrary coupling constants. In particular the gravitational coupling constant has to be regarded as a real-valued parameter equivalent to an infinite-bit ingredient in the theory, and there is nothing in the orthodox theory to explain why this parameter is not zero, i.e. why gravitation exists at all. However there is a recent programmatic theory which would overcome just this difficulty, and explain why gravitation exists at all and has the strength it does, namely Super-string theory. In fact Super-string theory not only eliminates various otherwise unavoidable infinities in the quantum theory of the other forces, but has no consistent solutions which do not include Einsteinian gravitation, and requires a non-zero gravitational coupling constant. The empirical value of this coupling constant should actually be calculable within this theory. Unfortunately the theory is not yet well enough understood for us to be able actually to carry out this calculation. Super-string theory remains to this extent still a merely programmatic theory. But should it prove right we will have to conclude that Newton was wrong in concluding that gravity was not an essential property of matter, and thus wrong in concluding on that basis that the presence of gravity in our world required a free creative act by an omnipotent Deity.

4. Discussion

What I want to emphasize about this brief review of the subsequent history of Newton's theory is that every change can in fact be viewed as one of formal simplification. Things do not seem like this to the layman because the layman does not realise how complicated the mathematical formalization has to be of all the background assumptions about the world which he takes for granted, and how much mathematical arbitrariness is implicit in naive formalizations of these background assumptions. But with the benefit of deeper mathematical understanding, we can see that arbitrarinesses here can be eliminated simultaneously with the elimination of arbitrarinesses in Newton's original explicit theory: such non-evident and evident arbitrarinesses can be made to cancel each other out, yielding what is overall a succession of mathematical simplifications of the theory of the world. This is what has happened in physics so far, and it is reasonable to suppose that it will continue in the future. (The most natural inductive inference here would be to the conclusion that Galileo and Einstein were right in thinking that our actual universe will ultimately turn out to be the simplest possible physical universe.)

This history creates, however, the following problem for deduction from the phenomena arguments, even for the most "impressive" deduction from the phenomena arguments. We would have liked these to yield actual high rational probabilities for the theories to which they lead. At first it seems they must do this because those theories are deduced from background assumptions which seem to have high probabilities. Any known rival theory can be shown to be inconsistent with one or other of these background assumptions and thus can be shown to be less probable in the then state of knowledge.

But the trouble is that such background assumptions must go well beyond the actual evidential data. And while they seem at the time to be the simplest mathematical generalizations consistent with that data (and were this so, this would entitle them to high Solomonoff-Levin rational probabilities), nevertheless subsequent mathematical investigation has always shown that there were formally simpler alternatives which had been overlooked, and which, had theorists known of them at the time, would have thus to have been assigned higher rational probabilities than the alternatives actually chosen.

Underlying this situation is a fundamental mathematical feature of the new inductive logic. Relative rational probabilities for any rival theories known to predict the same data can always be computed. It is enough to write those theories as programs and to count the number of program-bits required for each theory. Absolute rational probabilities are a different matter.

We would know these too if we knew that there were no simpler theories predicting the same data. For Solomonoff-Levin rational probabilities fall off fast enough with numbers of additional bits, for more complicated theories, even the disjunction of all more complicated theories, generally (though there are occasional exceptions) to get a low relative rational probability. So the problem is not the Popperian problem of the universe actually probably turning our more and more complicated.

The problem is the reverse. Absolute rational probabilities can only be assigned, or bounded from below, if we know there are no simpler theories predicting the same data. And we cannot ever know this because for any realistic data there can be no algorithm for determining the shortest program which will regenerate that data. Absolute rational probabilities are thus non-computable. This would not matter if we had meta-inductive evidence that we were actually good at finding the shortest possible encodings of realistic data. If we could show that modulo mathematical miracles concerning as yet uncomputed sequences of digits in the development of π, we had every reason to believe that we were often in practice actually succeeding in determining the shortest encodings of realistic data.

Unfortunately the existing meta-inductive evidence points in precisely the opposite direction. Each generation of theoretical physicists discovered that its predecessors had missed what were actually fewer-bit encodings of all the existing data. For this reason even the most impressive deduction-from-the-phenomena justifications, contrary to Newton's own inductive hopes, fail to yield high rational probabilities. At best we can take the conclusions to which such arguments lead as the most preferred of the available theories, and as theories which can be expected to yield correct predictions in the domain in which their background assumptions remain reasonable extrapolations from the data.

We indeed have some a priori mathematical guide as to when we are likely to be nearing the boundaries of such domains. Namely when the value of some physical quantity begins to approximate to the value of what appears to be a fundamental constant in the theory in question or in some related theory. For it is precisely at such points in a theory, or in the relation between two theories, that we expect deeper mathematical insights to yield future changes resulting in potential overall theoretical simplification.

Thus in practice it is reasonable to suppose that when an unexpectedly simple theory is deduced from the phenomena it will remain a very good approximation to the truth in a very extended domain. But the new inductive logic warns us that we cannot conclude that it is probably true, and meta-induction from the history of physics teaches us that it is almost certainly false, not because the truth will turn out to be more complicated, but because the truth will prove to be even simpler when all the relevant data that were available are taken into account.

The Newtonian strategy gives us a way of establishing theories which are rationally more probable than any rivals we are likely to be able to envisage given the current state of theoretical understanding. But at the same time the subsequent history of physics warns us that we will almost certainly be failing to envisage rationally even more probable theories.

The moral seems to be that physicists should spend more time reflecting on the foundations of current theoretical frameworks, than on tinkering with theories to explain particular recalcitrant experimental results. Every feature which is normally taken for granted in our currently successful theories needs to be repeatedly called in question, because we can be almost certain that there will be simpler future theories obtainable by abandoning some such features.

Extrapolating a little further from the history of physics, it is hard to avoid the inductive conclusion that our actual universe is likely ultimately to prove to be, in some meaningful sense, the simplest possible universe. This suggests that a more direct and a priori approach to characterizing the latter structure (e.g. by first laying down informal adequacy conditions on the class of mathematical structures which could count as characterizing possible physical universes, and then looking for the simplest mathematical structure which could meet those conditions) might eventually deliver the jackpot.

Thus the results of three centuries of reasoning from the phenomena are more surprising than philosophers seem to realize: for they seem to imply that we may be unduly neglecting a potentially viable alternative more aprioristic strategy.

References

Dorling, J. (1987), "Schrödinger's original interpretation of the Schrödinger equation: a rescue attempt," Schrödinger, *Centenary Celebrations of a Polymath*, C.W. Kilmister (ed.). Cambridge: Cambridge UP.

_____. (1991), "Einstein's methodology of discovery was Newtonian deduction-from-the-phenomena," *Scientific Discovery* (provisional title), J. Leplin (ed.), University of California Press (forthcoming 1991).

Li, M. and Vitányi, P.M.B. (1991), "Inductive reasoning and Kolmogorov Complexity," *Proceedings of the 4th Annual IEEE Structure in Complexity Theory Conference 1989*.

"From the Phenomena of Motions to the Forces of Nature": Hypothesis or Deduction?

Howard Stein

University of Chicago

There is a passage in Hume's *Enquiry concerning Human Understanding* that I have always found striking and rather charming. It concerns a metaphysical theory that Hume regards as bizarre; and he offers two philosophical arguments in its confutation. It is the first of these that I have in mind:

> *First*, [he says,] It seems to me, that this theory . . . is too bold ever to carry conviction with it to a man, sufficiently apprized of the weakness of human reason, and the narrow limits, to which it is confined in all its operations. Though the chain of arguments, which conduct to it, were ever so logical, there must arise a strong suspicion, if not an absolute assurance, that it has carried us quite beyond the reach of our faculties, when it leads to conclusions so extraordinary, and so remote from common life and experience. We are got into fairy land, long ere we have reached the last steps of our theory; and *there* we have no reason to trust our common methods of argument, or to think that our usual analogies and probabilities have any authority. Our line is too short to fathom such immense abysses. And however we may flatter ourselves, that we are guided, in every step which we take, by a kind of verisimilitude and experience; we may be assured, that this fancied experience has no authority, when we thus apply it to subjects, that lie entirely out of the sphere of experience (Hume 1777, pp. 59-60).

Hume was, of course, a great admirer of Newton, and took Newton's theory of gravitation in particular as the very paradigm of science. Can Hume have reflected seriously upon the chain of arguments by which Newton claims to establish—on the basis of phenomena accessible to everyone—that each particle of matter in the universe attracts each other particle by a force whose value he precisely states? By what line did Newton fathom *that* abyss?—Nothing in Hume's philosophy suggests that he did seriously consider this question; but the passage I have just cited sets the mood in which we ourselves, I think, ought to consider it.

Let me suggest two further points of general perspective. First, in the preface to the *Principia* Newton tells us that his subject in the book is what he calls *potentiae naturales*, or *vires naturae*—natural powers, or forces of nature. From the point of view of the physics of our own time, the discovery of universal gravitation was the discovery of the first of what we ourselves call the fundamental natural forces. In this sense, quite apart from the extraordinary scope of the law Newton stated—the enormous extrapolation involved in it—we have to see its discovery as one of astonishing *depth*; indeed, it was a discovery of a sort that the reigning epistemology of the circle of philosophers with whom Newton is most closely associated (and of which Hume is often considered the culmination) considered demonstrably beyond the scope of human capacities. There is, moreover, a second statement by Newton of his aim in the *Principia*: namely, to show how to distinguish the true from the apparent motions of bodies. This remark—which occurs at the end of the celebrated scholium to the Definitions—has occasioned some rather pointed comment on the blindness of scientists to the significance of their work; but one now understands pretty clearly that what Newton is talking about is his success in obtaining a decisive resolution of the issue posed by the competing geocentric and heliocentric cosmologies. Since that resolution is essentially a corollary of the theory of gravitation, we see that whatever argument leads from the phenomena to this theory must in some way implicate the deeper philosophical problems of space and time.

The law of universal gravitation is stated by Newton in Proposition VII of Book III of the *Principia* and its second corollary. Although I do not believe, as I shall explain presently, that what Newton calls the "deduction from the phenomena" of the law of gravitation is properly said to have been completed at that point, it is clearly the case that a "chain of arguments that conduct to" that proposition has occurred by then. It is thus our first task—and a great part of our main task—to examine that catenation of reasoning.

Let me review the state of affairs with respect to Propositions I-IV of Book III. It is almost (but not quite) true that the first three of these are *derived mathematically*—"mathematically demonstrated," Newton would say—from what are called *Phaenomena* in the introductory material to Book III. These latter are actually formulations of astronomical regularities, as regularities of the motions of the heavenly bodies (planets or their satellites), each referred to a suitable frame of reference: for each system of satellites of a central body, the motions are described from a perspective in which the fixed stars and the central body in question are taken to be at rest. In particular, then, Newton's statements of the *Phaenomena* carefully abstain from any commitment as to the "true motions." (In view of his statement of aim in the scholium to the Definitions, this is of course essential to his purpose: to "*collect* the true motions from their causes, effects, and apparent differences.")

Propositions I-III in effect translate the mathematical description of the astronomical regularities into a simpler but (nearly) equivalent form, relating the acceleration of the satellite to its position relative to the central body: the acceleration is directed always towards the central body, and its magnitude varies—both for a given satellite within its orbit, and from satellite to satellite within a single system—inversely as the square of the distance from the central body. To be sure, when one looks closely at the details, there are certain liberties in what I have called the "translation." I shall have occasion to refer later to the fact that the moon departs appreciably from Keplerian motion—a point that Newton not only acknowledges, but even accords a certain emphasis; he promises that the discrepancy will be shown to be due to the sun's action on the moon.

Proposition IV is another matter. What it claims to tell us is something about the *cause* of the behavior of a certain astronomical body. The proposition reads: *That the Moon gravitates towards the Earth, and by the force of gravity is drawn continually away from rectilinear motion, and retained in its orbit.* It is established by a calculation, with the help of the inverse square law, of what the orbital acceleration of the moon would become if the moon were brought down to the surface of the earth. The result agrees with the acceleration of terrestrial falling bodies; whence Newton concludes: "And therefore the force by which the Moon is retained in its orbit is that very same force, which we commonly call gravity."

It is most important to be clear about the content of such an assertion: that a certain force is the *very same force* as something-or-other. What notion of force is operative here—and what notion of identity? A parenthetic comment added in the second edition of the *Principia*, referring at this point to Rules I and II (of the *Rules of Philosophizing*), implies that what is asserted to be the same is "the cause" of the moon's behavior and that of falling bodies. By itself, however, this is not obviously helpful; for "same cause" is at least as problematic a concept as "same force." Within the bounds of the *Principia*, what seems the best guide to Newton's intention is contained in the section of Definitions prefatory to Book I. An action exerted upon a body [tending] to produce a change in its motion is called by Newton "a force impressed on the body." The acceleration of the moon, therefore, or that of a falling body, manifests such an "impressed force"; and so does the tendency of a heavy body (at any instant) to weigh down upon an obstacle that prevents it from accelerating downwards. But Newton says that such actions have in general diverse causes: "Impressed forces are of different origins; as from percussion, from pressure, from centripetal force." Since Propositions I-III of Book III in effect identify the forces on the planets and satellites as "centripetal forces," it is this last category that we are concerned with.

Now, it surely should seem odd to anyone who has had a course in Newtonian mechanics to read in Newton that an "impressed force" may be *caused by* a "centripetal force." Newton proceeds to define the latter as *that by which bodies are drawn or impelled, or any way tend, towards a point as to a center*. Is this not—one may wish to ask—*a special case*, rather than a cause, of "impressed force"?

The answer to this is that Newton himself uses the term "force" in a different way from that now customary in what we call Newtonian mechanics; a way resembling, rather, that in which physicists today use the word when they speak of the "four fundamental forces." The first force Newton actually defines is *materiae vis insita*, the "intrinsic force of matter," otherwise its vis *inertiae*, or "force of inactivity": committing the notorious crime of the freshman physics student, to call inertia a force. *Vis impressa*, I think, is best understood as a phrase denoting, not a kind of force at all, but as the *functioning* of a certain kind of force (the kind Newton elsewhere calls "active force," in contrast with the passive "force of inactivity"): thus, I am suggesting, "impressed force"="exerted force," and its "origin" or cause—e.g., a centripetal force—is "the force that is being exerted."

But this, if clarifying at all, is so only of (so to speak) the syntax of Newton's usage, not of its semantics. How are we to understand the notion of a centripetal force as the "cause" of an action on a body tending towards a center, and in particular to understand the notion of *the same* centripetal force, as *the same cause*?

Remaining within the text of the *Principia*, further instruction is to be found in the definitions Newton gives of three distinct "measures" of a centripetal force; and especially in his elucidating comments on these measures. Taking them in reverse order, the "motive measure" (which Newton also calls simply the "motive force")—Definition VIII—is the one we have all been introduced to in elementary physics. It is (in effect) the product of mass and acceleration; and Newton characterizes it as measuring the action *actually exerted upon a body*: "I refer the motive force to the body, as an endeavor and propensity of the whole towards a center, composed out of the propensities of all the parts." (Thus the motive measure is properly the measure of an impressed force—of the force *qua* acting on the body.) A second measure, the "accelerative measure" or "accelerative force," is just the acceleration engendered by the force; a notion that is a little puzzling, and far from innocent, as I shall explain in a moment. Finally—in Newton's sequence, initially—there is what he calls the "absolute quantity of a centripetal force." He defines it (Definition VI) as "the measure of the same, greater or less according to the efficacy of the cause that propagates it from the center through the surrounding regions." The definition is of considerable interest, but it certainly does not succeed in defining a quantity: rather, I should say, it gives us to understand what the absolute quantity of a centripetal force is supposed to be a measure *of*—so that we may, in favorable circumstances, come to recognize a particular quantity as of the sort required. And we do in fact succeed in this: Proposition VII of Book III allows us to say with precision what is the absolute measure of the centripetal force of gravity. Definition VI, therefore, rather identifies an aspect of the problem that confronts us, than contributes to its solution.

It is in a series of elucidatory paragraphs following Definition VIII that Newton goes to some length to try to convey to us what is on his mind in offering these several definitions of quantities of centripetal force. It is from that place that I have quoted his characterization of the motive force, as the action on—or the "endeavor or propensity of"—a particular body, composed of the propensities of all its parts. Of the absolute force he says that he refers it "to the center, as endowed with some cause, without which the motive forces would not be propagated through the surrounding regions; whether that cause be some central body (such as is the magnet in the center of the magnetic force or the Earth in the center of the gravitating force) or something else that does not appear. This concept is only a mathematical one. For I do not now consider the physical seats and causes of the forces." (In other words: the absolute quantity is in a certain sense referred "mathematically" to the center; and it is conceived as a measure of the "efficacy of the cause," *whatever the physical nature of that cause may be.*) This, once again, we shall later find of some interest; but it is of dubious assistance in clarifying Proposition IV.

For that, what Newton has to say here about the accelerative force turns out to be of most direct use. And what he does say is quite strange: he refers the accelerative force "to the place of the body, as a certain efficacy, diffused from the center to the several places around it, for moving the bodies that are in them." How, we may ask, does the acceleration of a body acted upon by a force towards a center serve appropriately as a measure of something or other that has been "diffused" to the *place* of that body? If the "motive force" measures something in which the body itself is involved, does not the acceleration—which is, after all, the acceleration *of that body*—do so equally? Most crucially: how does it make sense to speak of an "accelerative quantity" that characterizes *all the places* around the center of force—whether or not there happen to be bodies in them?

The answer, of course, is that it does not make sense in general. The notion that Newton has characterized qualitatively in his discussion of accelerative force as "referred to" place is the concept of what we call a "field"; but his quantitative definition does not in most cases accord with this notion—it does not, for instance, in the case, adduced by him, of magnetic force. What Newton has actually done is (a) to make it quite clear that the basic notion of centripetal force that he is concerned with is the notion of a *central force field*, and (b) to define, as measure of the field intensity, that quantity which happens to be appropriate to the force he intends to deal with in Book III.

But this does suggest an answer to our problem about the sense of the identification of the force on the moon with its weight: Newton draws the conclusion that the acceleration of the moon towards the earth (or rather, its accelerations, at all the points it traverses in its orbit), and the accelerations of freely falling terrestrial bodies, are all manifestations of a single *field of acceleration*: directed towards the earth at all points around it, and varying in magnitude inversely with the square of the distance from the earth's center.

Now it may seem that this puts Proposition IV into the same class, after all, as Propositions I-III. Is it not simply a matter of calculation from the data—a calculation that Newton carries out, and presents to establish the proposition—that the moon's acceleration is related to that of terrestrial bodies as Proposition IV claims? It is so indeed; but this is not the content of Proposition IV. The latter actually contains an extremely far-reaching physical implication that is not contained in the data, and not extrapolated in any straightforward way from the data; an implication that was hardly dreamt of before Newton published the Principia, and that was received with astonishment—and with assent—by the scientific community. For instance, it was regarded as a great and wholly unanticipated discovery by such distinguished opponents of the more general theory of Book III as Huygens and Leibniz.

The discovery I mean is that gravity—weight—varies inversely with the square of the distance from the center of the body towards which it tends. For note that the inverse square law for the acceleration *of the moon* is asserted in Proposition III on the basis of an argument from the phenomena, and this result plays a crucial role in the argument for Proposition IV; but before Proposition IV has been established, *no grounds whatever* are apparent for asserting a like law of variation for the weight of a terrestrial body. That the two "natural effects," the acceleration of the moon and that of falling bodies, are to be regarded as "the same" in the sense of Rule II, and thus as having "the same cause," has this as at least an important part of its effective meaning: the law governing the "diffusion from the center to the several places around it" of "an efficacy for moving the bodies" that are in those places is the same for terrestrial bodies and for the moon.

Proposition V of Book III is quite straightforward. What has been asserted of the moon in relation to the earth is now repeated for the other satellite systems: the satellites of Jupiter "gravitate"—have weight—towards Jupiter (and analogously, in the second and third editions, the satellites of Saturn towards Saturn); the circumsolar planets gravitate towards the sun. The argument is a simple appeal to Rule II: "The revolutions of the circumjovial planets about Jupiter, of the circumsaturnal about Saturn, and of Mercury and Venus and the other circumsolar planets about the sun are phenomena of the same kind as the revolution of the moon about the earth; and therefore (by Rule II) will depend upon causes of the same kind." Thus we are led to the conclusion that each of the bodies that has at least one satellite is a center of gravitational force. In the first corollary to Proposition V Newton extends this conclusion to the remaining planets: "Gravity therefore is to be conceded towards all the planets. For doubtless Venus, Mercury, and the rest, are bodies of the same sort with Jupiter and Saturn."

But although this argument presents no difficulties, there are two points about the conclusion so far reached that it is most important to understand—and that are easily overlooked. The first is this: As I have said earlier, Newton has formulated the astronomical regularities for each satellite system with respect to a frame of reference suitable for that system; and—in accordance with his program of *inferring* the true motions from their "causes, effects, and apparent differences"—with no commitment to a definitive statement about the "absolute" motions. (Indeed, the very word *Phaenomena*, introduced by Newton in the second edition as the heading for these formulations, indicates that they are concerned with the *apparent*—that is, the "relative"—motions.) It follows that the "accelerative forces" involved in the arguments for Propositions I-V have the same "relative" character. Thus the conclusion we are entitled to at this point is the following: For each satellite system, let us consider a rigid geometric frame F in which the central body A and the fixed stars are all at rest. Each point fixed relative to F will have, according to Newton's theory of absolute space and time, a certain "true" or "absolute" acceleration (not, of course, known to us). By Proposition V and its corollaries, there is a force of weight towards A, whose accelerative measure relative to F varies inversely with the square of the distance from the center of A; and the total (true) accelerative force on any body B at any moment will be the sum of the accelerative force of weight towards A, the resultant acceleration produced by any other forces that act on B relative to F, and the absolute acceleration of the point (fixed relative to F) at which B is located at that moment. Thus we have, at this point of the argument, a picture of several distinct fields of gravitational force, not only directed to as many distinct central bodies, but also characterized with respect to as many distinct frames of kinematical reference; we have so far no account of how this whole ensemble is to be organized into a single representation of the motions and forces.

One supplementary remark about this first point ought still to be made: namely, that the task of combining the several gravitational fields is essentially simplified by Newton's sixth corollary to the Laws of Motion. For the distances of any planet and its satellites from the sun are so large, in comparison to their distances from one another, that their "accelerative gravities" towards the sun may be regarded as very nearly parallel and equal; and under these circumstances, Corollary VI of the Laws assures us that the motions of the bodies among themselves will be the same as if that common additional acceleration were absent.

The second point to be made about the conclusion so far reached is less technical, but more startling: it has not, so far as my exposition has yet carried us, been claimed by Newton that the earth gravitates towards the sun (or towards anything at all); or that the moon gravitates towards the sun. Proposition V mentions, as gravitating, the circumjovial planets towards Jupiter; the circumsaturnal towards Saturn; and the circumsolar planets towards the sun. But we are not here justified in taking the earth to be a circumsolar planet. Indeed, in his formulation of Phaenomenon IV, which states Kepler's third law for the planets, Newton has very carefully referred to "the periodic times of the five primary planets, and of *either the sun about the earth or the earth about the sun*"; and he has said nothing so far to remove that uncertainty.

This issue—not only, that is, of the earth's motion, but of its standing subject to gravitation towards the sun, or subject to gravitation at all—turns out on close reading to present a minor crux in the interpretation of Newton's text. He seems to delay committing himself on the issue as long as he can; and the precise point at which the commitment is made, and the argument on which it there rests, is strangely hard to determine. There is a difference, small and puzzling, between the first and second editions of the *Principia* in their treatment of the matter; and a further quite interesting difference between both of them and the still earlier version of that book (Newton 1728), composed, as Newton tells us, "in a popular [rather than mathematical] method, that it might be read by many" (Newton 1687, introduction to Book III). But a discussion of these intricacies would be too long for the present occasion; and substantively they are of secondary importance. I shall confine myself to one or two points related to Newton's special arguments, but chiefly shall suggest how the knot can be cut.

First, however, one step that turns out to be relevant for his own approach to this question is taken by Newton in a portion of Corollary 1 of Proposition V that I have not yet quoted; and in this he introduces a theme that we shall find of pivotal importance for the reasoning that leads to the law of gravitation. Corollary 1, it will be recalled, has stated that there is a power of gravity tending to all the planets—on the grounds that these are doubtless all bodies of the same kind, and that we have already inferred the existence of such powers tending towards Jupiter and Saturn. What I have not yet quoted is this:

> And since all attraction (by the third law of motion) is mutual, Saturn reciprocally will gravitate towards the Huygenian Planet [that is, the satellite discovered by Huygens]. By the same argument Jupiter will gravitate towards all his satellites, the Earth towards the Moon, and the Sun towards all the primary planets.

Here, therefore, the assertion is made (on the basis of the third law) that the earth is subject to gravitation *at least towards the moon*; and a part of Newton's later argumentative strategy is to exploit this as a basis for the claim that the earth is the *kind* of body that is subject to gravity—and therefore should be supposed to gravitate towards the sun.

In the second edition of the Principia, there is a new corollary to Proposition V, which takes up the theme of mutual gravitation introduced by Corollary 1. This reads:

> All the planets do mutually gravitate towards one another, by cor. 1. and 2. And hence Jupiter and Saturn when near conjunction sensibly disturb each other's motions, by their mutual attractions; the sun disturbs the motions of the moon; and both sun and moon disturb our sea; as we shall hereafter explain.

The statements about disturbances of the motions are important, and I shall return to them; but at this point, they can only be regarded as parenthetical *anticipations* of matters to be discussed later (as Newton's concluding phrase indicates explicitly). The whole force of Corollary 3 is to allude to these later discussions, and to adduce them as evidence supporting the mutuality of the gravitation of the planets—which had itself been inferred already in Corollary 1.

Proposition VI states: *That all bodies gravitate towards each planet, and that their weights towards any one planet, at equal distances from the center of the planet, are proportional to their quantities of matter.* Of course this just says that gravity towards any planet is characterized by a field of "accelerative force," everywhere well-defined as a function of place alone, that affects all bodies without exception: that the weights are proportional to the masses precisely *means* that the "accelerative measures" are the same for all. The discussion offered by Newton in support of this proposition is somewhat lengthy, but its substance is very simple. It is tempting merely to say that we already know this proposition: for it is precisely the earth's acceleration-field, extending to all terrestrial bodies and to the moon, that has served to identify the force on the moon with its weight; and it is the character of their acceleration-fields, agreeing with that of the earth, that has led us to identify the sun, Jupiter, and Saturn—and then by induction the other primary planets—as gravitational centers. What Newton actually does in his proof of Proposition VI is, first, to cite additional and more precise evidence for the proposition that gravitational acceleration is the same for all terrestrial bodies; and then to argue, from the astronomical phenomena, that the *satellites* must have the same accelerations towards the sun, at equal distances, as the planets to which they belong. Finally, he offers a special argument that each *part* of a planet must tend to gravitate with the same acceleration at the same distance, on the grounds that otherwise the whole planet would gravitate more or less according as one or another kind of part predominates in its composition. All this serves, then, not as initial grounds for accepting Proposition VI—as I have said, its content had already been used in the earlier argument—but to provide strengthened support for it.

As to the question of the earth's gravitation towards the sun, Newton suggests the following argument: First, as for any other satellite and its central body, the moon and the earth must either both gravitate, or neither. But the moon is undoubtedly a body of the same kind as the planets; so it should be presumed to gravitate as they all do; and the earth is therefore carried along (not, of course, carried along by the moon, but by the argument).

We are now on the threshold of the chief step. Proposition VII states the law of universal gravitation: *That there is gravity towards all bodies* —gravitatem in corpora universa—*proportional to the quantity of matter in each.* The argument for this proposition is based entirely upon the conclusions already drawn, and it is very short. Newton's exposition refers back to Proposition LXIX of Book I; let me first state and prove a somewhat more general proposition, and then quote Newton's argument for Proposition VII in full.

The preliminary proposition is this: Suppose we have a system of bodies—A, B, C, etc.—in which A attracts all the others with accelerative forces depending only upon the distance, and B does likewise (in particular, then, each of the bodies A and B is subject to the accelerative force that tends towards the other); then (and I here quote Newton's words) *the absolute forces of the attracting bodies* A, B *will be to each other, as are those bodies themselves* A, B, *whose forces they are.*

In confronting this proposition, we are faced with the problem I mentioned earlier: we have been given no quantitative definition of "absolute [measure of a] force." But consider the bodies A and B, at any given distance d apart. According to the third law of motion, Newton argues, the motive forces on A and B will be equal; therefore the acceleration of B towards A will be to that of A towards B as the mass of A to the mass of B. But acceleration towards A at distance d is the same for all bodies; and likewise that towards B. So we have, quite simply:

The accelerative force towards A at any distance is to that towards B at the same distance, as the mass of A to the mass of B.

Clearly, then, mass serves as the appropriate measure of the efficacy of whatever cause makes the body a center of force.

Newton adds as a corollary that if (not only the first two listed, but) each of the bodies of the system attracts all the rest, with the corresponding condition (in effect: *well-defined fields of accelerative force*), then the absolute forces of all the bodies will be proportional to their masses.

With this prefaced, here is Newton's proof of Proposition VII of Book III:

That all the planets mutually gravitate towards one another I have proved before, as also that gravity towards each one of them considered separately is inversely as the square of the distance of places from the center of the planet. And thence it follows (by prop. lxix bk. i.) that gravity towards them all is proportional to the quantity of matter in them.

Further since all the parts of any planet A gravitate towards any planet B, and the gravity of any part is to the gravity of the whole, as the matter of the part is to the matter of the whole, and to every action there is (by the third law of motion) an equal reaction; the planet B will in turn gravitate towards every part of the planet A, and its gravity towards any one part will be to its gravity towards the whole, as the matter of the part to the matter of the whole. *Q. E. D.*

The proof is very simple indeed; yet it takes us directly to universal gravitation. What is the engine of this enormous step?

The answer is that the proof employs two crucial premises. The first is that weight is a force which acts independently upon all the parts of a body. This is a point that has already played a role in the argument for Proposition VI; it had been invoked by Galileo (1638, pp. 67-68) to resolve the apparent paradox that increasing the weight of a falling object—e.g., by putting one brick on top of another—does not increase its rate of fall. The other crucial premise is the third law of motion. Or, rather: it is an *application* of the third law of motion in *a very special form*.

Is there, then, some leeway in the application of the third law—so that the latter allows of being applied in alternative ways? The answer, in the present case, is that there indeed are alternatives. Back, for instance, in Corollary 3 to Proposition V, Newton, after concluding that Jupiter's satellites gravitate towards Jupiter, argued thus: "And since all attraction (by the third law of motion) is mutual Jupiter will gravitate towards all his satellites." But does this follow? What has been established about the satellites of Jupiter is that they are subject to forces directed towards Jupiter. The third law of motion does not tell us that whenever one body is urged by a force directed towards a second, the second body experiences an equal force towards the first; it tells us, rather, that whenever one body is acted upon *by* a second, the second body is subject to a force of equal magnitude and opposite direction. Therefore—putting the point in proper generality—what we may legitimately conclude, from the proposition that each planet is a center of gravitational force acting upon all bodies, is that for each body B there must be some body (or system of bodies) B' which, exerting this force on B, is subject to the required equal and opposite reaction. It must not be thought that the leeway implied by this formulation is one merely of far-fetched possibilities—that the only *plausible* subject of the reaction to gravitational force towards a planet is the planet itself. On the contrary, the very widespread view of Newton's time that one body can act upon another only by contact—a view that is well known to have had a powerful influence on Newton himself—makes for precisely the opposite assessment: that it is far-fetched to apply the third law in the way Newton does.

Here, then, is the shape of the whole main argument, from Proposition IV through Proposition VII: We have identified the force on the moon with its weight to the earth. Of weight, we have reason to believe that it acts on all bodies, and independently on all the parts of a body: the weight of the whole is simply the sum or resultant of the weights of the parts. Of the force on the moon, we have reason to believe that it varies inversely with the square of the distance from the earth. The latter holds as well for the various acceleration-fields about the astronomical bodies that have satellites; these too, then, ought to be considered to be fields of weight—gravity—acting, like weight to the earth, on all bodies. Thus we have a handful of centers of gravitational force, in each case producing accelerations determined by the inverse square law of variation with distance; in particular, accelerations the same for all bodies at any given place; producing, therefore, *weights*—*motive* forces—proportional to the masses of the bodies acted upon. *We now make a very far-reaching hypothesis*: noting that, in each case, there is a central body towards which the force of gravity tends, we ask, in effect,

What will be the consequence of *assuming* that the reactions called for by the third law are *exercised upon those central bodies*?

What in fact follows directly is that each central body experiences, towards every particle of matter whatsoever, a force proportional in each case to the mass of the particle and inversely proportional to the square of the distance from the particle. But Newton infers more: he concludes that each of these forces on each central body really acts *independently upon each part of the central body, proportionally to the mass of the part;* and then, further—on the grounds that any particle of matter whatever can be a part of a planet (e.g. just by falling on it)—that the force acts between all pairs of particles *simpliciter*. This consequence is not stated explicitly in Newton's argument. It is implicit, however, in his designation of the reaction-force as itself a *force of gravity*: of *weight;* and it is explicit in what he mentions as an objection that could be made to Proposition VII: namely, "that according to this law, all bodies about us must mutually gravitate one towards another." We must therefore still consider what justifies this step—the classing of the reaction-force as a kind of weight, with, therefore, the characteristic properties already assigned to weight.

On this point, there is a most illuminating passage—illuminating not for this alone, but for Newton's conception of the forces of nature in general—in the first version of what is now Book III of the *Principia*, composed "in a popular method." In §20 of that work we have the remark: "[A]ll action is mutual, and (by the third Law of Motion) makes the bodies approach one to the other, *and therefore must be the same in both bodies*" (Newton 1728, p. 568: emphasis added). He continues:

> It is true that we may consider one body as attracting, another as attracted; but this distinction is more mathematical than natural. The attraction really resides in each body towards the other, and is therefore of the same kind in both.

The ensuing §21 is then devoted entirely to the elaboration of this theme: the unity of the process involved in "action and reaction" (of *any* sort). The discussion is a full page long; and this is of interest, as showing how important it was to Newton to make his point clearly; but considerations of time and space prevent me from quoting it here in full, so I give only the central passage:

> It is not one action by which the sun attracts Jupiter, and another by which Jupiter attracts the sun; but it is one action by which the sun and Jupiter mutually endeavor to approach each the other. By the action with which the sun attracts Jupiter, Jupiter and the sun endeavor to come nearer together; and by the action with which Jupiter attracts the sun, likewise, Jupiter and the sun endeavor to come nearer together. But the sun is not attracted towards Jupiter by a twofold action, nor Jupiter by a twofold action towards the sun; but it is one single intermediate action, by which both approach nearer together. [Earlier in the passage, Newton has analogized this "one single intermediate action" to the contraction of a cord that is stretched between two bodies.] (Newton 1728, p. 569.)

We can now see clearly the answer to our question: what the engine is that drives the enormous step to the law of universal gravitation. It is a certain conception of the character of a force of nature, or natural power, such as gravitation: namely, that a force of nature is a *force of interaction*; and that such a force is characterized by a *law of interaction*: a law in which the interacting bodies enter altogether symmetrically. It is worth remarking that the conception of a force as characterized by a law is just what we really needed as far back as our discussion of Proposition IV: we can see the assertion that the moon's acceleration and that of falling bodies are *due to the same cause,* precisely as the assertion that these two classes of phenomena are governed by the same law of nature. With this interpretation of what the third law of motion requires—combined with what I have called the far-reaching hypothesis that gravity is such an interaction *between* the heavy body and the central body towards which it has weight—Newton's short and simple argument for Proposition VII leads directly to universal gravitation.

That, then, is how we have managed to get to fairy land. What are we to say—what could Newton have to say—in response to the strictures of Hume, applied to this chain of reasoning, and the reliability of our arguments in this realm?

I said at the outset that, in my opinion, what Newton calls the "deduction from the phenomena" of the law of universal gravitation is not properly said to be complete when Proposition VII has been formulated and the argument for it has been given. To discuss the issue with adequate reference to Newton's own statements about his view of proper method in natural philosophy would exceed the bounds placed on the present paper; I must therefore rather baldly state my own interpretation of Newton's view, with only one or two citations in support.

Baldly, then: In Newton's terminology, three terms describing kinds of argument are used in sharply distinguished fashion: namely, *demonstration, deduction,* and *proof.* The first of these is Newton's characteristic term for *purely mathematical reasoning.* The second—"deduction"—is used by him in a quite wide sense, for reasoning *competent to establish a conclusion as warranted* (in general, on the basis of *available evidence*). As for "proof," Newton typically means by it *the subjection of a proposition to test by experiment or observation* (with a successful outcome). Besides these, Newton uses the words "gather" or "collect"—corresponding to the Latin *colligere*—for any sort of inference, whether conclusive or merely tentative. In these terms, when a proposition has been "gathered" from evidence, it may or may not have thereby received adequate warrant—and so qualify as "deduced." If not, then the desideratum is to subject the proposition to further "proof," in the hope of achieving either a "deduction" or a refutation. Of course, *when* such proof is sufficient to provide adequate warrant, and so constitute the grounds for a proper deduction, remains a very difficult question indeed; but it is clear that Newton is no Popperian: he believes that this difficult question can be answered in practice, even in the absence of a general principle for deciding it. On the other hand, although a proposition may qualify as deduced from the phenomena, the issue may always be reopened by the discovery of new evidence from phenomena—this is explicit in Rule IV of the Rules of Philosophizing; and in this sense, the process of "proof" is in principle unending.

To know, then, whether the law of gravitation has been, properly, deduced from the phenomena when the argument for Proposition VII has been stated, we must ask whether that proposition has at that point been given adequate warrant. My own account suggests a negative answer. But what of Newton's procedure—what does it suggest he thought about this? Although I do not think one can feel entirely confident about Newton's state of mind, I believe it quite possible to arrive at a *definitive* judgment about the position that his procedure *objectively* entails. There are three main relevant points.

In the first place, it is essential to recognize that Proposition VII implies a vast range of consequences not implied by the propositions antecedent to it—and, in part, *contradictory* of the statements of "Phaenomena" on which the initial reasoning of Book III was based. That Newton understood this cannot possibly be called in question: the entire remainder of Book III of the *Principia* is devoted to the derivation of such consequences, *and to their confrontation*, so far as it was possible at the time, *with actual phenomena*. In short, in the formal terms I have suggested above as characteristic of Newton's usage, the remainder of Book III can be seen as devoted to the "proof by phenomena" of the law of gravitation; and furthermore, the proofs so obtained, in so far as they involve in part new (and confirmed) astronomical discoveries, and a great increase in both the scope and the precision of astronomical prediction, provide a kind of warrant for that law that can quite reasonably be seen as drawing the sting from the charge of "wild hypothesis" that could otherwise be levelled at Newton's way of applying the third law of motion.

In the second place (less important in principle, but perhaps more nearly decisive for the question of how Newton himself saw the structure of his own "deduction from the phenomena"), we should recall that the crucial argument for the moon had a slightly shaky relation to the data. Newton, in his own formulation of that argument, said of the discrepancy, "This in fact arises from the action of the sun (as will be shown later), and is therefore to be neglected." But *that* makes the argument leading from the Phaenomena through Proposition IV to Proposition VII itself *formally and explicitly dependent* upon consequences to be drawn from Proposition VII for its own proper completion.

In the third place, we do have at least two statements made later by Newton that bear directly upon the question of where he saw the main weight of evidence for the law of gravitation to lie. These agree with one another in resting the case for that law, not on the chain of arguments leading to Proposition VII, but on the fact that by means of it he has accounted for the behavior of the planets, the comets, the moon, and the sea. I shall here cite only one of them; it occurs in the *Principia* itself, in the third Rule of Philosophizing (which was added in the second edition). The passage reads as follows:

> Lastly, if it universally holds by experiments and astronomical observations that all bodies about the earth gravitate towards the earth, and that in proportion to the quantity of matter in each, and the moon gravitates towards the earth in proportion to her quantity of matter, and our sea in turn gravitates towards the moon, and all the planets gravitate mutually towards one another, and the comets in like manner gravitate towards the sun: it is to be asserted, by this rule [i.e., Rule III itself], that all bodies gravitate mutually towards one another. For the argument from the phenomena will be even stronger for universal gravitation, than for the impenetrability of bodies: for which among the heavenly bodies in particular we have no experiment, no observation whatever.

Since this rule was added expressly to clarify, and strengthen the force of, the argument for universal gravitation, it surely carries great authority as an indication of Newton's late, considered view. I think it clear that what in the text I have just quoted is called "the argument from the phenomena"—*argumentum ex phaenomenis*—for universal gravitation is indeed what I have claimed it to be: the entire third book of the *Principia*.

Something remains to be said about this whole "deduction from the phenomena"—or perhaps two closely related things. It may be asked, first: What has become of the question of the "true motions"—the program to "collect" these from their causes, effects, and apparent differences? The answer is that this, too, is carried out in Book III. The procedure is simple. From the separate pieces of the puzzle—the conclusions about the forces acting in the several satellite systems, *relative* in each case to *the appropriate frame of reference*—we have been led (on the basis also, as I have described it, of a daring speculative move) to a general law respecting a force of nature. We then consider a system of bodies let loose, in Newton's terminology, in "absolute space," and we ask how these bodies will behave *under the sole influence of this one natural power*. The answer is that if these bodies have the particular characteristics of the earth, moon, sun, planets, and comets, their behavior will be such as to produce just the phenomena we observe—that is the "argument from the phenomena" just characterized; and it leads to the conclusion, not only that the law of universal gravitation is to be affirmed, but that the force of gravity is the only one significantly affecting the observed behavior of those bodies. But this in turn settles the question of the "true motions," so far as that is at all possible in Newtonian physics: it determines those motions up to a common uniform motion of the entire system. The appropriate conclusion is drawn by Newton in his argument for Proposition XI of Book III; its further, and rather spectacular, "proof" can be seen in the later discussion—in Propositions XXI and XXXIX—of the precession of the equinoxes.

This means that the deduction from the phenomena in Book III can be regarded as not *only* a deduction of the law of universal gravitation, but *also* a deduction—or at any rate a contribution of evidence; a "proof" in Newton's sense—of a major metaphysical element of Newton's science: his theory of space and time. But one can say more than this. For clearly, in so far as the "deduction" validates what I have called Newton's speculative application of the third law of motion, it also contributes evidence for the cogency of the general conception of the natural powers that lies behind that application: that is, as I would put it, it "proves," besides the metaphysics of space and time, the *general metaphysics of nature* expressed in the introductory sections of the Principia and in the preface to the first edition. I believe that this whole conception of the constitutional frame of nature was actually developed by Newton *at the same time* that he was discovering the law of gravitation. In other words, as I see the situation, not only the "proof," but the discovery itself, of the background theory that made possible Newton's reasoning from the phenomena to the force of gravitation, occurred simultaneously and marched hand-in-hand with the latter.

Note

[1]Because of limitations of space, it has been necessary to abbreviate this paper and to omit all discursive notes and all but the most essential references; a fuller version will appear elsewhere.

References to Newton (1687) occur in the text *passim*. They are always obvious (since embedded in discussion of the work itself); and the source loci are always indicated by such structural references as Book and Proposition number. It has therefore been thought superfluous to flag them by author, date, and page.

This material is based on work supported, in part, by the National Science Foundation under Grant No. Dir-8808575.

References

Galileo (1638), *Discorsi e Dimostrazioni Matematiche intorno à due nuove scienze*. Cited from Galileo Galilei, *Two New Sciences*, trans. S. Drake. Madison, Wis.: University of Wisconsin Press, 1974.

Hume, D. (1777), *An Enquiry Concerning Human Understanding*. Cited from David Hume, *The Philosophical Works* (ed.), T.H. Green and T.H. Grose, vol. IV. London, 1882 (reprinted Darmstadt: Scientia Verlag Aalen, 1967).

Newton, I. (1687), *Philosophiae naturalis Principia Mathematica*. Cited from A. Koyré and I.B. Cohen (eds.), *Isaac Newton's Philosophiae Naturalis Principia Mathematica, the Third Edition (1726) with Variant Readings*. Cambridge, Mass.: Harvard University Press, 1972. (This edition records all significant differences among the first, second, and third editions—of 1687, 1713, and 1726, respectively.) In translating passages from this work, I have been guided by the English translation of Andrew Motte (1729; reprinted London: Dawsons of Pall Mall, 1968).

_____ . (1728), *De Mundi Systemate*. Cited from the English translation (presumably by Motte, revised by Cajori) in F. Cajori (ed.), *Sir Isaac Newton's Mathematical Principles of Natural Philosophy and his System of the World*. Berkeley, California: University of California Press, 1946.

Part VI

SCIENCE AS PROCESS

Phylogenetic Analogies in the Conceptual Development of Science[1]

Brent D. Mishler

Duke University

David Hull's approach to science, which culminated in his important book *Science as a Process* (1988), represents an unprecedented conjunction of philosophy of science with the results and concepts of a particular science. Hull takes an evolutionary approach to the conceptual development of science, importing much of his explanatory framework from comparative biology, the discipline where his empirical observations of scientists have been made. On the surface, such a cozy relationship between data, theory, and metatheory leads to worries about circular reasoning (Mishler 1989). Nevertheless, I will argue here that Hull's approach is basically sound, and that the strength of his arguments comes precisely from his recognition of key analogies between the evolution of organisms and the conceptual evolution of scientists; however, certain disanalogies must also be taken into account.

1. The Current Status of Systematic and Evolutionary Biology

Comparative biology can be divided into two distinct (although interrelated) halves based on differing orientations and types of questions asked. Systematic biology focuses primarily on patterns in the history of life. Evolutionary biology focuses primarily on process explanations for these patterns. The two are interconnected in a feedback loop, because the choice of proper systematic methods depends on at least a rough model of how evolution is proceeding, while defensible evolutionary explanations depend on a sound systematic framework. Such an arrangement needs careful examination, to prevent it from drifting on the one hand into vicious circularity or on the other hand into a sterile separation of theory from data. The recent cladistic revolution in systematics was spurred by Hennig's (1966) insights on how to connect these two areas of comparative biology. He initiated a body of method and theory that does allow us, for really the first time in the history of systematics, to walk the thin line between the two undesirable alternatives: circularity and sterility.

Hennig had one brilliant, central insight, from which several corollaries were derived. His key insight, sometimes known as the Hennig Principle, was that homologous similarities shared among a group of organisms are of two kinds. One kind of similarity is a feature shared by all and only members of an assemblage of organisms due to inheritance from their immediate common ancestor. He termed such features *synapomorphies* (i.e., shared, derived characters or "special" similarities). A second kind of similarity is a feature shared by all, but not only, members of an assemblage of organisms due to inheritance from a distant common ancestor. Hennig termed those features *symplesiomorphies* (i.e., shared, primitive characters or "general" similarities). His principle was that only synapomorphies are valid indicators of phylogenetic relationship (i.e., relative recency of common ancestry).

A major corollary to this principle is that only synapomorphies should be used as evidence for the naming of formal taxonomic groups (known in a general sense as "taxa" when the rank is unspecified). Hennig redefined *monophyly* so as to be compatible with this corollary: a monophyletic taxon is one that contains all and only descendants of a common ancestral species, recognized as such by the discovery of synapomorphies.

The meaning of *homology* has been clarified as well under Hennig's system. There are several distinct categories of homology, which when taken broadly, can be defined as a correspondence between two or more characteristics of organisms that is caused by a historical continuity of information (Roth 1988). *Iterative homology* is historical correspondence between different structures within a single organism. *Taxic homology* is a correspondence between features in different organisms due to inheritance from a common ancestor that possessed that feature (equivalent to synapomorphy; Patterson 1982). *Transformational homology* is a correspondence between two different features resulting from a historical modification of one into the other (equivalent to the relationship between an apomorphy and its plesiomorphy). Two or more features hypothesized to be related in this way are termed *transformation series*. Such transformationally related features are usually called "character states," whereas the whole series is called a "character."

The transformation series can be *polarized* if a defensible hypothesis can be made that one state is plesiomorphic (i.e., "primitive," or temporally prior). Evolutionary polarity of a transformation series can be hypothesized in several ways (Stevens 1980); the most favored criterion in cladistics is *outgroup comparison*, which is based on an examination of the distribution of character states outside the study group. That character state occurring widely outside the group is assumed to be plesiomorphic. Note that any state of a transformation series can be the plesiomorphic one, whether one of the "ends" of the series or an intermediate state.

Taxic homology (synapomorphy) is initially postulated based on detailed similarity between corresponding features in different organisms (the "similarity test" of Patterson 1982), but it is not accepted until the distributions of all putative homologies are compared using a parsimony criterion (see Sober 1988 for a recent discussion). Those putative homologies that are congruent with a *cladogram* (i.e., a branching diagram representing a hypothesis of relative recency of common ancestry) based on the most parsimonious arrangement of all putative homologies are accepted as homologies (the "congruence test" of Patterson 1982); those that are incongruent are termed *homoplasies*.

Such character incongruencies can be due either to simple mistakes in character study, or to various biological processes that are of evolutionary interest (even though they cause epistemological problems for systematics). Two identical features may arise independently in two different groups, a phenomenon known as parallelism or convergence. As long as sufficient evidence (in the form of other, independent synapomorphies) exists, parallelism can be discovered.

Homoplastic character distributions can also arise via "horizontal" transmission. Hennig assumed an underlying model of diverging evolution coupled with "vertical" descent with modification. Various reticulating processes, however, including hybridization, introgression, and even lateral transmission of genetic material by viruses are known to occur. The epistemological effects of horizontal transmission on biological systematics are as yet incompletely explored; however, the consensus among cladists seems to be that rampant reticulation can destroy our ability to reconstruct relationships (and should, because complete reticulation is an "information destroying" process in the sense of Sober 1988), but that partial or occasional reticulation can be discovered and accounted for in cladistic analysis by means of careful study of organisms.

These various concepts can be illustrated using Figure 1, which is a hypothetical cladogram of six species. The topology of this cladogram is assumed to be based on other characters than just the one discussed here (since a fully resolved cladogram must have synapomorphies supporting all nodes). The distribution of a single character is mapped onto the cladogram. This character is a three state transformation series between ○ (which is primitive, widely distributed outside this group of six species), ⊛, and ●. The latter two character states are transformational homologs of ○; ● is hypothesized to be transformed from ⊛ because elements of ⊛ are seen in ● (perhaps in ontogeny), yet the latter has elements unique to itself. Given this scenario, ● is a synapomorphy of species 5 & 6, relative to the plesiomorphy ⊛. Likewise ⊛ is a synapomorphy of species 3, 4, 5, & 6, relative to the plesiomorphy ○.

Figure 1

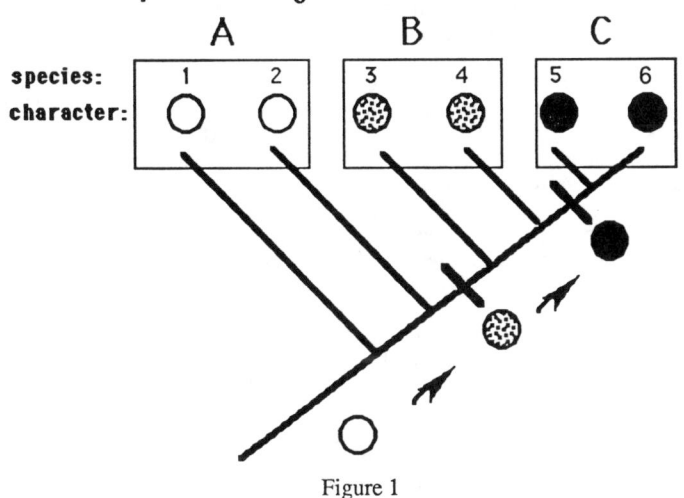

This example illustrates two important points: (1) The relational nature of the concept of synapomorphy; ⊛ can be referred to either as a symplesiomorphy or a synapomorphy, depending on the phylogenetic level (one reason why the apomorphy/plesiomorphy distinction is not identical to the older, derived/primitive distinction). (2) The abstract nature of synapomorphy; not all members of a taxon (e.g., a group composed of 3, 4, 5, & 6) need actually possess the synapomorphic feature. Several exceptions are possible. The original synapomorphy may have been transformed into a distinctive new feature in some derivative lineage (evidence for such transformation can often be discovered through ontogenetic studies), or it may even have been "lost" in a derivative lineage, giving rise to a character state that cannot be distinguished from the plesiomorphic state (such character "loss" or "reversal" is confusing epistemologically, but can be discovered if enough other synapomorphies are present to reconstruct the true relationships). The synapomorphy for a group may happen to only occur in adult females; the lack of the character in males and juveniles can be explained genetically and developmentally and thus is not taken as evidence to exclude such individuals from the group. Therefore, groups diagnosed using synapomorphies are not necessarily monothetic in an observational sense, but they are monothetic in a theoretical sense.

Continuing the example (Fig. 1), of the possible higher taxa shown, only C would be monophyletic. If named, A and B would be *paraphyletic* (i.e., a group of organisms including only some of the decendants of a common ancestor, thus having some included organisms that are actually more closely related to organisms outside the group). Such taxa are considered unnatural and therefore prohibited in Hennigian cladistics. Ontologically, groups such as C are monophyletic if they are hypothesized to include all and only descendants of a common ancestor. Synapomorphic characters have primarily an epistemological role; they do not, strictly speaking, *define* groups, but rather *diagnose* them. They are not the taxon, they are the evidence for the taxon.

The nature of species, as the basal taxon and perhaps a basic unit of evolution, is a subject of considerable controversy both within and without the cladistic school (see recent discussion by Mishler 1990). For the purposes of the present paper, it is particularly necessary to come to grips with the biological meaning of "basal" phylogenetic taxa, in order to investigate the model for Hull's classification of basal scientific research groups.

Hennig himself (1966), as well as many of his followers (e.g., Wiley 1981; Nixon and Wheeler 1990) argued that the concepts of synapomorphy and monophyly do not, and should not, extend to the species level. These authors have variously argued for the application of other criteria at the species level (e.g., reproductive compatibility or diagnosability using unpolarized similarities). Mishler and Donoghue (1982) and Mishler and Brandon (1987), however, developed a case for treating species like cladistic taxa at all other levels. The primary argument for this is logical consistency; the basal units in a phylogenetic system should be phylogenetic units.

In this phylogenetic approach, species are viewed as basal monophyletic taxa, into which organisms are grouped because of the presence of synapomorphies. Hennig's definition of monophyly was broadened and clarified by Mishler and Brandon (1987) to include "all and only descendants of a common ancestor, originating in a single event," where "ancestor" refers not to an ancestral *species* (as in Hennig's definition), but to a lower-level entity that is fully an *individual* in the sense of Hull (1976). The "event" referred to in this definition is the spatiotemporally localized action of one of a number of possible causes, including hybridization.

It was further argued in Mishler and Brandon (1987) that whereas all formally named taxa should be monophyletic, not all monophyletic groups should be formally named. Formal taxonomy should extend "down" (i.e., towards less inclusive groups) only so far as distinct, "important" lineages can be discovered. It is counterproductive (not to mention impossible) to attempt to name all monophyletic groups. Small lineages are constantly being produced, existing for a while, and going extinct (e.g., geographically localized kin groups in organisms with limited dispersability, multiple origins of sterile hybrids between two widespread parental types, and even cell lineages within clonal plants). Only a tiny fraction of these go on to be recognized as important lineages and to be formally named as species (e.g., those that have acquired a distinctive evolutionary novelty that alters selective regimes or patterns of interbreeding). Often this recognition is post facto, because it may not be clear early on that an important new lineage is arising.

Evolutionary theory, particularly the nature of natural selection, has also undergone considerable clarification and expansion in recent years. A recent book by Brandon (1990) codifies and integrates a rational approach to the related processes of selection and adaptation, that can be outlined briefly as follows. An adaptation, in a loose sense, is a match between some feature of an organism and a "problem" posed by its environment. The only known evolutionary process producing adaptations is natural selection. Several conditions are necessary for selection to occur: (1) There must be variation in some feature within a population (i.e., a spatiotemporally localized group of genealogically related individuals). (2) Variation in that feature must be heritable to some degree (i.e., offspring must preferentially resemble parents). (3) Some variants must endow their organisms with a propensity to leave more offspring in future generations (i.e., to have higher fitness) than other organisms. If these conditions hold, then competition in a common selective environment will lead to evolution of the population by natural selection, producing adaptations in the strict sense (Brandon 1990). Such adaptations are a match between organism and environment produced through a historical process of selection for certain heritable characters in the context of that environment.

The process of natural selection is a hierarchical one; the most familiar example of selection among individual organisms, given above, is by no means the only possibility. To understand, however, the nature of the "levels of selection" problem, one must realize that two hierarchies are involved (Brandon 1990).

Hull (1980) developed an important distinction between *replicators* and *interactors* in selection processes. The replicator is the unit of heredity and reproduction: an entity of which direct copies are made. Replicators thus form lineages; a population is a spatio-temporally isolated, integrated and/or cohesive, genealogically related, collection of replicator-lineages. The interactor is the unit involved in the competition process: an entity whose interaction with the environment causes replication to be differential. Interactors, unlike replicators, are not defined genealogically.

Brandon (1990) has shown that both sorts of entities are organized into hierarchies, and that these two hierarchies are not necessarily congruent. In other words, several levels of replication may be in operation in a particular group of organisms, none of which need be equivalent to the levels at which interaction is occuring. Controversy still rages over which levels actually *do* serve as replicators or interactors (whether organisms, populations, species, or higher taxa), but at least the properties necessary for a level to participate causally in a selection process have been clarified. For one of these higher-level entities to participate, it must either replicate itelf directly, or act as an interactor. For the latter, a "screening-off" relationship must obtain between it as a higher-level interactor and some lower-level replicator (such as in cases when the fitness of an organism in a particular selection process depends on its group membership rather than on its genotype; Brandon 1990).

Not all evolutionary change, at any hierarchical level, is due to the process of natural selection/ adaptation (Gould and Lewontin 1979). In finite populations (defined as above in a genealogical sense), *random drift* will occur, its importance depending upon population size and the strength of selection. Furthermore, natural selection/ adaptation is not the only constraint on observed phylogenetic patterns of character distribution. Not all conceivable variants are available to be input to selection. Some conceivable variants violate basic physical laws; such prohibitions are called *physical constraints*. Some features are physically possible, yet are never produced in a particular lineage because complex developmental processes give rise to an epigenetic homeostasis that biases against (or even prohibits) certain character combinations. Such historically contingent homeostatic mechanisms are called *developmental constraints*.

2. Hull's Evolutionary Approach to the Process of Science

The view of scientific process laid out by Hull, most fully in his 1988 book, relies heavily on analogies derived from systematic and evolutionary theory. His view can be summarized as follows (see also Mishler 1987, 1989). Scientists are grouped into basic lineages, called research groups, just as organisms are grouped into species. These research groups are bound together by various sociological ties, including: student/teacher and collegial relationships, shared language, peer reviewing of papers and grants, mutual use and citation of each other's work, and shared enemies, all analogous to the way in which species are bound together by interbreeding, ecology, and/or developmental constraints. Specific ideas function as traits; an analog of natural selection occurs as scientists use other scientists' ideas. Research groups evolve via this process, waxing and waning in size as they compete for new members (e.g., uncommitted scientists, graduate students), "speciating" sometimes as they give rise to splinter groups, and eventually going extinct.

A clear understanding of these putative analogies, and their efficacy, requires detailed comparisons between the biological concepts and scientific process. A summary of some proposed analogies is given in Table 1.

Any close analogy between the biological process of natural selection and the scientific process depends on the nature of replicators and interactors in the latter. The replicator, in the case of science, could be either the scientist or the idea, depending on the nature of transmission. Sometimes, scientific ideas are passed on wholesale, as in many professor/ graduate student relationships. In such a case, the "structure" of the scientist is being passed on directly. In many other cases (e.g., professional interaction among colleagues), individual ideas are taken up and added to an existing mix of other ideas.

Just as in biology, a distinction should be made between "genotype" and "phenotype" (see also Griesemer 1988). In terms of the evolution of scientific ideas, a distinction is necessary between the "meme" (sensu Dawkins 1976 — the basic structure of an idea-element as encoded in the brain of a scientist) and an idea as publicly communicated, applied, and understood in the scientific community. The "developmental" connection between individual memes and their assembly into functioning, public ideas can be complex, as is often the case in biology (and may lead to a sort of "intellectual inertia" — an analogy of developmental constraints in biological evolution). Unlike biology, however, the transmission of memes is via their phenotypic expression. There is no direct analog of genetic transmission in the conceptual evolution of science; students and colleagues do not have direct access to the memes in the brain of a scientist, only indirect access through writings, discussion, and observation of behavior.

Table 1. Proposed analogies between concepts derived from systematic and evolutionary biology and components of the scientific process. See text for discussion and explanation.

Concept From Comparative Biology	Corresponding Analogy in the Scientific Process
organism	scientist
species	research group
higher taxon	more inclusive research group/ school/ tradition
gene	basic structure of idea-element in mind of scientist = "meme"
trait	idea as communicated, applied, and perceived publicly
homologous trait	idea present in two scientists by descent from a common source
reproduction (birth)	production of "new" scientists with related set of ideas
death	scientist quitting science
fitness	relative number of scientific "offspring," related by descent, present in next generation
adaptive trait	idea that increses fitness by solving a perceived problem
developmental constraints	intellectual inertia
speciation	origin of new research group from old, by breakdown of sociological cohesion
extinction	termination of research group due to loss of members and lack of recruitment

Interaction in science clearly occurs at many levels of inclusion, thus presenting a pronounced "levels of selection" problem. Competition between ideas can occur within the mind of an individual scientist (analogous to somatic selection in biology;

Brandon 1990). Given that scientists are "fixed" in their ideas, competition can occur among them for converting or recruiting other scientists to their viewpoint. Higher fitness for a scientist would imply a propensity for propagating a relatively high number of copies of ideas to later "generations" of scientists. A group of scientists (Hull's "research groups") sharing homologous ideas can function as an interactor relative to other such groups (analogous to the models of group selection in biology discussed by Sober 1984; Brandon 1990), if the fitness of scientists is affected by group membership irrespective of their personal set of ideas. It is also at least conceivable that higher-order groups of scientists (e.g., entire research traditions, say systematic biology) could function as interactors in a selection process relative to other such groups (say molecular biology).

As emphasized by Brandon (1990), natural selection among several variants must occur within a common selective environment. The common environment in which the putative selection process among scientific ideas takes place, is composed of both sociological and empirical factors. This is analogous to the usual situation in biology, where the relevant environment consists of both biotic and physical factors (e.g., competitors, parasites, and climate). Other organisms (scientists) are part (but only part) of the environment that needs to be adjusted to. An adaptive idea is one that is perceived to "solve" a scientific problem (in the sense of Laudan 1977). This implies both "progress" in the eyes of other scientists, and "progress" in increasing our understanding of the natural world.

Hence, Hull's view of science does not boil down to sociological relativism (even though he seems to have missed the nice analogy with adaptation). Some ideas *are* intrinsically better than others in their fit to the empirical world (as it is currently known). The inescapable sociological milieu of science functions in both replication and interaction, yet competing scientists do continually attempt to match their concepts up with the real world. Just as in biological evolution, however, many ideas (traits) evolve by drift, rather than by natural selection. In addition, ideas that are adaptive are "tracking" an "environment" that may well be labile. Increasing fit between ideas and the real world occurs over time, but this progress is neither linear, precise, nor perfect. All this illustrates the way in which scientific evolution is and is not "progressive," and may help to assuage the concerns of Donoghue (1990), who rightly pointed out that Hull (1988) neglected the importance of "the worth of ideas" to scientists.

What is the nature of the patterns produced as a result of these various selection processes in science? Replicators, by their very definition, produce lineages. The biological concepts of taxic homology (synapomorphy), transformational homology, speciation, and monophyly have their analogs in recognizing and understanding lineages in the the history of science.

To count as the "same" (i.e., a taxic homology) an idea shared among several scientists must be identical by descent. As emphasized by Hull (1988), independent origins of an idea by an unappreciated precurser or by a contemporary working in isolation don't count as homology, but rather homoplasy. As a research group evolves, some (or all) of its diagnostic ideas will change to a greater or lesser extent. The new versions of old ideas, connected by descent, count as transformational homologies.

As with organisms, informative classifications of scientists should be based on homologies, take into account the polarity of transformation series of ideas, and attempt to group by synapomorphy. New research groups can be produced either by fission of a pre-existing one ("speciation"), or by fusion of two or more pre-existing ones ("hybridization"). The meaning of monophyly may seem even more problematic in reference to lineages of scientists than to organisms, because it is clear than an exceptionally large amount of horizontal transmission occurs in the former. Scientists are able to adopt ideas from diverse research groups (although the idea-flow among groups is much less than within a group), in a way that has no close analogy among the more complex organisms such as mammals or land plants.

Nonetheless, considerable horizontal transmission is suspected to have occurred in the phylogeny of simpler organisms such as the fungi or bacteria. As discussed above for organismal systematics, horizontal transmission causes epistemological problems for phylogeny reconstruction, yet not ontological problems so long as it is relatively infrequent compared to vertical transmission. Recognizable lineages can be maintained in the face of considerable horizontal gene-flow. If sufficient horizontal transmission occurs, then two lineages may blend to produce one.

The application of the concept of monophyly to research groups must involve a synchronic approach, called by Sober (1988) the "cut method" (similar to the approach taken for species monophyly by Mishler and Brandon 1987). At any given moment in time, one can make a horizontal slice across lineages and define monophyletic groups. At a later time, a horizonal slice might well imply a different decision. Furthermore, the synapomorphous characters marking a single lineage will be different (both in number and description) at different time-slices.

3. Conclusions

It appears that most of the relevant analogies that can be made between organismal and scientific evolution are sound. Given the complexity of comparative biology and its theoretical structure (which has only been rigorously formalized in the last two decades), it is not surprising that a rigorous view of evolution of science has been long in coming. This view, as it has developed in the works of Kuhn and Hull, among many others, is in need of further theoretical and empirical work.

In the theoretical realm, careful consideration is needed of the effects of various complicating factors known from the study of biological evolution. The models of selection discussed above are rather simplistic as compared to real situations known in biology wherein several levels of selection operate simultaneously and often in opposing directions. The presence of possible analogies to developmental or physical constraints should be investigated.

The meaning of heritability as applied to possible replicators in science similarly needs to be addressed, as does the process of "reproduction" itself. One disanalogy between science and evolutionary biology (mentioned above) is the lack of comparable transmission mechanisms for memes and genes. A second disanalogy is that in biology, organismic reproduction occurs only via production of new individual organisms, while in science, scientists reproduce either via production of "new" scientists (i.e., training of students) or via conversion of an existing scientist. The effects of such conversion on standard models of selection may be interesting; one could well imagine an accelerating effect on evolutionary rates.

In standard formulations of selection processes in evolutionary biology the raw genetic variation in populations is "random" or neutral with respect to possibly beneficial traits (Sober 1984). In the conceptual evolution of science, however, the origin of memes often occurs in a directed, "Lamarckian" manner (Boyd and Richerson

1985). Such a bias in production of raw memetic variation towards adaptive solutions would also be expected to have a accelerating effect on evolutionary rates.

In the empirical realm, well worked-out examples are needed. Despite Hull's descriptive efforts in the field of systematic biology itself, there remain no explicit applications of the analytical tools of cladistics or population biology to the evolution of science (or, for that matter, to other academic disciplines, where the same tools and concepts may be relevant). Such applications will present novel problems and opportunities. We have better sources of evidence for cohesion and interaction within and between lineages than are usually available for organisms (e.g., the written historical record), yet synapomorphous traits (ideas) marking lineages may be difficult to discover and describe given the speed with which transformation occurs in scientific ideas and the significant amount of inter-lineage borrowing that goes on in science. Selection processes in science should be relatively easy to discover, as long as care is taken to untangle causation at specific levels in the hierarchies of replicators and interactors.

It appears that a successful investigator of the process of science is going to need extensive training in the theory and methodology of comparative biology as well as in more standard training in philosophy and historical methods. Perhaps no one currently has sufficient strengths in all these areas to produce a rigorous study; extensive inter-lineage borrowing is clearly indicated. I just happen to have a place in my lab for any philosopher of science who wants to try.

Note

[1] I thank V. Albert, R. Brandon, A. Gutierrez, C. Horvath, D. Hull, and S. Rice for comments on the manuscript and discussion.

References

Boyd, R. and Richerson, P.J. (1985), *Culture and the Evolutionary Process*. Chicago: University of Chicago Press.

Brandon, R.N. (1990), *Adaptation and Environment*. Princeton: Princeton University Press.

Dawkins, R. (1976), *The Selfish Gene*. Oxford: Oxford University Press.

Donoghue, M.J. (1990), "Sociology, Selection, and Success: A Critique of David Hull's Analysis of Science and Systematics", *Biology and Philosophy* 5: 459-72.

Gould, S.J. and Lewontin, R. (1979), "The Spandrels of San Marco and the Panglossian Paradigm: A Critique of the Adaptationist Programme", *Proceedings of the Royal Society of London, B* 205: 581-98.

Griesemer, J.R. (1988), "Genes, Memes, and Demes", *Biology and Philosophy* 3: 179-84.

Hennig, W. (1966), *Phylogenetic Systematics*. Urbana: University of Illinois Press.

Hull, D.L. (1976), "Are Species Really Individuals?", *Systematic Zoology* 25: 174-91.

_____. (1980), "Individuality and Selection", *Annual Review of Ecology and Systematics* 11: 311-32.

_____. (1988), *Science as a Process: An Evolutionary Account of the Social and Conceptual Development of Science*. Chicago: Chicago University Press.

Laudan, L. (1977), *Progess and its Problems: Towards a Theory of Scientific Growth*. Berkeley: University of California Press.

Mishler, B.D. (1987), "Sociology of Science and the Future of Hennigian Phylogenetic Systematics", *Cladistics* 3: 55-60.

_____. (1989), [Untitled review of Hull, D.L. 1988, Science as A Process.] *Systematic Botany* 14: 266-268.

_____. (1990), "Species, Speciation, and Phylogenetic Systematics", *Cladistics* 6: 205-209.

_____ and Brandon, R.N. (1987), "Individuality, Pluralism, and the Phylogenetic Species Concept", *Biology and Philosophy* 2: 397-414.

_____ and Donoghue, M.J. (1982), "Species Concepts: A Case For Pluralism", *Systematic Zoology* 31: 491-503.

Nixon, K.C. and Wheeler, Q.D. (1990), "An Amplification of the Phylogenetic Species Concept", *Cladistics* 6: 211-23.

Patterson, C. (1982), "Morphological Characters and Homology", in *Problems of Phylogenetic Reconstruction*, K.A. Joysey and A.E. Friday (eds.). London: Academic Press, pp. 21-74.

Roth, V.L. (1988), "The Biological Basis of Homology", in *Ontogeny and Systematics*, C.J. Humphries (ed.). New York: Columbia University Press, pp. 1-26.

Sober, E. (1984), *The Nature of Selection: Evolutionary Theory in Philosophical Focus*. Cambridge, MA: MIT Press.

_____. (1988), *Reconstructing the Past: Parsimony, Evolution, and Inference*. Cambridge, MA: MIT Press.

Stevens, P.F. (1980), "Evolutionary Polarity of Character States", *Annual Review of Ecology and Systematics* 11: 333-58.

Wiley, E.O. (1981), *Phylogenetics: The Theory and Practice of Phylogenetic Systematics*. New York: John Wiley.

The Function of Credit in Hull's Evolutionary Model of Science

Noretta Koertge

Indiana University

1. Evolutionary Models and the Demarcation Problem

David Hull's book (1988) provides an evolutionary account of the development of science which pays attention to both the social and conceptual aspects of that process. Unlike most philosophers who only invoke Darwinian metaphors in a casual way, Hull takes the analogy between the biological evolution of species and the growth of scientific knowledge quite seriously and by providing abstract definitions of terms such as *replicator*, *interactor* and *lineage*, he makes it possible for us to see clearly the structural similarities between the two historical processes.

Other symposiasts will comment on how tight that analogy really is. I must remark in passing that I have never understood the intense interest which evolutionary epistemologists take in this comparison. Surely our major job is to understand how science works, perhaps by using evolutionary theory as a fallible heuristic, but nothing seems to hinge on the extent of the formal analogy. My main concern would be with the extent to which the evolutionary analogy illuminates what is distinctive about *scientific* development as opposed to other branches of intellectual history.

For example, Elaine Pagel's account (1979) of the struggle between the early Gnostics and what we would now consider the more orthodox variants of Christianity can be easily cast in quasi-Darwinian terms as follows: In the first few centuries after the death of Jesus, two important theological systems (cf. genotypes) existed. Each had its own gospels. The four orthodox gospels (Matthew, Mark, Luke, John) spoke of the bodily resurrection of Christ. The gnostic gospels (e.g. those of Mary Magdalene and Philip) on the other hand, spoke only of spiritual resurrection and implied that the post-crucifixation sightings of Jesus were really mystical visions, not observation reports. These two theologies were embodied in and had influence on the actions of two groups of Christians (cf. phenotypes). The orthodox group were more successful in gaining converts, not because of any intrinsic theological superiority, but because their doctrine of apostolic succession (passed down from Peter through ordination) gave their movement more stability. The leadership roles amongst the gnostics were more fluid because they depended on the charismatic and visionary powers of individuals, qualities which were much more difficult to operationalize. As a result, gnostic Christianity literally went extinct - there were no extant replications of many of the gnostic gospels until the discovery in 1945 of a bunch of manuscripts in a jar in Egypt.

There is an obvious moral to be drawn from this example. The mere fact that certain ideas have survival value does not tell us what those ideas were selected for. Urban legends or jokes survive because of their entertainment value. Myths survive because they allay fears and make us feel that we understand the meaning of life. There is no reason whatsoever to believe that those ideas which survive provide descriptions which "fit" the natural world. Quite the contrary. Soothing belief systems which are unfalsifiable often manifest an extraordinary degree of reproductive fitness!

Pagel's story also shows us why it is fruitful to incorporate a rough parallel to the phenotype/genotype distinction into an evolutionary account of the history of ideas. Although the gnostic system may have been theologically superior, it lost out to the orthodox system because of the superior political organization of the followers of Peter. I conclude that any adequate evolutionary account of the development of science will have to include an explicit account of which properties of theoretical ideas (e.g., empirical adequacy) play a crucial role in the process of selection. One also needs to discuss the extent to which scientific institutions provide mechanisms which insure that it is the most meritorious scientific ideas which in fact survive. What must be *added* to the general evolutionary account are the scientific norms which Hull summarizes by the slogan of *Curiosity, Checking, and Credit*. Theologians do not, in general, place high value on any of the three C's. But scientists are carefully trained to do so and it is these norms that determine *how* the struggle between competing scientific theories is to be conducted and differentiate the discussions of the Nobel Prize Committee from religious discussions about who should or should not be canonized. (I am not denying that politics plays a role in each, but I do claim that the standards of appraisal differ.)

So I now want to turn away from the Spencerian swamp of evolutionary epistemology and concentrate instead on what Hull says about the *distinctive* values and social practices of science, especially what he says about the roles of competition, cooperation and credit in organized science and how they contribute to the growth of scientific knowledge.

2. The Proximate Function of Credit

A significant portion of Hull's book is devoted to a detailed description of the public and private scheming and infighting which went on amongst various schools within systematic zoology. I read him as making two major claims: (i) Fights over credit and priority are an integral (maybe even an essential) part of scientific inquiry and (ii) his evolutionary model can help us understand why. I'll discuss these claims in turn.

Many commentators have taken note of the virulence of priority disputes and other forms of striving for recognition in science. Some attempt to explain it as a socially constructed response in a capitalist, patriarchal society. However, few have tried to argue that this is a functional aspect of science. Before I read Hull, I think if someone has asked me about the significance of the emphasis which scientists place on the ways in which credit is apportioned in science, I might have said something like the following: The "real" value/purpose of footnotes is to direct the interested reader to a place where they can get additional information on the subject. *Who* did the experiment or formulated the theory is "really" not very important, although it does provide a way of assigning responsibility for bad work. It is also sometimes important to know how many different laboratories have replicated a particular experiment. As far as credit is concerned, counting citations, etc. is something which only sociologists, tenure/promotion committees, funding panels, and other folks who haven't time to *read* the literature have any need to do. "Real" scientists are motivated by curiosity and the joy of problem solving; the essential ingredients of a scientific community are the traditional Mertonian norms of universalism, disinterestedness, and communalism. All of this striving for personal recognition is at best peripheral to the process of science.

As for priority disputes, I would have said that these are the legitimate concerns only for patent lawyers and hero-worshipping historians (still under the spell of romantic theories of genius). A "real" scientist is passionately concerned that the solutions to important problems be found and checked by others, but it can't possibly "really" matter to science whose discovery was first. Maybe sociobiologists can explain why (male) scientists have so much "paternity anxiety" or worry so much about who was the first to score the big "breakthrough". All of this is embarrassing nonsense which may tell us something about the pettiness or immaturity of scientists (which is exacerbated by today's funding squeeze) but has nothing to do with science *comme il faut*.

However, after reading Hull I am prepared to seriously entertain the idea that the credit practices of scientists cannot be ignored by any adequate theory of science and that credit does play an important role in the furthering of intellectual progress in science. I may be less optimistic than David, however, about how efficiently our present credit system functions. And I also think we need to have a good analysis of exactly how the present reward system benefits science. It is not good enough just to say science is doing pretty well so our present credit system must be O.K.

3. Why Curiosity and Checking Are Not Enough

To begin our analysis of how the reward system of science works, let us do a typical political philosophy thought-experiment in which we start out with isolated individuals and then assemble them into an efficient, scientific community. What new motivational ingredients would we need to inculcate? What special social norms would need to emerge? And since we are focusing on credit, let us also assume that our individual scientists are already well-equipped with curiosity and are already personally intrigued by the various types of problems which trigger scientific inquiry - problems arising from violated expectations, unexplained regularities, fragmented bodies of knowledge, etc. Let us also assume that they already have a propensity to subject proposals to critical scrutiny and realize that it is a good idea to have independent, skeptical collaborators to scrutinize observation claims, come up with cogent objections to other people's theories, etc. So, the present thought-experiment presupposes that the institutions which facilitate scientific curiosity and empirical checking are already in place.

We now ask, why isn't this enough? Why do we need to add a concern for individual credit in order to make our *New Atlantis* work? The general answer is, I think, fairly simple. We want scientists to solve *new* problems, ones which no one yet knows the answer to, and we want them to *publish* their solutions. Lest we take all of this for granted, we should remember that children routinely satisfy their curiosity and hone their problem-solving skills by re-discovering Archimedes' Principle or playing with Rubick's cubes. One tension in science education is how to teach students the skills necessary for and the satisfactions of solving problems by themselves (in which case the novelty of their solutions is unimportant) while also encouraging them to look up answers to questions in authoritative reference works and to devote their energies to working on new projects. And there are educated adults, many of them excellent college teachers, whose active curiosity makes them life-long readers of "Great Books" but who have few aspirations to make novel contributions. Satisfying one's personal curiosity does not insure communal progress. For the latter to occur, we need to make easily available combined communal information (e.g., libraries), we need to insure that people work on genuinely new problems (e.g., by requiring literature searches), and we need to reward people for actually publishing any solutions which they obtain, instead of secretly gloating that they know something which no one else does (hence, the publish-or-perish ethos).

Hull points out that in the past, scientists were often reluctant to make their results public. There was (and is) a strong tradition of passing on craft skills and secrets only to apprentices (e.g., alchemy, Stradavarius' violins) and until the patent system was developed it would be silly to divulge technological innovations. But gentlemen scientists also buried results in desk drawers out of laziness, or caution, or failure of nerve - or because the incentives and opportunities to publish were deficient. For example, here is Dijksterhuis' commentary on his countryman, Isaac Beeckman:

"Beeckman showed the same defects in the matter of science as Leonardo da Vinci. Both were deficient in the tenacity of purpose and powers of concentration required to systematize, finish, record, and publish their inquiries, even if only in one field. Of Faraday's motto: 'Work, Finish, Publish', they only took to heart the first injunction. In consequence they either did not advance science at all, or at least to a much smaller extent than they might have done..." "We shall see more of Beeckman's independent and frequently original way of thinking later: it is to be regretted that this candle never stood on a candle-stick." *(Dijksterhui's 1961, pp. 330-33)*

Although today we tend to think that it is "natural" to want to solve problems no one else has ever solved before and to get public credit for it, I think even a brief look at the early history of science and especially at traditional societies (cf. the description of resistance to new agricultural methods in Kemal's *Mehmet, My Hawk*, 1979) reminds us that such a drive is not to be taken for granted and must in fact be carefully shaped through scientific institutions. Every human being may be curious and want some kind of recognition from peers, but the kinds of things scientists get curious about and the kinds of credit that they find rewarding are both unusual tastes which are probably acquired.

4. Credit in *New Atlantis*

Let us now look in a little more detail at how the credit system in science works so that we can eventually ask how efficient it is in fostering scientific progress. Hull's description of the publishing/citation system quickly reveals just how complex the credit system is. Perhaps we can begin to analyze and evaluate it by looking at how credit considerations enter in at each step of the scientific process as philosophers would describe it. (Here I follow a quasi-Popperian schema.) Again I will adopt a thought-experiment strategy. Let us assume that scientists have the mundane *proximate* aim of maximizing personal recognition. How well will the behavior appropriate to such an aim coincide with the traditional *ultimate* aim of understanding the universe? In the sketch which follows I will emphasize the congruence between these goals (because that is what I found surprising).

(i) *Choice of problem*: If our immediate aim is to get published in scientific journals, we should choose problems which haven't been solved yet, but which are ripe for solution. (It is generally difficult to publish unsuccessful solution attempts or interim reports.) It may be wise to form a team so as to be able to tackle problems which others aren't equipped to solve and in order to solve problems more quickly, but this means we'll have to share credit with our co-authors. We also will need to be able to assess the competence of prospective teammates. We should also choose a problem whose solution will be of interest to our peers (otherwise they won't cite our work). This tends to lead to a *clustering* of research efforts around hot topics which means there is more data/theoretical speculations around that topic for everyone to use. But it also promotes a healthy *division of labor* because it encourages research teams to choose not just problems which they have a good chance of eventually solving, but ones which they also have a good chance of solving *first*.

(ii) *Working out a tentative solution*: Since the first publication often gets the most positive citations, we must work rapidly and secretly, especially if other individuals or teams are pursuing similar lines of inquiry. This is a time for team camaraderie, brainstorming, and the constructive criticism of conjectures. We will look for promising helpful hints while refereeing our competitor's grant-proposals or even their submitted journal articles (although note that in scientific journals submission dates are published). On the other hand, we will be reticent to share preliminary results with anyone who might scoop us. This will also protect our own reputations if the conjecture we're working on turns out to be way off-base.

Once we have a solution which has passed preliminary appraisals, we must decide when to publish it. This is a complicated choice. The reasons for publishing as soon as possible are obvious: if we are right, we want to get credit for being first. However, as Hull emphasizes, there are also lots of reasons not to rush into print. If we are quickly shown to be wrong, our reputations are likely to suffer somewhat. (This non-Popperian attitude towards refuted bold conjectures has the function of pruning the literature a little bit. Note that the greater the reward for being first, the greater should be the penalty for being wrong if the literature is not to deteriorate.) There is yet another consideration: if our conjecture is correct, it will generally lead to other lines of productive research. By temporarily delaying publication, we can explore these ramifications at our leisure and publish everything at once!

(iii) *Appraisal of the tentative solution:* Ignoring for the moment the pre-publication networks in science, the first hurdle that our tentative solution has to pass is the journal review process. In order to function well, the institutions which regulate publishing in science have to balance a variety of desiderata. Science (and the public) benefits when results are published, so there must be opportunities and incentives to publish. On the other hand, it is imperative to maintain quality control over what is published, so one needs to prevail on experts in each field to take time off from their own research to referee articles. Why should they consent to undertake these time-consuming and often unpleasant activites? Actually, as any journal editor knows, not everyone does consent. It is in every scientist's cognitive interest to keep the communal knowledge store as reliable as possible but are there any mundane (credit-related) reasons for doing so? Well, as I pointed out above, it's always nice to have advance knowledge of what other people working in your area are up to. Furthermore, an excellent way to make sure your own ideas are taken seriously (thus gaining you credit while increasing their fitness) is to eliminate, or at least point out the weaknesses in, rival viewpoints.

The form of appraisal most emphasized by philosophers is that of varied and severe empirical testing, but a scientist looking for professional credit will not spend time performing experiments which are unlikely to result in a significant number of citations. So routine replications are out, tests of theories which are of low interest are out, even refutations of other people's popular theories will be of rather low priority (because they will probably not cite your results except to explain them away) *unless* the refuted theory is in direct competition with your groups' own pet conjecture in which case your allies will cite it extensively. (One is reminded here of Lakatos' cognitive claim that there are no refutations, only superceded research programmes.) Under the credit system, bad theories don't die - no refutation of them may ever appear in print; they merely fade from view as their competitors get more citations.

5. The Problem of Tempering the Credit System

In his book Hull emphatically debunks the romantic myth of scientist as the objective, altruistic problem-solver whose only interest is that Nature be understood (and no matter who wins the Nobel Prize for being the first to probe her inner-most secrets). Scientists, like everyone else, want credit for their successes. However, one should not draw the cynical conclusion that since scientists *qua* scientists are motivated by mundane ambitions, the products of their inquiry have no special cognitive status. This would be like arguing that since business men and professional athletes are both "in it for the money", it makes no difference whether you put Donald Trump or Magic Johnson on the basketball court! The crucial question is not whether scientists want credit; what matters is *which activities* they in fact receive credit for. Do the proximate rewards reinforce the ultimate aims of science? Can scientists "do well by doing good" *science*?

In the above thought-experiment, I followed Hull in emphasizing the nice fit between what we might call the proximate mundane goals of professional success and the ultimate noble aims of the search for scientific understanding. Yet philosophy of biology reminds us how easy it is to make up "just so" stories which render any trait you like adaptive. And philosophers of social science have taught us to be skeptical of easy functionalist analyses which emphasize the beneficial effects of the potlach, cargo cults, sacred cows, primitive warfare and witch burning. Could we not also tell a pessimistic story about how the lust for quick publications and citations discourages scientists from tackling difficult problems which would take a long time to solve but which are nevertheless important? About how too much emphasis on credit can lead to the exploitation of graduate students, the mistreatment of laboratory animals, irresponsible methodological shortcuts, the practice of publishing virtually the same article in several places, unfair hiring practices, even outright fraud?

Even Hull's own optimistic account indicates that the balance between the cooperative and competitive aspects of science is a rather fine one. We should remember that some honorable professions are not so lucky as science seems to have been so far. The qualities and behaviors required to be a successful politician in an age of TV elections are almost contrary to those which contribute to statesmanship. And there are fewer professional incentives for doctors to stay abreast of new medical developments (unless their patients read about them in the popular press and demand them) than there are for scientists to keep up in their fields. When there is a dissonance between the success structure and the internal aims of the profession, such as exists in medicine and politics, we need to focus on institutional reforms, where the direction of the reform is dictated not by the selfish motives of individual practitioners but by the internal aims of the profession (which are why society values it in the first place).

Or consider the case of professional sports, which is like science in apparently having some congruence between mundane success (reflected in salaries) and internally defined excellence (extraordinary sporting performances). It would at first appear that even if athletes were only out for money, they would have to play just as well. So one might argue that mundane motivations are sufficient for and do not harm sports as long as there is a strong correlation between salary and batting averages. Yet perhaps it is not just romanticism which makes us suspicious of this cozy conflation of the sacred and the secular. What if coaches become reluctant to call for a sacrifice bunt (because players want to keep their averages up)? What if salaries come to depend on a player's charismatic box-office appeal, not just on box scores? Won't people playing primarily for money be easier prey for point-shaving deals with gamblers?

The general point is this: When a profession's reward system is consonant with the goals of that profession it is indeed possible to do good by doing well. But we should never forget the primary importance of doing good. Any congruence between careerism and love of the professional activity for its own sake is precarious enough that we are ill-advised to abandon our romantic-sounding rehearsals of the internal aims of science (or sports!) when we are educating our students - or representing science to a lay public. We sometimes forget that institutionalized invocations of rather high-sounding ideals such as truth-for-its-own-sake can also be functional. I once asked a seminar of graduate students how difficult it would be for them to completely fabricate their Ph.D. dissertation and get away with it. After they got over their initial shock, many of them answered that it would be quite easy. Well, why don't you do it, I asked. There was an embarrassed silence and finally the political scientist, whose survey research project we all agreed would be the easiestto fake, answered: "Because it wouldn't be any fun! I really want to know what my experimental subjects think!" Scientists want credit, yes. But what they want credit *for* is discovering interesting truths. It's the last part that most sociologists miss entirely. David Hull doesn't miss it — but perhaps he and I disagree on how important it is to keep harping on it. (But of course his book was published *before* Colorado used five downs to win a game!)

6. Conclusion

Hull hoped that by introducing the phenotype/genotype distinction he could improve on previous evolutionary models of science and give a unified account of both the conceptual and social aspects of the scientific process. Instead of viewing scientific ideas as disembodied propositions struggling for survival in a Popperian World - 3 (1972), Hull wanted to instantiate them in real flesh-and-blood scientists who competed for grants, graduate students, and glory.

In biological evolution, adaptation occurs when genes exert a causal influence on the reproductive success of the phenotypes which house them. The hope was that in science the cognitive merits of scientific ideas would directly influence the professional success of scientists which would in turn determine the extent to which those ideas would be accepted as true. To put it crudely, people who hit upon true theories would be more apt to make successful predictions, produce lots of experimental results and win Nobel prizes. It would then be their theories (possibly named after them) which would be cited in all subsequent science books. But as the example from the history of religion shows, the mere fact that the bearers of certain ideas are very successful in gaining converts who perpetuate these ideas through many generations tells us nothing about the empirical adequacy of those conceptual systems. Neither does the introduction of a citation/credit system help solve the demarcation problem. The Biblical interpretations of the Church fathers are much cited by later theologians and T.V. evangelists compete and get credit for numbers of souls saved or number of dollars raised.

Perhaps *artificial* selection would serve as a better model for epistemologists. Pigeon fanciers select for a wide variety of idiosyncratic properties - homing instincts, speed, or big neck ruffs. The pigeons which survive "fit" only the fancies of the breeders and are not at all adapted to any natural niche. People select ideas for all sorts of attributes - for their beauty, their whimsy, their ease of comprehension, their political correctness - and sometimes for their descriptive adequacy. (Post-modernists are right about one thing - science is not the only game in town!) So I conclude that Hull's evolutionary model does not describe what is distinctive about science. But what it does do is raise some very interesting questions about the nature of scientific institutions and how well adapted they are to the aims of the scientific enterprise.

References

Dijksterhuis, E.J. (1961), *The Mechanization of the World Picture*. Oxford: Oxford University Press.

Hull, D.L. (1988), *Science as a Process: An Evolutionary Account of the Social and Conceptual Development of Science*. Chicago: University of Chicago Press.

Kemal, Y. (1979), *Mehmet, My Hawk*. Denmark: Gyuldendalo Bogklub.

Pagels, E. (1979), *The Gnostic Gospels*. New York: Random House.

Popper, K.R. (1972), *Objective Knowledge: An Evolutionary Approach*. London: Oxford University Press.

The Evolution of Scientific Lineages

Michael Bradie

Bowling Green State University

1 Introduction

The fundamental dialectic of *Science as a Process* is the interaction between two narrative levels. At one level, the book is a historical narrative of one aspect of one ongoing problem in systematics - the dispute between cladists and more traditional evolutionary taxonomists and amongst the cladists themselves on the correct method of classifying species. This narrative is replete with details of the process whereby scientists promote and publish their ideas. It is an informative and somewhat 'racy' account of the rough and tumble battleground of ideas which puts the lie to the mythical ideal of the scientist as disinterested pursuer of the Truth. At the second level, Hull presents a theoretical model of the scientific process - a model which draws heavily on invoked similarities between biological and scientific change. The narrative serves as the evidence for the model. The model, in turn, helps shape the historical narrative. The triumph of Darwinism, according to Hull, is both the triumph of a particular view of nature and a particular view of the nature of science.

My remarks will focus on the theoretical model and one of its implications. I first want to situate the model as one alternative among several which loosely fit under the umbrella of 'evolutionary epistemologies.' Second, I want to explore one of the implications of Hull's model, namely, that insofar as scientific theories are [parts of] "conceptual lineages," they are "conceptual individuals." This has the rather unsettling consequence that "conceptual descent" turns out to be a more significant criterion of conceptual identity than structural similarity.

2. Evolutionary Epistemology and the Evolution of Science

Evolutionary epistemologies are broadly naturalistic approaches to the theory of knowledge which draw heavily upon evolutionary considerations to formulate models of conceptual growth.

There are two interrelated but distinct programs which go by the name "evolutionary epistemology." One is the attempt to account for the characteristics of cognitive mechanisms in animals and humans by a straightforward extension of the biological theory of evolution to those aspects or traits of animals which are the biological substrates of cognitive activity, e. g., their brains, sensory systems, motor systems, etc. I have labelled this the EEM (Evolution of Epistemic Mechanisms) program (see Bradie 1986). The other program attempts to account for the evolution of ideas, scientific theories and culture in general by using models and metaphors drawn from evolutionary biology. I have called this the EET (Evolution of Epistemic Theories) program. Both programs have their roots in 19th century biology and social philosophy, in the work of Darwin, Spencer, James and others. There have been a number of attempts in the intervening years to develop the programs in detail (see the bibliography and review in Campbell 1974). Much of the contemporary work in evolutionary epistemology derives from the work of Konrad Lorenz (1977, 1982), Donald Campbell (1960, 1974), Karl Popper (1968, 1972, 1976, 1978, 1984) and Stephen Toulmin (1967, 1972, 1974, 1981). Hull's concern in *Science as a Process* is with the changes in belief attendant upon human curiosity (an EET project) and not with the origin of human curiosity (an EEM project).

Although Hull wants to distance his project from the concerns of epistemologists, both traditional and evolutionary, his work is part of an ongoing lineage rooted in those concerns. It is a project inspired by the work of Toulmin.

Hull's complaint about most of what passes for evolutionary epistemology is that it tries to be "epistemology." For Hull, traditional epistemology is bankrupt and not to be emulated. On Hull's view, the contents and methods of science cannot be "justified" *a la* the aim of traditional epistemology. But evolutionary epistemologies, while in the tradition, do not necessarily address the traditional issues.

There are three possible configurations of the relationship between evolutionary and traditional epistemologies. (1) Evolutionary epistemologies might be conceived as addressing the problems posed by traditional epistemologies (justification, skepticism, the definition of "knowledge," etc.) and offering competing solutions to them. Riedl (1984) defends this position. Hull's model for the process of science does not qualify as evolutionary epistemology in this sense. (2) Evolutionary epistemology might be seen as complementary to traditional epistemology. That is, one might defend the legitimacy of traditional problems of epistemology but hold that evolutionary epistemology addresses different but complementary issues (concerning, e. g., the growth of knowledge). This appears to be Don Campbell's view. (3) Evolutionary epistemology might be seen as a successor discipline to traditional epistemology. On this reading, evolutionary epistemology does not address the questions of traditional epistemology because it deems them irrelevant or unanswerable or uninteresting. Many defenders of naturalized epistemologies fall into this camp and Hull's work also fits under this heading. In any case, Hull's model is a descendant of a selectionist model of scientific change first proposed by Stephen Toulmin.

The core thesis of Stephen Toulmin's Human Understanding is a commitment to what Toulmin considers a form of epistemological Darwinism.

> Darwin's populational theory of 'variation and natural selection' is one illustration of a more general form of historical explanation; and ... this same pattern is applicable also, on appropriate conditions, to historical entities and populations of other kinds. (Toulmin 1972, p. 135)

Science, according to Toulmin, develops in a two-step process analogous to biological evolution. At each stage in the historical development of science, a pool of competing intellectual variants exists along with a selection process which determines which variants survive and which die out. (Toulmin 1967, p. 465)

On Hull's view, neither biological evolution nor the growth of knowledge serves as the primarily model in terms of which we are to understand the other. Hull prefers to develop a general analysis of "evolution through selection processes which applies equally to biological, social and cultural evolution." (Hull 1982, p. 275; Hull 1988) Hull's rationale for treating both biological evolution and conceptual evolution as exemplifications of some common general selectionist model is to undercut objections to selectionist accounts of conceptual change which emphasize the disanalogies between biological and conceptual change. (Hull 1988, p. 418) Although the specific mechanisms of change are not the same in the two cases (Hull 1988, p. 431) and there is no clear evidence that there is any "significant correlation between genetic and conceptual inclusive fitness," (Hull 1988, p. 282f), Hull argues that both processes are exemplifications of a single selection model.

Hull's selection model is couched in terms of "interactors" and "replicators." Selection occurs as the differential proliferation (extinction) of replicators caused by the differential proliferation (extinction) of interactors. The abstract level of analysis is an attempt to avoid the misleading implications of particular selection models couched in terms of particular entities and processes. In sexually reproducing organisms, the interactors are the organisms themselves and the replicators are their genes or alleles. But, Hull argues, single cells (paramecium splitting) and multicellular organisms which undergo fission are candidates for being replicators as well. (Hull 1988, p. 414) Lineages are historical entities which result from replication. Thus, both genes (alleles) and organisms form lineages. Species are lineages if gradualism is true. If not, then species form lineages too. Other sequences of replicating entities, e. g., the HeLa cell line, form lineages as well. On Hull's view, all the crucial players in the model (interactors, replicators and lineages) are individuals. The interactors are the most ephemeral, the replicators are more long lived and the lineages are the longest lived of all.

Conceptual selection in science is one exemplification of the general selection model. The replicators are the ideas, themes, procedures, etc. that are passed on from one generation of scientists to another. The typical interactors are the scientists themselves and the books and articles that they write and publish. Scientists interact among themselves (when they read and criticize each other's work, check results and award or withhold credit) and with nature (when they make observations or perform experiments). The scientific process, for Hull, is driven by the "interplay between curiosity, giving and receiving credit for contributions, and the mutual checking of results." (Hull 1988, p. 431) As the results of a selection process, the resulting conceptual lineages are historical individuals.

How profitable is this analysis? This is, as Hull points out, an empirical question to be decided by more detailed empirical investigation into the processes of science. In conceding that "[t]he specific mechanisms involved in biological and conceptual evolution are quite different," Hull undermines some of the initial credibility of the similarity between biological and scientific evolution. For is this not the crucial point - that the specific mechanisms are not the same? In the case of specific theories such as Newtonian mechanics, the scope of the theory is extended by incorporating new phenomenological domains under the rubric of worked out examples or Kuhnian exemplars. In such cases, the same specific mechanisms, e. g., the force of gravitation or the spring force, are in play in both the old and new domains. However, in the present case, the situation is different. Either we are arguing by analogy from the biological to the conceptual or we are arguing from some common framework to both. In either case, as Hull admits, the specific mechanisms are not the same. This raises serious questions about the explanatory virtue of such a move. How strained do the connections between the mechanisms have to become before we concede that the "sameness" of process is *mere* "similarity?" Aristotle's characterization of motion as the actualization of that which is potential brought a wide diversity of phenomena under a single rubric but it did so at a price that, from our modern perspective and interests about motion, borders on the vacuous. In one sense, all (Aristotelian) motion exhibits the "same" features although the "specific mechanisms" can be quite different. But the important aspects of different kinds of motions, at least to our contemporary way of viewing things, lies in the details and the differences. Can Hull's view provide that? Only time will tell. Hull would, no doubt, agree that much significant conceptual work needs to be done along with the empirical case studies.

3. Species and Conceptual Lineages as Individuals

One of the main messages of the book is that "species, if they are to play the roles assigned to them in evolutionary theory, must be treated as historical entities." (Hull 1988, p. 79) It follows, Hull argues, that species are individuals and not natural kinds. The correlative implication for conceptual evolution is that "[j]ust as species cannot be treated simultaneously as historical entities and as eternal and immutable natural kinds, neither can concepts." (Hull 1988, p. 17)

I want to consider two questions in this regard. First, does the fact that species are historical entities entail that they are individuals and not natural kinds? And second, how plausible is it to construe conceptual lineages as individuals?

3.1 Species as Individuals

Consider the case of the elements. They are natural kinds structurally defined. Even if all the referents for hydrogen, e.g., disappeared, that is, if all the hydrogen atoms in the universe ceased to be, a "slot" for Hydrogen would remain and any new entity created with the appropriate structure would qualify as Hydrogen. Why not the same for species? On this view we would construe the sense-species of, say, human being as eternal and if all the current referents disappeared then it would remain as a "slot" etc., etc. Hull rejects this option and claims that other Darwinians (including Charles Darwin himself) now reject it as well. The ground for rejection is that species taxa are evolving individuals that are spatio-temporally limited existents. New "humans" would not have the appropriate genealogical connections and, thus, would not count as bona fide humans. But, why is this not simply a problem of re-qualification for human status? and not something essentially related to individuality?

Suppose Jones, after an imprint matrix of his being is taken, suddenly dies. A new "Jones" is constructed with appropriate memories, preferences, desires, personality, etc., but, of course, a lapse of memory for one crucial part of "his" life. Is it the same individual or not? If "physical" continuity is crucial, no. But, physical continuity is never enough as witness cases of multiple personalities. If "psychological" continuity is crucial, perhaps it is the same person. Again, psychological continuity is not everything either as cases of amnesia illustrate.

Why not say that the 'genealogical requirement' just makes it that much harder for 'new' members to qualify as members of the club but does not necessitate that species taxa have a different ontological status from physical elements? Which comes first? Do we first note the genealogical connections and then infer species taxa are individuals or do we argue that since they are individuals, so the genealogical requirement must be met? Presumably the former, in which case if we can make out a case for satisfying the 'genealogical requirement' and not being an individual, then the inference from 'X is a historical entity satisfying a genealogical requirement' to 'X is an individual' will be blocked.

If, indeed, there is no difference in principle between human beings and hydrogen atoms, we should say that just as there are atom particulars and atom kinds, so there are human particulars and human kinds. It is just part of the vagaries of biological existents that they are less easily come by than their atomic counterparts. Hydrogen atoms come about "spontaneously" -they are not spawned by other Hydrogen-atoms. But, suppose spontaneous generation amongst biological organisms were more widespread than we now think it is? Wouldn't this raise problems for Hull's account? At the very least it would show the tight interconnection between metaphysical and scientific issues if a matter of fact can determine which ontological category a given entity belongs to. This reminds us that the descent and genealogical nexus of organisms is an empirical fact (or theory) about natural entities. These natural entities would still exist even if we turn out to have had fundamentally false beliefs about them.

3.1.1. A Science Fiction Fantasy in Four Scenarios and a Coda

Scenario 1: "Aristotle Redux"

Spontaneous generation is more widespread than we currently believe. The net effect is that "species" or kinds may or may not have the appropriate lineages (depending upon whether sex is obligatory or facultative).

Scenario 2: "Howdy, pardner"

The biospecies concept and the ethos of the Old West: It's bad form to ask a Stranger about his/her past. If they mind their own business and don't cause any trouble, then they're OK. The Stranger comes, woos, wins, weds, beds and fertile offspring ensue. The biospecies concept sorts individuals into different species according to whether they can (same species) or can not (different species) produce fertile offspring. Now, is Hull going to deny those children their birthright by challenging the pedigree of the Stranger? And, even if he did, wouldn't we call the Stranger "human" if it looked like, sounded like, tested out as, etc., etc., even if it didn't have the appropriate pedigree?

Scenario 3: "The Artificial Human"

Plot A: In virtue of advances in biotechnology, the parts of a "real" human being are replaced one by one with ingenious plastic substitutes which, through the miracles of modern science, work just as well as the originals. First the left thumb, then . . . Finally, the time comes for the final original part to be replaced. The new product looks like, walks like, talks like, etc.. Isn't it still human?

Plot B: As the transplantation process described above is being performed, 'baby' replicas are being assembled and maintained. At the last stage, we have a perfectly formed "human[?]" baby, capable of growth, development, etc. Suppose it does grow, develop, move west, woo, win, wed ... Isn't it human? No? Why not? Because it doesn't have the appropriate genealogical connections?? BUT, now we have to ask: is having the appropriate lineage so important as to outweigh all the other respects in which it does seem to qualify as human?

Scenario 4: "Gothian visions"

We are, alas, merely parochial existents in one bubble out of an infinity of bubble universes, each having its own 'Big Bang' and subsequent expansionist evolution. The laws of nature being what they are and the initial distributions being what they are, we need not assume that all these bubble universes will evolve in exactly the same way. Some collapse before they barely get started, others expand indefinitely, others oscillate forever. In some of those universes, an element that looks like our Hydrogen and reacts like our Hydrogen and has the 'same' structure as our Hydrogen exists. Is it Hydrogen? Natural kinds being what they are, we are supposed to say yes! But, these "Hydrogen atoms" don't share the appropriate genealogy with our Hydrogen. They didn't come from our Big Bang. Perhaps, but genealogy does not play a role for being Hydrogen like it does for being a Human Being. Aha, but that's because we didn't realize before that there was more than one universe and more than one historical lineage for elements. Now that we do know, history should make a difference, shouldn't it? If not, why not?

Why should we count the (hypothesized) ability of our Hydrogen to combine with their Hydrogen to form molecules of H_2 in such a way the two components are structurally indistinguishable as evidence that their Hydrogen really is Hydrogen while we are expected to discount the (hypothesized) ability of our human beings to mate with their "human beings" and produce fertile offspring as evidence that their "human beings" are "really" human? Is it because it is part of the meaning of "human being" (or of being a biological taxa) that its "parts" have the appropriate lineage? How true to Darwin is this? And don't we have essences back again?

Coda: "The Counterfactual Defense"

Of course, in general, Hull rejects such science fiction scenarios as philosophically irrelevant (Hull 1988, p. 28). Science and the philosophy of science are too interconnected, he holds, to allow for the relevance of "unconstrained science fiction" examples.

Hull rejects the use of contrary to fact conditionals to test conceptual limits: but philosophical claims about science, if they are to be testable [!?], must submit to empirical tests of physical possibility. Conceptual/physical possibility must be treated together and counterfactuals, to the extent they are used, must be structured within the possibilities of science as we now know it. But, aren't the scenarios described above within the bounds of the empirically possible? And don't they show that the categorical difference between lineages and kinds is not as sharp as one might suppose?

3.2. Conceptual lineages as individuals

Insofar as Hull's model takes conceptual systems to be analogous to biological species, it endorses the view that conceptual systems are evolving lineages. This has implications for the criteria of identity and individuation of conceptual systems. For species, the (evolutionary) criterion for conspecificity is descent not similarity of morphology. If we are to take the selectionist model of scientific change seriously, the criterion for being the same system or theory should be likewise descent not similarity in logical or conceptual structure. This is a radical thesis. It entails, among other things, that two individuals who hold structurally similar views but who do not share the appropriate causal nexus cannot properly be said have the same views. On the other hand, two individuals who do stand to one another in the appropriate causal relationships, but whose views may be structurally quite different, can be said to share the same theory, program or tradition. So, "Darwinians" are not individuals grouped together because of a common core of beliefs (which would constitute the "essence" of "Darwinism" - as an individual, the lineage of "Darwinians" has no essence) but rather the "Darwinians" constitute a coterie of individuals who have learned from, and interacted with, each other in appropriate ways. At the very least, such an approach promises to reshape the intellectual landscape of the history of science. It is a very interesting way to do history and I admit to being somewhat partial to it. However, it does produce counterintuitive results. Consider a doctrine D1 put forth by a scientist S1. S1 transmits his doctrine to S2 and in the process D1 is slightly modified to become D2. Imagine this transfer to continue for some time in whatever ways are causally appropriate to preserve continuity of tradition. All the Si's can proudly claim to D-ists. Now suppose doctrine D1 has the simple structure "p" where "p" is some declarative assertion. On the grounds that descent is (almost) everything for determining lineage identity, it does not seem inconceivable that, for some Sn, the form of the doctrine espoused could be "not-p." Now I, for one, would find it extremely odd to say that both S1 (who believes "p") and Sn (who believes "not-p") are endorsing the same view or even that they can properly be said to be in the same (intellectual) tradition.2 What blocks immediate assent is the radical structural dissimilarity between the views of S1 and Sn. Of course, on Hull's view, this is not relevant if the criterion of identity is the existence of some appropriate causal chain.

We have the following situation. There is a phenomena called the "scientific process" and we have competing models put forward to account for it. On what we may call the traditional view, the criterion of conceptual individuation is structural similarity of views. Two individuals hold the "same" views just in case the views they endorse are structurally similar in relevant respects. The traditional view has in its favor the commonplace that we distinguish Believers from non-Believers in terms of whether or not they accept the existence of God, regardless of how they come to hold such beliefs. On Hull's view, the criterion of conceptual individuation is causal descent. Two individuals hold the "same" views just in case they have interacted in appropriate ways regardless of any structural dissimilarity between their respective views. This, as I have suggested, is prima facie, counterintuitive. I do not suggest it is wrong, just a bit fishy.

The question is, how is the neutral observer supposed to decide which way to go? Do the cases where structurally dissimilar views which nonetheless stand in a relation of descent count against the view of conceptual systems as lineages or is it merely some aspect of our intuitions that we need to adjust in the light of or acceptance of the truth about conceptual systems? This is an open question to defenders of the "intellectual traditions as entities" view. A correlative question is what difference does it all make? We wind up reclassifying some scientists whom we took to be Darwinians to be otherwise and some others whom we took not to be Darwinians to be so, but so what? Is there more, and if so, what is it?

4. Conclusion

The selectionist model of scientific change advanced by Hull has a number of important and problematic consequences. In section 2, I raised some questions about the appropriateness of the selectionist model for understanding scientific change. In section 3, I raised some questions about the appropriateness of the distinction between "lineages" and "natural kinds," endemic to many contemporary interpretations of Darwinian theory and proceeded to explore some of the implications of treating scientific concepts and traditions as lineages.

With respect to the question of whether scientific concepts are "lineages" or "natural kinds," I suspect there may be something to both claims. There is an ambiguity about terms like "Darwinian" (or "Newtonian," "Lamarckian," "Dadaist," etc.). On the one hand, such terms label individuals who have been appropriately influenced by the named individuals or movements. In this sense, "Christians," e. g., are those who share a relationship of "apostolic descent" to the founding fathers of the faith, regardless of the disparities of their respective beliefs. On the other hand, such terms label individuals who share structurally similar views. In this sense, individuals who never heard of Darwin or anyone who was associated with Darwin, could be called a Darwinian, given that they held relevantly similar views. With respect to the historiography of science and the taxonomy of scientific doctrines it remains to be seen which reading is most appropriate or whether there is indeed room for both. To the extent that there is room for both, concepts can be construed as both "historical entities or lineages" and "kinds" structurally defined. But, if so, and if Hull is right about the fundamental similarity between biological and conceptual evolution, then we may suspect that biological species have a mixed pedigree as well.

The distinction that Hull and others draw between ecological and evolutionary perspectives in studying species may be helpful here (Hull 1988, ch. 11). Considered from an ecological perspective, a tradition or view is a kind. Seen from an evolutionary perspective, a tradition or view is a lineage. The case where a view evolves into its negation shows that the two ways of cutting up conceptual reality do not map onto one another in a one to one fashion. This, I think is both true and an interesting observation. There is, however, the following point. Scientific views or traditions function as individuals in both perspectives. A tradition, conceived ecologically, has the power to mold and change opinions and the views of others. To the extent this is correct, it points up an important difference between traditions and species. Species function as individuals when conceived from an evolutionary point of view but they do not do so when conceived from an ecological perspective (cf. Eldredge's analysis of the dual hierarchies in Eldredge 1985).

Finally, Noretta Koertge's paper raises the question of whether Hull's view helps to solve the demarcation problem. As Hull remarks in his response, the demarcation problem, which was a central problematic of the logical empiricist and Popperian philosophies of science, is nowadays somewhat out of fashion. The onslaught of social constructivist interpretations of science in the wake of Kuhn's historicist analysis has blurred the distinction between what is science and what is not. This is not surprising. The meaning criterion of the logical empiricists, which was to serve as the demarcation principle was essentially ahistorical. Even Popper, with his emphasis on the importance of the problem of conceptual change, relied on the fundamentally ahistorical criterion of falsifiability. Hull's analysis, which focuses on the evolutionary dynamics of conceptual change, holds no promise of a quick and dirty resolution of the problem of what distinguishes science from non-science. Indeed, insofar as science stands to other intellectual traditions as one species stands to another we should not expect there to be any sharp delineation. But this brings to the fore an unresolved tension in Hull's analysis. Whereas he wants to treat particular scientific traditions as evolving, non-essentialistic individuals, he is prepared to defend the view that science, as such, has an essence (Hull 1988, ch. 2). His response to Koertge reflects this: "other sorts of conceptual change . . . [might] . . . turn out to exhibit some of the characteristics of conceptual change in science . . . [and] . . .if a sort of conceptual change turns out to have all the characteristics of science fully developed [my emphasis], then it is science, common conceptions to one side." More work needs to done in spelling out the details of Hull's conceptualization of scientific process before we can be satisfied with this.

Notes

[1] Robert Richards, in his recent book, *Darwinism and the Emergence of Evolutionary Theories of the Mind and Behavior*, reaches a similar conclusion on the basis of a selectionist model of scientific change which he calls the NSM. For a discussion of Richards's book, see Bradie (forthcoming).

[2] Consider the following dialogue: She: I believe in God. He: I do not. She: Oh, I am so glad we share the same religious tradition.

References

Bradie, M. (1986), "Assessing Evolutionary Epistemology", *Biology & Philosophy*, 1: 401 - 459.

_____ . (forthcoming), "Darwin's Legacy", *Biology & Philosophy*.

Campbell, D.T. (1960) "Blind Variation and Selective Retention in Creative Thought and in Other Knowledge Processes", *Psychological Review* 67: 380-400.

_____ . (1974), "Evolutionary Epistemology", in *The Philosophy of Karl Popper*, Vol. I, P.A. Schilpp (ed.). LaSalle, Ill: Open Court: pp. 413-463.

Eldredge, N. (1985), *Unfinished Synthesis*. New York: Oxford University Press.

Hull, D. (1988), *Science as a Process: An Evolutionary Account of the Social and Conceptual Development of Science*. Chicago: University of Chicago Press.

Lorenz, K. (1977), *Behind the Mirror*. London: Methuen.

_____. (1982), "Kant's Doctrine of the a priori in the Light of Contemporary Biology", in *Learning, Development, and Culture*, H.C. Plotkin (ed.). Chichester: John Wiley & Sons, pp. 121-143.

Popper, K.R. (1968), *The Logic of Scientific Discovery*. New York: Harper.

_____. (1972), *Objective Knowledge: An Evolutionary Approach*. Oxford: The Clarendon Press.

_____. (1976), "Darwinism As A Metaphysical Research Programme", *Methodology and Science* 9: 103-119.

_____. (1978), "Natural Selection and the Emergence of Mind", *Dialectica* 32: 339-355.

_____. (1984), "Evolutionary Epistemology", in *Evolutionary Theory: Paths into the Future*, J.W. Pollard, (ed.). London: John Wiley & Sons Ltd, pp. 239-255.

Riedl, R. (1984), *Biology of Knowledge: The Evolutionary Basis of Reason*. Chichester: John Wiley & Sons.

Richards, R. (1987), *Darwinism and the Emergence of Evolutionary Theories of the Mind and Behavior*. Chicago: Chicago University Press.

Toulmin, S.E. (1967), "The Evolutionary Development of Natural Science", *American Scientist* 55: 456-471.

_ _ _ _ _ _ _ . (1972), Human Understanding: The Collective Use and Evolution of Concepts. Princeton, N.J.: Princeton University Press.

_ _ _ _ _ _ _ . (1974), "Rationality and Scientific Discovery", in *Boston Studies in the Philosophy of Science XX*, K. Schaffner and R. Cohen (eds.). Dordrecht: Reidel, pp. 387-406.

_ _ _ _ _ _ _ . (1981), "Evolution, Adaptation, and Human Understanding," in *Scientific Inquiry and the Social Sciences: A Volume in Honor of Donald T. Campbell*. M.B. Brewer and B.E. Collins (eds.). San Francisco: Jossey-Bass, pp. 18-36.

Conceptual Evolution: a Response[1]

David L. Hull

Northwestern University

1. Introduction

Each of the commentators on my *Science as a Process* has emphasized a different part of my book: Mishler concentrates on the relevant biology, Koertge the sociological mechanism, and Bradie conceptual and biological lineages as historical entities. Bradie (1986) has made the important distinction between two sorts of evolutionary epistemology—the evolution of cognitive mechanisms (EEM) and the evolution of theories (EET). On the first program, a biological, gene-based theory of evolution is extended to explain the development of cognitive structures. On the second program, a meme-based theory of conceptual evolution is modeled on the biological theory. In order for people to be capable of producing conceptual evolution, they have to be sensient, social, able to communicate, curious, etc. These characteristics evolved through a biological, gene-based selection process. As a result, EET takes place within the constraints imposed by EEM. As Bradie correctly points out, I have concerned myself almost exclusively with EET.

Koertge remarks in passing that she has never understood the intense interest that evolutionary epistemologists take in the comparison between biological evolution and the growth of scientific knowledge. From the perspective of traditional epistemology, I share Koertge's perplexity. If the issue is justification, EEM does not provide the sort of foundation for knowledge that epistemologists have sought. If cognitive abilities of our species have evolved via the mechanisms of biological evolution, then certain things follow about these abilities. First, they are likely not to be universally distributed among human beings. Variability is essential to evolution. Our blood type, eye color, resistance to disease, etc. vary. Why not cognitive abilities? And in the context of evolutionary theory, translating talk of *variation* into talk of *deviation* is illicit. Hence, epistemologists are denied reference to one of the staples of their literature—the "normal" observer.

Second, to the extent that reference to biological evolution can provide warrant for our beliefs, this warrant applies only to middle-sized entities and processes, when science deals with the universe at large. EEM cannot provide support for relativity theory or quantum theory. Even within the tiny slice of middle-sized entities and processes, biological evolution can lead us to hold false beliefs. For example, apparently the psychological imposition of sharp boundaries and internal homogeneity on biological species has been advantageous in our evolution, even if this belief is itself incompatible with our understanding of the evolutionary process. Evolution has made the understanding of evolution quite difficult.

I do not want to deny the relevance of biological evolution to our cognitive capacities and predispositions. I do deny that biological evolution can provide a warrant for our beliefs *tout court*. Holding false beliefs can be evolutionarily advantageous. Koertge is willing to entertain using evolutionary theory as a "fallible heuristic." It is at least this much. For example, the analogs to species in science are scientific groups and conceptual systems at the social and conceptual levels respectively. Species are internally variable. Sometimes there are greater differences within species than between closely related species. As a fallible, heuristic guide, we might try treating groups of scientists and conceptual systems in the same way. Perhaps scientists working together in the same group can disagree with each other even over essentials, their own claims to universal agreement notwithstanding.

My own goal is somewhat more ambitious. I think that various sorts of phenomena are all instances of basically the same sort of process, in particular biological evolution, the reaction of immune systems to antigens, the development of the nervous system, learning in general, and conceptual change in science in particular. I analyze selection processes in terms that are sufficiently general that they apply to more than just gene-based biological evolution. The issue then becomes the discovery of which of these (and possibly other) processes have what it takes to count as selection processes. Bradie reasons that if the specific mechanisms involved in a putative selection process are different (as they are in biological evolution and the reaction of immune systems to antigens), then they both cannot be instances of the same sort of process. I hate to see what effect this line of reasoning would have on the huge philosophical literature on the nature of functional organization. Bradie also points to the danger of making an analysis so general that it applies to anything and everything. This is a real danger, but I do not think that this point has yet to be reached with respect to analyses of selection processes.

As out of fashion as the principle of demarcation is nowadays, Koertge is interested in "what is distinctive about *scientific* development as opposed to other branches of intellectual history." I would not be surprised to discover that other sorts of conceptual change (e.g., the struggle between the Gnostics and more "orthodox" variants of Christianity) turn out to exhibit some of the characteristics of conceptual change in science. They might even exhibit all of them but developed insufficiently. However, if a sort of conceptual change turns out to have all the characteristics of science fully developed, then it is science, common conceptions to one side.

2. Selection Processes

In my book I analyze selection in terms of two constituent processes: replication and interaction. Replication is the transmission of information in ancestor-descendant sequences of entities largely intact. Entities are included in the same token trees or networks because of descent. Secondarily but *not* incidentally, these entities will also be similar. In interaction, entities interact with their environments so that the causally relevant replication sequences are differential. Interactors evolve properties (adaptations) that allow them to interact with their environments in very special ways. Ecologists group these adaptations into kinds and attempt to discover regularities in the processes within which they function. For example, wings are the things that can be used to fly, and eyes are the things that can be used to see. Ecologically these are the same characteristics even though genealogically they may have evolved independently several times. Selection is an alternation of genealogy and ecology. Ecological entities interact with their environments as *instances of kinds*, but then immediately they must reproduce, and these kinds are dismantled and integrated genealogically.

For example, organisms belonging to two species of mammal may live in the same valley, each rooting about in the dirt for their food. Ecologically they are indiscriminately rooters, but genealogically they are distinct (see Dupré 1990 and Hull 1990). Because inter-specific competition differs significantly from intra-specific competition, the selective fates of these two groups of organisms will differ even if they fill exactly the same ecological niche. This alternation of genealogical tokens and ecological kinds makes selection possible. It also makes selection processes very difficult to understand. This contrast may be what Ettinger, Jablonka, and McLaughlin (1990, p. 504) have in mind when they remark, "While it is true that only concrete entities (for example, individuals) reproduce, we shall argue that it is only abstract entities such as types, genotypes and alleles that can be fit."

One of the main messages of my book is that we tend to treat conceptual change as if concepts were ecological kinds, ignoring genealogy. The issue is not two *alternative perspectives*, but two *alternating processes*. Cannot we conceptualize species both as genealogical lineages and as kinds? Of course we can. However, we slip back and forth between the two perspectives so easily as it is that we hardly need to legitimate such equivocation. I have nothing against pluralism. Species can be conceived of consistently in a variety of ways. What I would like to see is a scientific theory in which such species play an important role. Might we not have a theory in which species function as natural kinds? Until such a theory has at least been sketched, I for one am not interested in such idle speculations. Parallel observations hold for concepts, but here the onus is on those of us who propose to treat conceptual change genealogically.

Koertge questions how closely connected my EET in terms of replication and interaction is to the social mechanism of curiosity, credit, and checking that I set out. I must admit that these two wheels do not mesh as nicely as I would like, but they do have at least one point of intersection—genealogy. If the social mechanism that I describe is to have the effects I claim it has, link-on-link conceptual connections are necessary. The most significant influence of the evolutionary part of my EET on the social mechanism I propose is the genealogical dimension it imposes.

3. Realism and Idealism

One of the distinctive features of science is the social mechanism that makes the pursuit of certain sorts of self interest on the part of individual scientists contribute to the realization of the traditional goals of science. By and large, the best way for scientists to further their own careers is to behave the way that they are supposed to. Koertge thinks I am too optimistic about how robust the coincidence between individual and general good in science actually is and that I should emphasize more strongly the idealism so common among scientists. How optimistic one is depends on one's contrast class. In comparison to how well the savings and loan industry has worked during the past few years in the United States, science looks very good indeed. In fact, science looks very good when compared to any other social institution, but it can go awry. I agree that we need to foster the idealism that so many young scientists bring with them as they enter science, but idealism alone is not good enough. Many doctors, police officers, and politicians may well have been extremely idealistic when they entered their professions, but in the absence of any social mechanism to encourage right behavior, the results are anything but heartening. Most scientists really are interested in their subject matter, almost pathologically so, many also have very high professional morals, but the point I wish to stress is that the general coincidence between individual and institutional goods adds another force to lead scientists to behave the way that they should. It also allows scientists who are not quite so saintly to contribute to the traditional goals of science.

A common complaint of viewing social institutions realistically is that a thick layer of hypocrisy is needed in order for society to work. If the agents were aware of what was actually going on, the system would cease to function. Without denying this general principle, I think that it does not apply to science. Scientists can come to understand how science works without disrupting the scientific process. In fact, periodic self-awareness might even make the system work better. For example, it is important that scientists run their experiments carefully and describe the results accurately, but it is also important that they present their results in ways that other scientists are likely to find relevant to their own work. Pick a dissertation topic that you think is important and interesting, but also pick a topic that at least one of the professors in your department thinks is important and interesting—preferably a productive, established professor. Viewing a life in science realistically does not preclude an idealistic perspective as well. It also increases the likelihood of survival.

4. Biological and Conceptual Evolution

What are the replicators in biological evolution? G. C. Williams' (1966, p. 25) answer is chunks of the genetic material, "any hereditary information for which there is a favorable or unfavorable selection bias equal to several or many times its rate of endogenous change." What are replicators in conceptual change? Various conceptual chunks that fulfill these same requirements. But must not conceptual replicators be absolutely discrete, all of the same size, and come in pairs? No. Since biological replicators frequently lack these characteristics, there is no reason to insist that conceptual replicators possess them.

What are the interactors in biological evolution? Everything from genes through cells and organisms to colonies, demes and possibly entire species. What are the interactors in conceptual evolution? To answer this question, we need something like the distinction between genome and phenome, homocatalysis and heterocatalysis, and transcription and translation (Griesemer 1988), and on this issue Mishler and I parse the two contexts differently. Perhaps Mishler's alternative will pan out better than mine. We will have to wait to see.

On my view, like produces like in replication (transcription). Germ-line genes produce other germ-line genes, but periodically they also produce genes that are shunted off into somatic cells. In translation, a gene contributes to the production of something that is different from but systematically related to itself—a protein. One of the noteworthy features of translation is the amount of information that is lost. Proteins are produced by sequences of genes, each interacting in complicated ways with its environment, including other genes, to produce the protein. Of the numerous proteins that could have been produced only one is actually produced. A genome gets tested in terms of only one of the potentialities included in its reaction norm.

Interaction occurs in conceptual evolution when testing takes place. Ideas can be transmitted from one agent to another in a variety of ways, but periodically, when all else fails, we evaluate our beliefs by seeing how well they fit nature. Theories in their totality are never tested. Instead, limited consequences are derived from them. These consequences are then "operationalized" to make them testable. Of all the consequences of a particular theory, only a very few will ever be tested. Scientists are interactors in the sense that they orchestrate these comparisons. Conceptual replicators do not produce scientists. Rather, the beliefs that scientists hold are elements in conceptual replication sequences. Because scientists hold certain beliefs, they perform certain tests and not others. The relevant traits (phenotypes) are the results of these tests. The litmus paper changes color. To put it as crudely as possible, the conceptual genome is made up of the ideas transcribed in a scientist's brain while the conceptual phenome results from scientists' interacting with nature as they test their beliefs. Scientists do not fit nature; their beliefs do.

5. The Social Dimension of Science

In my book I emphasize that the interplay between conceptual replication and interaction does not occur in isolation from all else. Scientists interact with each other as well as with the parts of the natural world that they are studying. Professional relations among scientists are also important. The chief reward in science is the use of your work by other scientists, and it is this mutual use that drives the process of self-correction that is so central to science. What matters in science is not individual fitness, but inclusive fitness—conceptual inclusive fitness. I spent so much time in my book on the professional interactions among scientists because I had not expected them to have so much epistemic relevance. The system of credit for contributions that characterizes science not only *is* epistemically relevant, it *should* be. It is one of the main reasons why the knowledge claims that scientists make are credible. Because of all the attention that I pay to the professional relations among scientists, some people have read me as advocating social relativism. Nothing could be further from my intentions. As Mishler puts it, on my theory science does not boil down to social relativism because "some ideas are intrinsically better than others in their fit to the empirical world."

6. Conceptual Systems as Historical Entities

Bradie is somewhat partial to viewing conceptual systems extended through time as historical entities. He agrees that there is something right about treating Christianity as having historical roots in Christ no matter how much it has split, merged, and changed in the meantime. But he is much less attracted to treating species in the same way. My line of argument through the years has been that, since species as the results of biological evolution must be viewed as historical entities, then so must conceptual systems in conceptual evolution. In actual fact, the influence was reciprocal. Because treating conceptual systems as historical entities seemed so right, I was led to get over some of my early intuitions about how biological species must be conceived. However, some of the consequences of treating conceptual systems as historical entities are not all that intuitively appealing. In a footnote, Bradie asks us to consider the following dialogue:

She: I believe in God.
He: I do not.
She: Oh, I am so glad that we share the same religious tradition.

She is right to be so happy. He and she have come to the opposite conclusion but within the same conceptual tradition. The God under discussion is the same God.

They would have a much harder time disagreeing if they were working in significantly different conceptual traditions. Such gaps can be bridged, but it is not easy.

In biological evolution, one trait can evolve into its opposite. Some organisms have legs. Other organisms do not. Of the organisms with legs, some eventually lose these legs. Worms never had legs; snakes are descended from organisms that once had legs but lost them. But "no legs" is not the same trait in the two instances. *Absence* of legs is not the same trait as the *loss* of legs. As Mishler points out. the loss of a character belongs as a derived character in the same transformation series as the possession of that character. The absence of this character does not. The correct sequence is: absence of legs/development of legs/loss of legs.

This distinction is crucial in phylogenetic reconstruction, but it is also important in selection processes themselves because of the alternation of replication and interaction. When organisms interact with their environments, the absence of legs functions as exactly the same character as the loss of legs. Neither worms nor snakes can run and for the exactly the same reason. But when it comes to replication, the difference between the two becomes apparent. Because of descent, loss of legs is conjoined with numerous other traits, some of which are the way that they are because the ancestors of these organisms once had legs. Extant organisms can even retain the information necessary to produce legs. Typically different organisms within the same or closely related species will have retained different vestiges of the information necessary for the production of characters that are no longer produced. That is why crossing these organisms can result in atavism, the reappearance of a trait that has been lost, e.g., stripes on the legs of horses. The distinction between homologies and homoplasies is absolutely essential for historical reconstruction, but it is also relevant to how selection processes work in the here and now.

Parallel observations hold with respect to conceptual evolution. Two groups of scientists might hold a belief that is the same in its assertive content, but from the perspectives of both historical reconstruction and EET, these beliefs do not count as the "same." For instance, according to Hennig's principle of dichotomy, taxonomic groups should be subdivided dichotomously whenever possible. At one time, all of Hennig's intellectual descendants accepted his principle of dichotomy, while his opponents rejected it. As time went by, some of Hennig's descendants abandoned this principle. For anyone wanting to reconstruct the history of these disputes on the basis of who believed what, the distinction between conceptual loss and absence is quite obviously crucial, but it is also important in understanding the dynamics of these disputes.

One source of this importance is that conceptual systems exhibit an internal coherence. Although the connections within conceptual systems are rarely deductive, some claims mutually support each other while others conflict. As Mishler explains, on the principles of cladistic analysis, merger is an information destroying process. It produces unresolvable polychotomies. Merger here and there in the phylogenetic tree can be handled, but wholesale merger makes the recovery of phylogenetic relations on the basis of character distribution very difficult if not impossible. For that reason, those scientists interested in reconstructing phylogeny have tended to play down its prevalence. Among animals species at least, genuine merger is supposedly quite rare. However, it is quite common among plants. Roughly 47% of angiosperms and 95% of ferns and their allies are polyploids, and most of these polyploids resulted from interspecific hybridization (Grant 1985). One often hears that the frequency of merger in conceptual evolution is one more reason why EET is bound to fail. The contrast is spurious, stemming from a very limited view of biological evolution. Selection processes can produce both splitting and merger. Merger does pose certain difficulties in both biological and conceptual change, but these are the same difficulties for both.

A second reason for the importance of the distinction between homologies and homoplasies in the ongoing process of science turns on the social nature of scientific disputes. Some of Hennig's opponents emphasized his principle of dichotomy because they found it to be one of the most vulnerable parts of his system. Hennig's disciples reluctantly added to its significance by responding to these criticisms. The net effect was that abandoning the principle of dichotomy threatened the entire program. Hence, a Hennigian rejecting Hennig's principle of dichotomy was of much greater moment than a non-Hennigian rejecting this very same principle. From the perspective of conceptual transformation series, these are two distinct beliefs, even if they have the same assertive content.

If all you are interested in is truth or falsity, then "snow is white" will be true or false depending on the color of snow and nothing else. But if you are interested in science as a dynamic process, what is termed the "same proposition" in Old Speak may well function as different propositions depending on its genealogy. The process by which we come to understand the world in which we live is crucially dependent on our studying conceptual trees and networks. Perhaps one might admit that such considerations actually *do* intrude into science but complain that they *should* not. Science should be a function of subjects interacting with the world in total isolation from all other subjects—the model that informed traditional epistemology. All powerful evil demons, inverted spectra, and brains in a vat are epistemically relevant, but the social structure of science is not—even though it is one of the most important sources of scientific claims having the credibility that they have. I think that we need to reevaluate our understanding of epistemic relevance if epistemology is to have anything to do with how we come to understand the world in which we live.

7. Species as Individuals One More Time

If species are taken to be the entities that result from biological evolution, then they cannot possibly be kinds of the sort that function in laws of nature but come closer to having the characteristics assigned to spatiotemporally localized individuals. This view has come to provide a superabundant source of philosophical disputes. As a good philosopher should, Bradie confronts this position with a battery of counter-examples. In the past I have refused to discuss what I take to be casual, superficial, often downright silly counter-examples Too many real problems exist without worrying about bug-eyed monsters. Thought experiments can be useful, especially if they are specified in detail and are raised in connection with explicitly stated contexts. For example, Koertge sets out several thought experiments about the social structure of science. She specifies which characteristics of science are to be considered in place and which ones are to be modified or replaced. Although such examples do not settle anything, they do help us to probe the functional organization of science. Which parts are likely to serve which functions? Of course, empirical investigations are necessary to determine which of these surmises are actually correct (for further discussion, see Hull 1989).

I do not find Bradie's thought experiments with respect to species all that helpful. He asks lots of questions about species that arise in the context of a series of thought experiments, but frequently he does not describe his examples in sufficient detail for me to be able to answer his questions. Nor was I able to discern the contexts of these examples. As far as I am concerned, "ordinary experience" or how "ordinary people might conceive of the issue" do not afford sufficiently determinate contexts to be of any use. Even if they did, I am a philosopher of science, not an ordinary language philosopher. If scientific usage departs from ordinary usage, so be it. In general, Bradie asks questions about single organisms of a sort that cannot be answered with respect to species as units of evolution. Specieshood is a populational matter. It all depends on how many organisms are involved and the relations among them. From an evolutionary perspective, the significant question is the individuation of species, not whether a particular organism belongs or does not belong to a particular species. Because the boundaries between species are frequently equivocal, certain organisms do not belong unequivocally to one species rather than another, and the relevant "boundaries" exist in real space, not the conceptual space of cluster analysis.

What if the spontaneous generation of organisms turns out to be much more widespread than we now think it is? Wouldn't this raise problems? If the world were very different from how it is, our current theories would be false. My argument for species being treated as spatiotemporal particulars is based on our current understanding of biological evolution. If we are wrong, then all bets are off. For me at least, ontological status is a function of our current understanding, not some a priori prescription. If organisms sprang up all over the place spontaneously, they would not form more inclusive historical entities, but they might form natural kinds. There is no way to know in advance.

What if the proverbial stranger from another planet landed and mated successfully with humans, and these hybrid offspring were fertile and mated with other human beings? We do not have to invent science fiction examples. Introgression occurs among species right here on Earth. As Mishler explains, whether introgression results in the two lineages merging into one or staying sufficiently distinct depends on how much introgression occurs. As long as the gene flow is minimal, the two species remain distinct. If it becomes too extensive, the two species merge into one. Obviously, a single visitor from outer space would scarcely touch the human species. If species are treated genealogically as historical entities, then descent matters. Species must have single origins, but these origins need not be from single species. Hybrid species are as much species as those that arise from a single ancestral population splitting off from the main body of a single parent species. Prejudices against merger to one side, most sexual organisms result from merger—the merger of a sperm and an ovum. They are nonetheless single historical entities.

But what of the stranger himself? Is *he* human? I assume that he was part of a non-human species on his home planet. By mating successfully with humans here on Earth, he merged his genes with human genes. But does he belong to *Homo sapiens*? One consequence of treating species as individuals is that it really does not matter whether or not a single organism is considered as belonging to a particular species. The principles of individuation for species are not determinate enough to apply unequivocally to every single organism, nor need they. Species are not that sort of thing.

What about individuating people as persons? Is physical continuity necessary and/or sufficient? How about psychological continuity? As far as I can see, contemporary biological theories have no implications for the problem of individuating persons (see Wilkes 1988 for a position comparable to mine on science-fiction counterexamples and her own suggestions for the individuation of persons). What if Theseus, after replacing the parts of his ship bit by bit, did the same for himself? Here the problem is the individuation of organisms. Although current biological theory might have implications for individuating organisms, I do not think that "organism" plays a sufficiently important role in biological theory for the resolution of such problems to be worth the effort. Is a Portuguese man-of-war one organism or several hundred? The most common answer is several hundred. A more fruitful question is whether, in the evolutionary process, it functions as one interactor or several hundred.

Bradie goes on to ask if the replaced Theseus would be human. This issue is not the replacement of parts. That goes on all the time. But the replacement of parts generated from the original zygote with manufactured plastic parts. As in the case of the visitor from another planet, such questions cannot be answered with respect to single organisms, nor does it matter. Consider them human; consider them non-human; it does not matter. Only if enough of these organisms were introduced would the integrity of the human species be brought into question, and given Bradie's science fiction example, I have not even a clue what the answer might be.

Why is interbreeding so important in sexual species? Because that is the way that sexual species produce ancestor-descendant sequences.

Why are such genealogical connections so important? Because they are necessary for selection.

But why is selection so important? Because that is how species evolve.

In the case of Bradie's bubble universes, I do not know what to say because I do not know the necessary physics. So I have been told, all the things that we think are laws of nature might actually be contingencies due to singularities in the Big Bang that gave rise to our universe. What if there were an infinite number of such universes? Would hydrogen be a natural kind in any or all of them? I simply do not know. All these universes have histories. Shouldn't history make a difference? In selection processes, the issue is not just histories but genealogies, and genealogy requires replication. Bradie says not a word in this thought experiment about replication sequences. Do we have essences back again? In my view of the world, they have never left. Just because particular species do not have essences, it does not follow that there are no essences. For example, I hope that replicator, interactor, and lineage turn out to be natural kinds with essences of the old fashioned sort.

I am certainly willing to submit my philosophical views to "empirical tests of physical possibility." The world *could* be different from the way that it is, but a lot of hard work is necessary to unpack these possibilities. For thought experiments like those sketched by Bradie to be really useful, they have to be spelled out in much greater detail.

Notes

[1] This paper was written in part under NSF grant DIR-9012258.

References

Bradie, M. (1986), "Assessing Evolutionary Epistemology", *Biology & Philosophy* 1: 401-60.

Dupré, J. (1990), "Scientific Pluralism and the Plurality of the Sciences: Comments on David Hull's 'Science as a Process'", *Philosophical Studies* 60: 61-76.

Ettinger, L, Jablonka, E. and McLaughlin, P. (1990), "On the Adaptations of Organisms and the Fitness of Types", *Philosophy of Science* 57: 490-513.

Grant, V. (1985), *The Evolutionary Process*. New York: Columbia University Press.

Griesemer, J.R. (1988), "Genes, Memes and Demes", *Biology & Philosophy* 3: 179-84.

Hull, D.L. (1989), "A Function for Actual Examples in Philosophy of Science", in *What Philosophy of Biology Is: Essays for David Hull*, M. Ruse (ed.). Dordrecht: D. Reidel, pp. 313-24.

_ _ _ _ _. (1990), "Conceptual Selection", *Philosophical Studies* 60: 77-87.

Wilkes, K. (1988), *Real People: Personal Identity without Thought Experiments*. Oxford: Oxford University Press.

Williams, G.C. (1966), *Adaptation and Natural Selection*, New York: Princeton University Press.

Part VII

SELF-ORGANIZATION, SELECTION AND EVOLUTION

Form and Order in Evolutionary Biology: Stuart Kauffman's Transformation of Theoretical Biology[1]

Richard M. Burian and Robert C. Richardson

Virginia Polytechnic Institute and State University and University of Cincinnati

1. Introduction

Stuart Kauffman's forthcoming book, *The Origins of Order: Self Organization and Selection in Evolution* (1991), is a large and ambitious attempt to bring about a major reorientation in theoretical biology and to provide a fundamental reinterpretation of the place of selection in evolutionary theory. Kauffman offers a formal framework which allows one to pose precise and well-defined questions about the constraints that self-organization imposes on the evolution of complex systems, and the relation of self-organization and selection. He says at the outset that he wants to "delineate the spontaneous sources of order, the self organized properties of simple and complex systems" and to understand how they "permit, enable and limit the efficacy of natural selection" (Introduction, p. 2).[2] As he says somewhat later, the central theme running through his book is that "the order in organisms may largely reflect spontaneous order in complex systems" (ch. 6, p. 224). These evolutionary constraints are general and universal constraints intended to apply across the spectrum defined by biological specializations (see Maynard Smith, et. al., 1985 for a description of the range of developmental constraints). If Kauffman is right, he offers a vindication of D'Arcy Thomson's "principle of discontinuity," emphasizing that "nature proceeds 'from one type to another' among organic as well as inorganic forms; and these types vary according to their own parameters, and are defined by physico-mathematical conditions of possibility" (1952, p. 1094). The reorientation required has strong philosophical overtones. Kauffman often speaks of complex, integrated, systems as "emergent," and is committed to a novel form of holism together with a revised view of the interrelationship between the biological and the physical sciences. The scope of his analysis is remarkable. He treats at some length the origin of life, neural and genetic networks, the interplay of proteins in metabolism, evolutionary theory, developmental biology and ontogeny, genetics, immunology, and a number of other disciplines. In each case, he looks to generic sources or order. In each case, it is surprising how much he finds.

We intend to focus particularly on the ways in which Kauffman aims to restructure the foundations of—and the ongoing work in—a number of biological disciplines, including his treatment of the sources of order (section 2) and the relation between adaptation and organization (section 3). With this general groundwork in place, we will raise a few of the issues that must be faced in understanding his enterprise. These include more specific questions about the role of adaptation in explaining order (section 4), and about the limitations of Kauffman's formal approach in explaining the distinctively biological (section 5). These are exploratory inquiries rather than objections to Kauffman's enterprise or approach. We hope to stimulate further discussion into the scope and limitations of self organization in determining the course of evolution. Similarly, it is important to examine the relation of Kauffman's universal, or 'generic,' constraints on evolution to more specific local constraints, such as those imposed by the characteristic materials out of which organisms are constructed, the manifestly accidental features characteristic of the *Bauplan* of a lineage, or the even more local vicissitudes of adaptation. We offer, as one would expect, no answers to these large questions. We hope only to focus the questions, and to bring out their importance for Kauffman's enterprise and for theoretical biology generally.

2. Generic Properties as a Source of Order

The central novelty in Kauffman's approach to theoretical biology is his use of "self organized collective properties" or "self organizing systems" (ch. 2, p. 20). These are formally characterized properties of complex systems, determined by the mode of aggregation of their constituent entities. Kauffman argues that these self organized collective properties are sufficient to determine "generic properties of complex systems." *Generic properties,* in turn, are those that are statistically common within an array of systems for whatever reason, whether the consequence of some uniform selective pressure or because of the regularities governing the behavior of the constituent entities and their organization. If they are due to self organization, then these generic properties must be largely independent of selection or of the limitations imposed by phylogeny. They are the natural and expected consequences of simple components interacting in simple ways. These generic properties both create and limit the possibilities available to development, evolution, and metabolism. They are, to employ a useful expression from Donna Haraway, "enabling constraints."

The first thing to note is that there is nothing *distinctively biological* about the properties of complex systems, or the properties of their components, on which Kauffman draws in delineating these enabling constraints. Kauffman examines the higher level, abstract, regularities of complex systems. His focus is on what is typical of classes, or ensembles, of systems specified in terms of their components and organization. The components are relatively simple, and their modes of combination are formally characterized. In contrast, the raw material of closed metabolisms is complex and interaction is highly specific; so is that of organismic development and of evolution. If it is possible to give a minimal—and perhaps minimalist—but correct, mathematical characterization of the features of these complex raw materials and their rules of combination, one then can hope to provide useful information about the ways in which they *can be combined,* the relative ease of combining them in certain ways, the possibilities which their combinations open up, and the behavior of large ensembles of such entities. One might hope to derive properties of the behavior of a "typical" system of such entities from an adequate description of the components of the system and their rules of combination. One might even hope to predict some features of the resulting processes, or some patterns in the resulting taxonomic array. The features to be predicted will be, of course, generic rather than specific. The predictions would govern what would be common, rather than what would be universal; nonetheless, such predictions would be ambitious and testable. They would reveal, for example, something about what is likely to happen in the course of evolution even in the absence of selection. Just as appeals to genetic drift (Lande 1976; Lande and Arnold 1983) and the neutral theory (Kimura 1968, 1980, 1983) provide useful and important tools for evaluating the significance of selection by providing benchmarks from which to measure adaptive change, so likewise a serious understanding of the generic sources of order would provide an analysis of what is likely to happen in the absence of selection. The establishment of null hypotheses of this sort is a major accomplishment. If well established and relatively precise, they provide expectations against which to test the workings of selection. Deviations of observed genomic architectures or actual distributions of evolutionary trees, compared with Kauffman's null hypotheses, could be used to detect perturbing effects of selection or other 'agents' of evolutionary change. Such null hypotheses would also allow us to evaluate the importance of more local phylogenetic constraints by seeing what aspects of form might be expected even in their absence. Local effects can be seen only against a backdrop that defines what is generally to be expected.

Among the "predictions" which Kauffman derives from his formal models are the following: typically, evolutionary radiations, even in the absence of selection, will behave like that in the Cambrian explosion, with most of the radiation coming early, followed by pruning and refining among a reduced number of basic body plans. Again, he offers this: evolutionary modifications in early ontogeny will exhibit a "locking in" of processes in early development; consequently, a similarity in early ontogeny would be expected across a wide array of taxonomic groups. Yet again, we have this: in the development of multicellular organisms with a large number of cell types, the cells of any particular lineage (if they can switch type at all) will normally be competent to switch to only one, two, or (rarely) three of the cell types in the body. Finally, we have this: the number of cell types across lineages will increase roughly as the square root of the number of structural genes.

The predictions just listed may sound cheap, because they represent claims already accepted by many biologists. The crucial point is that they are derived, *without* parameter fixing or other formal trickery, from elementary mathematical models basic to Kauffman's work—models that *do not incorporate selection or drift*. Kauffman uses such models to make many more predictions, some of which are controversial. For now, let these stand as illustrations of the sorts of results he can be expected to achieve.

Complexity of certain sorts can be mathematically characterized. For example, a soup of prebiotic molecules that interact with one another will have certain probabilities of forming substances which enter into mutually catalytic interactions and of forming a collection of molecules that can produce yet more catalytically interacting molecules of increasing complexity and stability. Such a system, once stabilized, will evolve. This does not require that any single molecule be autocatalytic or that there be any genetic information or template; it simply requires that the density of molecules that interact—in part by catalyzing reactions involving other molecules—be sufficiently great that the molecules act collectively as an autocatalytic set. Such a scenario—which is, indeed, a moderately plausible one with the additional feature that it includes a path by which template functions and autocatalysis might enter the picture—is developed at some length in Kauffman's chapter 6 and amplified in chapters 7 and 8. That work has been elaborated in extensive simulation studies, now leading to experimental work. For present purposes, the conceptual basis for these studies yields a take-home moral even without paying serious attention to the details: the likelihood of forming autocatalytic sets and the likelihood that any that *are* formed will have certain properties is determined by the statistics of the interactions among high numbers of organic molecules within certain ranges of dilutions and boundary conditions.

Molecules may be specified abstractly by their relationships to one another—including, for example, the effects they have on other molecules and their reactions. If one can get a fix on the right boundary conditions, the rules of organic chemistry determine the transition probabilities between chemical species and the changes in those probabilities caused by the addition of particular molecules at particular densities. These probabilities, in turn, determine the likelihood of forming autocatalytic sets and of those sets having certain properties; for example, they determine the likelihood of properties that do or do not favor formation of RNA or RNA-mediated catalytic reactions. Once autocatalytic sets of the sorts here hinted at are formed, some of their behaviors (for example, their interactions as wholes) are not readily derivable from the rules of organic chemistry. Such sets are thus "emergent" entities of a new sort, distinct from that of any simple collection of organic molecules. They could, in short, be the stuff of life.

The treatment of these matters in Kauffman's book is far more serious than in our simple fable. But the *strategy* of argumentation is characteristic of the actual cases Kauffman discusses. For there to be a metabolism, an immune system, evolution of adaptations among competing organisms, or orderly modes of development, there must be a substrate of complexly organized entities. Kauffman's theoretical strategy, typical of that of physicists working on spin glasses and like matters, is to analyze this substrate as an ensemble of relatively simple and uninteresting entities sharing some intrinsic characteristics and modes of interaction. This procedure allows one to abstract from the particularities that distinguish one member of the ensemble from another. The focus is then shifted to what is common among elements of the ensemble: the interaction between these constituents, and the effects of their organization on the evolution of the systems they constitute.

Other things being equal, the characteristics of specialized components and their modes of interaction determine what sorts of metabolism, immune system, developmental program, or patterns of evolution are likely and what sorts unlikely; systems consisting of such specialized components will be prone to form complexes of some kinds and not of others, to form networks of some kinds, but not of others, to have histories of some kinds, but not of others. Kauffman systematically abstracts from differences between and among constituents of the systems he is considering. For example, in modelling metabolism, Kauffman ignores differences among polypeptides, instead analyzing "typical" interactions and relationships between them. Similarly, he applies formally identical models in different and disparate domains; for example, a model appropriate to the interaction of polypeptides could also apply to genomes or cell types, or even to the evolution of species within an ecosystem. Thus, the differences among constituents typically do not enter into Kauffman's characterization of higher level entities or their evolution. Much of his work involves the extraction of "generic properties" of higher-level systems through the use of mathematical modelling and computer simulations—but always also with an eye to what is biologically reasonable and realistic. In brief, Kauffman holds that there is a sort of statistical mechanics governing ensembles of complex systems, depending only on formally characterizable properties of underlying entities. These statistical features describe many interesting and important features of major biological systems and their evolution.

3. Adaptive Evolution and Organization

The method of reasoning that Kauffman employs to determine which collective properties are generic, and the consequences of the organizational properties of complex systems, is of quite general interest. We will illustrate them in this section with two central examples that play a crucial role in his book: fitness landscapes and genomic networks. These examples illustrate, respectively, the relative autonomy of self-organizing properties of complex systems from the limitations of selection and the sources of spontaneous order. We will see that both embody similar abstract and formal characterizations which, in the terms we have used, contain nothing distinctively biological. Nonetheless, their biological interpretations are of sweeping importance.

3.1. Adaptive Landscapes

The general problem Kauffman poses in examining adaptive landscapes can be understood this way. We can suppose there is an intrinsically favored generic order for some class, or ensemble, of systems. Suppose they are naturally blue. We can also suppose that there is selection favoring some member or subset of that ensemble, which deviates from the favored generic order. Selection favors red. Is it likely or even possible that the generic order will be discernible within the population even when it is selected against? Is it, conversely, likely or even possible that the selected subset will constitute a significant, or a dominant, part of the observed order? Will we see blue or red? If the generic order will be discernible even when it is selected against, then its presence when it is not selected against is unremarkable. As Kauffman says, "if selection can only slightly displace evolutionary systems from the generic properties of the underlying ensembles, those properties will be widespread in organisms, not *because* of selection, but *despite* it" (ch. 1, p. 27). Kauffman demonstrates that the generic order will be discernible across a wide variety of differences in parameter values—blue will shine through—by exploring the statistical structure of fitness landscapes as population size (N) and integration (K) vary. This is what he calls the NK model. We can see the general point fairly easily by looking at the two limiting cases.

For the sake of concreteness and simplicity, consider a simple haploid system with three genes and two alleles at each locus. Any one genotype can be transformed into three others by single mutations. In figure 1 the genotype G_1 (A,B,C) can be transformed into any of three others, G_2 (A,B,c), G_3 (A,b,C), or G_4 (a,B,C) by modifying exactly one allele to its alternative form. There will be eight possible genotypes, in all, in an ordered space illustrated in figure 2.

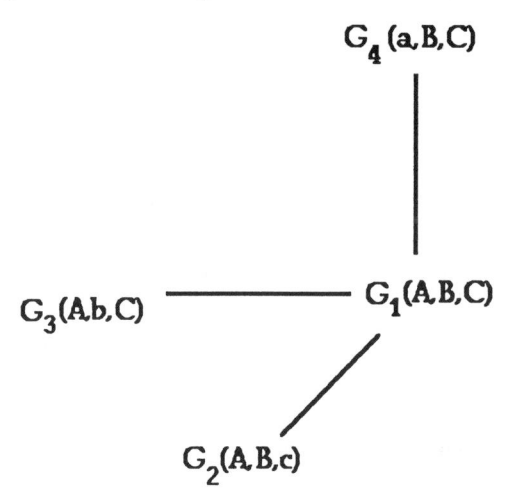

Figure 1. One step transformations in a haploid model with two alleles at each locus. G_1 can be transformed into any one of three different genotypes (G_2, G_3, or G_4) by altering a single locus (A,B, or C) to an alternative form (a,b, or c).

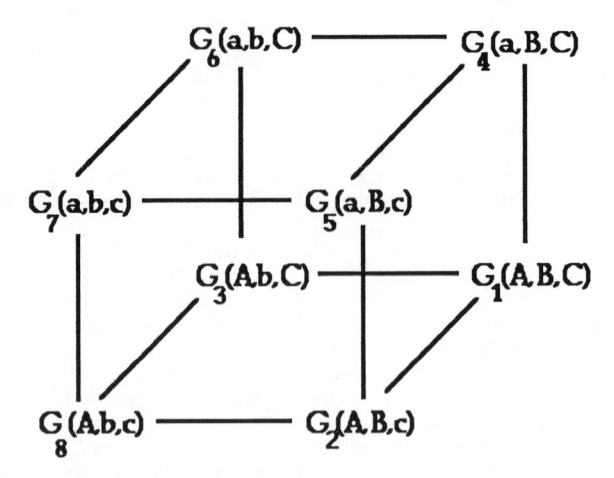

Figure 2. Structure of a space with 3 dimensions:. Once again, each of the three genotypes can be modified by single steps into exactly three other forms (D=3). There are three loci (N=3), and two alleles at each locus (A=2).

The structure of the space determines the number of mutational steps needed to transform any one genotype into any other. To find an adaptive landscape, we need only calculate the fitness to be assigned to each genotype. The contribution of each gene to the fitness of a genotype will, in general, depend on other genes present at other loci. The fitness contribution of A to the genotype G_1 will not, in general, be the same as the fitness contribution of A to the genotype G_2. Kauffman tells us the fitness w_i of an allele i is a vector composed of the K contributions to w_i, plus i's own contribution. If genes at the second and third loci in our simple model affect the fitness of A in G_1, then w_A can be treated as composed of the independent contribution of A to its own fitness, and the modifications induced by B and C. The fitness of the genotype is then the total of the fitnesses of the N genes in the system. Normalizing to take account for the number of genes, the result Kauffman gives us this:

$$W_G = (\Sigma w_i)/N$$

What appears initially complex turns out to have a simple form. Having determined the fitness for each genotype—that is, each node—we end up with a directed graph. The simple structure in figure 2 becomes a three dimensional fitness landscape in figure 3. We have an adaptive landscape with two local optima (G_4 and G_7) and two minima (G_3 and G_5).

Notice that, though we have described this as if it were a simple haploid system, nothing depends on that interpretation. N might be the number of genes in the genotype, but it might equally be the number of amino acids in a protein, the number of traits composing a phenotype, the number of conspecifics in a population, the number of species in a community, or, for that matter, the number of firms in a local economy. Whether the problem is one of explaining evolutionary dynamics as a consequence of genomes, ecosystem dynamics, or economics does not matter to the general model. There is nothing distinctively biological in the explanation.

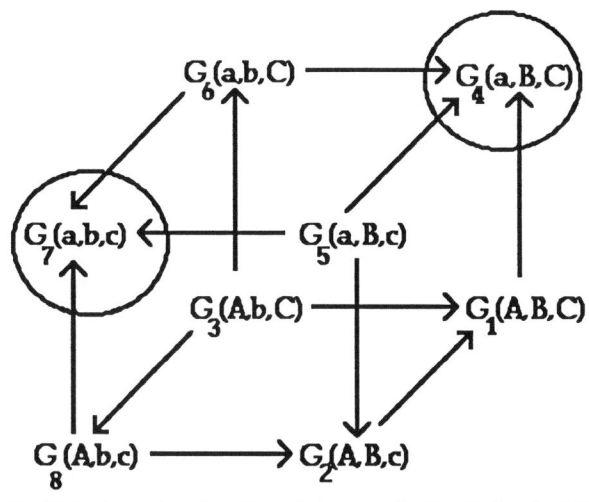

Figure 3. An adaptive landscape in a three dimensional space (D=3) with three loci (N=3) and two alleles at each locus (A=2). Arrows indicate the direction of higher fitness. So, for example, W_{G4} > W_{G1} > W_{G2}. Optima are indicated by circles.

To return to our simple example, the problem is to understand how adaptive walks vary, and how adaptive landscapes vary, as a function of other variables. Kauffman tells us that the variables that matter most are N, the number of components, and K, the connectivity of the system, defined in terms of the number of interactions *for each component*. There are two limiting cases. If K=0, then there is no interaction among components and fitness is simply additive. The total fitness of a genotype is simply the sum of the intrinsically determined fitnesses of the N genes which constitute it. If K=(N-1), then every component interacts with every other, and interaction is at a maximum. Kauffman explores each in considerable detail.

Let us begin with K=0. In this case there will be exactly one optimum genotype, composed of the the aggregate of the N more fit alleles. All other genotypes will be less fit. Moreover, for any genotype with m less fit alleles, any alteration among those m alleles to its alternative form will bring us closer to the optimum. That is, any suboptimal genotype can climb to W_{max} by m steps, each providing some incremental increase in fitness. This is a smooth, highly correlated landscape. If G_1 and G_2 differ in only one allele, then with N alleles, they can differ in fitness by, at most, 1/N. If N is large, this will be relatively small. The landscape will be smoothly graduated. If N is small, any increment will be large. The landscape will be steep. Selection in the first case will have relatively small differences to work with, and is likely to be overcome by disruptive pressures such as mutation. In the second, relatively small differences will yield large effects under disruption. In either case, we should expect shifts away from the optimum. If red occurs only at the peaks, we will see lots of blue.

It is possible to quarrel with the details. As Kauffman notices, normalization is largely responsible for the resulting correlated landscape. It is true that the K=0 case yields a correlated landscape when N is large: the fitness of any genotype will be a relatively good—that is, a better than chance—predictor of the fitness of any of its one mutant neighbors. But this does not mean that all changes are equally good. If G_1

differs from both G_2 and G_3 by only one step, then either change can result in a maximum improvement of 1/N, but the magnitude of the changes could also differ by several orders of magnitude. A change from G_1 to G_2 might result in a change in fitness of 1/N, while a change from G_1 to G_3 might result in a change in fitness of 1/1000N. At a smaller scale, the fitness landscape might look very rugged indeed.

Now consider the other extreme, when K=(N-1). These are fully integrated networks, where each component interacts with every other. Any change in any unit changes everything everywhere else too. These are Cuvier's organisms, and Leibniz' world. The differences between this and the K=0 case are striking. Even if there is only one *global* optimum, there will be a plethora of local optima, and it will be hard to find the global optimum in the space of possibilities. Kauffman shows that the expected number of local optima is, in fact $2^N/(N+1)$. As N increases the number of local optima increases drastically. Moreover, fitness values—even for neighbors removed by one step—are entirely uncorrelated; that is, a mutant differing by only one locus differs randomly from its parent. This "rugged fitness landscape" has two important dynamic consequences. First, adaptive walks will reach only a small fraction of the local optima, and, therefore, they are unlikely to reach the global optimum. Once again, we should be able to see any generic order in the ensemble. Blue will show through. Second, since there is massive interaction, there will be conflicting constraints on every gene; satisfying conflicting constraints will cause adaptive peaks to fall toward the mean fitness of the ensemble. It is harder to satisfy two people than one, and impossible to satisfy everyone. "Accessible optima," Kauffman tells us, "become ever poorer. Fitness peaks dwindle." (ch. 3, p. 47). As a consequence "Optima become hardly better than chance genotypes in the space of possibilities" (ch 3, p. 49). Correlation between the fitness of alternative genotypes therefore increases. Heritability of fitness decreases, and the response to selection slows. As the fitness peaks dwindle, approaching the mean fitness, alternative local optima will become more readily accessible, and disruptive forces should spread the population across the landscape. We see more blue.

Kauffman says this reveals a "fundamental limitation in adaptive evolution" (ch. 3, p. 49). Everything depends on N and K. If K is much smaller than N, then there will be high optima, but any genotype will be only marginally better than its neighbors in the space of genotypes. If N is approximately the same order as K, then there will be a host of local optima, but any one genotype will be only marginally better than the average. In both cases, sub-optimal forms will be common: "... powerful spontaneously ordered properties typical of most ensemble members will almost certainly still be found [even] in the presence of strong selection" (ch. 3, p. 49).

Kauffman goes on to consider various wrinkles in this basic model and their implications for adaptive evolution and organization. He considers the effects of accumulated mutations, longer jumps in the adaptive space, recombination, variable environments, sex, and much more. In every case, the implications are explored with a careful eye to realistic development, and rigor.

The power of the resulting vision is considerable. Let us briefly point toward two implications. First, consider what is often called "Von Baer's law" (Gould 1977; Wimsatt 1986; Wimsatt and Schank 1988): there is generally greater resemblance among embryonic forms than adult forms, and greater resemblance among early embryos than later ones.[3] Differentiation occurs only later in ontogeny. Why would modifications of earlier stages be less common?[4] Kauffman observes that if early modifications affect many traits, then mutations affecting (early) embryonic stages will be adapting on highly uncorrelated fitness landscapes. It will be harder to find viable mutants. Correspondingly, mutations affecting later stages will be adapting on relatively correlated fitness landscapes. Finding viable alternatives will be relatively easy. Second, Kauffman offers an explanation of the pattern of rapid proliferation and subsequent constriction found in the Cambrian explosion. The initial explosion at higher taxonomic levels occurs because even jumps of a broad scale will find locally adapted peaks (or comparable peaks); but after this initial success, the landscape will become more differentiated. Finding improvements will get harder and harder.

3.2 Genomic Regulatory Networks

If the story we have just recounted from Kauffman is correct, the origins of generic order do not lie in adaptive evolution. One source of such order, Kauffman claims, lies in intrinsically organized mechanisms, or in self organization. The general strategy Kauffman employs in exploring the importance of self organization can be illustrated by turning to genomic regulation (which occupies Kauffman's attention in chapters 9-13). The problem Kauffman poses is *how* genetic regulatory systems can be maintained in the face of genetic mutation and recombination, both of which should tend to disrupt and disaggregate them. Kauffman describes his project this way:

> The genome is a system in which a large number of genes and their products directly and indirectly regulate one another's activities. The proper aim of molecular and evolutionary biology is not merely to analyze their structure and dynamical behavior, but also to comprehend *why* they have more or less the architecture and behavior observed and *how* they may evolve in the face of continuing mutations. I shall suggest that we must build *statistical theories of the expected structure and behavior* of such networks. Those expected properties then become testable predictions of the body of theory. If those properties are discovered in organisms, they then find their explanation as the *typical or generic* properties of the *ensemble* of genomic regulatory systems which evolution is exploring (ch. 11, p. 254)

The problem is one of understanding how to maintain an optimal or nearly optimal network in the face of mutational pressure, which will tend to create suboptimal genotypes. Mutation pressure can be thought of, as we have said, as tending to distribute genotypes more broadly over a fitness landscape, whereas selection would tend to concentrate the distribution of genotypes within a population over a narrower and more optimal region of the space. In genomic regulatory systems, mutations alter the structure of connections within the system. However, an additional problem derives from research on the mechanisms of gene expression. There are, as is well known, complex sets of genes which regulate the expression of other genes. The more complex the regulatory system, the more unstable and delicate it would seem to be; for the more complex it is, the more targets there are which can be disrupted by mutation. Moreover, mutations or transpositions in regulatory genes would alter the expression of genes as much—and in fact more—than would mutations in non-regulatory genes. High selection pressures would be required to counter the mutation rate and maintain the regulatory system. Given the actual complexity of the genome, Kauffman concludes it is implausible that selection could maintain order.

Kauffman suggests, alternatively, that the stability of the regulatory system is not something that requires a special explanation in terms of selection. If the genome consists of large ensembles of genes, mutually influencing and regulating one another, then, he argues, there will be a natural, generic, order to the system which is maintained independently of the influence of selection. The genome is a *spontaneously self organizing system*. To support this suggestion, he turns to the statistical features of systems with multiple interconnections. Kauffman looks at the patterns of interactions between genes instead of the specific effects of any single gene or set of genes. In many cases, he claims, the critical determinant of the genome's behavior is the self organizing pattern of connections; no particular gene or set of genes must be maintained by selection.

Kauffman models genetic networks abstractly, just as he models adaptive change abstractly. He assumes as a first approximation that genomic networks can be modelled as Boolean switching networks (see ch. 5, pp. 140 ff.). Each gene will "regulate" other genes, either directly or indirectly, and either alone or in conjunction with other genes. When a gene directly or indirectly regulates its own behavior, we have a feedback loop. Since we are limited to Boolean operations and a finite number of nodes, a finite number of states (defined in terms of configurations of nodes)will be possible. It follows that such a network must eventually return to a previous state. Since a Boolean system is deterministic, these are cycles in the state space. The cycles constitute stable *dynamical attractors* for the network: once a system enters a cycle, it will repeat that cycle indefinitely.

As a simple illustration, we can consider a system consisting of three units, each sending activation signals to the other two, in which the response of each unit to the incoming signals is a Boolean AND or OR operation. A system with three 'genes' has eight possible patterns of activation (arrayed in a space isomorphic to the genotype space of figure 2). In the specific network shown in Figure 4, the value of unit 1 is + if the values of both units 2 and 3 were + on the previous cycle. It is governed by an AND function. Units 2 and 3 will record a + value if any other unit assumed a + value in the previous cycle, and otherwise will assume a - value. They are OR Gates. We assume the network is synchronously updated: a pattern at time t completely determines the new pattern to be found at t+1. Since there are a finite number of states, and deterministic transitions, the system will inevitably encounter cycles, which it will repeat indefinitely. These are dynamical attractors, or attractor state cycles. For example, the alternation between <--+> and <-+-> is a stable cycle. Which cycle a given network will settle into depends totally upon the starting point.

Kauffman assumes that the genes within a genomic regulatory network can be treated as nodes in the model. The connections between nodes provide the vehicle for one 'gene' to regulate the behavior of another. He then analyzes the statistical properties of such networks, by looking at the effects of different connectivity, different rules, different numbers of genes, and the effects of mutation on the resulting networks. As with adaptive landscapes, the ratio of connections to nodes turns out to be highly important: the ratio of the number of connections within the system (M) to the number of genes (N) affects the expected number of genes any gene influences directly and the connectivity of the network.[5] This in turn affects how many steps are required for a gene to communicate to all the genes it regulates. It likewise affects the size of feedback loops, and the chances of a given gene receiving feedback, via a loop, from itself. If M is significantly less than N, then there will be a number of relatively independent sets of nodes connected with one another in a tree structure. As M increases, the average size of these trees grows and they tend to overlap and merge; moreover, feedback loops form. Once M exceeds N, we have a richly interconnected tree with multiple feedback loops (ch. 11, passim.).

Kauffman also assumes that genetic regulatory networks exhibit a relatively low average connectivity. Given a physical interpretation, this means that the regulatory functions of genes are highly specific: any given gene will regulate, and be regulated by, relatively few genes. Kauffman develops (ch. 12) models that specify some of the details of interactions within such networks. He considers Boolean networks where each unit (gene) can have one of two values (on, off) and sends out signals to other

```
        1
       ╱│╲              1   2 │ 3
      ╱ │ ╲            ─────────────
     ╱  │  ╲            +   + │ +
    ╱   │   ╲           +   - │ +
   2 ⇄ 3            -   + │ +
                         -   - │ -


  2   3 │ 1              1   3 │ 2
─────────────          ─────────────
  +   + │ +              +   + │ +
  +   - │ -              +   - │ +
  -   + │ -              -   + │ +
  -   - │ -              -   - │ -
```

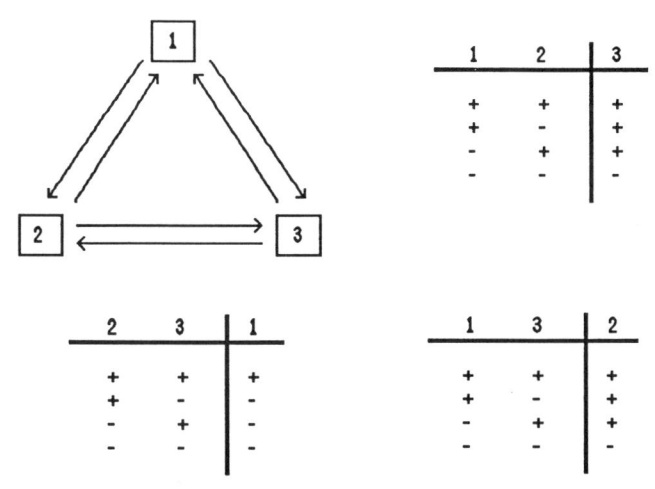

Figure 4. A Simple Boolean Network with Three Nodes and Eight Possible States. Unit 1 is an AND gate, assuming a + value iff both units 2 and 3 assumed a + value on the previous cycle. That is, we have a function from <...++> to <+......>. Units 2 and 3 are OR gates, assuming a + value if any other unit assumed a + value on the previous cycle, and otherwise assuming a - value. That is, we have a function from either <+......> or <......+> to <...+...> and from either <+......> or <...+...> to <......+>.

units. The inputs to other units determine the next value of these units, again using Boolean operations. He focuses especially on systems in which the value of one regulating unit can suffice to guarantee that the regulated unit assumes one of its two values. He calls these *canalyzing functions,* and says that the majority of known genes in bacteria and viruses are governed by such Boolean functions. (The AND function governing unit 1 above is a canalyzing function, since a negative value for either unit 2 or unit 3 is sufficient to insure that 1 will be negative on the next cycle. Likewise OR is a canalyzing function, since a positive value for either input unit is sufficient to insure a positive value in the target unit on the next cycle.) Kauffman's procedure is then to look at the total set, or ensemble, of large regulatory systems (10,000<N<250,000) under the constraints of a low K and a limitation on the allowable Boolean operations. This tells us what to expect, statistically, given just these local constraints. It is a Kauffmanian null hypothesis governing the expected features of genomic systems (ch. 12, p. 287).

Once again we have a general and simple formal model, depending on no assumptions that are distinctively biological. Kauffman suggests that each distinct attractor state cycle can be interpreted as a distinct cell type in the repertoire of the genomic system. When the system settles into one cycle, it produces one cell type. When it settles into another cycle, it produces another cell type. These are the cell types into which the system settles naturally. To support the claim that his network provides a model of such cell types, Kauffman identifies a number of similarities between the properties of the attractor state cycles and the cell types found in eukaryotes: (1) The number of cell types in an organism and the number of attractor state cycles in a network both roughly equal the square root of the number of constituent genes; (2) A large core of units in networks, as well as genes in organisms, are ubiquitously active in all attractor states or cell types in an organism; (3) the patterns of activation in attractor states as well as the patterns of gene expression in different cell types are highly similar; (4) a network attractor state or a cell type can be induced to differentiate directly only to a few neighboring attractor states or cell types; and (5) both networks and cell types are highly stable and resistant to change in the face of alteration in the network.

Kauffman is able to explore the conditions under which such networks maintain their stability and the role selection might play in maintaining such stability. He begins by examining the behavior of highly interconnected systems undergoing random mutations, investigating the effects of these mutations on fitness and the selection strength necessary to maintain the network in a high fitness state. He stipulates, arbitrarily, that some specific set of connections is to count as optimal, and specifies a measure of fitness, in terms of the percentage of connections which are 'correct' by comparison with the standard. The fitness for a given network, W_x, is defined thus (ch. 11, p. 265):

$$W_x = b + (1-b)(G_x/T)^a$$

where T is the total number of regulatory connections in the network, G_x is the number of 'good' connections in x, and b is a residual basal fitness in the absence of any 'good' connections. The value of a corresponds to the three broad ways fitness might correlate with the fraction of good connections; that is, it determines the shape of the fitness curve. If a=1, fitness falls off linearly in proportion to the percentage of bad connections; if a>1, fitness falls off sharply as G falls below T and then levels off; if a<1, fitness initially drops off slowly, and then more rapidly.

The general qualitative results are straightforward. If we assume a fixed mutation rate μ, and a basal fitness of zero (b = 0), then the equation above reduces to this:

$$W_x = (G_x/T)^a$$

There are three cases, depending on the value of a. When a=1, the fitness of a network is a linear function, proportional to the number of good connections and inversely proportional to the number of bad connections. A mutation in the network can increase or decrease the fitness of the network only incrementally. Fitness, in short, is additive. When a<1, the fitness curve is convex. W_x falls off slowly at first, and then increases. A limited number of bad connections is tolerated, but there will be a threshold (depending on μ) beyond which W_x declines precipitously. When a>1, the fitness curve is concave. W_x falls off steeply at first, and then levels off with 0 as the limit. In this case, networks are relatively intolerant to error, but networks with errors are not significantly different from one another.

These results generally replicate the NK model discussed in section 3.1. The case in which a>1 corresponds to high K, and a<1 to relatively low K. Just as in the previous case, the complexity of the network matters greatly. Complexity here is essentially a function of T, the number of regulatory connections in the network. As T increases, mutation is more able to drive networks off the adaptive peaks. Once again, we assume μ is constant. Since the power of selection is inversely proportional to the number of connections, as the number of connections increases the impact of selection on any given node decreases. Increasing T thus decreases the relative strength of selection. Depending on the specific value of a, as T increases, the population will move away from the optimum more or less quickly. This is the error catastrophe revisited.

A high basal fitness value for b means that even a randomly ordered network can approach the optimum. In the limiting case in which b=1, there is no difference at all between networks. With intermediate values of b (1>b>0) and given that a>>1, two alternative stable states emerge—one with low, and one with high, mean numbers of good connections. Which state a network ends up in depends on its history, just as with state cycles in Boolean networks. If one starts an evolving population of genetic networks as a small network with high fitness and gradually increases the number of connections, the population will remain near optimal fitness even as it becomes more complex. It resists mutation pressure. If one starts with low fitness values, it will tend to stabilize at the lower value, and is unlikely to achieve the higher fitness value. Moreover, there is a threshold for the number of connections in such a network above which the system, even if started in an optimal condition, will lose fitness. There will be values of T and a at which μ will overwhelm selection. Once again, we have a complexity catastrophe.

Kauffman offers his models as useful characterizations of the null state against which selection must be assessed. Kauffman contends that without a characterization of the null state, evolutionists are prone to attribute too much power to natural selection, viewing it as capable of generating almost any possible genetic state that is highly adaptive. Kauffman contends that the null state has quite powerful properties and that selection is generally powerless to overcome them. In particular, Kauffman contends that the usual treatments of the means by which selection maintains complex regulatory systems rest on mistaken presuppositions. Selection is not able by itself to produce (rather than merely maintain) patterns of connectivity radically different from the base line—and this base line connectivity turns out to be relatively stable. Once a system settles into this kind of pattern, even massive selection pressure is powerless to alter it. Again, we have what appears to be a fundamental limitation on adaptive evolution.

4. Coevolution and Complex Adaptation

We now turn to two questions concerning Kauffman's treatment of adaptation and complexity. One concerns the impact of coevolution on models of evolutionary dynamics. The other focuses on the generic character of Kauffman's constraints on evolution.

4.1. Coevolution

As we explained in section 3.1, Kauffman's appeal to the NK model motivates his skepticism about the power of selection. Recall that by increasing the connectivity, K, one increases the ruggedness of the adaptive landscape. When K=0, we have a highly correlated landscape with a single adaptive peak, and fitness is additive. When K=(N-1), we have an uncorrelated landscape with multiple adaptive peaks, and fitness is non-additive. Since, in the K=0 case, the landscape is highly correlated and the selective effect of a single allelic change is correspondingly low, disruptive pressures will be able to counteract selection. Since, in the K=(N-1) case, there are multiple constraints, local optima will tend to be depressed, and, once again, disruptive pressures will be able to counteract selection. At one point, Kauffman makes the point this way:

Since increasing epistatic interactions simultaneously increases the number of conflicting constraints, increased multipeaked ruggedness of the fitness landscape as K increases reflects those increasingly mutual constraints (ch. 3, p. 43)

The conclusion appears to be robust. As we move from cases with additive fitness through to highly integrated systems by increasing K, we simultaneously increase the number of adaptive constraints and reduce the relative advantage. The tradeoff bodes ill for the power of selection.

But we should be reluctant to embrace this conclusion without further consideration. One possibility which must be taken seriously is the prospect of increasing K without increasing conflicting adaptive constraints. Flightless birds, for example, are often neotenic relative to their nearest relatives. They retain a longer leg length, reduced breast muscles, and a less developed sternum. Or to take a slightly different example, regressive evolution among cave fauna affects a complex group of traits. Among other things, adaptation to the cave environment means reducing optic sensory organs and increasing non-optic sensory organs. There is reason to think that there is selection acting on both traits, but probably not independently (cf. Culver, et. al., 1990; Kane, Richardson and Fong 1990). These are not, from a functional perspective, independent traits; rather they form an adaptive complex. Clearly, there is interaction, but no constraints that conflict. The same is true in general for mosaic evolution. Evolution operates at different rates on different characters, but often affects character-complexes rather than simple characters. Character sets may provide the "compromise between these bracketing limits of too smooth and too rugged landscapes" (ch. 5, p. 126) required to achieve adaptive evolution.

The same general concern can be raised at a higher level. Consider the analogous case a level or two up. Van Valen (1973) found that there was a roughly constant rate of extinction among taxa. Ammonoids, Echinoids, and Foraminifera have similar survivorship curves, indicating that survivorship is independent of age (contrary to the expectation of, among others, Hyatt). To explain this observation, Van Valen proposed what he called the "Red Queen Hypothesis." The environment constantly degrades because of interactions with other species, whether they are predators or competitors. As a consequence, extinction rates are essentially stochastic. The more general point, already emphasized by Darwin (1859), is that the relevant selective environment of an organism, or a species, is largely biotic. Correspondingly, evolution is largely co-evolution. This may lead to a kind of "arms race" with a constantly deforming adaptive landscape (Dawkins 1982) or to an equilibrium state (Stenseth and Maynard Smith 1984). In either case, fitness landscapes are largely formed and informed by other species. The connectivity here will be limited, but, more importantly, it will be localized.

At whatever level we may address the problem, the question is relatively simple. Granted that the extreme cases of no connectivity, ($K=0$), and maximal connectivity, $K=(N-1)$, leave adaptive evolution only a secondary role, will the intermediate cases suffer from the same dilemma? Or, put positively, under what conditions of connectivity will the dilemma be escaped? The general outline of Kauffman's answer is clear. Coevolution will be understood in terms of the interaction of fitness landscapes in a hyperspace. Kauffman says that "... in a *coevolutionary process,* the very adaptive landscape of one actor itself heaves and deforms as other agents make their own adaptive moves" (ch. 5, p. 165). What is needed is to find parameter values which allow improvement without falling into chaotic fluctuations. The solution is to couple the fitness landscapes in modelling coevolution (ch. 5, pp. 169 ff.). In a multiple species system, this means introducing a parameter C for the interaction effect. Fitness depends not only on the K interactions within the system but on the C interactions between systems. The NK model becomes an NKC model. Again, everything depends on the details. The relative values of K and C matter; so does the number of coevolving species. Kauffman argues that coevolution requires that the interaction between coevolving species be weaker than the interaction within either population. In Herbert Simon's (1981) sense, the systems must exhibit *near decomposability*. In consequence, evolution will occur on *correlated* rather than *uncorrelated* fitness landscapes—close to, but not within, the chaotic regime. Adaptation, in Kauffman's delightful phrase, occurs "at the edge of chaos" (ch. 5, p. 175). This allows us to pose a difficult question, one with considerable mathematical and biological depth: Can one avoid the 'complexity catastrophe' with the right pattern of connectivity? Can selective interconnections structure fitness spaces and their interactions so that an increase in K need not depress the fitness peaks? The answer, to us at any rate, is not clear.

4.2. Development and Evolution

It is worth at least remarking on how radical Kauffman's vision is. Kauffman proposes that intrinsic organizational constraints are critical in the evolutionary process, and that at least the 'generic' properties will be unaltered by selection. Others, such as Alberch (1982), hold similar views. For many of us, this requires a kind of gestalt shift. One way to think of developmental mechanisms such as heterochrony or allometry is to think of these as the *targets* of selection. So, for example, there may be selection for genes that attenuate development. Selection may favor genes that affect antennal length and which only incidentally affect limb length; or there may be selection for larger body size, only incidentally affecting length (cf. Kane, et. al., 1990). Evolution in which developmental stages are dissociated is apparently quite common. This provides for at least two important ontogenetic changes (cf. Raff and Kaufman 1983): one of these is mosaic evolution; the other is a dissociation of larval from adult development. Approached another way, developmental mechanisms are additional constraints on evolution. Thus, Alberch says that evolution is "the product of two independent processes, where diversification ... precedes adaptation" (1982, p. 19). Phenotypic evolution is controlled by processes regulating variation in body plan and by adaptive processes. Kauffman carefully argues that generic constraints typically dominate selection. Form rules, function follows.

5. The Distinctively Biological

We have repeatedly emphasized that Kauffman's models focus on more formal constraints, and pay limited attention to biological contingencies. This does not mean that he pays no attention to biological details. He often does, as we illustrated above in discussing genomic regulatory networks. In emphasizing the "distinctively biological," we mean to emphasize the difference between local and global constraints, including both what is characteristic of biological as opposed to abiotic systems and what is characteristic of some biological systems but not others. To what extent is Kauffman's approach capable of capturing what is distinctively biological? Is it, like many other attempts to develop physics-based analyses of biological subject matter, fated to yield serious misrepresentations of the systems to which it is applied, or does it provide a central core from which more specifically biological analyses can comfortably proceed? What does it do well and what does it leave for others to do better? In the end, these questions must be answered by detailed examination of attempts to apply the new apparatus to biological problems, but, as we will now argue, some general considerations let us localize issues that are particularly sensitive to these questions.

5.1. Historical contingency

Kauffman's approach, by and large, does not pretend to answer questions about historical contingencies and is ill-equipped to do so. Thus, his work speaks to the statistical distribution of patterns of evolutionary radiation, but says virtually nothing about which organisms or which lineages will play a role, if any, in a radiation. Kauffman's results are based more on abstract rules of combination than on the details of chemistry or biology. As such, they might yield, for example, the expected architecture of genetic networks, the likely patterns of differentiation in cell lineages, or the likely branching patterns in evolutionary trees, but not the particular functions of, or connections among, particular genes, the sorts of cells that will appear in particular lineages, or the likely tree for a particular monophyletic group. Similarly, Kauffman argues that evolutionary change and adaptation are most likely to occur when ecological systems are "poised on the edge of chaos," but his apparatus can say little or nothing about the concrete details of the changes that move an ecology closer to or further from the edge of chaos.

To put the point generally, Kauffman's theoretical apparatus is not intended to handle the contingent details of the structure or function of particular genes, proteins, cells, stages of ontogeny, and so on, or to resolve questions about the particular ecological or evolutionary interactions of particular populations or species. There may be important interactions between his results and the investigations of biologists interested in such issues, but, in general, they will turn up only after independent investigations of the contingencies of structure, function, and history.[6] In this sense, there is a very large set of issues of central concern to many biologists to which Kauffman's work speaks only indirectly or not at all.

There is a sense of contingency which Kauffman's approach can accommodate, as he shows using another formal model—this time based on random grammars—in chapter 10. The point he makes is actually quite general: history is sensitive to small perturbations. What might appear to be minor details can have effects which are amplified with time. When history is treated as a walk in an adaptive space, it is clear that small differences in initial conditions or minor perturbations of parameter values can significantly affect the result. Thus, Kauffman can explain how it is that there is contingency to history. However, the particular contingencies which govern our world remain beyond the reach of his formal apparatus. For example, it appears to be simply accidental that terrestrial life relies on L-amino acid isomers and D-ribose rather than D-amino acids and L-ribose. We would expect it to rely on one or the other even if this reliance is a consequence of a random walk, but we know, at least as yet, of no general reason to expect one rather than the other or to explain why terrestrial life actually relies on L-amino acids and D-ribose. It is contingencies such as this that appear beyond the reach of Kauffman's formal apparatus, even though the general fact of contingency is not.

On the other hand, there are many traditional biological questions on which his work, if sound, has immense impact. Take two seemingly simple questions: To what extent does selection affect the general patterns of evolution? To what extent are the patterns of ontogeny the result of selection? These questions have typically been posed in ways that make attempts to answer them misleading and confused. To ask after causal responsibility is to ask what difference the presence of the putative causal factor makes in the relevant context. To resolve such a question, one must know the way the world *would have been* in the absence of the factor in question in order to compare it with the way it *is* in the presence of that factor. It is obvious that the relevant comparisons required to answer our questions about selection are inaccessible to biologists. We do not have access to systems in which there has been no selection and we are rarely if ever in a position to compare systems virtually free of selection with others, otherwise similar, in which selection plays a major role. Correlation and causation are confounded. Kauffman's work at least allows an intelligible formulation of such comparisons by providing a theoretical (re)construction of what would happen to systems that meet the combinatorial conditions with which he works in the absence of selection. To the extent that such theoretical reconstructions are sound, his work allows us to pose clearly some crucial questions about evolutionary patterns that were already raised obscurely by Darwin but which, frustratingly, have remained unclearly formulated and unresolvable to this day.

5.2. Biological Laws

In our pursuit of the distinctively biological, let us turn from historical contingencies to the place of laws and regularities in the biological sciences. We have in mind, for example, the regularities of biochemistry as they pertain to organisms, and not the chimerical, logically universal, laws of philosophers. For lack of space, we shall simply embrace an admittedly controversial claim, namely that *there are no universal and distinctively biological laws of nature*. Such biological laws as there are are, in a strong sense, contingent.[7]

To provide some content for this claim, note that the principle of natural selection is not a law, but is a schema requiring specification by means of an account of the physico-chemical and ecological factors affecting survival and reproduction (cf. Brandon 1978 and 1990). Note also that such evolutionary "laws" or rules as Allen's and Bergmann's rules and such "laws of development" as those that specify the consequences of spiral cleavage are nothing like universal generalizations under any plausible interpretation. Note, finally, that all biochemical rules that are not narrowly chemical—perhaps, the rules connecting nucleotide triplets with amino acids via the genetic code or connecting certain sequences of amino acids with protein conformation (or with catalytic roles)—are contingent on the physiological contexts secured in the course of evolution. This fact provides strong grounds for denying that such rules have the generality over evolutionary time that would be required of laws as these are often understood.

To the extent that the contingencies of evolution underlie biochemical, ontogenetic, and ecological regularities, Kauffman's devices for capturing generic properties and regularities cannot be expected to produce the specific regularities with which biologists have typically been concerned. First order biological regularities are, in the end, historical contingencies of just the sort from which Kauffman abstracts in defining his generic regularities. If historical contingencies—whether they have their source in adaptation or in phylogeny—underlie the regularities of biology, a statistical mechanics based on abstract combinatorial rules will not be able to (re)produce those regularities. To the extent that biological "laws" are the products of evolution, then, Kauffman's abstract mechanics will produce patterns of a higher order, but not biological laws; in full abstraction, the underlying combinatorics employed in his mechanics must be independent of biological evolution. In the case of biochemistry, for example, the underlying combinatorial rules are those of organic chemistry, and they do not change in the course of biological evolution.[8]

5.3. Functional claims

The most obvious remaining bastion of distinctively biological claims concerns the functions of various entities and processes—the various functions of DNA and RNA molecules (and of specific segments of such molecules), of particular enzymes, proteins, lipids, organelles, organs, etc., and also of behaviors, of circadian and seasonal cycles that affect the physiological states of organisms, and so on. Briefly put, the point is that the functions of a molecule—say an arbitrary hormone or an arbitrary gene product—depend on its physiological context. They depend on whether there are receptors for that molecule, and if so where, on whether there are enzymes that process it as part of a reaction chain, and if so on whether the substrates for that activity are available. No intrinsic analysis of the molecule can reveal its actual functions. Parallel claims are true, though perhaps to a lesser degree, for organs—where, as John Beatty (1980) shows, engineering analyses are sometimes possible which, arguably, are somewhat less context dependent—and for behaviors. Here, too, *the specific functions of specific entities or processes are not amenable to the sort of statistical mechanics that typifies Kauffman's approach to biological regularities.*

It will help to anchor these claims if we return to one example for a very brief discussion. There is a generic genomic architecture that, according to Kauffman's analysis, is to be expected in typical cells of typical organisms merely on grounds that genes produce products and, directly or indirectly, influence other genes while retaining stable and heritable interrelationships. In this architecture, some genes are turned on in most cell types, the activity of most genes is affected by a small number of other genes, and whole batteries of genes are switched on in a cascade if one—or, more likely, a few—critical genes are turned on. Kauffman's results, as we have said, are rather more complicated than this, but this captures the spirit of the enterprise. His account of the architecture follows from minimal assumptions that say nothing (and allow no conclusions) about the functions of particular genes. This is not a shortcoming in Kauffman's approach, nor do we mean to suggest that it is. It simply characterizes his enterprise. To the extent that the characterization is fair, it makes it clear that there is no easy way for him to provide a functional analysis of particular genes.

Strongly analogous claims apply to Kauffman's treatment of metabolic networks, of ontogenies, and of fitness landscapes. The power of his constructional principles rests on the fact that they are built by reference to the statistics of *typical* genes (or proteins, cells, organisms, or species), taken as indifferently alike for the analysis of the systems of which they are components. The weakness of his constructional principles is that they abstract from the particular case. This makes it an open question to what extent such constructional principles can capture the distinctive features of metabolisms, genomes, ontogenies, and evolutionary patterns that have been of central interest to biologists. It also leaves open the question whether Kauffman's transformative vision will prove so powerful that the center of biological interest shifts to the novel questions placed at the center of investigation by his formal apparatus.

6. Conclusion

Kauffman's approach to a large suite of major biological issues is that of a physicist studying complex materials and complex systems. His approach opens up new vistas. It suggests that the behaviors of metabolisms, immune systems, genetic networks, and ecological communities are driven by dynamics different from those that most orthodox biologists recognize or accept. He puts forward important hypotheses about the origin of life, the power of selection, the conditions under which evolution will be rapid, the structure of the genome and immune systems, and much more.

One consequence of thinking like a physicist here is that one is primarily, and perhaps solely, concerned with higher level regularities. That implies that one no longer focuses on the details of the systems of concern; those details are mere contingencies, buried and inaccessible to analysis. Kauffman is of two minds here. In his conciliatory moments, he sides with Darwin, treating the higher level results as providing null hypotheses against which we might judge the importance of more local constraints, including those of selection. Thus, he says in his Epilogue (p. 408): "I have tried to take modest steps toward characterizing the interaction of selection and self organization." In this mood, self organizational properties are simply formal constraints on evolution.

When he is in a more radical mood, the generic characteristics overwhelm the particularities. Here he sides with D'Arcy Thompson—as he himself acknowledges one page earlier than the last remark quoted. There he says that his book "is an effort to continue in D'Arcy Thompson's tradition, with the spirit now animating parts of physics. It seeks the origins of order in the generic properties of complex systems" (Epilogue, p. 407). On one standard view, biology studies features of physical systems that contribute to the functioning of living systems (Bechtel, in preparation). Biology is a science of function rather than form. When Kauffman sides with D'Arcy Thompson's program, his denies this characterization of biology. Form dominates, and, to that extent, Kauffman's *The Origins of Order* seeks to bring about a transformation of biology into a science of form, a transformation which, we believe, will be a focus of debate for some time to come.

Notes

[1] This began as two papers, and evolved into a joint enterprise. The order of authors is only alphabetical. We are grateful to Marjorie Grene for arranging the symposium in which this work was presented, and for her encouragement in this project as well as everywhere else. We also thank Stu Kauffman for his patient help and discussion as we struggled to understand his views. We both benefited greatly from a workshop at the Santa Fe Institute, and from the participants there. RCR is indebted to the National Science Foundation (DIR-8921837) for supporting this work.

[2] Since the book is not published, we cannot supply proper page references. We refer to the page numbers in the penultimate draft, kindly supplied to us by Stuart Kauffman. We also identify chapters to assist in locating the cited texts in the published version of the book.

[3] Historically, this is only a rough and misleading approximation. Von Baer did hold that there was greater resemblance among embryonic forms than adult forms, and greater resemblance among early embryos than later ones; but this was because he thought earlier forms were relatively *unformed and undifferentiated,* Evolution, like development, he thought of as a process of differentiation, and thus not a matter of divergence so much as specialization.

[4] But see Raff (1987), Raff and Wray (1989), and Raff, et. al, (1990) for a discussion of ontogenetic processes in sea urchins which suggests this is simply false.

[5] It is worth emphasizing that M is the number of connections within a network, whereas K is the number of connections per component. That is, K is just M/N where N is the number of components. When faced with systems that exhibit relatively low connectivity (K), Kauffman tends to look at the consequences of variations in M.

[6] A possible exception concerns Kauffman's suggestions for experimental design of novel peptides. At the end of Chapter 7, he argues on general grounds that the best path to follow in such work may well be to establish sets of interacting novel enzymes under appropriate constraints; the efficiency of the search for an enzyme that can perform a novel catalytic task may be easier from such starting points than from the positions currently occupied in "catalytic task space."

[7] See J. Beatty, R. Brandon, and R. Burian in progress. We owe the initial suggestions that led us to undertake this treatment of laws here to John Beatty.

[8] As Marjorie Grene and Norman Gilinsky have pointed out to RB in discussion, the combinatorial rules of organic chemistry may well depend on cosmological evolution. Our claim is relative: given the laws of organic chemistry, the distinctively bio-

logical laws of biochemistry (which deal, for example, with the products and functions of particular molecules in physiological settings) are contingent on the pathways of biological evolution.

References

Alberch, P. (1982), "The Generative and Regulatory Roles of Development in Evolution", in *Environmental Adaptation and Evolution,* D. Mossakowski and G. Roth (eds.). Stuttgart: Gustav Fischer, pp. 19-36.

Beatty, J. (1980), "Optimal-Design Models and the Strategy of Model Building in Evolutionary Biology", *Philosophy of Science* 47: 532-561.

Beatty, J., Brandon, R., and Burian, R. (in progress), "The Evolutionary Contingency Thesis".

Bechtel, W. (in preparation), "Integrating Sciences by Creating New Disciplines: The Case of Cell Biology".

Brandon, R. (1978), "Adaptation and Evolutionary Theory", *Studies in History and Philosophy of Biology* 9: 181-206.

_____. (1990), *Adaptation and Environment.* Princeton, NJ: Princeton University Press.

Culver, D., Kane, T. C., Fong, D. W., Jones, R., Taylor, M. A., and Sauereisen, S. C. (1990), "Morphology of Cave Organisms — Is It Adaptive?", *Memoires de Biospeologie* 17: 13-26.

Darwin, C. (1859), *On the Origin of Species.* First Edition. London: John Murray.

Dawkins, R. (1982), *The Extended Phenotype: The Gene as the Unit of Selection.* Oxford and San Francisco: W. H. Freeman.

Gould, S. J. (1977), *Ontogeny and Phylogeny.* Cambridge: Harvard University Press.

Kane, T. C., Richardson, R. C., and Fong, D. (1990), "The Phenotype as the Level of Selection: Cave Organisms as Model Systems", in *PSA 1990*, Volume 1, A. Fine, M. Forbes and L. Wessels (eds.). East Lansing: Philosophy of Science Association, 151-164.

Kauffman, S. (1991). *The Origins of Order: Self Organization and Selection in Evolution.* Oxford: Oxford University Press.

Kimura, M. (1968), "Evolutionary Rate at the Molecular Level",*Nature* 217: 264-626.

_____. (1980), "A Simple Method for Estimating Evolutionary Rate of Base Substitutions Trhough Comparative Studies of Nucleotide Sequences", *Journal of Molecular Evolution* 16: 111-120.

_____. (1983), *The Neutral Theory of Molecular Evolution.* Cambridge: Cambridge University Press.

Lande, R. (1976), "Natural Selection and Random Genetic Drift in Phenotypic Evolution", *Evolution* 30: 314-334.

Lande R., and Arnold, S. (1983), "The Measurement of Selection on Correlated Characters", *Evolution* 37: 1210-1226.

Raff, R. (1987), "Constraint, Flexibility, and Phylogenetic History in the Evolution of Direct Development in Sea Urchins", *Developmental Biology* 119: 6-19.

Raff, R., and Kaufman, T. C. (1983), *Embryos, Genes, and Evolution*. New York: MacMillan.

Raff, R., and Wray, G. A. (1989), "Heterochrony: Developmental Mechanisms and Evolutionary Results", *Journal of Evolutionary Biology* 2: 409-434.

Raff, R., Parr, B. A., Parks, A. L., and Wray, G. A. (1990), "Heterochrony and Other Mechanisms of Radical Evolutionary Change in Early Development", in *Evolutionary Mechanisms*, M. Nitecki (ed.). Chicago: University of Chicago Press, pp. 71-98.

Simon, H.: 1981. *The Sciences of the Artificial*, Second Edition. Cambridge: M.I.T. Press.

Smith, J. Maynard, Burian, R., Kauffman, S., Alberch, P., Campbell, J., Goodwin, B., Lande, R., Raup, D., and Wolpert, L. (1985), "Developmental Constraints and Evolution", *Quarterly Review of Biology* 60: 265-287.

Stenseth, N. C., and Smith, J. Maynard (1984), "Coevolution in Ecosystems: Red Queen Evolution or Stasis", *Evolution* 38: 870-880.

Thompson, D. W. (1952), *On Growth and Form*. Second edition. Cambridge: Cambridge University Press.

Van Valen, L. (1973), "A New Evolutionary Law", *Evolutionary Theory* 1: 1-30.

Wimsatt, W. (1986), "Developmental Constraints, Generative Entrenchment, and the Innate-Acquired Distinction", in *Integrating Scientific Disciplines,* W. Bechtel (ed.). Dordrecht: Martinus-Nijhoff, p. 185-208.

Wimsatt, W., and Schank, J. C. (1988), "Two Constraints on the Evolution of Complex Adaptations and the Means for their Avoidance", in *Evolutionary Progress,* M. Nitecki (ed.). Chicago: University of Chicago Press, pp. 231-273.

Self Organization and Adaptation in Insect Societies[1]

Robert E. Page, Jr. and Sandra D. Mitchell

University of California, Davis and University of California, San Diego

1. Introduction

The social organization of insect colonies has fascinated biologists and natural historians for centuries. Aristotle wrote in *History of Animals* about a division of labor among workers within the hive that is based on age. He observed that the field bees foraging for nectar and pollen have less "hair" on their bodies than the hive bees that care for young larvae and tend the nest. He concluded that the more pubescent hive bees must be older. We now know that, in fact, the field bees are older and have less hair because the hairs break off as the bees age. The phenomenon of age related changes in behavior, age-polyethism, is now well documented for many social insects (Oster and Wilson 1978).

Evidence of the ecological success of social insects is inescapable. Virtually everywhere you look you see them or the results of their activities. The main features of insect societies that are believed to be adaptations responsible for their tremendous ecological success are (Oster and Wilson 1978; Wilson 1985a, 1985b):

division of labor—between reproductives and workers, and a further division among workers that is often, perhaps usually, based on age and/or anatomical differences;
specialization—some individuals perform some tasks with a significantly greater frequency than do other individuals;
homeostasis—colonies regulate internal conditions, such as food stores, temperature, humidity, etc.;
plasticity and resiliency—colonies are able to change the numbers of workers engaged in different tasks in response to changing internal and external colony environments;
mass action responses— colonies are able to mobilize large numbers of workers for specific emergency needs, then return to their original state.

Much of the classical and contemporary research in insect sociobiology has been dedicated to studying these phenomena with the assumption that they represent colony-level functional adaptations and demonstrate the processes of evolution by natural selection. For example, division of labor and specialization among workers within societies have been explained as adaptations that optimize energy use of colonies (Oster and Wilson 1978). Seeley (1985, p. 1) opens the preface of his book with " The honey bee is a wonderful example of adaptation..." and the main reason to study them is because of their "... wealth of adaptations associated with group living". However, any explanation of adaptation depends on historical conditions of heritable variation in the trait and the causal process of selection resulting in differential reproductive success among the variants (Mitchell 1987). Hence, in this example, for division of labor to be an adapted trait, we must presuppose the prior competition between colonies, some with division of labor, some without, and a colony level selection process acting to preserve the trait through differential colony reproduction.

Alternative explanatory strategies that replace pan-selectionism have not been implemented by evolutionary biologists even though uncritical selectionist explanations for the complex traits we observe have been criticized for some time (Gould 1978; Gould and Lewontin 1979). We believe that Kauffman's (Kauffman 1984, forthcoming) approach to the study of complex systems provides an important component in developing alternative explanatory frameworks. That is, Kauffman shows what features one would expect to be typical of complex systems irrespective of, or even in spite of, selection operating on that system. Specific traits of complex systems may be self-organized consequences emerging from the basic structure of the system itself and not the direct product of natural selection. Given a framework for identifying those typical features, the questions of how and on what aspects of the system selection operates can be addressed to generate an explanation which takes into acount both development and selection.

In this paper we show that some features of insect societies, formerly explained as direct products of natural selection, are expected outcomes of self organization. Using Kauffman's approach to studying complex systems as networks of connected binary elements with boolean switching functions, we demonstrate how the typical features of insect societies may emerge from these simple models by self organization.

3. Natural History of Honey Bees

Throughout this paper we will try to interpret the parameters and dynamical properties of theoretical network models for a real social insect, the honey bee *Apis mellifera*. A "typical" honey bee colony consists of roughly 10—40 thousand female workers, zero to several thousand reproductive male drones, and a single egg-laying queen (see Winston 1987). The queen is the mother of all workers and nearly all of the drones. The nest is composed of several vertically oriented, parallel combs made of wax secreted from special exocrine glands of workers. Each comb may contain thousands of individual hexagonally-shaped cells that serve to store food reserves of honey and pollen, and are used to raise the immature life stages of workers and drones. The spatial orientation of colony resources is roughly that of concentric hemispheres with the brood at the center surrounded by stores of pollen that are then surrounded by stores of honey.

The female workers do virtually all of the "work" necessary for colony function. Each worker changes the set of tasks that she performs as she ages, a phenomenon called age polyethism. Currently four "age castes" of workers are recognized, each with associated task sets that are generally dominated by one or two activities: 1) workers that are 0—2 days old clean cells; 2) workers that are 2—11 days old care for immature bees and attend to the queen; 3) 11—20 day old workers process and store incoming food; and 4) workers more than 20 days old forage for pollen and/or nectar for about 1—3 weeks, then die (see Seeley 1985 and Winston 1987).

4. The Model

At any instant in time a colony exists in an organizational state that can be described as the number of individual workers engaged in each of the possible tasks in a worker's behavioral repertoire (males do not work so we do not include them in this state description). Using Kauffman's boolean dynamical theory, assume that individual worker honey bees are nodes, or elements, of a boolean switching network. Individuals can be either *on* or *off* with respect to performing a particular task. For this model, assume that the number of members of the network, N, is all individuals that belong to a particular behavioral state (age caste) that are behaviorally competent to perform a particular task. Also assume that K, the "connectance" (Gelfand and Walker 1980, p. 51), is large, close to or equal to N, with elements receiving directed inputs from and outputs to nearly all other elements. An input line is *on* or *off* depending on the state of the element that initiates it. Each element is in an *on* or *off* state according to the number of *on* or *off* inputs coming from all N individuals. Thus, each element has a threshold of response. Individual threshold functions, f_i, are assigned randomly to individuals from the set {F'} which is a subset of the set of all boolean switching functions, {F}, that has as its members all and only those functions that switch the element *on* or *off* when the number of *on* or *off* inputs exceed some specific value. For this specific model, individuals are switched *off* when the number of *on* inputs is equal to or exceeds some given threshold value, and switched *on* when the number of *on* inputs falls below that threshold.

To make this model represent a honey bee society, assume that individuals receive information of the state status of other individuals via a common perceived stimulus, rather than by individual input connections. Each *on* individual performs a given task

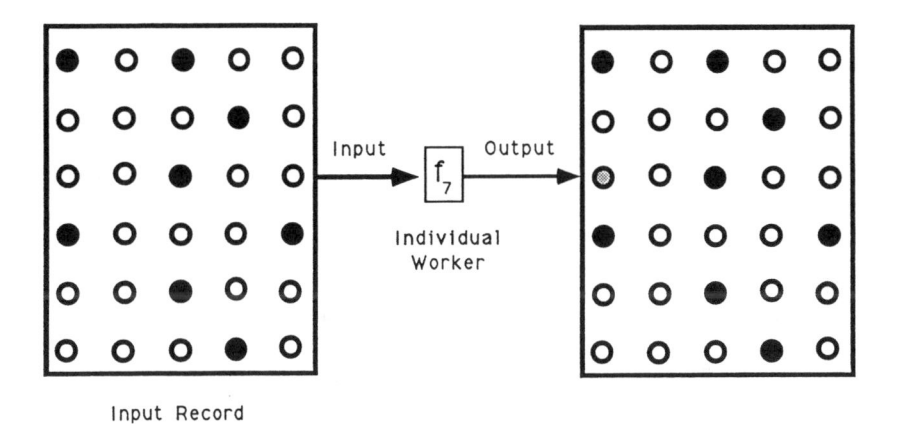

Figure 1. A diagram illustrating how a switching network can be transformed to represented a network of honey bees sharing information through cues provided by the amount of food, say pollen, stored in a comb. The left box represents a section of comb with 30 cells. The 22 open circles are empty while the 8 closed circles are full. An individual worker assesses the number of empty cells and compares it with its threshold function, f_7. Because the number of empty cells exceeds the threshold value of 7, the worker forages and fills the stippled cell represented in the right-hand box. The record now changes as a consequence of the behaviour of the worker.

and by doing so decrements the stimulus level by one stimulus unit (negative feedback). When an individual switches *off*, she stops performing the task and the stimulus level is incremented one stimulus unit. For example, a worker honey bee of foraging age may be *on* for pollen collecting when she successfully forages and *off* if she does not forage or forages for nectar instead. The stimulus in a honey bee colony may be the number of empty food storage cells that stimulates individuals to forage. As more individuals forage, storage cells fill up and, with a constant food consumption rate, the number of empty or full cells provides a record of foraging activities of all N individuals.

For these simulations, assume that all individuals are initially off and the initial stimulus level, S_0, is set equal to N. The order of events for the simulations is as follows:

1. All "individuals" (elements) of the network are assigned a threshold of response f_i randomly drawn from a discrete uniform distribution represented by the set of threshold functions $F' = \{f_i, f_j, ...f_n\}$.

2. Individuals are sampled from the network either simultaneously or randomly one at a time. For the simultaneous model, all individuals check their inputs and respond at the same time. For the random model, an individual is sampled, its inputs checked, the response recorded, then the next individual is sampled.

3. After being sampled, each individual is checked to determine if it is currently *on*. If it is *on*, it is turned *off* and the residual stimulus incremented one unit. (The residual is the initial stimulus level minus the number of individuals currently engaged in that task [*on*] because each individual decrements the stimulus 1 unit while she performs the task. We assume that an individual is not performing a given task while it is collecting and assessing stimulus information.) If it is *off*, the residual stimulus is not changed.

4. The threshold value for that individual (or each individual in the case of the simultaneous model) is then compared with the current residual stimulus level. If the stimulus level exceeds the randomly assigned threshold value, then the individual is turned *on* and the residual stimulus level is decremented one unit. The individual is then recorded as being *on* or *off* for that sampling event.

Case 1: Variable Data Sampling, N = 100, K = N, $F' = \{f_1, f_2, ..., f_N\}$, $S_0 = N$

Simultaneous Sampling—We investigated dynamical behavior of model systems over 100 simultaneous sampling events. Simultaneous sampling leads to highly ordered dynamical behavior with great instability (Fig. 2). At time **t**, all N individuals are *off* and the residual stimulus level exceeds the thresholds of all individuals except those that have thresholds of 100 (individuals randomly assigned threshold values of 100 are never *on*). All other individuals turn *on* simultaneously and the stimulus level goes toward zero at time **t + 1**. Now the stimulus level is below all thresholds so individuals turn *off*, and the stimulus level goes back up to 100 at **t + 2**. The system oscillates between nearly all individuals on and all individuals *off*.

Random Sampling— For this model system we assume that only one individual samples and changes the residual data base stimulus at a time. In this case, connectivity of the network can be asymmetrical, those that are sampled more frequently, due to chance, have more outputs than those that are sampled less often. Results of the simulations show that there is an initial negative feedback phase followed by a search for an attractor (Fig. 2).

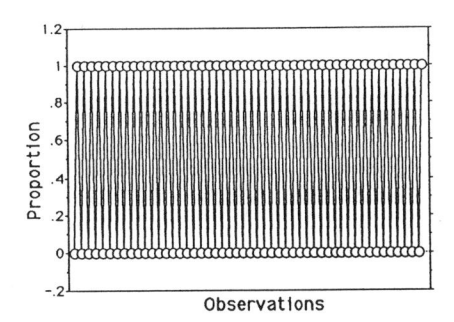

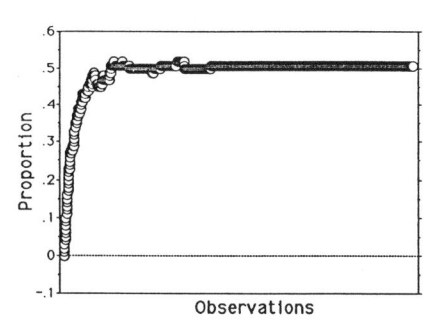

Figure 2. A comparison of the results of simulations using simultaneous (left) and random (right) data access sampling of individual network elements. The proportion of individuals that are on is shown for each of 100 simultaneous sampling observations and for each of 2000 observations for the random sampling model.

The random model behaves differently each time it is run but ultimately ends up at the same steady state value with respect to the proportion of individuals turned *on* (density). We don't know if the same individuals are always fixed *on* or *off*, but it seems unlikely. Attractors in these systems represent a typical property of the network, the average of the distribution of thresholds among elements.

Case 2: Variable Threshold Distributions, Random Sampling, N = 100, S_0 = N

Uniform threshold distributions used for random assignment of individual elements can vary with respect to their mean values and their variances. We examined the effects of varying each on the dynamical behavior of model systems.

Variable Mean Thresholds—We compared the dynamical behavior of networks with thresholds drawn randomly from discrete uniform distributions with ranges 1—50, and 51—100. Each showed the characteristic negative feedback phase followed by random searching near an equilibrium point, then each located a steady state attractor. Systems with higher mean thresholds end up in attractors with lower densities — in other words, fewer individuals work.

Variable Threshold Variances—Changing the variance in thresholds around a given mean affects both the rate at which the system locates an attractor and the stability of the system to external perturbation — homeostasis. We constructed a network where all individual elements had the same threshold value of 50.5, the expected value for other distributions we tested. Systems with no variance in thresholds locate their steady state attractors much more quickly than those with variance.

Case 3: Variable Stimulus Level, Variable Threshold Distributions, Random Sampling, S_0 = N

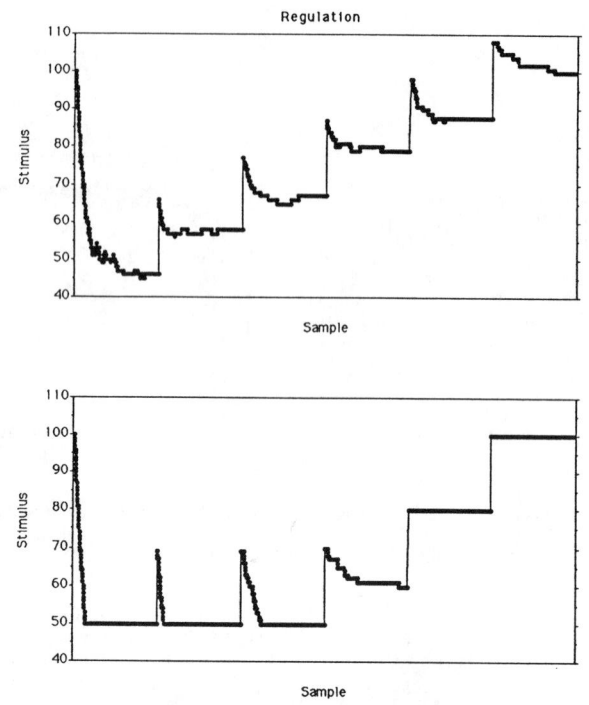

Figure 3. Regulation of stimulus level by model networks containing N = 100 elements with thresholds drawn at random from a uniform distribution with integer values of 1 to 100 (upper graph) and fixed threshold values of 50.5 (lower graph). Y-axis represents residual stimulus level (see text) for each of 3000 sample observations (X-axis).

The variance in thresholds among elements of our model systems affects dynamical behavior resulting from external perturbations to the stimulus level. To test these effects we constructed networks of N = 100 with thresholds drawn at random from a discrete uniform distribution with range 1—100 and networks where all elements had thresholds of 50.5. We started with an initial stimulus, S_0, equal to N, then increased or decreased the residual stimulus level 20 units after every 500 random individual samples. We measured the residual stimulus level at each of 3000 sample events.

Networks with no variance in thresholds show strong homeostatic properties, they return the residual stimulus to predisturbance levels following each increment or decrement of stimulus until either all individuals are *on*, in the case of incrementing the stimulus, or all individuals are *off*, in the case of decrementing the stimulus (Fig. 3). Networks with variable thresholds do not return precisely to predisturbance levels but do regulate the number of individuals turned *on* or *off* in response to each increase or decrease in stimulus (Fig. 3). They buffer individuals from changing stimulus levels providing a more integrated system response to external changes in conditions.

5. Discussion

It is apparent that highly ordered dynamical behavior can emerge from these model networks of individual elements. Even in the simplest systems we see the kinds of behavior long marvelled at in insect societies: **1. homeostasis** resulting from negative feedback reducing residual stimuli to near an equilibrium determined by the mean of the threshold distribution; 2. **mass action responses** when all elements turn on then off when simultaneous sampling is employed; 3. system **plasticity and resiliency** in response to increases and decreases in external stimuli; 4. **division of labor and specialization** demonstrated most dramatically by steady state attractors where some individuals are frozen on and others are frozen off.

What does this mean for social insects? Perhaps many or even most of the phenomena that we hold up as examples of adaptive functional organization are self-organized typical properties of complex social systems and not a direct consequence of natural selection. The minimal requirements for complex social organization may be only mutual tolerance and variability among individuals for thresholds of response to task inducing stimuli; the rest may be self organized.

Our models display emergent behavior via the interactions of individual elements governed by restricted threshold functions drawn from a subset of boolean functions. One could argue that we get order out because we put order in with the set of threshold functions we use. However, it is likely that both solitary and individual social insects vary similarly with respect to thresholds of response to specific stimuli that elicit the release of behavior. Individuals are expected to vary with respect to specific thresholds of response because they have different learning experiences that modify thresholds, they are in different physiological states that have correspondingly different behavior, and because they have different genotypes.

Many solitary female bees and wasps display ordered sequences of behavior associated with egg laying: 1) initiate a nest by burrowing in the ground, boring a hole in wood, or building with mud or paper pulp; 2) construct a special cell or chamber; 3) forage for food or prey to provision the cell or chamber; 4) lay an egg; 5) close the cell or chamber and construct a new cell or chamber to begin the cycle over again. There is evidence that response thresholds to specific stimuli change during this cycle as a consequence of feedback from activities and changes in physiology associated with the the maturation of eggs (Evans 1966; Michener 1974; West-Eberhard 1987).

If we look at honey bee workers we see the same kinds of changes occuring in behavior, only they are played out in a single sequence over the lifetime of the worker rather than repeated cyclicly. Workers first perform duties as cell cleaners followed by brood care then food processing and handling. Then, lastly they make the transition to foraging for pollen and nectar, and die. They change physiologically with age with specific exocrine glands turning on and off specific glandular products that are appropriate for the specific tasks performed.

Genotypic variability for the likelihood that individual worker honey bees will perform specific tasks has also been demonstrated (see Page and Robinson, in press). Workers of the same age that have different fathers (honey bee queens are polyandrous) have different likelihoods of performing certain tasks. This has been interpreted as the result of genotypic variability for response thresholds. Therefore, we believe that protosocial insect systems may have been organized in ways analogous to our model systems.

We are not advocating that the process of natural selection cannot or has not played a role in shaping the organization of complex insect societies. However, we believe that our models provide a framework for identifying those functional components that typically emerge from the social structure and don't require natural selection and those parameters on which natural selection can operate to "fine tune" social organization. Natural selection for colony organization may operate on the dynamical systems parameters **N**, **K**, and **F'**. Individual colony members have behavioral and developmental properties that are selectable that affect these parameters (Page and Mitchell, unpublished manuscript).

6. Conclusions

The application of Kauffman's boolean network model of complex systems to the behavior of social insects clarifies the relationship between developmental, or "emergent", "self-organizing" processes and natural selection. Darwin himself faced the problem of explaining how a complex, integrated system like the eye could arise by the gradual progression of natural selection on minor variations. To do so, he required the assumption that each stage on that gradual trajectory be relatively better adapted to the environment than the other available variants. To Darwin, the only process capable of creating such complexity was natural selection. Kauffman's theory allows us to identify specific ordered features of complex systems that are not the direct object of natural selection but rather the spontaneously generated order arising from the coupling of lower level ordered features. His abstract model goes further by showing how order emerges at a higher level in the least ordered systems—randomly constructed networks. If order can occur in unordered systems, then it is not surprising that it can emerge in more ordered systems. In our application of the boolean network model to honey bee colonies we have shown that some of the basic orderly features of insect societies emerge as typical features of integrated systems.

A specific concern for our model, where **K = N**, is the avoidance of the complexity catastrophe predicted by Kauffman's theory. Kauffman shows that complexity catastrophes occur as N and K get large and represent constraints for natural selection on building and maintaining the complex patterns of development needed for the ontogeny of complex individual organisms. This occurs because the "typical" or average properties of the system become inescapable. Our model systems exhibit the characteristics of one complexity catastrophe presented by Kauffman: the system equilibrates at a level, with respect to the number of individuals on, that is determined by the average properties of the set of threshold functions used. The density attractors of our systems have enormous basins of attraction. This may be a catastrophic condition for natural selection of development but yields predictability and reliability for dynamical social systems and may facilitate the evolution of age and caste structure of colonies (see Oster and Wilson 1978).

Another kind of complexity catastrophe occurs with our models when individuals simultaneously sample the environment and respond. This leads to great instability of the system. However, there are times when this kind of "catastrophe" may be a desirable outcome of a connected network — say in the mass action responses required for colony defensive behavior.

System modularity can effectively reduce **N** and **K** avoiding catastrophe and restoring stability. Modularity can be achieved either structurally or functionally. The physical structure of the nest or the environment can limit the exchange of information among individuals, thereby reducing **K**. Functional modularity can be generated by restricting the functions available for processing the **N** inputs into a given activity. Individuals with low thresholds may form information feedback loops that restrict the information flow to small regions of the network.

We can thus see how selection can operate on the parameters **N**, **K**, and **F'** to tune the network in ways that are adaptive at the colony level. However, what selection may not be able to do, or at least is unlikely to achieve, is to move the entire system outside the domain of its typical features. If division of labor is a property that emerges under almost all systems that have the basic features of insect societies, then natural selection is not the most obvious explanation for this feature. If certain forms of specialization within a general framework of division of labor are not typical of most such systems, then natural selection becomes the framework for explaining them.

We have attempted to show how Kauffman's theory of coupled networks can be used to address concrete biological phenomena. It provides a much needed framework that recognizes the contributions to colony phenotypes of both developmental (emergent) properties of complex systems, and natural selection. Distinguishing between self-organized and adapted traits is important to avoid falling into either uncritical selectionist or anti-selectionist explanations. The incorporation of Kauffman's ideas of biological system dynamics and self organization into insect sociobiology represents a major conceptual advance and transforms into a workable, testable program, the plea that biological explanations take develomental as well as selective processes into account. We think that this "new view" of insect societies as dynamical, at least partially self-organized systems will guide future empirical studies of social insect behavior.

Note

[1]This work was funded by National Science Foundation Grants BNS-8719283 and BNS-9096139 to R.E. Page. S. D. Mitchell thanks the Population Biology Center, University of California, Davis, and the University of California Chancellor's Summer Faculty Fellowship for partial support.

References

Evans, H.E. (1966), "The Behavior Patterns of Solitary Wasps", *Annual Review of Entomology* 11: 123-154.

Gould, S.J. (1978), "Sociobiology: the Art of Storytelling", *New Scientist* 80: 530-533.

Gould, S.J. and Lewontin, R.C. (1979), "The Spandrels of San Marco and the Panglossian Paradigm: a Critique of the Adaptationist Program", *Proceedings of the Royal Society of London* 205: 581-598.

Kauffman, S.A. (1984), "Emergent Properties in Random Complex Automata", *Physica* 10D: 145-156.

_ _ _ _ _ _ _ _ . (forthcoming), *Self Organization and Selection in Evolution.* Oxford: Oxford University Press.

Michener, C.D. (1974) *The Social Bees.* Campbridge: Belknap Press of Harvard University Press.

Mitchell, S.D. (1987), "Competing Units of Selection? A Case of Symbiosis", *Philosophy of Science* 351-367.

Oster, G. F. and Wilson, E.O. (1978), *Caste and Ecology in the Social Insects.* Princeton: Princeton University Press.

Page, R.E. and Robinson, G.E. (in press), "The Genetics of Division of Labor in Honey Bee Colonies", *Advances in Insect Physiology*, Volume 23.

Seeley, T.D. (1985), *Honey Bee Ecology: a Study in Adaptation in Social Life.* Princeton: Princeton University Press.

West-Eberhard, M.J. (1987), "Flexible Strategy and Social Evolution", in *Animal Societies: Theories and Facts*, J. L. Brown and J. Kikkawa (eds.). Tokyo: Japan Sci. Soc. Press, pp. 35-51.

Wilson, E.O. (1985a), "The Sociogenesis of Insect Colonies", *Science* 228: 1489-1495.

——————. (1985b), "The Principles of Caste Evolution", in *Experimental Behavioral Ecology and Sociobiology*, B. Hölldobler and M. Lindauer (eds.). Sunderland: Sinauer Associates, Inc., pp. 307-324.

Winston, M.L. (1987), *The Biology of the Honey Bee.* Cambridge: Harvard University Press.

Winston, M.L. and Katz, S.J. (1982), "Foraging Differences between Cross-fostered Honeybee Workers (*Apis mellifera*) of European and Africanized Races", *Behavioural Ecology and Sociobiology* 10: 125-129.

The Sciences of Complexity and "Origins of Order"

Stuart A. Kauffman

University of Pennsylvania

A new science, the science of complexity, is birthing. This science boldly promises to transform the biological and social sciences in the forthcoming century. My own book, *Origins of Order: Self Organization and Selection in Evolution*, (Kauffman, 1992), is at most one strand in this transformation. I feel deeply honored that Marjorie Grene undertook organizing a session at the Philosophy of Science meeting discussing *Origins*, and equally glad that Dick Burian, Bob Richardson and Rob Page have undertaken their reading of the manuscript and careful thoughts. In this article I shall characterize the book, but more importantly, set it in the broader context of the emerging sciences of complexity. Although the book is not yet out of Oxford press's quiet womb, my own thinking has moved beyond that which I had formulated even a half year ago. Meanwhile, in the broader scientific community, the interest in "complexity" is exploding.

A summary of my own evolving hunch is this: In a deep sense, E. coli and IBM know their respective worlds in the same way. Indeed, E. coli and IBM have each participated in the coevolution of entities which interact with and know one another. The laws which govern the emergence of knower and known, which govern the boundedly rational, optimally complex biological and social actors which have co-formed, lie at the core of the science of complexity. This new body of thought implies that the poised coherence, precarious, subject to avalanches of change, of our biological and social world is inevitable. Such systems, poised on the edge of chaos, are the natural talismen of adaptive order.

The history of this emerging paradigm conveniently begins with the "cybernetic" revolution in molecular biology wrought by the stunning discoveries in 1961 and 1963, by later Nobelists Francoise Jacob and Jacques Monod that genes in the humble bacterium, E. coli, literally turn one another on and off (Jacob and Monod, 1961, 1963). This discovery laid the foundation for the still sought solution of the problem of cellular differentiation in embryology. The embryo begins as a fertilized egg, the single cell zygote. Over the course of embryonic development in a human, this cell divides about 50 times, yielding the thousand trillion cells which form the newborn. The central mystery of developmental biology is that these trillions of cells become radically different from one another, some forming blood cells, others liver cells, still other nerve, gut, or gonadal cells. Previous work had shown that all the cells of a human body contain the same genetic instructions. How, then, could cells possibly differ so radically?

Jacob and Monods' discovery hinted the answer. If genes can turn one another on and off, then cell types differ because different genes are expressed in each cell type. Red blood cells have hemoglobin, immune cells synthesize antibody molecules and so forth. Each cell might be thought of as a kind of cybernetic system with complex genetic-molecular circuits orchestrating the activities of some 100,000 or more genes and their products. Different cell types then, in some profound sense, calculate how they should behave.

1. The Edge of Chaos

My own role in the birth of the sciences of complexity begins in the same years, when as a medical student, I asked an unusual, perhaps near unthinkable question. Can the vast, magnificent order seen in development conceivable arise as a spontaneous self organized property of complex genetic systems? Why "unthinkable"? It is, after all, not the answers which scientists uncover, but the strange magic lying behind the questions they pose to their world, knower and known, which is the true impulse driving profound conceptual transformation. Answers will be found, contrived, wrested, once the question is divined. Why "unthinkable"? Since Darwin, we have viewed organisms, in Jacob's phrase, as bricolage, tinkered together contraptions. Evolution, says Monod, is "chance caught on the wing". Lovely dicta, these, capturing the core of the Darwinian world view in which organisms are perfected by natural selection acting on random variations. The tinkerer is an opportunist, its natural artifacts are ad hoc accumulations of this and that, molecular Rube Goldbergs satisfying some spectrum of design constraints.

In the world view of bricolage, selection is the sole, or if not sole, the preeminent source of order. Further, if organisms are ad hoc solutions to design problems, there can be no deep theory of order in biology, only the careful dissection of the ultimately accidental machine and its ultimately accidental evolutionary history.

The genomic system linking the activity of thousands of genes stands at the summit of four billion years of an evolutionary process in which the specific genes, their regulatory intertwining and the molecular logic have all stumbled forward by random mutation and natural selection. Must selection have struggled against vast odds to create order? Or did that order lie to hand for selection's further molding? If the latter, then what a reordering of our view of life is mandated!

Order, in fact, lies to hand. Our intuitions have been wrong for thousands of years. We must, in fact, revise our view of life. Complex molecular regulatory networks inherently behave in two broad regimes separated by a third phase transition regime: The two broad regimes are chaotic and ordered. The phase transition zone between these two comprises a narrow third complex regime poised on the boundary of chaos. (Kauffman 1969, 1989; Fogleman-Soulie 1985; Derrida and Pomeau 1986; Langton 1991; Kauffman 1991, 1992). Twenty five years after the initial discovery of these regimes, a summary statement is that the genetic systems controlling ontogeny in mouse, man, bracken, fern, fly, bird, all appear to lie in the ordered regime near the edge of chaos. Four billion years of evolution in the capacity to adapt offers a putative answer: Complex adaptive systems achieve, in a lawlike way, the edge of chaos.

Tracing the history of this discovery, the discovery that extremely complex systems can exhibit "order for free", that our intuitions have been deeply wrong, begins with the intuition that even randomly "wired" molecular regulatory "circuits" with random "logic" would exhibit orderly behavior if each gene or molecular variable were controlled by only a few others. Notebooks from that period mix wire-dot diagrams of organic molecules serving as drugs with wire dot models of genetic circuitry. The intuition proved correct. Idealizing a gene as "on" or "off", it was possible by computer simulations to show that large systems with thousands of idealized genes behaved in orderly ways if each gene is directly controlled by only two other genes. Such systems spontaneously lie in the ordered regime. Networks with many inputs per gene lie in the chaotic regime. Real genomic systems have few molecular inputs per gene, reflecting the *specificity* of molecular binding, and use a *biased* class of logical rules, reflecting molecular simplicity, to control the on/off behavior of those genes. Constraint to the vast ensemble of possible genomic systems characterized by these two "local constraints" also inevitably yields genomic systems in the ordered regime. The perplexing, enigmatic, magical order of ontogeny may largely reflect large scale consequences of polymer chemistry.

Order for free. But more: The spontaneously ordered features of such systems parallels a host of ordered features seen in the ontogeny of mouse, man, bracken, fern, fly, bird. A "cell type" becomes a stable recurrent pattern of gene expression, an "attractor" in the jargon of mathematics, where an attractor, like a whirlpool, is a region in the state space of all the possible patterns of gene activities to which the system flows and remains. In the spontaneously ordered regime, such cell type attractors are inherently small, stable, and few, implying that the cell types of an organism traverse their recurrent patterns of gene expression in hours not eons, that homeostasis, Claude Bernard's conceptual child, lies inevitably available for selection to mold, and, remarkably, that it should be possible to predict the number of cell types, each a whirlpool attractor in the genomic repertoire, in an organism. Bacteria harbor one to two cell types, yeast three, ferns and bracken some dozen, man about two hundred and fifty. Thus, as the number of genes, called genomic complexity, increases, the number of cell types increases. Plotting cell types against genomic complexity, one finds that the number of cell types increases as a square root function of the number of genes. And, outrageously, the number of whirlpool attractors in model genomic systems in the ordered regime also increase as a square root function of the number of genes. Man, with about 100,000 genes should have three hundred seventy cell types, close to two hundred and fifty. A simple alternative theory would predict billions of cell types.

Bacteria, yeast, ferns, and man, members of different phyla, have no common ancestor for the past 600 million years or more. Has selection struggled for 600 million years to achieve a square root relation between genomic complexity and number of cell types? Or is this order for free so deeply bound into the roots of biological organization that selection cannot *avoid this order*? But if the latter, then *selection is not the sole source of order in biology*. Then Darwinism must be extended to embrace self organization and selection.

The pattern of questions posed here is novel in biology since Darwin. In the NeoDarwinian world view, where organisms are ad hoc solutions to design problems, the answers lie in the specific details wrought by ceaseless selection. In contrast, the explanatory approach offered by the new analysis rests on examining the statistically typical, or generic, properties of an entire class, or "ensemble" of systems all sharing known local features of genomic systems. If the typical, generic, features of ensemble members corresponds to that seen in organisms, then explanation of those features emphatically *does not* rest in the details. It rests in the general laws governing the typical features of the ensemble as a whole. Thus an "ensemble" theory is a new kind of statistical mechanics. It predicts that the typical properties of members of the ensemble will be found in organisms. Where true, it bodes a physics of biology.

Not only a physics of biology, but beyond, such a new statistical mechanics demands a new pattern of thinking with respect to biological and even cultural evolution: Self organization, yes, aplenty. But selection, or its analogues such as profitability, is always acting. We have no theory in physics, chemistry, biology, or beyond which marries self organization and selection. The marriage consecrates a new view of life.

But two other failures of Darwin, genius that he was, must strike us. How do organisms, or other complex entities, manage to adapt and learn? That is, what are the conditions of "evolvability". Second, how do complex systems coordinate behavior, and more deeply, why are adaptive systems so often complex?

Consider "evolvability" first. Darwin supposed that organisms evolve by the *successive accumulation of useful random variations*. Try it with a standard computer program. Mutate the code, scramble the order of instructions, and try to "evolve" a program calculating some complex function. If you do not chuckle, you should. Computer programs of the familiar type are not readily "evolvable". Indeed the more compact the code, the more lacking in redundancy, the more sensitive it is to each minor variation. Optimally condensed codes are, perversely, minimally evolvable. Yet the genome is a kind of molecular computer, and clearly has succeeded in evolving. But this implies something very deep: Selection must achieve the kinds of systems which are able to adapt. That capacity is not Godgiven, it is a success.

If the capacity to evolve must itself evolve, then the new sciences of complexity seeking the laws governing complex adapting systems must discover the laws governing the emergence and character of systems which can themselves adapt by accumulation of successive useful variations.

But systems poised in the ordered regime near its boundary are precisely those which can, in fact, evolve by successive minor variations. The behavior of systems in the chaotic regime are so drastically altered by any minor variation in structure or logic that they cannot accumulate useful variations. Conversely, systems deep in the ordered regime are changed so slightly by minor variations that they adapt too slowly to an environment which may sometimes alter catastrophically. Evolution of the capacity to adapt would be expected, then, to achieve poised systems.

How can complex systems coordinate behavior? Again, complex adaptive entities achieve the edge of chaos because such systems can coordinate the most complex behavior there. Deep in the chaotic regime, alteration in the activity of any element in the system unleashes an avalanche of changes, or "damage", which propagates throughout most of the system (Stauffer 1987). Such spreading damage is equivalent to the "butterfly effect" or sensitivity to initial conditions typical of chaotic systems. The butterfly in Rio changes the weather in Chicago. Crosscurrents of such avalanches unleashed from different elements means that behavior is not controllable. Conversely, deep in the ordered regime, alteration at one point in the system only alters the behavior of a few neighboring elements. Signals cannot propagate widely throughout the system. Thus, control of complex behavior cannot be achieved. Just at the boundary between order and chaos, the most complex behavior can be achieved.

Finally, computer simulations suggest that natural selection or its analogues actually do achieve the edge of chaos. This third regime, poised between the broad ordered regime and the vast chaotic regime, is razorblade thin in the space of systems.

Absent other forces, randomly assembled systems will lie in the ordered or chaotic regimes. But let such systems play games with one another, winning and losing as each system carries out some behavior with respect to the others, and let the structure and logic of each system evolve by mutation and selection, and, lo, systems do actually adapt to the edge of chaos! No minor point this: Evolution itself brings complex systems, when they must adapt to the actions of other, to an internal structure and logic poised between order and chaos, (Kauffman 1991).

We are lead to a bold hypothesis: Complex adaptive systems achieve the edge of chaos.

The story of the "edge of chaos" is stronger, the implications more surprising. Organisms, economic entities, nations, do not evolve, they *coevolve*. Almost miraculously, coevolving systems, too, mutually achieve the poised edge of chaos. The sticky tongue of the frog alters the fitness of the fly, and deforms its fitness landscapes that is, what changes in what phenotypic directions improve its chance of survival. But so too in technological evolution. The automobile replaced the horse. With the automobile came paved roads, gas stations hence a petroleum industry and war in the Gulf, traffic lights, traffic courts, and motels. With the horse went stables, the smithy, and the pony express. New goods and services alter the economic landscape. Coevolution is a story of coupled deforming "fitness landscapes". The outcome depends jointly on how much my landscape is deformed when you make an adaptive move, and how rapidly I can respond by changing "phenotype".

Are there laws governing coevolution? And how might they relate to the edge of chaos? In startling ways. Coevolution, due to a selective "metadynamics" tuning the structure of fitness landscapes and couplings between them, may typically reach the edge of chaos (Kauffman 1992). E.coli and IBM not only "play" games with the other entities with which they coevolve. Each also participates in the very *definition or form* of the game. It is we who create the world we mutually inhabit and in which we struggle to survive. In models where players can "tune" the mutual game even as they play, or coevolve, according to the game existing at any period, the entire system moves to the edge of chaos. This surprising result, if general, is of paramount importance. A simple view of it is the following: Entities control a kind of "membrane" or boundary separating inside from outside. In a kind of surface to volume way, if the surface of each system is small compared to its volume it is rather insensitive to alterations in the behaviors of other entities. That is, adaptive moves by other partners do not drastically deform one partner's fitness landscape. Conversely, the ruggedness of the adaptive landscape of each player as it changes its "genotype" depends upon how dramatically its behavior deforms as its genotype alters. In turn this depends upon whether the adapting system is itself in the ordered, chaotic, or boundary regime. If in the ordered, the system itself adapts on a smooth landscape. In the chaotic regime the system adapts on a very rugged landscape. In the boundary regime the system adapts on a landscape of intermediate ruggedness, smooth in some directions of "genotype" change, rugged in other directions. Thus, both the ruggedness of one's own fitness landscape and how badly that landscape is deformed by moves of one's coevolving partners are *themselves possible objects of a selective "metadynamics"*. Under this selective metadynamics, tuning landscape structure and susceptibility, model coevolving systems which mutually know and interact with one another actually reach the edge of chaos. Here, under most circumstances, most entity optimizes fitness, or payoff, by remaining the same. Most of the ecosystem is frozen into a percolating Nash equilibrium, while coevolutionary changes propagate in local unfrozen islands within the ecosystem. More generally, alterations in circumstances send avalanches of changed optimal strategies propagating through the coevolving system. At the edge of chaos the size distributions of those avalanches approach a power law, with many small avalanches and few large ones. During such coevolutionary avalanches, affected players would be expected to fall transiently to low fitness, hence might go extinct. Remarkably, this size distribution comes close to fitting the size distribution of extinction events in the record. At a minimum, a distribution of avalanche sizes from a common size small cause tells us that small and large extinction events may reflect endogenous features of coevolving systems more than the size of the meteor which struck.

The implications are mini-Gaia. As if by an invisible hand, coevolving complex entitites may mutually attain the poised boundary between order and chaos. Here, mean sustained payoff, or fitness, or profit, is optimized. But here avalanches of change on all length scales can propagate through the poiised system. Neither Sisyphis, forever pushing the punishing load, nor fixed unchanging and frozen, instead E.coli and its neighbors, IBM and its neighbors, even nation states in their collective dance of power, may attain a precarious poised complex adaptive state. The evolution of complex adaptive entities itself appears lawful. How far we come from Darwin's genius.

This strand in the birth of complexity theory, here spun, has its history. The first stages were set in the mid 1960s by the discovery of spontaneous order, as well as the expected chaos, in complex genomic systems. The discovery was not without attention among scientists the day. Warren McCulloch, patriarch of cybernetics, author with Pitts of "The Logical Calculus of Ideas Imminent in the Mind", step-child of Bertrand Russell's logical atomism, and ancestor to todays neural connectionist tony flowering, invited me to share his home with his remarkable wife Rook. "In pine tar is. In oak none is. In mud eels are. In clay none are", sang this poet of neural circuitry, demonstrating by dint of a minor Scots accent that no hearer could unscramble four simple declarative sentences. Mind, complex, could fail to classify. "All Cambridge excited about your work", wrote McCulloch to this medical student who, thrilled, was yet to decode Warren's style.

Yet the time was not ripe. McCulloch had said twenty years would elapse before biologists took serious note. He was right, almost to the hour. And for good reason had he made his prediction. The late 1960s witnessed the blunderbuss wonderful explosion of molecular biology. Enough, far more than enough, to thrill to the discovery of the real molecular details: How a gene is transcribed to RNA, translated to protein, acts on its neighbors. What is the local logic of a bacterial genetic circuit controlling metabolism of lactose? Of a bacterial virus, or phage? What of the genes in a higher organism like the heralded but diminutive fruit fly? What of mouse and man? Enveloped by the Darwinian world view, whose truths run deep, held in tight thrall by the certainty that the order in organisms resides in the well wrought details of construction and design, details inevitably ad hoc by virtue of their tinkered origins in the wasteland of chance, molecular biologists had no use for heady, arcane, abstract ensemble theories. The birth of complexity theory, or this strand of it, though noted, received no sustaining passion from its intended audience.

Twenty years, indeed. Rebirth of this strand was midwifed by the physicists. An analogue ensemble theory, called "spin glasses", had been developed starting in the mid 1970s by solid state physicists such as Philip Anderson, Scott Kirkpatrick, Bernard Derrida, Gerard Toulouse, were struggling with an odd kind of dilute magnet material. Unlike the familiar ferromagnet, captured in the famous Ising model, where magnetic spins like to orient in the same direction as their neighboring spins, hence the magnetized state with all spins oriented in the same direction arises, in these bewildering spin glasses, adjacent spins might like to orient in the same or in the opposite direction, depending sinusoidally on the distance between the spins. What a mess. Edwards and Anderson started an industry among their brethren, and legitimized the new class of ensemble theories, by building mathematical models of spin glasses on two or three dimensional lattices,. Here each vertex houses a spin. But, to capture the bizarre logic of their magnetic materials, Edwards and Anderson assumed that each adjacent pair of spins "chose", once and forever, whether they wanted to point in the same or opposite direction, and how much they cared, given by an energy for that bond. Such messy models meant two major things. First, since couplings are assigned at random, any one model spin glass is a member of a vast ensemble governed by the same statistics. This is an ensemble theory averaging, not over the states of one system as in the familiar statistical mechanics of gases, but over billions of systems in the same ensemble. One seeks and characterizes the typical, or generic features of these systems. Second, such systems have tortuous and rugged "energy landscapes". This is due to "frustration". Consider four spins around a square, where three pairs wish to point in the same direction, the fourth does not. All cannot be satisfied. Each configuration of the many spins in the lattice of a spin glass has a total energy. The distribution of energies over the configurations is the energy landscape, the analogue of a fitness landscape. Frustration implies that the landscape is rugged and multipeaked.

Later, the structures of these spin glass landscapes would provide new models of molecular evolution over rugged multipeaked fitness landscapes. Molecular evolution turns out to be much like an electron bouncing on a complex potential surface at a small temperature. At too low a temperature, the electron remains trapped in poor potential wells. At too high a temperature, the electron bounces all over the potential surface and has a high, unhappy, average energy. On any defined time scale, energy is minimized at a specific fixed temperature at which the electron is just "melting" out over the energy landscape, sliding gracefully over low saddles in the surface separating wells such that it finds good potential wells rather easily, then does not hop out of them too rapidly. The analogue in molecular evolution or other biological evolution over a fixed fitness landscape, or one deforming at a given mean rate, is to tune the parameters of adaptive search over the space such that an adapting population is just "melting" out of local regions of the space. Again: The edge of Chaos!

By 1985 many of the physicists had tired of their spin glasses. Some turned to models of neural networks, sired by McCulloch, where neurons turn one another on and off rather like genes, or like spins for that matter. Hopfield found further fame by modeling parallel processing neural networks as spin systems, (Hopfield 1982). Attractors of such networks, rather than modeling cell types as I had suggested, were taken to model memories. Each memory was an attractor. Memories were content addressable, meaning that if the network were begun in the "basin of attraction" drained by one whirlpool attractor, the system would flow to that attractor. Partial data, corresponding to an initial state in a basin of attraction but not on the attractor itself, could be reconstructed to the stored memory. (All scientists regret the article not written. Jack Cowan and I had sketched an article in 1970 arguing against the logical atomism implicit in McCulloch and Pitts, an atomism melted by Wittgenstein's Investigations. In contrast, we wanted to suggest that concepts were attractors in neural networks, hence a collective integrated activity. From Wittgenstein we knew that language games are not reducible to one another, law to human action to physical phenomena. We wanted to argue that new concepts, new language games, arose by bifurcations yielding new attractors in the integrated activity of coupled neurons. One such attractor would not be reducible in any obvious way to another attractor. Grandmother cells be damned, concepts are collective properties.) Toulouse, brilliant as Hopfield, followed with other spin glass like models whose basins of attraction were, he said, more like French than English gardens. Many have followed, to the field's flowering.

Not all the physicists who tired of spin glasses turned to neurobiology. In the way of these things, French physicist Gerard Weishbuch was romantically involved with French mathematician Francoise Fogleman-Soulie. Francoise chose, as her thesis topic, the still poorly understood order found in "Kauffman nets", (Fogleman-Soulie 1985). Many theorems followed. Gerard's interest extended from Francoise and spin glasses to this strange hint of order for free. Summers in Jerusalem and Haddasah hospital with Henri Atlan, doctor, theoretical biologist, author of Crystal and Smoke with its search for order and adaptability, led to more results. Put these bizarre genetic networks on lattices, where any good problem resides. See the order. Scale parameters. Find phase transitions and the scaling laws of critical exponents. A new world to a biologist. And Gerard shared an office with Bernard Derrida, nephew of deconstructionist Jacques. Bernard looked at at these "Kauffman nets", the name is due to Derrida, and leaped to an insight no biologist would ever dare. Let the network be randomly rewired at each moment, creating an "annealed" model. Theorem followed theorem. No genome dances so madhatterly. But the mathematics can. Phase transition assured. Order for free in networks of low connectivity. Analysis of sizes of basins of attraction, and of overlaps between attractors, (Derrida and Pomeau 1986). I lost a bottle of wine to Derrida, shared over dinner, on the first theorem.

Even I chimed in with a few theorems here and there: a mean field approach to attractors, the existence of a connected set of elements which are "frozen" and do not twinkle on and off, that spans or percolates across the system. This frozen component, leaving behind isolated twinkling islands, is the hallmark of order. The phase transition to chaos occurs, as parameters alter, when the frozen component "melts", and the twinkling islands merge into an unfrozen, twinkling, percolating sea, leaving behind small isolated frozen islands. The third, complex regime, the boundary between order and chaos, arises when the twinkling connected, percolating sea is just breaking up into isolated islands. Avalanches of changes due to perturbations, which only propagate in the twinkling unfrozen sea, show a characteristic "power law" distribution at the phase transition, with many small avalanches and a few enormous ones, (Kauffman 1989).

Now the reader can see why systems on the boundary between order and chaos can carry out the most complex tasks, adapt in the most facile fashion. Now too, I hope, you can see the intrigue.at the possibility that complex adaptive systems achieve the edge of chaos in their internal structure, but may also coevolve in a selective metadynamics to achieve the edge of chaos in the ecosystem of the mutual games they play! The edge of chaos may be a major organizing principle governing the evolution and coevolution of complex adaptive systems.

Other themes, again spawned by physicists, arose in America, and lead quasi-independently, quasi-conversing, to the growth of interest in complexity. "Kauffman nets", where the wiring diagram among "genes" or binary elements, is random, and the logic governing each element is randomly assigned, hence differs for different "genes", are versions of a mathematical structure called "cellular automata". Cellular automata were invented by von Neuman, whose overwhelming early work, here and on the existence of self reproducing automata, filters down through much that follows. The simplest cellular automata are lines or rings of on/off sites, each governed by the same logical rule which specifies its next activity, on or off, as a function of its own current state and those of its neighbors to a radius, r. Enter young Stephen Wolfram, quick, mercurial, entreprenurial. The youngest MacArthur Fellow, Wolfram had begun publishing in high energy physics at age 16. While a graduate student at Cal Tech, he earned the mixed admiration and emnity of his elders by inventing computer code to carry out complex mathematical calculations. Cal Tech did not mind his mind. It minded his marketing the products of his mind. Never mind. Thesis done, Wolfram packed off to the Institute for Advanced Study and fell to the analysis of cellular automata. He amazed his audiences. The world of oddball mathematicians, computer scientists, wayward physicists, biologists soon twiddled with CA rules. Four classes of behavior emerged, stable, periodic, and chaotic, of course. And between them, on the edge between order and chaos, capable of complex computation, perhaps universal computation? A third "complex class". Among the most famous of these CA rules is Conway's "Game of Life", provable capable of universal computation, demonstrably capable of capturing gigabits of memory and gigaseconds of time among amateurs and professionals world wide. The game of life, like true life itself according to our bold hypothesis, also lies at the edge of chaos.

Paralleling Derrida is the lineage flowing from Chris Langton. Langton, a computer scientist and physicist, elder graduate student, survivor of early hang gliding and an accident relieving him of most unbroken bone structure in his mid-twenties body, thought he could improve on von Neuman. He invented a simple self reproducing automaton and littered computer screens from Los Alamos to wherever. Then Langton, following von Neuman again, and fired up by Wolfram, began playing with cellular automata. Where I had shown that the transition from order to chaos was tuned by tuning the number of inputs per "gene" from 2 to many, Langton reinvented Derrida's approach. Derrida, like Langton after him, in turn reinvented a classification of logical rules first promulgated by Crayton Walker. This classification marks the bias, P, towards the active, or inactive state, over all combinations of activities of the inputs to an element. Derrida had shown that the phase transition occurred at a critical value of this bias, Pc. At that bias, frozen components emerge. Langton found the same phase transition, but measured in a different way to focus on how complex a computation might be carried out in such a network. This complexity, measured as mutual information, or what one can predict about the next activity of one site given the activity of another site, is maximized at the phase transition (Langton 1991).

The poised edge reappears, like a new second law of thermodynamics, everywhere hinted, but, without Carnot, not yet clearly articulated, in the recent work of physicist Jim Crutchfield. "Symbolic dynamics" is a clever new tool used to think about complex dynamical systems. Imagine a simple system such as a pendulum. As its swings back and forth, it crosses the midpoint where it hangs straight down. Use a 1 to denote times when the pendulum is to the left of the midpoint, and 0 to denote times when the pendulum swings to the right. Evidently, the periodic pendulum gives rise to an alternating sequence of 1 and 0 values. Such a symbol sequence records the dynamics of the pendulum by breaking its state space into a finite number of regions, here two, and labeling each region with a symbol. The flow of the system gives rise to a symbol sequence. Theorems demonstrate that, with optimally chosen boundaries between the regions, here the midpoint, the main features of the dynamics of the real pendulum can be reconstructed from the symbol sequence. For a periodic process, the symbol sequence is dull. But link several pendulums together with weak springs and again denote the behavior of one pendulum by 1 and 0 symbols. Now the motion of each pendulum is influenced by all the others in very complex ways. The symbol sequence is correspondingly complex. The next step is to realize that any symbol sequence can be generated as the output of a finite automaton, a more or less complex "neural" or "genetic" network of on off elements. Further, theorems assure us that for any such symbol sequence, the smallest, or minimal automaton, with the minimal number of elements and internal states, can be found. Thus, the number of elements, or states, of such a system is a measure of the complexity of the symbol sequence. And now the wonderful surprise. The same three phases, ordered, chaotic, and complex, are found again. That is, such automata, like Kauffman nets and neural nets, harbor the same generic behaviors. And, as you will now suspect, the complex regime again corresponds to the most complex symbol sequences, which in turn arise in dynamical systems themselves on the boundary between order and chaos.

If one had to formulate, still poorly articulated, the general law of adaptation in complex systems, it might be this: Life adapts to the edge of chaos.

2. The Origin of Life and its Progeny

This story, the story of the boundary between order an chaos achieved by complex coevolving systems, is but half the emerging tale. The second voice tells of the origin of life itself, a story both testable and, I hope, true, a story implying vast stores of novel drugs, vaccines, universal enzymatic tool boxes, a story latent with the telling of technological and cultural evolution, of bounded rationality, the coemergence of knower and known, hence at last, of telling whether E. coli and IBM do, in fact, know their worlds, the worlds they themselves created, in the same deep way.

Life is held a miracle, God's breath on the still world, yet cannot be. Too much the miracle, then we were not here. There must be a viewpoint, a place to stand, from which the emergence of life is explicable, not as a rare untoward happening, but as expected, perhaps inevitable. In the common view, life originated as a self reproducing polymer such as RNA, whose self complementary structure, since Watson and Crick remarked with uncertain modesty, suggests its mode of reproduction, has loomed the obvious candidate urbeast to all but the stubborn. Yet stubbornly resistant to test, to birthing in vitro is this supposed simplest molecule of life. No worker has yet succeeded in getting one single stranded RNA to line up the complementary free nucleotides, link them together to form the second strand, melt them apart, then repeat the cycle. The closest approach shows that a polyC polyG strand, richer in C than G, can in fact line up its complementary strand. Malevolently, the newly formed template is richer in G than C, and fails, utterly, to act as a facile template on its own. Alas.

Workers attached to the logic of molecular complementarity are now focusing effort on polymers other than RNA, polymers plausibly formed in the prebiotic environment, which might dance the still sought dance. Others, properly entranced with the fact that RNA can act as an enzyme, called a ribozyme, cleaving and ligating RNA sequences apart and together, seek a ribozyme which can glide along a second RNA, serving as a template that has lined up its nucleotide complements, and zipper them together. Such an ribozyme would be a ribozyme polymerase, able to copy any RNA molecule, including itself. Beautiful indeed. And perhaps such a molecule occurred at curtain-rise or early in the first Act. But consider this: A free living organism, even the simplest bacterium, links the synthesis and degradation of some thousands of molecules in the complex molecular traffic of metabolism to the reproduction of the cell itself. Were one to begin with the RNA urbeast, a nude gene, how might it evolve? How might it gather about itself the clothing of metabolism?

There is an alternative approach which states that life arises as a nearly inevitable phase transition in complex chemical systems. Life formed by the emergence of a collectively autocatalytic system of polymers and simple chemical species.

Picture, strangely, ten thousand buttons scattered on the floor. Begin to connect these at random with red threads. Every now and then, hoist a button and count how many buttons you can lift with it off the floor. Such a connected collection is call a "component" in a "random graph". A random graph is just a bunch of buttons connected at random by a bunch of threads. More formally, it is a set of N nodes connected at random by E edge. Random graphs undergo surprising phase transitions. Consider the ratio of E/N, or threads divided by buttons. When E/N is small, say .1, any button is connected directly or indirectly to only a few other buttons. But when E/N passes 0.5, so there are half as many threads as buttons, a phase transition has occurred. If a button is picked up, very many other buttons are picked up with it. In short, a "giant component" has formed in the random graph in which most buttons are directly or indirectly connected with one another. In short, connect enough nodes and a connected web "crystalizes".

Now life. Proteins and RNA molecules are linear polymers build by assembling a subset of monomers, twenty types in proteins, four in RNA. Consider the set of polymers up to some length, M, say 10. As M increases the number of types of polymers increases exponentially, for example there are 20 M proteins of length M. This is a familiar thought. The rest are not. The simplest reaction among two polymers consists in gluing them together. Such reactions are reversible, so the converse reaction is simply cleaving a polymer into two shorter polymers. Now count the number of such reactions among the many polymers up to length M. A simple consequence of the combinatorial character of polymers is that there are many more reactions linking the polymers than there are polymers. For example, a polymer length M can be formed in M 1 ways by gluing shorter fragments comprising that polymer. Indeed, as M increases, the ratio of reactions among the polymers to polymers is about M, hence increases as M increases. Picture such reactions as black, not red, threads running from the two smaller fragments to a small square box, then to the larger polymer made of them. Any such triad of black threads denotes a possible reaction among the polymers; the box, assigned a unique number, labels the reaction itself. The collection of all such triads is the chemical reaction graph among them. As the length of the longest polymer under consideration, M, increases, the web of black triads among these grows richer and richer. The system is rich with crosslinked reactions.

Life is an autocatalytic process where the system synthesizes itself from simple building blocks. Thus, in order to investigate the conditions under which such a autocatalytic system might spontaneously form, assume that no reaction actually occurs unless that reaction is catalyzed by some molecule. The next step notes that protein and RNA polymers can in fact catalyse reactions cleaving and ligating proteins and RNA polymers: trypsin in your gut after dinner digesting steak, or ribozyme ligating RNA sequences. Build a theory showing the probability that any given polymer catalyzes any given reaction. A simple hypothesis is that each polymer has a fixed chance, say one in a billion, to catalyze each reaction. No such theory can now be accurate, but this hardly matters. The conclusion is at hand, and insensitive to the details. Ask each polymer in the system, according to your theory, whether it catalyzes each possible reaction. If "yes", color the corresponding reaction triad "red", and note down which polymer catalyzed that reaction. Ask this question of all polymers for each reaction. Then some fraction of the black triads have become red. The red triads are the catalyzed reactions in the chemical reaction graph. But such a catalyzed reaction graph undergoes the button thread phase transition. When enough reactions are catalyzed, a vast web of polymers are linked by catalyzed reactions. Since the ratio of reactions to polymers increases with M, at some point as M increases at least one reaction per polymer is catalyzed by some polymer. The giant component crystalizes. An autocatalytic set which collectively catalyzes its own formation lies hovering in the now pregnant chemical soup. A self reproducing chemical system, daughter of chance and number, swarms into existence, a connected collectively autocatalytic metabolism. No nude gene, life emerged whole at the outset.

I found this theory in 1971. Even less than order for free in model genomic systems did this theory find favor. Stuart Rice, colleague, senior chemist, member of the National Academy of Science asked, "What for?" Alas again. When famous older scientists say something warrants the effort, rejoice. When famous older scientists are dismissive, beware. I turned to developmental genetics and pattern formation, the beauty of Alan Turing's theory of pattern formation by the establishment of chemical waves, the quixotic character of homeotic mutants in the fruit fly, Drosophila melanogaster, where eyes convert to wings, antennas to legs, and heads to genitalia. Fascinating disorders, these, called metaplasias, whose battered sparse logic hinted the logic of developmental circuits. But experimental developmental genetics, even twelve years and surgery on ten thousand embryos, is not the central thread of the story.

In 1983 interest in serious theories of the origin of life was rekindled. In 1971 and the ensuing decade, Nobelist Manfred Eigen, together with theoretical chemist Peter Schuster, developed a well formulated, careful model of the origin of life, called the "hypercycle". In this theory, the authors begin by assuming that short nude RNA sequences can replicate themselves. The hooker is this: During such replication, errors are made. The wrong nucleotide may be incorporated at any site. Eigen and Schuster showed that an error catastrophe occurs when RNA sequences become too long for any fixed error rate. The RNA population "melts" over RNA sequence space, hence all information accumulated within the "best" RNA sequence, culled by natural selection, is lost. The "hypercycle" is a clever answer to this devastation: Assume a set of different short RNA molecules, each able to replicate itself. Now assume that these different RNA molecules are arranged in a control cycle, such that RNA 1 helps RNA 2 to replicate, RNA 2 helps RNA 3, and so on until RNA N closes the loop by helping RNA 1. Such a loop is a hypercycle, "hyper" because each RNA itself is a tiny cycle of two complementary strands which copy one another. The hypercycle is, in fact, a coevolving molecular society. Each RNA species coevolves in company with its peers. This model has been studied in detail, and has strengths and weakness. Not the least of the latter is that no RNA sequence can yet replicate itself.

But other voices were lifted, from the most intelligent minds. Freeman Dyson, of the Institute of Advanced Studies, an elegant scientist and author of lyric books such as "Disturbing the Universe", suggested, in Origins of Life, that life arose as a phase transition in complex systems of proteins. Philip Anderson, with Daniel Stein, and Rothkar, borrowed from spin-glass theory to suggest that a collection of template replicating RNA molecules with overlapping ends and complex fitness functions governing their survival might give rise to many possible self reproducing sequences.

Lives in science have their peculiar romance. I heard of these approaches at a conference in India. Central India, Madya Pradesh, sweats with the sweet smell of the poor cooking over fires of dried buffalo dung. The spiritual character of India allows one to speak of the origin of life with colleagues such as Humberto Maturana, riding in disrepair except for his glasses and clear thoughts, in a bus of even greater disrepair among the buffalo herds to Sanchi, early Buddest shrine. The Budda at the west portal, thirteen hundred years old, ineffably young, invited only a gentle kiss from the foreigners in time, space, culture. Dyson's and Anderson's approaches appeared flawed. Dyson had assumed his conclusion, hidden in assumption 7. Life as an autocatalytic crystalization was trivially present in his model, slipped in by hand, not accounted for as a deeply emergent property of chemistry. And Anderson, overwhelmingly insightful, proposed nothing truly deep not already resting on RNA self complementarity. The romance continues with a flurry of theorems and lemmas, simple to a real mathematician.

This hiccup of creativity, I hoped, warranted investigation. Doyne Farmer, young physicist at Los Alamos, and his childhood friend Norman Packard, and I began collaborating to build detailed computer simulations of such autocatalytic polymer systems. Six years later, a Ph.D. thesis by Richard Bagley later, it is clear that the initial intuitions were fundamentally correct: In principle complex systems of polymers can become collectively self reproducing. The routes to life are not twisted backalleys of thermodynamic improbability, but broad boulevards of combinatorial inevitability.

If this new view of the crystalization of life as a phase transition is correct, then it should soon be possible to create actual self reproducing polymer systems, presumably of RNA or proteins, in the laboratory. Experiments, even now, utilizing very complex libraries of RNA molecules to search for autocatalytic sets are underway in a few laboratories.

If not since Darwin, then since Weisman's doctrine of the germ plasm was reduced to molecular detail by discovery of the genetic role of chromosomes, biologist have believed that evolution via mutation and selection virtually requires a stable genetic material as the store of heritable information. But mathematical analysis of autocatalytic polymer systems belies this conviction. Such systems can evolve to form new systems. Thus, contrary to Richard Dawkin's thesis in "The Selfish Gene", biological evolution does not, in principle, demand self-replicating genes at the base (Dawkins 1976). Life can emerge and evolve without a genome. Heresy, perhaps? Perhaps.

Many and unexpected are the children of invention. Autocatalytic polymer sets have begotten an entire new approach to complexity.

The starting point is obvious. An autocatalytic polymer set is a functional integrated whole. Given such a set, it is clear that one can naturally define the function of any given polymer in the set with respect to the capacity of the set to reproduce itself. Lethal mutants exist, for if a given polymer is removed, or a given foodstuff deleted, the set may fail to reproduce itself. Ecological interactions among coevolving autocatalytic sets lie to hand. A polymer from one such set injected into a second such set may block a specific reaction step and "kill" the second autocatalytic set. Coevolution of such sets, perhaps bounded by membranes, must inevitably reveal how such systems "know" one another, build internal models of one another, and cope with one another. Models of the evolution of knower and known lay over the conceptual horizon.

Walter Fontana, graduate student of Peter Schuster, came to the Santa Fe Institute, and Los Alamos. Fontana had worked with John McCaskill, himself an able young physicist collaborating with Eigen at the Max Planck Institute in Gottingen. McCaskill dreamt of polymers, not as chemicals, but as Turing machine computer programs and tapes. One polymer, the computer, would act on another polymer, the tape, and "compute" the result, yielding a new polymer. Fontana was entranced. But he also found the autocatalytic story appealing. Necessarily, he invented "Algorithmic Chemistry". Necessarily, he named his creation "Alchemy", (Fontana 1991).

Alchemy is based on a language for universal computation called the lambda calculus. Here almost any binary symbol string is a legitimate "program" which can act on almost any binary symbol string as an input to compute an output binary symbol string. Fontana created a "Turing gas" in which an initial stock of symbol strings randomly encounter one another in a "chemostat" and may or may not interact to yield symbol strings. To maintain the analogue of selection, Fontana requires that a fixed total number of symbol string polymers be maintained in the chemostat. At each moment, if the number of symbol strings grows above the maximum allowed, some randomly strings are lost from the system.

Autocatalytic sets emerge again! Fontana finds two types. In one, a symbol string which copies itself emerges. This "polymerase" takes over the whole system. In the second, collectively autocatalytic sets emerge in which each symbol string is made by some other string or strings, but none copies itself. Such systems can then evolve in symbol string space evolution without a genome.

Fontana had broken the bottleneck. Another formulation of much the same ideas, which I am now using, sees interactions among symbol strings creating symbol strings and carrying out a "grammar". Work on disordered networks, work which exhibited the three broad phases, ordered, chaotic and complex, drove forward based on the intuition that order and comprehensibility would emerge by finding the generic behavior in broad regions of the space of possible systems. The current hope is that analysis of broad reaches of grammar space, by sampling "random grammars", will yield deep insight into this astonishingly rich class of systems.

The promise of these random grammar systems extend from analysis of evolving proto-living systems, to characterizing mental processes such as multiple personalities, the study of technological coevolution, bounded rationality and non-equilibrium price formation at the foundations of economic theory, to cultural evolution. And the origin of life model itself, based on the probability that an arbitrary protein catalyzes an arbitrary reaction, spawned the idea of applied molecular evolution the radical concept that we might generate trillions of random genes, RNA sequences, and proteins, and learn to evolve useful polymers able to serve as drugs, vaccines, enzymes, and biosensors. The practical implications now appear large.

Strings of symbols which act upon one other to generate strings of symbols can, in general, be computationally universal. That is, such systems can carry out any specified algorithmic computation. The immense powers and yet surprising limits, trumpeted since Gödel, Turing, Church, Kleene, lie before us, but in a new and suggestive form. Strings acting on strings to generate strings create an utterly novel conceptual framework in which to cast the world. The puzzle of mathematics, of course, is that it should so often be so outrageously useful in categorizing the world. New conceptual schemes allow starkly new questions to be posed.

A grammar model is simply specified. It suffices to consider a set of M pairs of symbol strings, each about N symbols in length. The meaning of the grammar, a catch-as-catch-can set of "laws of chemistry" is this: Wherever the left member of such a pair is found in some symbol string in a "soup" of strings, substitute the right member of the pair. Thus, given an initial soup of strings, one application of the grammar might be carried out by us, acting Godlike. We regard each string in the soup in turn, try all grammar rules in some precedence order, and carry out the transformations mandated by the grammar. Strings become strings become strings. But we can let the strings themselves act on one another. Conceive of a string as an "enzyme" which acts on a second string as a "substrate" to produce a "product". A simple specification shows the idea. If a symbol sequence on a string in the soup, say 111, is identical to a symbol sequence on the "input" side of one grammar pair, then that 111 site in the string in the soup can act as an enzymatic site. If the enzymatic site finds a substrate string bearing the same site, 111, then the enzyme acts on the substrate and transforms its 111 to the symbol sequence mandated by the grammar, say 0101. Here, which symbol string in the soup acts as enzyme and which is substrate is decided at random at each encounter. With minor effort, the grammar rules can be extended to allow one enzyme string to glue two substrate strings together, or to cleave one substrate string into two product strings.

Grammar string models exhibit entirely novel classes of behavior, and all the phase transitions shown in the origin of life model. Fix a grammar. Start the soup with an initial set of strings. As these act on one another, it might be the case that all product strings are longer than all substrate strings. In this case, the system never generates a string previously generated. Call such a system a jet. Jets might be finite, the generation of strings petering out after a while, or infinite. The set of strings generated from a sustained founder set might loop back to form strings formed earlier in the process, by new pathways. Such "mushrooms" are just the autocatalytic sets proposed for the origin of life. Mushrooms might be finite or infinite, and might, if finite, squirt infinite jets into string space. A set of strings might generate only itself, floating free like an egg in string space. Such an egg is a collective identity operator in the complex parallel processing algebra of string transformations. The set of transformations collectively specifies only itself. The egg, however, might wander in string space, or squirt an infinite jet. Perturbations to an egg, by injecting a new string, might be repulsed, leaving the egg unchanged, or might unleash a transformation to another egg, a mushroom, a jet. Similarly, injection of an exogenous string into a finite mushroom might trigger a transformation to a different finite mushroom, or even an infinite mushroom. A founder set of strings might galvanize the formation of an infinite set of strings spread all over string space, yet leave local "holes" in string space because some strings might not be able to be formed from the founder set. Call such a set a filligreed fog. It may be formally undecidable whether a given string can be produced from a founder set. Finally, all possible strings might ultimately be formed, creating a pea soup in string space.

Wondrous dreamlike stuff, this. But more lies to hand. Jets, eggs, filligreed fogs and the like are merely the specification of the string contents of such an evolving system, not its dynamics. Thus, an egg might regenerate itself in a steady state, in a periodic oscillation during which the formation of each string waxes and wanes cyclically, or chaotically. The entire "edge of chaos" story concerned dynamics only, not composition. String theory opens new conceptual territory.

3. New Territory

Models of mind, models of evolution, models of technological transformation, of cultural succession, these grammar models open new provinces for precise thought. In "Origins of Order" I was able only to begin to discuss the implications of Fontana's invention. I turn next in this essay to mention their possible relation to artificial intelligence and connectionism, sketch their possible use in the philosophy of science, then discuss their use in economics, where they may provide an account, not only of technological evolution, but of bounded rationality, non-equilibrium price formation, future shock, and perhaps most deeply, a start of a theory of "individuation" of coordinated clusters of processes as entities, firms, organizations, so as to optimize wealth production. In turn, these lead to the hint of some rude analogue of the second law of thermodynamics, but here for open systems which increase order and individuation to maximize something like wealth production.

Not the least of these new territories might be a new model of mind. Two great views divide current theories of mind. In one, championed by traditional artificial intelligence, the mind carries out algorithms in which condition rules act on action rules to trigger appropriate sequences of actions. In contrast, connectionism posits neural networks whose attractors are classes, categories, or memories. The former are good at sequential logic and action, the latter are good at pattern recognition. Neither class has the strengths of the other. But parallel processing symbol strings have the strength of both. More broadly, parallel processing string systems in an open coevolving set of strings, wherein individuation of coordinated clusters of these production processes arise, may be near universal models of minds, knower and known, mutually creating the world they inhabit.

Next, some comments about the philosophy of science by an ardent amateur. Since Quine we have lived with holism in science, the realization that some claims are so central to our conceptual web that we hold them well neigh unfalsifiable, hence treat them as well neigh true by definition. Since Kuhn we have lived with paradigm revolutions and the problems of radical translation, comparability of terms before and after the revolution, reducability. Since Popper we have lived ever more uneasily with falsifiability and the injunction that there is no logic of questions. And for decades now we have lived with the thesis that conceptual evolution is like biological evolution: Better variants are cast up, never mind how conceived, and passed through the filter of scientific selection. But we have no theory of centrality versus peripherality in our web of concepts, hence no theory of pregnant versus trivial questions, nor of conceptual recastings which afford revolutions or wrinkles. But if we can begin to achieve a body of theory which accounts for both knower and known as entities which have coevolved with one another, E. coli and its world, I.B.M. and its world, and understand what it is for such a system to have a "model" of its world via meaningful materials, toxins, foods, shadow of a hawk cast on newborn chick, we must be well on our way to understanding science too as a web creating and grasping a world.

Holism should be interpretable in statistical detail. The centrality of Newton's laws of motion compared to details of geomorphology in science find their counterpart in the centrality of the automobile and peripherality of pet rocks in economic life. Conceptual revolutions are like avalanches of change in ecosystems, economic systems, and political systems. We need a theory of the structure of conceptual webs and their transformation. Pregnant questions are those which promise potential changes propagating far into the web. We know a profound question when we see one. We need a theory, or framework, to say what we know. Like Necker cubes, alternative conceptual webs are alternative grasped worlds. We need a way to categorize "alternative worlds" as if they were alternative stable collective string production systems, eggs or jets. Are mutually exclusive conceptual alternatives, like multiple personalities, literally alternative ways of being in the world. What pathways of conceptual change flow from a given conceptual web to what "neighboring" webs, and why? This is buried in the actual structure of the web at any point. Again, we know this, but need a framework to say what we know. I suspect grammar models and string theory may help. And conceptual evolution is like cultural evolution. I cannot help the image of an isolated society with a self consistent set of roles and beliefs as an egg shattered by contact with our supracritical Western civilization.

Now to economic webs, where string theory may provide tools to approach technological evolution, bounded rationality, non-equilibrium price formation, and perhaps individuation of "firms". I should stress that this work is just beginning. The suspected conclusions reach beyond that which has been demonstrated mathematically or by simulations.

A first issue is that string theory provides tools to approach the fundamental problem of technological evolution. Theoretical economists can earn a living scratching equations on blackboards. This is a strange way to catch dinner. One hundred thousand years ago, their grandfathers and grandmothers scratched a living in a more direct way. Economists survive only because the variety of goods and services in an economy has expanded since Neanderthal to include the services mathematical economists. Why? And what role does the structure of an economic web play in its own growth?

The central insight is that, in fact, the structure of an economic web at any moment plays the central role in its own transformation to a new web with new goods and services and lacking old goods and services. But it is precisely this central fact that the economists have, until now, no coherent means to think about. The richness of economic webs has increased. Introduction of the automobile, as noted, unleashes an avalanche of new goods and services ranging from gas stations to motels, and drives out horse, buggy, and the like. Economists treat technological evolution as "network externalities". This cumbersome phrase means that innovation is imagined to occur due to causes "outside" the economy. While innovation has cascading consequences of the utmost importance, traditional economic theory is not to account for technological evolution, but to note its history and treat such innovation as exogenous. Strange, since the bulk of economic growth in the current century is driven by innovation.

There is a profound reason why economics has had a difficult time building a theory of the evolution of technological webs. They lack a theory of technological complementarity and substitutability without which no such web theory can be built. String theory offers such a framework. Economists call nut and bolt, ham and eggs, "complements". That is, complements are goods and services which are used together for some purpose. Screw and nail are "substitutes", each can replace the other for most purposes. But the growth of technological niches rests on which goods and services are complements and substitutes for one another. Thus, the introduction of the computer led to software companies because software and hardware are complements. Without a theory of which goods and services are complements and substitutes for one another, one cannot build a decent account of the way technological webs grow autocatalytically.

String theory to the rescue. Any random grammar, drawn from the second order infinite set of possible grammars, can be taken not only as a catch-as-catch-can model of the "laws of chemistry", but of the unknown "laws of technological complementarity and substitutability". Strings which act on strings to make strings are tools, or capital goods. The set of strings needed as inputs to a tool to made product strings is itself a set of complements. Each string is needed with the rest to make the products. Strings which can substitute for one another as inputs to a tool to yield the same products are substitutes for one another. Such complements and substitutes constitute the "production functions" of the economist, or, for consumption goods, the consumption complementarities and substitutions, ham and eggs, salt and potassium chloride.

We have no idea what the laws of technological complementarity and substitutability are, but by scanning across grammar space we are scanning across possible models of such laws. If vast regimes of grammar space yield similar results, and we can map the regimes onto real economic systems, then those regimes of grammar space capture, in an "as if" fashion, the unknown laws of technological complementarity and substitutability which govern economic links. An ensemble theory again. Catch-as-catch-can can catch the truth.

These economic string models are now in use with Paul Romer to study the evolution of technological webs. The trick is to calculate, at each period, which of the goods currently produced, or now rendered possible by innovation based on the current goods, are produced in the next period and which current goods are no longer produced. This allows studies of avalanches of technological change.

In more detail, an economic model requires production functions, an assignment of utility to each good or service, and budget constraints. An economic equilibrium is said to exist if a ratio of the amounts of goods and services produced is found which simultaneously optimizes utility and such that all markets clear no bananas are left rotting on the dock, no hungry folk hankering for unproduced pizzas. At each period those goods, old and now possible, which are profitable are incorporated into the economy, those old and new goods which would operate at a loss will not. Thus, these models are literally the first which will show the ways avalanches of goods and services come into and leave economic systems. Heretofore, the economists have lacked a way to say why, when the automobile enters, such and such numbers of other new goods are called into existence, while old goods are rendered obsolete. Now we can study such transformations. The evolution of economic webs stands on the verge of become an integral feature of economic theory. Since such evolution dominates late 20th century and will dominate early 21st century economic growth, the capacity to study such evolution is not merely of academic interest.

String theory provides new insights into economic take off. The 21st century will undoubtedly witness the encroaching struggle between North and South, developed and underdeveloped economies, to learn how to share wealth and, more essentially, learn how to trigger adequate economic growth in the South. For decades economists have sought adequate theories of economic take off. Typically these rest on the idea of accumulation of sufficient surplus to invest. But string theory suggests this picture is powerfully inadequate. Summon all the surpluses one wishes, if the economic web is too simple in variety to allow the existing niches to call forth innovation to create novel goods and services, the economy will remain stagnant and not take off. In short, string theory suggests that an adequate complexity of goods and services is required for phase transition to take off.

These phase transitions are simply understood, and can depend upon the complexity of one economy, or the onset of trade between two insufficiently complex economies. Think of two boxes, labeled France and England. Each has a forest of types of founder strings growing within it. If, in either country, the goods and services denoted by the founder strings is too simple, then the economy will form a faltering finite jet, which soon sputters out. Few novel goods are produced at first, then none. But if the complexity of goods and services within one country is great enough, then, like an autocatalytic set at the origin of life, it will explode into an indefinitely growing proliferation of goods and services, a mushroom, filligreed fog, etc. Thus, takeoff requires sufficient technological complexity that it can feed on itself and explode. It will occur in one economy if sufficiently complex. Or onset of trade can trigger take off. Let France and England be subcritical, but begin to exchange goods. The exchange increases the total complexity, thus the growth of new opportunities, new niches for innovation, thus may catapult the coupled economies to supracritical explosion. Takeoff.

Price formation is an unsolved problem. There is no established mechanism which assures equilibrium price formation in current economic theory. We hope to show using string theory that an adequate account requires a radical transformation. Price formation is not an equilibrium phenomenon. The proper answer rests on and optimally, but boundedly, rational economic agents, who may jointly approach a price equilibrium as best as can be achieved, but typically do not reach it. Among other implications, arbitrage opportunities must typically exist.

Here is the issue. Price equilibrium is meant to be that ratio of prices for goods, denominated in money or some good, such that if all economic agents simultaneously optimize their utility functions, all markets clear. But there is a sad, if brilliant, history here. In days of old one envisioned supply and demand curves crossing. Bargaining at the bazaar between buyer and seller was to drive price to equilibrium where supply matched demand and markets cleared. Helas, ordinary folk like butter with their bread, sometimes marmalade. Unfortunately, this linkage of consumption complementarities can mean that alteration in the price of bread alters the demand for butter. In turn, theorems show that price adjustment mechanisms at the bazaar or by an auctioneer do not approach equilibrium where markets clear, but diverge away from equilibrium given any price fluctuation. Panic among the economists, if not the markets.

General equilibrium theory, the marvelous invention of Arrow and Debreu, is odd enough. Posit all possible conditional goods, bananas delivered tomorrow if it rains in Manitoba. Posit the capacity to exchange all possible such oddities, called complete markets. Posit infinitely rational economic agents with prior expectations about the future, and it can be shown that these agents, buying and selling rights to such goods at a single auction at the beginning of time, will find prices for these goods such that markets will clear as the future unfolds. Remarkable, marvelous indeed. But it was easier in the days of hunter-gatherers.

Balderdash. In the absence of complete markets the theory fails. Worse, infinite rationality is silly, we all know it. The difficulty is that if there were a "smart" knob, it always seems better to tune that knob towards high smart. The problem of bounded rationality looms as fundamental as price formation. Since Herbert Simon coined the term all economists have known this. None, it appears, has known how to solve it.

I suspect that non-equilibrium price formation and bounded rationality are linked. Using "string theory", we hope to show for an economic web whose goods and services changes over time, that even infinitely rational agents with unbounded computer time should only calculate a certain optimal number of periods into the future, a "Tc", to decide how to allocate resources at each period. Calculation yet further into the future will leave the infinitely rational agents progressively less certain about how to allocate resources in the current period. In short, even granting the economist complete knowledge on the part of agents, there is an optimal distance into the future to calculate to achieve optimal allocation of resources. Bounded rationality not only suffices, but is optimal. There is an optimal tuning of the "smart knob". If this obtains, then the same results should extend to cases with incomplete knowledge in fixed as well as evolving economies. Thank goodness. If evolution itself tunes how smart we are, perhaps we are optimally smart for our worlds.

Our ideas can be understood by comparing the task of an infinitely rational Social Planner in an economy with a fixed set of goods and services, or production technologies, from a Social Planner in evolving economic web with new goods and services becoming possible over time. In a standard economy with a fixed set of goods and services, the Social Planner can calculate precisely how he should allocate resources in the first period, or any period thereafter, by thinking infinitely far ahead. In the case of an economy with new goods and services, if the Planner thinks too far ahead he becomes progressively more confused about how he should allocate resources in the first period. He should only think optimally far ahead: He should be optimally boundedly rational.

In a standard economic model, with unchanging goods and services, each modeled as a symbol string and endowed with a utility, the Social Planner proceeds as follows: In order to allocate resources in the first period, he calculates 1 period ahead and assesses allocation of resources in the first period. Then he calculates 2 periods ahead to see how this further calculation changes the optimal allocation of resources in the first period. Then he calculates 3, 4, T periods ahead. At each such calculation he obtains an optimal ratio or allocation of economic production activities for the first period which optimizes the utility attained by action in the first period. The most important result is this: As he calculates ever further ahead, this ratio of activities at first jumps around, then settles down to a steady ratio as T approaches infinity. Two features are important. First, the further out he calculates, the larger T is, the higher the utility achieved by allocation in the first period, since he has accounted for more consequences. Second, because the ratio settles down asymptotically, that asymptotic ratio of activities at T = infinity is the optimal allocation of resources in the first period. Given this, he carries out the allocation and passes to the second period.

The central conclusion of the standard problem is that the Social Planner should tune the smart knob to maximum. A next standard step is to assume a large number of independent economic agents, each infinitely rational, each carrying out the same computation as the Social Planner. All calculate the same optimal ratio of economic activities, each does the calculated amount of his activity, utility is optimized, and because each has computed the same ratio of all activities, those activities are coordinated among the independent agents such that markets clear.

In this context, the major approach taken by economists to the fact of bounded rationality is to assume a cost of computation such that it may not be worth thinking further. The increase in utility is balanced by the cost of computing it. Such a cost is a trivial answer to why bounded rationality occurs. The deep answer, I think, is that too much calculation makes things worse. The Social Planner can be too smart by half, indeed by three quarters, or other amounts.

In an economic web where goods and services evolve over time due to innovation and replacement we hope to show that the ratio of activity calculated by the Social Planner generically will not settle down to a fixed asymptote. Rather, the further out he calculates, the more the ratio thought to be the optimal allocation of activities for the first period should jump around. Consequently, if independent economic agents carry out the same calculation as the Social Planner, the further out they calculate the harder it will become to coordinate activities. Thus, individual agents should only calculate an optimal time ahead, when the jumpiness of the optimal ratio of activities is minimized.

Here it is more slowly. As the Planner calculates for periods, T, ever further into the future in order to allocate resources in the first period, at first, as T increases, the optimal ratio will appear to settle down, but then as the hoards of new goods and services which might enter proliferate, the ratio should begin to change ever more dramatically as he calculates ever further into the future. At period 207 just the good which renders our current good 3 utterly critical makes its appearance. We would miss a gold mine had we not considered things until period 207. But upon studying period 208 we find a substitute for good 3 has become possible. We should not make 3 in the first period then. Alas again. The more we calculate the less we know.

But if the calculated ratio of activities producing goods and services first starts to settle down then becomes more variable as the Planner calculates further into the future, how in fact should he allocate resources in the first period? Every deeper calculation he becomes more confused. If he continues to calculate to infinity, even with discounting of future utilities, he may change his mind every further period he calculates.

The problem is not overwhelming for the planner, however, for he is the single commander of the entire economy, hence suffers no problems in coordinating activities across the economy. If he picks any large future T to calculate, say 1000 periods, he will make a very good allocation of resources at the current moment. Where, then, is the profound problem?

The profound problem is that there is no Social Planner. Let there be an economic agent in charge of each production function, and N such agents in the economy. Suppose, as economists do, that these agents cannot talk to one another, that they know as much as the Social Planner, and can only interact by actions. How should they coordinate their mutual behaviors? Each makes the same calculations as does the Social Planner. Each realizes that the further out he calculates, the more the ratio of activities varies. He must choose some T and act on it. But if he tries to optimize utility by choosing a large T, and others in the economy choose even slightly different T values, then each will elect to produce levels of outputs which do not mesh with the inputs assumed by others. Vast amounts of bananas will rot on the dock, hunger for apples will abound. Massive market disequilibrium occurs.

Optimally bounded rationality provides the answer. There is some period, Tc, in the future, say 7 periods ahead, when the ratio of activities is the most settled down it shall be as T varies from 1 to infinity. Here, slight differences in T chosen by other agents minimizes the bananas left on the dock and hunger for apples. Near here, then, a finite calculation ahead, is the best guess at how to allocate resources so that markets nearly clear. Bounded rationality, I believe, is linked to non-equilibrium price formation.

More should fall out. Among these, future shock. As the goods and services in the web explode in complexity, Tc should become smaller.

The time horizon for rational action would become crowded into the present.

Perhaps most importantly, we may achieve a theory of something like "optimal firm size". In a webbed economy, let local patches of production functions, vertically and horizontally integrated, count as "firms". Coordination within a firm on the choice of T can be obtained by the C.E.O. We hope that there will be an optimal distribution of firm sizes which optimizes utility. Tiny firms must remain close to Tc to minimize waste. Larger firms can push beyond Tc. But, if the economy has too few firms, hence each too large, then fluctuations over successive periods of play will drive them into bankruptcy. Thus, an intermediate number of firms, of intermediate size, should optimize average self and mutual wealth production over time.

But a theory of firm size as a cluster of production processes which optimizes the distribution of firm sizes such that each "patch" optimally maximizes growth of utility is no small conceptual step. It is a start toward a theory of individuation of clustered sets of production processes as "entities " which optimally coevolve with one another in the economic system. As such, it seems deeply linked to coevolution to the edge of chaos. In both cases, tuning something like the surface to volume ratio of an "individual" such that all individuated entities optimize expected success, is the key . More, such a theory of individuation hints an analogue of the second law of thermodynamics for open thermodynamic systems. Such a law is one ultimate focus of the "sciences of complexity".

The root requirement is the primitive concept of "success", taken as optimizing utility in economics, or optimizing reproduction success in biology. Let red and blue bacteria compete, while the former reproduces more rapidly. Soon the Petri plate is red. Let the bacteria not divide but increase in mass. Again, soon the Petri plate is red. Increase of mass is the analogue of increase of wealth. It is not an accident that biology and economics borrow from one another. The wealth of nations is, at base, the analogue of the wealth of a species. More is better.

Given a primitive concept of success, then a theory of parallel processing algorithms, alive in the space of possible transformations, should yield a theory of individuals as clumps or patches of processes which coordinate behavior such that each optimizes success, while all optimize success as best they can in the quivering shimmering world they mutually create. Economics is the purest case, for coordination into firms is a voluntary process. But Leo Buss has stressed the puzzle of the evolution of multicellular individuals since the Cambrian. Why should an individual cell, capable of dividing and passing its genome towards the omega point of time, choose to forego that march and enter into a multicellular organism where it shall form somatic tissue, not gonadal tissue, hence die. Think it not so? The slime mold Dicteostylium discoidum coalesces thousands of starving amoebae, each capable of indefinite mitotic division, into a crawling slug in which many cells later form stalk, not spore, hence die. Their progeny are cut off. Yet they have opted to associate into a biological firm, replete with specialization of labor and termination with extreme disfavor, for some form of profit. It is, at base, the same problem. What sets the size, volume, and boundary membrane of an individual.

Given a theory of individuals, patches of coordinate processes optimizing success in a coevolving world mutually known, then the argument above and its generalizations in the case of incomplete knowledge and error amplification with excessive calculation, yield a bound to rationality. A coevolving individual does not benefit, nay, does worse, by calculating too far into the future. Or too far into the web away from each. Or, equally, too far into the future light cone of events. But, in turn, a bound on rationality, better, an optimally bounded rationality, implies a bound on the complexity of the coevolving individual. No point in being overcomplex relative to one's world. The internal portrait, condensed image, of the external world carried by the individual and used to guide its interactions, must be tuned, just so, to the ever evolving complexity of the world it helped create.

We draw near an analogue of the second law of thermodynamics. The latter fundamental law states that closed systems approach a state of disorder, entropy increases to a maximum. Living systems, E.coli or IBM, are open systems. Matter and energy range through each as the precondition for their emergence. The investigations sketched here intimate a law of increasing order and differentiation of "individuals", packages of processes, probably attaining the boundary of chaos, a wavefront of self organizing processes in the space of processes, molecular, economic, cultural, a wavefront of lawful statistical form governed by the generalized insight of Darwin and Smith. While the attainment of optimally sized "individual" which optimize coevolution may be constrained by the means allowing individuals to form, aggregation of cells, antitrust laws, the direction of optimization, like the direction of entropy change, will govern the emerging structure.

4. Closing Remark: A Place for Laws in Historical Sciences

I close this essay by commenting on Burian and Richardsons' thoughtful review of *Origins of Order*. They properly stress a major problem: What is specifically "biological" in the heralded renderings of ensemble theories? This is a profound issue. Let me approach it by analogy with the hoped for use of random grammar models in economics as discussed above. As emphasized, economists lack a theory of technological evolution because they lack a theory of technological complementarities and substitutes. One needs to know why nuts go with bolts to account for the coevolution of these two bits of econo-stuff. But we have no such theory, nor is it even clear what a theory which gives the actual couplings among ham and eggs, nuts and bolts, screws and nails, computer and software engineer, might be. The hope for grammar models is that each grammar model, one of a non-denumerably infinite set of such grammars since grammars map power sets of strings into power sets of strings, each such grammar model is a "catch-as-catch-can" model of the unknown laws of technological complementarity and substitutability. The hope is that vast reaches of "grammar space" will yield economic models with much the same global behavior. If such generic behaviors map onto the real economic world, I would argue that we have found the proper structure of complementarity and substitutability relationships among goods and services, hence can account for many statistical aspects of economic growth. But this will afford no account of the coupling between specific economic goods such as power transmission and the advent of specific new suppliers to Detroit. Is such a theory specifically "economics"? I do not know, but I think so.

Grammar models afford us the opportunity to capture statistical features of deeply historically contingent phenomena ranging from biology to economics, perhaps to cultural evolution. Phase transitions in complex systems may be lawful, power law distributions of avalanches may be lawful, but the specific avalanches of change may not be predictable. Too many throws of the quantum dice. Thus we confront a new conceptual tool which may provide a new way of looking for laws in historical sciences. Where will the specifics lie? As always, I presume, in the consequences deduced after the axioms are interpreted.

References

Bagley, R. (1991), *A Model of Functional Self Organization*. Ph.D Thesis, University of California, San Diego.

Dawkins, R. (1976), *The Selfish Gene*. Oxford University Press, Oxford, N.Y.

Derrida, B. and Pomeau, Y. (1986), "Random networks of automata: a simple annealed approximation." *Europhys. Letters*. 1(2): 45-49.

Edwards, D.F., and Anderson, P.W. (1975), *Journal of Physics* F 5:965.

Eigen, M. and Schuster, P. (1979), *The Hypercycle: A Principle of Natural Self Organization*, Springer Verlag, N.Y.

Fogleman-Soulie, F. (1985), "Parallel and sequential computaiton in Boolean networks." In *Theoretical Computer Science 40*, North Holland.

Fontana, W. (1991), "Artificial Life II", in Langton, Farmer, Taylor (eds.) in press. Addison Wesley.

Hopfield, J.J. (1982), *Proceedings of the National Academy of Science*. U.S.A. 79: 2554-2558.

Jacob, F., and Monod, J. (1961), "On the regulation of gene activity." Cold Spring Harbor Symposium. *Quantum Blology* 26: 193-211.

_____. (1963), "Genetic repression, allosteric inhibition, and cellular differentiation." In E. (M. Locke, ed.) 21st Symposium for the Society for the Study of Development and Growth. Academic Press, N.Y. pp. 30-64.

Kauffman, S.A. (1969), "Metabolic stability and epigenesis in randomly connected nets." *Journal of Theoretical Biology* 22: 437-467.

_____. (1986) "Autocatalytic sets of Protiens." *Journal Theoretical Biology* 119: 1-24.

_____. (1989), "Principles of Adaptation in Complex Systems." In *Lectures in the Sciences of Complexity* in Dan Stein (ed.) The Santa Fe Institute Series. Addison Wesley.

_____. (1991), "Antichaos and Adaptation." *Scientific American*, August 1991.

_____. (1992), *Origins of Order: Self Organization and Selection in Evolution*. In press, Oxford University Press.

Langton, C. (1991), in *Artificial Life II*, (eds.) Langton, Farmer, Taylor, in press, Addison Wesley.

Smolensky, P. (1988), "On the proper treatment of connectionism." *Behavioral and Brain Science* 11: 1-74.

Stauffer, D. (1987), "Random Boolean networks: analogy with percolation." *Philosophical Magazine* B, 56 no. 6: 901-916.

Part VIII

STATISTICAL INFERENCE AND THEORY CHANGE

The Appraisal of Theories: Kuhn Meets Bayes

Wesley C. Salmon

University of Pittsburgh

Can statistical inference shed any worthwhile light on theory change? For many years I have believed that the answer is "Yes." Let me try to explain why I think so. On my first reading of Thomas S. Kuhn's *The Structure of Scientific Revolutions* (1962) I was so deeply shocked at his repudiation of the distinction between the context of discovery and the context of justification that I put the book down without finishing it. By 1969, when a conference was held at the Minnesota Center for Philosophy of Science on the relations between the history of science and the philosophy of science, I had returned to *Structure* and formed the view that Kuhn's rejection of this fundamental distinction resulted from his adoption of an inadequate conception of scientific justification. It appeared that he saw scientific confirmation in terms of the traditional hypothetico-deductive (H-D) schema, according to which a scientific hypothesis (or theory) is confirmed by observing the truth of its logical consequences. More precisely, given a hypothesis T, some initial or boundary conditions I, and auxiliary hypotheses A, an observational consequence is deduced. If the consequence turns out to be true that fact constitutes at least a bit of support for T.

The H-D method seemed to me at the time (and still does) grossly inadequate as a schema for the characterization of scientific confirmation (Salmon 1967, chap. VII). Moreover, this opinion had been shared by a number of leading experts on the subject. In his (1949), Hans Reichenbach explicitly rejected the H-D method and advocated the use of Bayes's theorem as the proper schema. In his (1950), Rudolf Carnap adopted a logical interpretation of probability, the structure of which conforms to Bayes's theorem, as an explication of degree of confirmation. Personalists, such as L. J. Savage (1954), were so wedded to Bayes's theorem that they took its name and called themselves *Bayesians*. Karl Popper, who repudiated confirmation altogether, a fortiori rejected the H-D schema. The fact that Popper accepted the use of the same schema as modus tollens when the observational prediction turns out to be false in no way commits him to allowing confirmation when the observational prediction is true.

In my contribution (1970) to the proceedings of the Minnesota conference I adopted Bayes's theorem (see equation (1) below) as the basic schema for scientific confirmation, but without presupposing any particular interpretation of probability. Quite plainly, I argued, if Bayes's theorem is to be used as a schema for confirmation it is necessary to take account of the prior probabilities that appear therein. The most natural construal of the prior probabilities is as plausibility considerations. To anyone who, like Kuhn, does not think of confirmation in terms of Bayes's theorem, plausibility arguments seem completely heuristic, strongly suggesting that they belong to the context of discovery rather than the context of justification. Nevertheless, Kuhn makes a striking case for the thesis that plausibility considerations play an indispensable role in the choice among scientific theories. Therefore, he seems to conclude, the distinction between discovery and justification is misconceived.

It seems to me that Kuhn is quite right regarding the indispensability of plausibility arguments; however, from the standpoint of Bayes's theorem, these plausibility considerations belong squarely within the context of justification. They appear as explicit terms in the logical schema for the confirmation of scientific hypotheses. Thus, I argued, since we can locate plausibility considerations in the context of justification, there is no need to give up the distinction between the contexts of discovery and justification. In an era in which sharp distinctions had been dropping like flies, I did not succeed in generating much enthusiasm for this one.

The 1983 program of the Eastern Division of the American Philosophical Association contained a Symposium on the Philosophy of Carl G. Hempel. On that occasion I had the singular honor to share the platform with Kuhn and Hempel, the most distinguished representatives, respectively, of the historical approach to philosophy of science and of logical empiricism. On this occasion Kuhn chose to speak about rationality in science, a topic that he and Hempel had been discussing for some time. This symposium struck me as an appropriate opportunity to suggest that a bridge could be built between these two approaches to the philosophy of science — at least insofar as rationality is concerned — by means of Bayes's theorem. As in the Minnesota paper, my attention here was focused on the status of plausibility considerations.

Near the beginning of his paper, Kuhn expressed his appreciation to Hempel. "More than other philosophers of his persuasion, Hempel has examined my views in this area with care and sympathy: he is not one of those who suppose that I proclaim the irrationality of theory choice. But he sees why others have supposed so" (1983, p. 563). Indeed, in (1977) Kuhn had already expressed his dismay at the accusation of irrationality. Although he acknowledged remarks to the effect that theory choice involves persuasion and judgment and is not a matter of proof, and that theory choice goes beyond observation and logic, he insisted that these aspects do not make science irrational — that he did not make theory choice "a matter for mob psychology" (1977, pp. 320-21).

Kuhn's attitude toward rationality takes, as its point of departure, the supposition that mature physical science constitutes *the prime example* of a rational enterprise. If we want to understand the nature of rationality, it is better to examine the practice of that domain of science, in order to try to learn how that practice proceeds, than to lay down a priori conditions of rationality.

What, I ask to begin with, are the characteristics of a good scientific theory? Among a number of quite usual answers I select five, not because they are exhaustive, but because they are individually important and collectively sufficiently varied to indicate what is at stake. First, a theory should be accurate: within its domain, that is, consequences deducible from a theory should be in demonstrated agreement with the results of existing experiments and observations. Second, a theory should be consistent, not only internally or with itself, but also with other currently accepted theories applicable to related aspects of nature. Third, it should have broad scope: in particular, a theory's consequences should extend far beyond the particular observations, laws, or subtheories it was initially designed to explain. Fourth, and closely related, it should be simple, bringing order to phenomena that in its absence would be individually isolated and, as a set, confused. Fifth — a somewhat less standard item, but one of special importance to actual scientific decisions — a theory should be fruitful of new research findings: it should, that is, disclose new phenomena or relationships previously unnoted among those already known. These five characteristics — accuracy, consistency, scope, simplicity, and fruitfulness — are all standard criteria for evaluating the adequacy of a theory. (Kuhn 1977, pp. 321-22).

Kuhn makes it quite clear that the foregoing criteria cannot be applied in any mechanical way. Different scientists may place different relative weights upon them, and, in any case, their application requires judgment. At times, moreover, the criteria may conflict with one another. Simplicity, for example, might conflict with accuracy or scope.

When we look at this list, it seems to me, at least two of the items stand out as having a direct bearing on prior probabilities. Simplicity is often invoked as a plausibility consideration in the physical sciences; a nice example, involving proliferation of fundamental particles, can be found in Harari (1983). In the behavioral sciences, however, simplicity is not always prized. For example, an archaeological explanation in terms of drought alone of the abandonment of a sizable dwelling at Grasshopper, Arizona, near the close of the 14th century would be rejected as too simple — as an oversimplification. Certainly the drought was part of the story, but many other factors are required in a satisfactory theory. Practicing scientists in any given domain must exercise their judgment regarding the degree of simplicity or complexity appropriate in their fields.

In his remarks about consistency, Kuhn states explicitly that he is not concerned solely with the internal consistency of a theory. Clearly, internal consistency is important and desirable, but that is not the whole story. Another major consideration is how well a given theory fits with what we already accept in related domains. On this criterion, Immanuel Velikovski's *Worlds in Collision* (1950) receives an extremely low rating (see Gardner 1957). In contrast, Coulomb's inverse square law of the electrostatic force scores well on this desideratum, in view of Newton's law of gravitation and the then accepted Euclidean structure of physical space.

Kuhn's criterion of consistency is closely related, it seems to me, to arguments by analogy that are invoked to establish plausibility. Let me give three examples, one each from the physical, biological, and behavioral sciences. In physics, Louis de Broglie's hypothesis of wave-particle duality for material particles on the basis of analogy with the same duality for light is a perfect example. In the biological sciences, inference to the carcinogenic nature of saccharin for humans on the basis of the results of experiments with rats constitutes a strong plausibility claim. In archaeology, ethnographic analogy is used to support plausible hypotheses about the use of an artifact in a prehistoric culture on the basis of the observed use of a similar artifact in an extant culture.

There can be little doubt that plausibility considerations are ubiquitous in the sciences; only their status can be open to question. As long as the H-D schema is used to characterize scientific confirmation, plausibility arguments are relegated to heuristics as an aid in the generation of interesting and promising hypotheses. Bayes's theorem, in contrast, shows that they play an indispensable role in the appraisal of theories.

Having discussed two of Kuhn's criteria, I must briefly remark on the remaining three — accuracy, scope, and fruitfulness. All of them can be construed in terms of likelihoods. Consider accuracy. Inasmuch as Kuhn refers explicitly to deducible consequences, the likelihood of the results, given the truth of the theory, is one. If there are results of many observations and experiments, the likelihood of an accurate agreement, if the theory were false, would be small. Similar remarks apply to scope and fruitfulness. If the theory were false it is unlikely that it would extend successfully beyond "the particular observations, laws or subtheories is was initially designed to explain," or that it would "disclose new phenomena or relationships previously unnoted among those already known." Accordingly, it appears that Bayes's theorem provides a logical rationale for the kinds of considerations Kuhn sees as guiding actual scientific practice.

During the discussion in the 1983 APA Symposium, a problem was raised concerning the likelihoods in Bayes's theorem. One useful way of formulating Bayes's theorem is

$$Pr(T_1|E.B) = [Pr(T_1|B)Pr(E|T_1.B)] / [Pr(T_1|B)Pr(E|T_1.B) + ... + Pr(T_k|B)Pr(E|T_k.B)] \quad (1)$$

where B is background knowledge, E is a new piece of evidence, and $T_1, \ldots, T_k$ is a mutually exclusive and exhaustive set of theories. In this equation, $Pr(T_1|E.B)$ is the posterior probability of T_1, $Pr(T_i|B)$ are prior probabilities, and $Pr(E|T_i.B)$ are likelihoods. A likelihood is the probability that the evidence actually found would occur, given our background knowledge and the truth of the the theory T_i mentioned in that expression.

Likelihoods are sometimes unproblematic. Consider Galileo's observation of the phases of Venus. On the Copernican theory the probability of this evidence is one; on the Ptolemaic theory the likelihood is zero. Sometimes, however, they are quite problematic. Consider the fact that no annual stellar parallax was observed at the time of Galileo. According to the Ptolemaic theory stellar parallax does not exist; that is why it is not observed. According to the Copernican theory stellar parallax must occur; the fact that we do not observe it can only be explained on the supposition that the fixed stars are unspeakably distant from us — a blatantly ad hoc hypothesis. How probable is it that the Copernican theory is correct and that the stars are so far away? We can easily understand why people rated that likelihood low.

The cosmological problem during the scientific revolution brings out another problem associated with Bayes's theorem (1), namely, the fact that the enumeration of theories in the denominator of the right hand side must be exclusive and exhaustive. Typically, when we have competing theories they are incompatible with one another, but seldom are they exhaustive. In addition to the Copernican and Ptolemaic theories there are other possibilities, e.g., the Tychonic system, in which the earth is stationary, the sun and moon move around the earth, but the other planets move around the sun. On the Tychonic system, the likelihood of absence of observed stellar parallax is one; in addition, the likelihood of the full set of phases of Venus later observed by Galileo is also one.

The Ptolemaic, Copernican, and Tychonic systems are obviously not the only logical possibilities, so if we want the exhaustive enumeration required by equation (1) we must add some other possibilities. The only plausible candidate to complete the set is what Abner Shimony (1970) called "the catchall" — i.e., the hypothesis that says "none of the above." Looking again at equation (1), we can always take T_k to be the catchall. At any given stage of scientific investigation, the catchall is the disjunction of all of the hypotheses we have not yet conceived. What is the likelihood of any given piece of evidence with respect to the catchall? This question strikes me as utterly intractable; to answer it we would have to predict the future course of the history of science. No one is ever in a position to do that with any reliability.

In a recent paper (Salmon 1990) I have offered a solution to the problem of the likelihood on the catchall. Borrowing an idea often emphasized by Kuhn, we may consider what happens if, instead of trying to evaluate one theory in isolation, we make only a comparative appraisal of competing alternatives. The catchall, incidentally, is never one of those in the competition, for it cannot be considered a bona fide scientific theory. Suppose, then, that we have two theories, T_1 and T_2, that we wish to compare. In a way completely parallel to equation (1) we can write

$$Pr(T_2|E.B) = [Pr(T_2|B)Pr(E|T_2.B)] / [Pr(T_1|B)Pr(E|T_1.B) + \ldots + Pr(T_k|B)Pr(E|T_k.B)] \quad (2)$$

Comparing the two equations, we note that the denominators on the right hand sides are identical. Assuming that the neither of the numerators is zero — if either were zero we would have no interest in the hypothesis involved in it — we divide (1) by (2) with the result

$$Pr(T_1|E.B) / Pr(T_2|E.B) = [Pr(T_1|B)Pr(E|T_1.B)] / [Pr(T_2|B)Pr(E|T_2.B)] \quad (3)$$

We note that the likelihood on the catchall, as well as its prior probability, have vanished. What remain are only the prior probabilities, posterior probabilities, and likelihoods of the theories we are explicitly comparing. If the ratio of the posterior probabilities is greater than one, we prefer T_1 to T_2; if the ratio equals one, we prefer neither to the other; if the ratio is less than one, we prefer T_2 to T_1.

Even though we have eliminated the most intractable probability that occurs in equations (1) and (2) — the likelihood on the catchall — we are still not in a position to say that the likelihoods are unproblematic. The likelihoods on the theories being compared may still pose difficulties. Remember the Copernican theory and the absence of observed stellar parallax. There is, I think, a rather common strategy for dealing with such cases. By the invocation of suitable auxiliary hypotheses A, one can construct a *plausible scenario* according to which the problematic observation is made into a likely, or even necessary, consequence of the theory T in conjunction with the auxiliary A. For the Copernican system, the plausible scenario places the fixed stars at a huge distance from the earth. The numerator on the right hand side of (3) becomes $Pr(T_1.A_1|B)Pr(E|T_1.A_1.B)$. Although the likelihood $Pr(E|T_1.A_1.B) = 1$, we are left with the prior probability of the *plausible scenario* $Pr(T_1.A_1|B)$ — namely, with the question of just how plausible the proffered scenario actually is.

Consider another example (see Worrall 1990). In the 1830s David Brewster, a distinguished British optician, still supported the corpuscular theory of light, and refused to adopt the by then widely accepted wave theory. Although he recognized the difficulty for the corpuscular theory in accounting for various diffraction phenomena such as the Poisson bright spot, he believed that the wave theory encountered problems that were equally difficult if not more so. For example, he did *not* find plausible the hypothesis that the universe is completely filled with an aether of suitable mechanical properties to transmit starlight across vast distances of apparently empty space. Moreover, a phenomenon that he discovered — the selective absorption of light passing through a gas — seemed to pose insuperable problems for the wave theory. If you believe that light is a wave phenomenon, then how could you explain why light of a given wavelength passes freely through a gas, while light differing only slightly in wavelength is almost completely absorbed by the same gas? (Brewster avoided talk of wavelengths by referring to indices of refraction.) At the same time, he had no plausible scenario to offer on behalf of the corpuscular theory to deal with such phenomena as the Poisson bright spot. From our late 20th century vantage point, we can see that both the corpuscular and the undulatory theories of the 19th century faced insuperable difficulties. Plausible scenarios were not available.

A contemporary example of the use of plausible scenarios can be seen in the case of dinosaur extinction. The discovery in 1979 of an iridium anomaly — an extraordinarily high concentration of iridium — at the Cretaceous-Tertiary (K-T) boundary near Gubbio, Italy, led to the postulation of a collision of an asteroid or comet with the earth 65 million years ago. This event, it has been claimed, coincided with the extinction of dinosaurs and many other living species. Walter Alvarez and his father, Luis W. Alvarez, advanced a scenario designed to explain how the extinction was caused by the impact (see Alvarez, et al., 1980). This hypothesis has generated an enormous amount of controversy; other scientists maintain, for example, that massive volcanic activity was responsible. They too have fashioned scenarios they find plausible. A historical project studying the development of this controversy is reported by William Glen (1989).

Possible scenarios are especially successful if they enable the theorist to deduce the evidence, thus making the likelihood one, and if they are, indeed, plausible. If this result is achieved for theories T_1 and T_2 by means of auxiliaries A_1 and A_2, equation (3) assumes the following simplified form,

$$\Pr(T_1.A_1|E.B) / \Pr(T_2.A_2|E.B) = \Pr(T_1.A_1|B) / \Pr(T_2.A_2|B) \qquad (4)$$

in which the ratio of the posterior probabilities equals the ratio of the prior probabilities. Since theory choice does not hinge ultimately on plausibility considerations (prior probabilities) alone, new evidence will be sought to discriminate between the two plausible scenarios. If a *crucial experiment (or observation)* can be found, yielding a likelihood of zero on one of the scenarios and a likelihood of one on the other, eliminative induction can be successfully practiced (see Earman forthcoming, chap. 7).

Two major objections have often been raised against the foregoing account of theory appraisal. First, it has been questioned whether two scientists, differing in their preferences between two theories, always share the same background knowledge B. While it may be true that they sometimes differ, the problem does not arise for the individual scientist who is comparing the merits of two theories for herself or himself; both must be evaluated in light of the individual's total body of background information. Thus, we can reasonably assume that, for any given individual, "B" is univocal in equations (3) and (4).

Second, a far more difficult challenge has been raised regarding the status of the prior probabilities that occur in Bayes's theorem. At the outset of this discussion I mentioned the views of Carnap, Reichenbach, and Savage. As representatives of the logical, frequency, and personalist interpretations of probability, they differ regarding the nature of the priors. For Carnap, they are a priori measures. I do not see how it can reasonably be maintained that we have a priori prior probabilities for the hypotheses seriously considered in science. Reichenbach required that prior probabilities be related to classes of similar theories in terms of the relative frequency of success of hypotheses in such classes. Although Reichenbach never made a convincing case for frequencies in this context, he did, at least, insist that prior probabilities be related to scientific experience. For Savage and other personalists, prior probabilities are degrees of belief subject only to coherence requirements. The extreme subjectivism of this view has been a severe obstacle.

I am inclined to think that a compromise can be made between the frequentist and personalist approaches, along a line suggested by Patrick Suppes (1966, pp. 202-3). Suppes points out that we generally bring to any scientific hypothesis a heterogeneous body of information that cannot be stated explicitly in full. A judgment of prior probability gives at least a rough assessment of the way that knowledge applies to a given hypothesis. It is, to be sure, a subjective judgment, but it also reflects objective experience. It is important that any such prior probability assessment be made in the light of scientific experience, to the exclusion of idiosyncrasies, prejudices, ideologies, and emotions. Moreover, as personalists have often noted, it is unnecessary to have precise numerical values for the priors.

The moral to be drawn from the preceding discussion is twofold. First, the logical empiricists, it seems to me, should take seriously Kuhn's point that in most cases, if not always, scientists are concerned to make comparative evaluations among theories. Seldom, if ever, is a hypothesis judged in isolation from all potential competitors. Second, philosophers with a historical disposition should look closely at the nature of scientific confirmation. They should relinquish the H-D schema and consider the merits of Bayes's theorem as a more adequate account. In this way, I believe, can a significant degree of rapprochement be achieved between these two important schools of philosophy of science.

References

Alvarez, L.W., et al. (1980), "Extraterrestrial Cause for the Cretaceous-Tertiary Extinction", *Science* 208: 1095-1108.

Carnap, R. (1950), *Logical Foundations of Probability*. Chicago: University of Chicago Press.

Earman, J. (forthcoming), *Bayes or Bust? A Critical Examination of Bayesian Confirmation Theory*. Cambridge, Mass.: MIT Press/Bradford Books.

Gardner, M. (1957), *Fads and Fallacies in the Name of Science*. New York: Dover.

Glen, W. (1989), "Meteorites, Volcanoes and Dinosaurs: Update on a Historical Project on the Current Debates", *Center for the History of Physics Newsletter* 21: 2-3.

Harari, H. (1983), "The Structure of Quarks and Leptons", *Scientific American* 248, no. 4 (April): 56-68.

Kuhn, T.S. (1962), *The Structure of Scientific Revolutions*. Chicago: University of Chicago Press.

_____. (1977), "Objectivity, Value Judgment, and Theory Choice", in *The Essential Tension*. Chicago: University of Chicago Press, pp. 320-39.

_____. (1983), "Rationality and Theory Choice", *Journal of Philosophy* 80: 563-72.

Popper, K.R. (1959), *The Logic of Scientific Discovery*. New York: Basic Books.

Reichenbach, H. (1949), *The Theory of Probability*. Berkeley & Los Angeles: University of California Press.

Salmon, W.C. (1967), *The Foundations of Scientific Inference*. Pittsburgh: University of Pittsburgh Press.

_____. (1970), "Bayes's Theorem and the History of Science", in *Historical and Philosophical Perspectives of Science,* vol. 5, *Minnesota Studies in the Philosophy of Science,* Roger H.Stuewer (ed.). Minneapolis: University of Minnesota Press, pp. 68-86.

_____. (1983), "Carl G. Hempel on the Rationality of Science", *Journal of Philosophy* 80: 555-62.

_____. (1990), "Rationality and Objectivity in Science, *or* Tom Kuhn Meets Tom Bayes", in *Scientific Theories* vol. 14, *Minnesota Studies in Philosophy of Science,* C. Wade Savage (ed.). Minneapolis: University of Minnesota Press, pp. 175-204.

Savage, L.J. (1954), *The Foundations of Statistics*. New York: John Wiley & Sons.

Shimony, Abner (1970), "Scientific Inference", in *The Nature and Function of Scientific Theories,* Robert G. Colodny (ed.). Pittsburgh: University of Pittsburgh Press, pp. 79-172.

Suppes, Patrick (1966), "A Bayesian Approach to the Paradoxes of Confirmation", in *Aspects of Inductive Logic,* Jaakko Hintikka and Patrick Suppes (eds.). Amsterdam: North-Holland Publishing Co., pp. 198-207.

Velikovski, Immanuel (1950), *Worlds in Collision*. Garden City, NY: Doubleday.

Worrall, John (1990), "Scientific Revolutions and Scientific Rationality: The Case of the Elderly Holdout", in *Scientific Theories* vol. 14, *Minnesota Studies in Philosophy of Science,* C. Wade Savage (ed.). Minneapolis: University of Minnesota Press, pp. 319-54.

Giving up Certainties[1]

Henry E. Kyburg, Jr.

University of Rochester

1. Choosing Among Theories

People have worried for many years — centuries — about how you perform large changes in your body of beliefs. How does the new evidence lead you to replace a geocentric system of planetary motion by a heliocentric system? How do we decide to abandon the principle of the conservation of mass?

The general approach that we will try to defend here is that an assumption, presupposition, framework principle, will be rejected or altered when a large enough number of improbabilities must be accepted on be basis of our experience. If I think that all swans are white, and a student claims to have a counterexample, I will assume that he has made some observational error. I will reject his result, and continue to accept the generalization. When a lot of people claim to have seen counterexamples, I will come around: to continue to accept the generalization would require me to accept too many improbabilities. This is a discontinuous process as we will construe it: it is not a matter of a general statement becoming less probable, while certain reports become more probable. We cannot accept the generalization and even one of the observation reports: that would be a simple inconsistency.

One suggestion, due to Karl Popper, is that we invent Bold Conjectures, and Put Them to the Test. (Popper, 1959) Bold conjecture: the Earth is the Center of the Solar System. Test... what? Bold conjecture: Mass is conserved. Test: weigh a mass of plutonium and its by products before and after. Obviously things are a little more complicated than the slogans suggest.

Alternatively, gather evidence, and accept the hypothesis that is most probable, relative to that evidence. So far, so good (maybe). But then what? How do you change from that hypothesis to one inconsistent with it when the evidence so indicates? For as soon as a hypothesis is accepted, it has probability 1; and as soon as a hypothesis has probability 1, its contraries have probability 0; and as soon as a contrary hypothesis has probability zero, its probability can never leave zero — at least not by Bayes' theorem.

A natural response to this observation is to say, as Carnap (1950) did, that "acceptance" is just an approximation to the real truth, and that no hypothesis ever achieves literal acceptance, which would entail its having a probability of 1. What we really have (as opposed to the approximate way we talk) is a probability blanket over a field of empirical propositions, none of which is ever assigned a probability of 0 or 1 unless it is a mathematical or logical truth, or the denial of one.

This latter approach presents us with serious problems. We will consider the problem of assigning prior probabilities to the sentences of a reasonably rich language later, but already we are faced with a difficult computational problem. Gil Harman (1989) has pointed out that in a language win n basic sentences there are 2^n assignments to make. But of course we can get by with wholesale assignments; if we decide that each conjunction of basic sentences or their negations is to have the same measure assigned to it, there is in fact only one assignment to make: one simple algorithm that provides the measure for any sentence.

In general, however, a useful and realistic language will have an infinite number of sentences, and this procedure breaks down. It is still possible to assign measures systematically, without assigning zero to any sentence representing a possibility. The number of sentences in any ordinary language is denumerable, and we can find a denumerable number of finite numbers that add up to 1. But the rationale of the system is hard to find.

It is, at any rate, worth exploring alternatives to either of these approaches to to rational acceptance. One of the first to offer a systematic procedure for this was Isaac Levi. In *Gambling with Truth* and *The Enterprise of Knowledge,* Levi proposes a rule for *adding* to your body of knowledge. Given such a rule, one can obtain a rule for replacing one conjecture, law, theory, hypothesis by another by proposing that when faced with a choice, one simply deletes both candidates from one's body of knowledge, and then adds the one indicated by the application of the rule for addition.

The rule is just this: (1967, p.86) Let U be a set of most specific possible hypotheses — i.e., a set of which exactly one member is true. Let M be an "information determining probability" (1980, p. 48): M(g) represents the informational value of rejecting g, and let p be an expectation forming probability (a degree of belief, a credibility). Let q in [0,1] be an index of caution. The rule (Rule A, of 1967) is to reject all and only those elements g of U such that $p(g) < qM(g)$, and to accept, with deductive closure, the disjunction of the remainder.

Given a rule for acceptance, we can construe contraction as suspending belief in a proposition and then failing to add it back under subsequent expansion; and we can construe replacement as suspending belief in one proposition, and arriving at another on subsequent expansion.

We can accomplish a change of framework of "accepted facts" this way, and we can be sure of maintaining consistency in the process. There are some knotty problems, however. When and how do we decide to suspend belief in a framework proposition? There are clear cases: when observations render our corpus inconsistent, for example. "For the sake of argument," in a friendly social context. In the context of a debate. Levi can afford to be casual about this, since a proposition erroneously deleted will be easily recaptured, and he is not concerned with real time changes. But there are other questions. How should q, the index of caution, be chosen? Where does the information measure M come from? How do we arrive at the credal probability p? More fundamental: How is the "abductive" step — the step in which the ultimate partition U is formed — to be controlled and rationalized?

One can always raise questions, of course. But these questions are disturbing because the rule presupposes a framework (a language, an information measure, a credibility measure, a set of most specific answers), and thus to be not even potentially capable of providing guidance in the choice of a framework. But let us look further.

An approach similar to Levi's has been developed in various ways by Makinson, Alchurron, and Gardenfors (1982, 1985), Gardenfors and Makinson (1988) and Gardenfors (1986, 1988). While Levi approaches the question from a constructive, analytic angle, and seeks to provide formal analysis of what goes on in changes in a corpus of knowledge, Gardenfors and the others approach the question from a logical point of view: they seek to explore axioms that may be taken to characterize the change of a body of knowledge, construed as a set of propositions. Thus, for example, it is clear that if we add the proposition A to our body of knowledge K, then A should belong to that expanded body of knowledge. As is the case with Levi, it is assumed by these writers that a body of knowledge K should be construed as a deductively closed set of propositions.

An excellent examination of these logics of theory change is provided by Gardenfors' book (1988). It is from that source that I take the following axioms. A *belief set* here is construed as a deductively closed set of propositions.

If we denote by K_A^+ the expansion of a body of knowledge K by the addition of the consistent proposition A, then we may express the the properties of the expansion of a belief set by the following relatively uncontroversial axioms.

(K+ 1) K_A^+ is a belief set.
(K+ 2) $K_A^+ \supseteq A$
(K+ 3) If $\sim A \notin K_A^+$, then $K_A^+ \supseteq K$
(K+ 4) If $A \in K$, then $K_A^+ = K$
(K+ 5) If $H \supseteq K$, $H_A^+ \supseteq K_A^+$
(K+ 6) For all belief sets K and all sentences A, K_A^+ is the smallest belief set that satisfies (K+ 1) – (K+ 5).

What is not so uncontroversial is the question of the principles according to which a body of knowledge should be contracted. This is not a terribly serious question for Levi: any proposition in our body of knowledge can be doubted with relative impunity. It can be doubted with relative impunity, since, if it belongs in our corpus of knowledge, it will be reinstated on reflection. One can thus suspend belief in a proposition A on quite casual grounds.

A serious reason to suspend belief in something arises from the circumstance that our corpus of knowledge is inconsistent. For example, if there are observational routines that warrant our acceptance of the statement that a is a crow and a is not black, then when we practise those routines, we should accept the corresponding statement. (Or proposition.) But if we already accept the generalization that all crows are black, this renders our corpus inconsistent.

With an inconsistent corpus, we are clearly *obligated* to suspend belief in something. Levi says that we should shrink our corpus of knowledge in such a way as to retain the most "information." But it is clear that no simple-minded construal of "information" will lead to the right results. In some sense it is clear that the information content of "all crows are black" is greater than that of "a is a crow and a is not black," but of course on any standard construal of hypothesis testing it is the former that will be suspended and the latter that will be retained.

While Levi offers us no logic of contraction, that is the main concern of Gardenfors et al. Gardenfors offers a number of axioms characterizing the contraction operation, denoted by K_A^-. Most of these axioms are relatively uncontroversial, as in the case of expansion. We have:

(K$^-$ 1) For any sentence A and any belief set K, K_A^- is a belief set.
(K$^-$ 2) $K \supseteq K_A^-$
(K$^-$ 3) If $A \notin K$, then $K_A^- = K$.
(K$^-$ 4) If not $\vdash A$, then $A \notin K_A^-$.
(K$^-$ 5) If $A \in K$ then $K_A^- \supseteq K$.
(K$^-$ 6) If $\vdash A \leftrightarrow B$, then $K_A^- = K_B^-$
(K$^-$ 7) $K_{A \& B}^- \supseteq K_A^- \cap K_B^-$.
(K$^-$ 8) If $A \notin K_{A \& B}^-$, then $K_A^- \supseteq K_{A \& B}^-$

These axioms may be more controversial than those for the expansion of a body of knowledge, but there is still nothing obviously wrong with them. It is possible to provide intuitively plausible axioms for theory replacement, and to show that in general replacement can be construed as a contraction followed by an expansion.

What becomes controversial is the procedure for conducting contractions. The contraction $\supseteq$ is not uniquely determined by these axioms, in contrast to K_A^+ (under the assumption of deductive closure). We must thus consider *how* to perform the contraction. One possibility is the following. Consider a subset of K that is deductively closed, that does not contain A, and that is such that if any other sentence of K is added to it, A will be a consequence of it. The set of all such sets of sentences is denoted by $K \perp A$. Clearly the result of contraction should be a member of this set (if it isn't empty; if A is a theorem, then we can take the contraction of K by A to be K itself. All we need to do is to devise a "selection function" S that will pick one set out of $K \perp A$. But, as Gardenfors shows, this yields contractions that are "too big." If $A \in K$ then this procedure will yield a K_A^- that for any proposition B contains either $A \vee B$ or $A \vee \sim B$.

The next idea one might have is to consider the intersection of all the sets of sentences in $K \perp A$. (This is called the "full meet contraction.") This is too small: K_A^- will consist only of the logical consequences of $\sim A$.

Finally, we may consider a selection function S that picks *some* of the members of $K \perp A$, intuitively the most *epistemically entrenched* members, and take K_A^- to be the intersection of these.

But what does epistemic entrenchment come to? That seems to be where the real controversy lies. Levi seeks to preserve information; he can be thought of as construing epistemic entrenchment in terms of information. But the epistemic entrenchment ranking of sets of propositions can plausibly be taken to reflect a system of beliefs, and thus to be sensitive to scientific or conceptual revolutions, whether these be understood in the dramatic Kuhnian sense or not. Gardinfors says that "...the fundamental criterion for determining the epistemic entrenchment of a sentence is how useful it is in inquiry and deliberation." (p.87) (Note that the selection function S is originally defined over sets of sentences, rather than sentences. This reflects a difference that could be exploited.)

One idea for representing such factors is provided by Wolfgang Spohn (1987). Spohn defines an "ordinal conditional function" that maps possible worlds into ordinals. The value of the function represents a degree of implausibility, or a degree of unwillingness to accept, or a degree of potential surprise (Levi, Shackle).

This function is can be extended to propositions in general by taking the value of the function for a proposition, to be the minimum value of the function over the set of worlds in which that proposition is true. Thus, since it is assumed that there is some world with value 0, either $k(A) = 0$ or $k(\sim A) = 0$, and $k(A \cup B) = \min\{k(A),k(B)\}$, where k is Spohn's ordinal conditional function.

Spohn's approach is more general than Gardenfors' since it takes as epistemic input a pair (A,a) consisting of a proposition A and an ordinal a. This yields a new ordinal function on possible worlds, and thus a new ordinal function. In the extreme cases, however, the treatment yields results parallel to those of Gardenfors (p. 73).

2. The Probabilistic Alternative

To be contrasted with this approach in terms of deductively closed sets of propositions, we may consider a purely probabilistic construal of knowledge: We take a statement as acceptable in our knowledge base when it becomes overwhelmingly probable. This is in accord with the nearly universal agreement that when it comes to empirical matters of fact, there is nothing (or almost nothing) that is certain. Almost any of the things we take for granted "could" turn out to be wrong. Nothing is incorrigible. Not even "observation" statements: without knowing how to handle errors of observation, modern science could hardly get off the ground. Of course, very crude observation statements, e.g., "the sun is shining now," are very unlikely to require correction. (They could be wrong: my "observation" may result from post–hypnotic suggestion, rather than the state of the weather.)

One way of dealing with an approach to knowledge that takes nothing empirical to be incorrigible is to become a thoroughgoing Bayesian: Represent knowledge as a probability function defined over the whole algebra of propositions in the language we are using for knowledge representation. Of course, as Carnap observed (1950), we must suppose that all refinements have been made in the language: we cannot introduce new terms without risking having to change our probability function. Then when experience causes us to shift the probability of some proposition, that change in probability propagates through the algebra in accord with some rule of propagation. (One possibility is "Jeffrey conditionalization," Jeffrey, 1965)

This approach to corrigibility has a number of drawbacks. The main one is computational. In language capable of representing some piece of common sense knowledge, or of reasoning about even quite a limited domain, the computational resources needed mount dramatically. The number of possible worlds, describable in even a constrained language, is LARGE. There is also the problem of the source of the original probability measure. Experts? There is the problem of soliciting consistent opinions. Generalize to sets of probability measures? This might be some help, but perhaps not much. There is the problem of updating: No set of probability assessments is likely to be consistent; adjustments will have to be made to achieve conformity with the probability calculus; and one of the items most natural to adjust is the ratio of probabilities $P(A \& B)/P(B)$; but this is just the important probability of A given B. And supposing a collection of agents with a common goal, sharing knowledge: how are disagreements concerning probabilities among these agents to be resolved?

These are difficult questions, and while one cannot be certain that plausible answers can't be found, it seems at least worth while to explore an alternative strategy. An alternative that has been explored for some years is that of adopting a purely probabilistic rule of acceptance: In general, "Accept *P* when its probability is high enough." (Kyburg, 1961)

One question rises immediately: how probable is "high enough?" A tentative answer to this question ("It depends on how much is at stake in using the corpus of knowledge in question") has been outlined in (Kyburg, 1988).

A less immediate question arises when we reflect that probability itself — especially evidential probability — depends on evidence. What is probable depends on what we know; and we are proposing that what we know depends on what is probable. Can we have it both ways? In particular, can evidential probability be serve both functions?

We answer yes. It has been proposed (Kyburg, 1984, Kyburg, 1988, Kyburg, 1974) that having fixed on practical certainty, we can introduce evidential certainty as the square root of practical certainty. (This stems from the fact that, using a probabilistic rule of acceptance, the conjunction of a pair of statements that do not appear conjoined in a higher level corpus will appear in a lower level corpus.)

A purely probabilistic rule of acceptance does not yield what Gardenfors has called "belief sets." The set of accepted statements is not closed under deduction, nor — what comes to the same thing in a logic with compactness — is it closed under conjunction. In general, it is not the case that if *A* and *B* are in our corpus of knowledge, their conjunction will also be in it. Of course it does not follow that the conjunction of a pair of statements in our corpus of knowledge will *not* be in it! There may be large conjunctions of statements whose probability is high enough to qualify for acceptance, and every conjunct of such a set of statements will also be in the corpus. In fact, every logical consequence of *each* statement in our body of knowledge will also be in it.

An immediate consequence is that there is an axiomatic representation of our body of knowledge. That is, there is a (presumably finite) set of statements from which the entire contents of our body of practical knowledge follows. This fact has useful consequences when it comes to talking about revisions of our body of knowledge.

The failure to embody deductive closure is not entirely unintuitive. Our confidence in the conclusion of an argument that involves many premises tends to decrease, even though we cannot put our finger on a specific doubtful premise, as the number of premises decreases. There are good intuitive grounds, even, for thinking that the set of statements that I am well justified in accepting is inconsistent; if it is inconsistent, to apply deductive closure to it would be a disaster. One particularly natural example concerns measurement. Suppose the method M yields errors that are distributed approximately normally with a mean of 0 and a variance of .04. Consider a set of applications of that method, from which we infer, in each case, that the length measured lies in the interval r ± .8 (i.e., within four standard deviations of the observed value.) Surely, by any ordinary standard, these results are acceptable. But if we accept a large number n of these results, it will also be overwhelmingly probable that at least one of them is wrong — according to the same distribution. The resulting body of knowledge is inconsistent.

The picture we work with so far is this: There are two sets of sentences we use to represent our bodies of knowledge. One, the practical corpus, contains the other, the evidential corpus, as a part. Everything in our evidential corpus is also in the practical corpus, since an item is a member of the practical corpus if and only if the lower bound on its probability (since we are using evidential probability), relative to the evidential corpus, exceeds some fixed probability p.

Statements may come and go, in the practical corpus, according as their probabilities vary with the contents of the evidential corpus. Thus there is no direct problem of revision, expansion, or contraction: all are taken care of by the probabilistic rule of acceptance. This applies to statistical statements, as well as other statements. So we will have such statistical statements in our practical corpus as "about 95% of birds fly," "less than 2% of penguins fly," etc.

Now how about the corpus of evidential certainties? How do statements get in this corpus? By being probable enough, if we are to have a uniform treatment of acceptance and corrigibility. But we can't (for reasons pointed out in Kyburg 1963) just consider simultaneously a sequence of bodies of knowledge. So we must construe a question about the contents of the evidential corpus as shifting context: now we are thinking of a different and higher level as the "evidential" corpus, and what was the evidential corpus as a practical corpus.

3. Probabilistic Inference

Statistical inference is no problem for evidential probability, but there is no ordinary way that empirical generalizations ("All Crows are Black," "Length is additive under collinear juxtaposition," etc) can be given probabilities. And it is just such items of knowledge that we would like to be able to correct. A related fact is that epistemological probability is defined only relative to a fixed language: the definition is syntactical, and depends on the recursive specification of potential reference classes and potential target classes. How do we handle generalization? And how do we deal with the relativization of probability to a language?

The key notion is that of error. We do not suppose that we have a clear cut distinction between "observational" predicates and "non-observational" predicates. We suppose instead that there is a metalinguistic corpus, parallel to our evidential corpus, that contains a representation of our knowledge concerning observational error. For example, it is there that we store the knowledge that method M for measuring length yields errors approximately normally distributed with a mean of 0 and a variance of .04.

The details of this construction are to be found in (Kyburg, 1984). The general idea is that empirical generalizations and theories are construed as features of the language we choose to use. But each of those possible languages will have going along with it, based on a given stock of actual experience, a corpus of knowledge concerning observational error. Good "observational" predicates are those that can be used with little chance of error; "non-observational" predicates will be those that have significant errors associated with them.

Observational error is generated by the interaction of our experience and a language in the following way: We know that error has occurred when we make a set of judgments that cannot all be true. Thus if we were content to live in the flowing sensuous moment, we need never suppose we made an observational error. But our bodies of knowledge would be empty of predictive content, communication would be useless, and language would be impossible. Alternatively, if we were willing to disregard experience, we could hold any theories we pleased; observations that conflicted with what our theories led us to could be dismissed as erroneous, and, like the seers of old, we would have achieved TRUTH.

What we need, then, is a way of choosing between candidate languages on the basis of the consequent errors associated the languages. In earlier work (1984, 1990) we approached this question in a very abstract framework, with a view to obtaining treatments of error in both direct and indirect measurement. Here we will adopt the same general standpoint, but examine a variety of replacements of framework assumptions (and expansions and contractions) that are rather more specific.

4. New Observations

There are a number of ways in which new data can impinge on our old body of knowledge. The most common is simply to have new observations added to our body of knowledge. This has an impact on what we believe even when it does not contradict anything we already belief. This impact has two forms. To accept the observation that A is a crow and that A is black entails, in our body of knowledge that A is a bird, since we know that all crows are birds. What is entailed by our background knowledge, and the new observation, becomes part of our background knowledge. (Subject to some caveats we'll get to later: the consequences of long conjunctions of premises may not be in our body of knowledge.)

The other form, more interesting in this context, is the impact that the observation has on our general statistical background knowledge. If we have statistical beliefs concerning the frequency with which A's are B's —e.g., that it is between p and q — and we observe an A that is not a B, that should change our body of knowledge, but not very much. If we had earlier accepted our statistical knowledge on the basis of an observation of n A's, of which m were observed to be B's, we now have, as a basis for our statistical knowledge about A's and B's a sample of $n + 1$, of which m are B's. It is clear that our body of knowledge will change relatively gradually as new observations come in: we will not, in this context, find the discontinuities that we observed earlier.

There is also the possibility that our background knowledge, even though statistical, is based on more than observation. For example, my belief that the chances of a birth being the birth of a male is about in [.50,.52] is based on lore obtained from sources that I regard as reliable. To learn that my daughter just gave birth to a boy will not only have little impact on that statistical generalization: it will have *no* impact. But if my source of knowledge were impugned, that would have a large effect. And it is conceivable that I could myself acquire such a large database of sex observations that my own data would impugn the authority on which I had accepted the conventional interval.

This also applies to the sort of statistical knowledge based on physical principles and assumptions. If a die is well balanced, then the velocities and momenta that characterize its trajectory will lead to its landing on each side with very nearly equal frequency in the long run, in view of the fact that very small changes in these momenta will lead to discontinuously different outcomes. If I roll a die and get a '1', my beliefs concerning its statistical characteristics will be unchanged. (Contrary to the Bayesian view, which would demand a tiny change.) If I roll the die a lot, and get a disproportionate frequency of '1's, then at some point I will question my assumptions — in particular, the assumption that the die is well balanced — and replace (not modify) my belief that the long run relative frequency of 1's is 1/6, by a statistical belief determined by my experience. (This will not be a very exact statistical belief, since I may well make this replacement on the basis of a fairly small sample. Thus I might come to believe that the frequency of 1's is in [.5,1.0].)

Thus even in the case of statistical knowledge, augmented by some more instances, there may be discontinuities. We have continuity (and, strictly, even this is not usually continuity in the mathematical sense) only when our evidential knowledge base contains representations of all the data on which the statistical law in our practical corpus is based, and when, in addition, we obtain additional statistical evidence by a procedure which is evidentially reliable. These conditions are almost invariably met when we are philosophers in our study making up examples out of moonbeams. They are rarely met otherwise.

5. Conflicting Observations

It is useful here to make a distinction between 'observation reports' — what is said to have been observed, and 'observation statements' — what is alleged in the report to have been observed. Observation reports cannot really conflict. If I report the weight of body W on one weighing as 23.654 grams, and on another weighing as 23.655, there need be nothing wrong with my observations, although the observation statements, "W weighs 23.654 grams," and "W weighs 23.655 grams" are inconsistent. This is why the natural and appropriate observation *statement* is rather, "W weighs $23.65 \pm .02$ grams." Note that this statement is not certain: It is acceptable, because the chance of error is negligible, not because the assertion cannot be wrong. On the usual treatment of errors, under which they are treated as normally distributed, an error of any magnitude is *possible*.

We treat the interval statement as *evidence*, however. We take it to be a statement that we can use in designing machinery, in engineering, in prediction, etc. It is not a statement to which we merely assign a high probability.

Even so, it is corrigible. We may weigh W twice again, and conclude (with the same degree of justification as we had before) that it weighs $23.60 \pm .02$ grams. The two interval statements are strictly incompatible. They are contraries. One *replaces* the other.

There are various possibilities. First, we may suppose that we simply have made somewhat unusual errors of measurement. If it is evidentially certain that W weighs between 23.63 grams and 23.67 grams, then W cannot weigh as little as 23.62 grams. But if W can't change weight, the discrepancy *must* be due to errors of measurement. If this is the case, then there are two impacts of our conflicting observations: The observations should be combined; and the discrepancy between the two sets of measurements should be taken as evidence concerning the distribution of errors of measurement for the measuring device(s) involved.

Merely combining the measurements would yield $23.62 \pm .015$, if we assume that all four measurements are simply taken from the same normal population of measurements. But the discrepancy might suggest that we should regard the measurements as coming from two distinct populations (corresponding to two instruments, say), or as coming from a population with a larger variance than we had thought.

In general, the conflict among observation reports must be taken as evidence concerning the reliability of the observer, or of the apparatus, of both. We will find that this is true also in the case of more basic conflicts.

6. Conflict between Observations and an Accepted Framework

This is the most interesting sort of conflict. In the example of weighing just described, it can arise. The conflict of measurements may be taken as evidence throwing suspicion on our framework assumptions: "Who *says* the object can't change weight? This is the sort of conflict that is most likely to be noticed, since we often make relatively local assumptions that we take for granted, act on the basis of, until and unless they lead us into difficulty. Good judgment consists in knowing when to abandon an assumption. But can good judgment be codified, reduced to mechanical rules? In some respects, we will argue, it can.

The simple-minded view of belief change is this: You have a generalization (general assumption) that you have taken for granted that leads you to infer that observational circumstances C will be followed by or accompanied by observational outcome O. You observe C. You observe some contrary of O. You reject your assumption.

But things are almost never this simple. Even when (rarely the case) a qualitative generalization is understood to be strict, to admit of no exceptions, there are alternatives to rejecting the generalization in the face of apparently conflicting observation. We may take the alleged observations to have been in error. Illusion, hallucination, are always available to explain away apparent refutations. And this is not irrational. In fact it has been argued (Kyburg, 1984) that this is the source of our knowledge of the qualitative errors of observation. The identification of an object or observation as belonging to a given *kind* is subject to error. The frequency of such errors is given by two principles: One is the conservation principle:

> We should not attribute more error to our observations than we are obliged to by the model of the world we accept.

The other principle guiding our assessments of error is the distribution principle:

> *Given* the satisfaction of the conservation principle, we should distribute the errors we are obliged to attribute to our observations as evenly as possible among the *kinds* of errors we might have made.

Thus if our model of the world assumes (presupposes) that all crows are black, and we have some observations of blue crows, we would assume that those observations contain errors. And further that the errors (other things being equal) are distributed equally between judgments of blueness and judgments of crowness. The metalinguistic fact that we must assume that we have made these errors of observation provides evidence about the *reliability* with which blueness and crowness can be identified.

7. Quantitative Observations Conflicting with Laws

Suppose in general that we assume the quantitative law, $y = f(x,z)$ in our body of knowledge. Then we observe a series of measurements of the quantities X, Y, and Z. No set of measurements can contradict the law in question, since any measurement is subject to error, and indeed, on the usual theories of measurement error, subject to error that can *possibly* be arbitrarily great. But of course large discrepancies, relative to a body of knowledge that contains the law in question, are extremely improbable.

The same general approach makes sense: The very improbable happens all the time (the particular set of measurements we make would be improbable even if they agreed with our assumed law, and the law *were* true), but if there is an alternative that renders the improbable not so improbable, the observations support that alternative. To put a quantitative measure on this is not trivial. One way, in terms of the framework we have already talked about, is the following: Anomalous observations can have two effects: they can provide new data concerning the errors of observation of a certain sort, or they can be taken at face value, and thus provide grounds for the rejection of general formulas. What we need are principles that will explain why one approach (admit to more errors of observation) or the other (reject a quantititive law) is to be preferred. Note that treating the law as an "approximation" or "idealization" is simply a way of taking it to be false, a way of rejecting it. A full discussion of this would amount to a general discussion of scientific inference. A more detailed treatment will be found in Kyburg (1990).

8. Fundamental Assumptions

Before going on to consider the grounds on which one would choose to give up an assumption in favor of attributing errors to one's observations, it is worth looking at one more extreme cases. This is that of measurement, and has been discussed more fully in (Kyburg, 1984). We suppose that length is additive: that the length of the collinear juxtaposition of two bodies is the sum of their lengths. Our measurements, of course, do not support this supposition; less dramatically: we can maintain the additivity of length only by attributing error to almost all our measurements.

Is this the alternative? To suppose that we can measure accurately, but that length is not additive, on the one hand, or, on the other, to suppose that length is additive, but that all our measurements are infected with error? Put this way it seems odd that one would ever opt for the second alternative. But we do.

Here is a possible explanation. The errors of measurement we need to introduce are very rarely large. They therefore do not deprive us of much useful knowledge. But the additivity of length is an enormously powerful predictive device. Knowing the length of two rigid bodies, we know, without even measuring, the *approximate* length of their collinear juxtaposition.

The choice between attributing error to observations and maintaining a generalization, as opposed to taking observations to be accurate and to refute the generalization, lies in the predictive observational content of the whole body of knowledge involved.

9. Choosing Between Assumptions and Errors

Suppose we consider two bodies of knowledge, one that embodies among its evidential certainties (among other things) the assumption A, the other of which does not. We make a set of observations (add to our evidential certainties a set of observation reports). We have in our background knowledge statistical information about errors in observations of this sort. Given the assumption A, the observation reports must be taken to embody unusually (improbably) large errors. These errors are not without observational consequences. They render observational predictions less dependable, since the correspondence between what is predicted and what *probably* going to be observed is only approximate, and reflects our knowledge of errors of observation.

How do we weight the advantages of one choice or the other? In order to have an actual measure that will yield an answer in these cases, we must focus on a class of predictive statements — that is, a class of statements that is of interest to us in the circumstances at hand. It is in this class that the predictions of the two cases are to be found. Let this class be **C**. We also need a measure of the precision of the predictions: thus if a prediction has the form "Bird B is Blue," the amount of content of that prediction should reflect the chance of an error in the observation that would test that prediction. If we can't accurately tell blue things, there is less content to the prediction that something is blue. If the prediction has the form, "Object O will be observed at an angle between $a - d$ and $a + d$," then its content will reflect the distribution of errors of observation of angle in the circumstances under consideration.

The class **C** of predictive statements about which we are concerned should be finite. It can be large, but we want to ensure that ratios are well defined in it. Next we need a measure **m** of accuracy, or predictive usefulness. What we need are only:

(i) a (finite) set of sentences **C** that include all those that may be of predictive interest in a given context, and
(ii) a measure **m** of how important errors of various kinds are.

We get the frequencies of error from our background knowledge of the observation reports we have had, together with assumptions of our body of knowledge. When we change the assumptions (or eliminate one) we change the statistical representation of these errors that we have reason to accept. If, for example, we eliminate an assumption, we can replace a number of predictions (those that stemmed from that assumption) by no predictions. If we increase the error of a certain kind of observation, we decrease the value of what we can predict.

Let **B** be a set of atomic sentences from **C** reflecting their historical proportion in our experience. Add **B** to our body of knowledge. Let **P** be the set of sentences that then become newly practically certain. The *predictive content* of a body of knowledge, relative to **C** and **m**, might be measured by the sheer number of predictions in **P**, each weighted by its reliability. This is a crude measure for determining the replacement of one general scientific theory by another, but for many purposes, it might be illuminating. In limited circumstances, we can find a class **C** that includes the statements that concern us; what is at issue is itself relatively straight-forward (Shall we assume that instrument I is working correctly, or shall we assume that it is broken?); and in these cases predictive content provides an appropriate criterion.

When we have quantitative statements in **P**, the natural measure of predictive content is $1/|u - l|$ — the measure of the precision which which we can confidently (with evidential certainty) predict. Of course the interval reflects the scale on which the quantity is measured. To alleviate the problem of artificial changes of scale, we can take the maxium and minimum values of a quantity in **B**, normalized to [0,1], to determine the scale. The measure **m** will then consist of the sum over **P** of the lengths of all the predictive intervals, reduced to that common scale.

10. Relation to Other Procedures

Can we relate this approach to replacement to other replacement formalisms that have been used? There is no direct reduction, obiously, since we are looking here at only a small segment of the statements of the language. Furthermore, these statements are not even statements that we have (now) reason to believe: what we have is a set **C** of statements that we are using a a *test instrument* for determining the relative desirability of two alternative frameworks. The procedure offered here is far less global than the procedures offered by Levi, Gardenfors, Spohn, and the others. It only takes into account the usefulness of one alternative, compared to another, when they make a difference to the test consequences **C** under the information measure **m**.

This seems quite natural (and perhaps even useful) when it comes to weighing relatively local assumptions. But it is not clear how far it can be extended, and how generally plausible the procedure can be made.

On the other hand, the procedure outlined is more general than the global replacement schemes previously suggested, since it allows us to compare two quite different languages, so long as they have the same test consequences **C**. We do not need an information measure on the languages themselves, nor do we need a global probability measure. Straight forward evidential probabilities will suffice.

Furthermore, our approach opens up the feasibility of treating framework assumptions as if they were infallible — that is, as genuine assumptions. The problem that has always surrounded talk of "presuppositions," "local assumptions," "framework assumptions," and the like has been their imunity from critical control. How do you weigh one against another? How do you tell when an assumption is dumb, compared to another that might have been made? It is hoped that we have offered an approach which allows these comparisons to be made in a rational way.

11. Summary

Global approaches to replacing one theory by another require relatively universal conventions: an ordering of all the sets of sentences in a formal language, for example, as well as, a Bayesian probability measure over all the sentences in the language. Approaches to eschewing acceptance, and therefore replacement, such as proposed by "Bayesian probabilists" tend to be impractical for simpler reasons: too much computation is devoted to issues that are at best peripheral to the often relatively simple question at hand, e.g. "Should we assume that instrument I is operating correctly?"

We have proposed instead an approach characterized by a set of sentences **C** (sentences that could, in principle, be construed as predictive observational sentences in the sense characterized above), and also by a measure of informational value **m** determined by a distribution of errors for these sentences. Suppose we are given a pair (**C**,**m**) consisting of a set of sentences and a measure of the importance of errors. Suppose we are given a body of knowledge. Then the relative value, in the face of a given body of observation reports, of two assumptions, or of one assumption as opposed to none, is determined. It is determined by machinery of evidential probability that we already have in hand.

There is, of course, the problem of determining the pair (**C**,**m**) to fit a given context. We have not yet dealt with this problem. We observe only that it is a far less overwhelming problem than that of determining informational content of all the sentences of a language (Levi) or of associating with each sentence of the language an ordinal number (Spohn). It can be done for a specific class of circumstances when certain kinds of predictions or anticipations are the kinds at issue. When the "assumptions" about which we are talking are relatively limited in scope ("Instrument 47 is working correctly"), it is not at all unreasonable to suppose that in fact we can isolate such a useful set of sentences. The question of deriving such a set of sentences from our concerns in a given context, and the question of deriving the importance of various kinds of error from the utilities of the outcomes possible in a given context, are questions that must be reserved for another time.

Note

[1]Research on which this work is based was supported by the Signals Warfare Center of the United States Army.

References

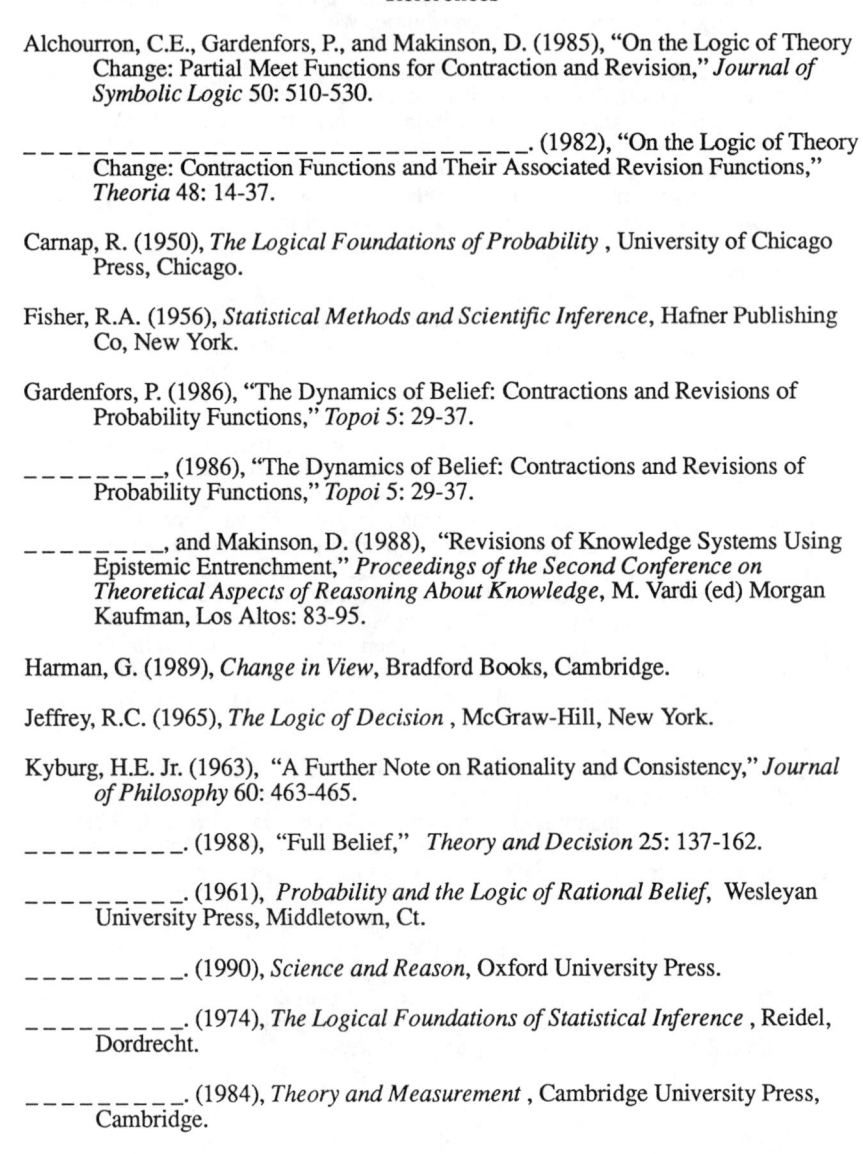

Alchourron, C.E., Gardenfors, P., and Makinson, D. (1985), "On the Logic of Theory Change: Partial Meet Functions for Contraction and Revision," *Journal of Symbolic Logic* 50: 510-530.

_____. (1982), "On the Logic of Theory Change: Contraction Functions and Their Associated Revision Functions," *Theoria* 48: 14-37.

Carnap, R. (1950), *The Logical Foundations of Probability* , University of Chicago Press, Chicago.

Fisher, R.A. (1956), *Statistical Methods and Scientific Inference*, Hafner Publishing Co, New York.

Gardenfors, P. (1986), "The Dynamics of Belief: Contractions and Revisions of Probability Functions," *Topoi* 5: 29-37.

_____, (1986), "The Dynamics of Belief: Contractions and Revisions of Probability Functions," *Topoi* 5: 29-37.

_____, and Makinson, D. (1988), "Revisions of Knowledge Systems Using Epistemic Entrenchment," *Proceedings of the Second Conference on Theoretical Aspects of Reasoning About Knowledge*, M. Vardi (ed) Morgan Kaufman, Los Altos: 83-95.

Harman, G. (1989), *Change in View*, Bradford Books, Cambridge.

Jeffrey, R.C. (1965), *The Logic of Decision* , McGraw-Hill, New York.

Kyburg, H.E. Jr. (1963), "A Further Note on Rationality and Consistency," *Journal of Philosophy* 60: 463-465.

_____. (1988), "Full Belief," *Theory and Decision* 25: 137-162.

_____. (1961), *Probability and the Logic of Rational Belief*, Wesleyan University Press, Middletown, Ct.

_____. (1990), *Science and Reason*, Oxford University Press.

_____. (1974), *The Logical Foundations of Statistical Inference* , Reidel, Dordrecht.

_____. (1984), *Theory and Measurement* , Cambridge University Press, Cambridge.

Levi, I. (1967), *Gambling with Truth*, Knopf, New York.

____. (1980),*The Enterprise of Knowledge*, MIT Press, Cambridge.

Popper, K.R. (1959), *The Logic of Scientific Discovery* , Hutchinson and Co., London. First German Ed. 1934.

Spohn, W. (1987), "Ordinal Conditional Functions: A Dynamic Theory of Epistemic States," in *Causation in Decision, Belief Change, and Statistics*, W. Harper and B. Skyrms (eds) Reidel, Dordrecht, pp. 105-134.

Belief Revision and Relevance[1]

Peter Gärdenfors

University of Lund

1. Belief Revisions as Minimal Changes of Relevant Beliefs

The theory of belief revision deals with models of states of belief and *transitions* between states of belief. The goal of the theory is to describe what should happen when you update a state of belief with new information. In the most interesting case, the new information is *inconsistent* with what you believe. This means that some of the old beliefs have to be deleted if one wants to remain within a consistent state of belief. A guiding idea is that the change should be *minimal* so that as few of the old beliefs as possible are given up.

A central problem for the theory of belief revision is what is meant by a minimal change of a state of belief. The solution to this problem depends to a large extent on the *model* of a state of belief that is adopted. In the literature, two types of models dominate: One where a state of belief is described by a *set of sentences* from a given language, sometimes called a *belief set*, and another where states of belief are modelled by *probability functions* defined over the language. I shall briefly outline these models of belief revision in Section 3.

The criteria of minimality used in these models have been based on almost exclusively *logical considerations*. However, there are a number of non-logical factors that should be important when characterizing a process of belief revision. The focus of this article will be the notion of *relevance*. The key criterion to be developed here is the following:

(I) If a belief state K is revised by a sentence A, then all sentences in K that are *irrelevant* to the validity of A should be retained in the revised state of belief.

In my opinion, (I) has a solid intuitive support. However, a criterion of this kind cannot be given a technical formulation in a model based on belief sets built up from sentences in a simple propositional language because the notion of relevance is not available in such a language. In models based on probability functions, there is a standard definition of relevance that could be used to formulate the desired principle.

However, as shall be shown below, the traditional definition suffers from some shortcomings that make it unsuitable to use in a more precise formulation of criterion (I).

So, before we can proceed to a version of criterion (I) that could be added to a theory of belief revision, based on either belief sets or probability functions, we must analyse the notion of relevance itself. This will be the purpose of Section 2. Only after this can we return to a theory of belief revision that incorporates the principle (I).

2. On the Logic of Relevance

In the traditional treatment of the notion of relevance, it is defined in terms of a probability function. However, since we want to develop an analysis of 'relevance' that can be used in various forms of models of belief states, a characterization that does not rely on probabilistic notions would be useful. This will be one of the aims of this section. Much of the material to be presented here is adopted from Gärdenfors (1978).

2.1 The Standard Definition

The traditional way of introducing the relevance relation is to define it with the aid of a given probability measure P in the following way:[2]

(D1) (a) A is *relevant* to C iff
$P(C/A) \neq P(C)$

(b) A is *irrelevant* to C iff
$P(C/A) = P(C)$

More general versions of this definition, but with the same basic idea have been studied by David (1979), Geiger (1990) and Pearl (1988). The implications for problems within philosophy of science are investigated by Salmon (1971, 1975) among others. Carnap (1950) points out that the theorems on irrelevance become simpler if the following definition of irrelevance is adopted instead:[3]

(b') A is *irrelevant* to C iff
$P(C/A) = P(C)$ or A is logically false.

If it is assumed that only logically false sentences have zero probability, then this definition has the consequence that any sentence A is either relevant or irrelevant to C. In the following sections I will adopt Carnap's suggestion, so when (D1) is mentioned, the conjunction of (a) and (b') is referred to.

There are two problems connected with (D1). One was already pointed out by Keynes (1921) who was among the first to discuss the concept of relevance. He observes that, intuitively, there is a stronger sense of 'relevance' which is not covered by (D1). In connection with his discussion of the 'weight' of arguments, he writes:[4]

> "If we are to be able to treat 'weight' and 'relevance' as correlative terms, we must regard evidence as relevant, part of which is favourable and part unfavourable, even if, taken as a whole, it leaves the probability unchanged. With this definition, to say that a new piece of evidence is 'relevant' is the same thing as to say that it increases the 'weight' of the argument."

Here Keynes is referring to the case when P(C/A & B) = P(C), even though P(C/A) ≠ P(C) and P(C/B) ≠ P(C), which, according to (D1), means that A & B is irrelevant to C, while both A and B, taken as separate pieces of evidence, are relevant to C.

In order to capture this stronger sense of 'relevance', Keynes proposes the following definition which, he believes, "is theoretically preferable":[5]

(D2) (a) A is *irrelevant* to C iff
there is no sentence B, which is derivable from A such that P(C/B) ≠ P(C).

(b) A is *relevant* to C iff A is not irrelevant to C.[6]

This definition has the consequence that if A is relevant to C, then, for any sentence B such that A & B is not logically contradictory, A & B is also relevant to C and thus it blocks the seemingly counterintuitive feature of (D1) mentioned above.

Carnap shows that the definition (D2) leads to the following trivialization result:[7] if neither C or ¬C are logically valid, then A is irrelevant to C iff A is logically valid. This is certainly absurd. For most sentences C there are many other sentences that we judge as irrelevant to C. (D2) is therefore not the appropriate way to define the relevance relation in the stronger sense hinted at by Keynes. The question is now whether it is possible to give a definition of this relation that satisfies Keynes's requirement.

The second problematic feature of (D1) is that as soon as P(A) = 1, it follows that A is irrelevant to B, *for any sentence* B. This is highly counterintuitive because if P is taken to model the current state of belief, and A is a contingent sentence, then P(A) = 1 means that A is held to be a true fact, but this does not entail that there are no sentences that are relevant for the fact that A. We will return to this drawback of (D1) in Section 4.

I will now first show that Carnap's trivialization result is not dependent on the definition (D2) or any other definition in terms of probability measures. I will formulate some general criteria for the relevance relation and show that if Keynes's requirement is added, then the trivialization result will follow. When formulating the criteria, it will not be assumed that the relevance relation is to be explicated in terms of probability measures.

Because of the trivialization result, I conclude that Keynes's requirement has to be abandoned. However, this should not prevent us from seeking a definition of 'relevance' that is stronger than (D1) and that follows Keynes's (and our) intuitions as far as possible. I will present two criteria for the relevance relation which are weaker than Keynes's requirement but which are not satisfied by (D1). Their logical consequences will be investigated. Finally, I will propose a new definition of the relevance relation that satisfies one of these criteria and briefly investigate its properties.

2.2. Basic Criteria for the Relevance Relation

In this paper, relevance is taken to be a relation between sentences. I therefore assume that there be a given language $\mathcal{L}$ where the sentences are taken from. This language is assumed to be closed under standard truth-functional operations. I will use A, B, C, etc. as symbols for sentences. If A is provable, I will write ⊢A. A sentence A is said to be *contingent*, if neither ⊢ A, nor ⊢ ¬A. The expression 'A is relevant to C' will be abbreviated A $\mathbb{R}$ C, and similarly, 'A is irrelevant to C' will be written A ⊥⊥ C.

I will now proceed to formulate some general criteria for the relevance relation. The criteria are not intended to be a complete characterization of the logic or 'relevance', but are rather meant to be as weak as possible.

(R0) If ⊢ A ↔ B, then A ℝ C iff B ℝ C.

This is a simple rule of replacement of logical equivalents.

(R1) A ℝ C iff not A ⊥ C.

Relevance and irrelevance are complementary and mutually exclusive relations. Carnap saw this criterion as an argument for changing (b) in (D1) to (b').

(R2) A ℝ C iff ¬A ℝ C.

If one obtains some new information about the sentence C when learning that A, then one also learns something about C when ¬A is added.

From (R1) and (R2) we can derive

(1) A ⊥ C iff ¬A ⊥ C.

(R3) (A ∨ ¬A) ⊥ C.

Counting A ∨ ¬A as new evidence does of course not affect our judgement of the degree of truth of C. From (R3) and (R2) we can derive

(2) (A & ¬A) ⊥ C.

This is in accordance with Carnap's changing (b) in (D1) to (b'), and it enables us to formulate (R2) without restrictions.

A consequence of (R0), (R1), (R3) and (2) is

(3) If A ℝ C, then A is contingent.

The following condition is introduced in order to secure that relevance is a non-empty relation.

(R4) If C is contingent, then C ℝ C.

If it is assumed that only sentences which are logical consequences of the evidence have probability one, then (D1) fulfills the requirements (R0) – (R4). I take these criteria to be necessary for any explication of the relevance relation.

2.3. A Trivialization Result

We next turn to Keynes's requirement. In connection with his definition, which I call (D2), he gives the following argument:[8]

> "Any proposition which is irrelevant in the strict sense [i.e., according to (D2)] is, of course, also irrelevant in the simpler sense [i.e., according to (D1)] but if we were to adopt the simpler definition, it would sometimes occur that a part of evidence would be relevant, which taken as a whole was irrelevant."

This quotation motivates the following criterion:[9]

(R5)   If A $\mathbb{R}$ C and not ⊢ ¬(A & B), then (A & B) $\mathbb{R}$ C.

As we have already observed, (D1) does not satisfy (R5) for any non-trivial probability measure P. The following simple lemma will show the connection between (R5) and (D2) and throw some light on why Keynes chose this definition for his stronger concept of relevance.

LEMMA: If (R0) is assumed, then the following criterion is equivalent to (R5):

(4)   If B $\mathbb{R}$ C, A → B, and not ⊢ ¬A, then A $\mathbb{R}$ C.

The proof of the lemma and the following three theorems can be found in Gärdenfors (1978).

I will now show that (R5) leads to strongly counterintuitive consequences, if combined with the criteria (R0) – (R4).

THEOREM 1: If the relations $\mathbb{R}$ and $\mathbb{U}$ satisfy (R0) – (R5), then every contingent sentence is relevant to every other contingent sentence.

This theorem presents us with a dilemma. On the one hand, there seems to be some truth in the observation that (D1) does not cover our intuitive conception of 'relevance', and, on the first impression, Keynes's requirement seems acceptable. On the other hand, the remaining criteria for the relevance relation, needed to derive the theorem, are seemingly innocent. However, the consequence that all non-trivial sentences are relevant to any contingent sentence is strongly counterintuitive.

In my opinion, the only reasonable way out of the dilemma is to reject the assumption that (R5) is valid. This does not mean, however, that (D1) has to be accepted as the correct definition of the relevance relation.

The unsatisfactory feature of (D1) is, roughly, that it makes *too few* sentences relevant. This view is supported by the quotations from Keynes (1921) given above. One way to find a more appropriate definition of the relevance relation is therefore to investigate further general criteria that may be added to the basic criteria (R0) – (R4) and that enlarge the set of relevant sentences.

2.4. Two Further Criteria

In this section I will investigate the logical consequences of the following criteria:

(R6)   If A $\mathbb{R}$ C, B $\mathbb{R}$ C, and not ⊢ ¬(A & B), then (A & B) $\mathbb{R}$ C.

(R7)   If A $\mathbb{U}$ C and B $\mathbb{U}$ C, then (A & B) $\mathbb{U}$ C.

These criteria will be called 'the conjunction criterion for relevance' and 'the conjunction criterion for irrelevance' respectively. Neither of these criteria is fulfilled by (D1). (R6) is a special case of (R5) and thus trivially derivable from (R5). A consequence of Theorem 1 is that the sentences that are irrelevant to a sentence C are those that are logically valid or invalid. From this it is easy to see that (R7) too is derivable from (R0) – (R5). Thus (R6) and (R7) are consequences of (R5) in the presence of (R0) – (R4). In fact, the converse is also true.

THEOREM 2: (R5) is derivable from (R6) and (R7) together with (R0) – (R4).

This theorem shows that (R6) and (R7) can not both be acceptable since Theorem 1 would then be derivable. In the sequel, it will be shown that neither (R6) nor (R7) is alone sufficient for (R5).

From (R6) and (R2) it is easy to derive the following condition:

(5)   If A $\mathbb{R}$ C, B $\mathbb{R}$ C and not ⊢A ∨ B, then (A ∨ B) $\mathbb{R}$ C.

For a fixed sentence C, we see by (R2), (R6) and (5) that the set of sentences relevant to C is closed under truth-funcional operations, as long as these operations do not yield sentences that are logically valid or contradictory.

Using (R0) – (R4) one can show that (R7) is equivalent to

(6)   If (A & B) $\mathbb{R}$ C, then A $\mathbb{R}$ C or B R C.

In words, this condition could be interpreted as saying that if a sentence is relevant, then some of its parts are also relevant. In a sense, this is the converse of (R5) which says that if a part of a sentence is relevant, then the sentence as a whole is relevant. As we have seen, (6) is derivable from (R0) – (R5).

Analogous to the case above is the possibility of deriving the following condition from (R7) and (1):

(7)   If A ⊥ C and B ⊥ C, then (A ∨ B) ⊥ C.

For a given sentence C, we conclude from (1), (R7) and (7) that the sentences relevant to C will be closed under truth-functional operations (with no restrictions). And, conversely, if the irrelevant sentences are closed under truth-functional operations, (1), (R7) and (7) will be fulfilled. This connection will be utilized in Section 5.

These results provide us with some ideas of the power of conditions (R6) and (R7). But, as we have seen, we cannot require both to be satisfied for a reasonable relevance relation. It is argued in Gärdenfors (1978) that (R7) is valid, but there are good counterexamples to (R6). Thus, the appropriate conditions for a relevance relation on this level seems to be (R0) - (R4) together with (R7). Next I would like to show that it is possible to improve the definition of irrelevance so that these conditions will be satisfied.

2.5 An Amended Definition of Irrelevance

In Gärdenfors (1978, p. 362) it is argued that in order to establish that A is irrelevant to C, it is not sufficient that $P(C/A) = P(C)$, but we must also know that if we learned that A, then no sentences that are now irrelevant to C would become relevant to C on the new evidence A. This is in accordance with the earlier idea that (D1) makes too few sentences relevant. This argument motivates the following definition:

(D3)   (a) A ⊥ C iff
$P(A) = 0$ or $P(C/A) = P(C)$ and for all B such that $P(C/B) = P(C)$ and $P(A\&B) \neq 0$, it also holds that $P(C/A\&B) = P(C)$.

(b) A $\mathbb{R}$ C iff not A ⊥ C.

Note that the condition that P(C/A) = P(C) is a special case of "for all B such that P(C/B) = P(C) and P(A&B) ≠ 0, it also holds that P(C/A&B) = P(C)", namely, the case when B is a tautology. This means that (D3) can be simplified to:

(D3') (a) A ⊥⊥ C iff
P(A) = 0 or for all B such that P(C/B) = P(C) and P(A&B) ≠ 0, it also holds that P(C/A&B) = P(C).

(b) A ℝ C iff not A ⊥⊥ C.

THEOREM 3: (D3) satisfies (R0) - (R4), and (R7).

It is easy to show by a small finite example that (D3) can be satisfied nontrivially so that no trivialization result is possible for the set of conditions (R0) - (R4), and (R7).[10] But (D3) still suffers from the second drawback mentioned for (D1), i.e., the property that if P(A) = 1, then A ⊥⊥ C for all C. In order to get around this problem, we need yet another amendment of the definition. This will be the topic of Section 4.

Furthermore, the following feature of (D3) is worth noticing. It is easy to verify that (D1) satisfies the following principle:

(8)   A ℝ C iff C ℝ A

However, this symmetry principle is, in general, not satisfied by (D3). The following kind of example might be a counterexample to (8): Let A be the proposition that a mother is blond and C that her daugther is blond. Even though probability calculus tells us that P(A/C) ≠ P(A) if and only if P(C/A) ≠ P(C), our intuitions seem to be that A ℝ C but not C ℝ A since a mother's being blond can be a *cause* of her daughter's being blond, but not the other way around (cf. the results obtained by Tversky and Kahnemann 1982). I conclude that the fact that (D3) does not satisfy (8) need not be a drawback of the definition in relation to (D1). On the contrary, this feature may be in full accordance with our intuitions about relevance.

3. Belief Revision Models

The definitions (D1) - (D3) all have the drawback that if P(A) = 1, then A ⊥⊥ C for all C. What one would like to have is that even if A is known, i.e., if P(A) = 1, C can be relevant to A for some sentences C. One way of capturing this idea is to say that *if A had not been known*, the information that C would have affected the probability of A. However, in order to formulate this idea more precisely, we need an account of belief revision and contraction processes.

In this section I will outline two models of belief revision. The first is based on belief sets as models of epistemic states, as developed in, for example, Alchourrón, Gärdenfors, and Makinson (1985) and Gärdenfors (1988). The second is based on probability functions as models of epistemic states. Models of this kind can be found in, for example, Harper (1975) and Gärdenfors (1986, 1988).

3.1 Belief Revision Models Based on Belief Sets

One way of modelling epistemic states is to describe them by *belief sets* which are sets of sentences from a given language. The interpretation is that if a sentence A belongs to a belief set K, this means that A is accepted as true in the state of belief modelled by K. Belief sets are assumed to be closed under logical consequences (classical logic is generally presumed), which means that if K is a belief set and K logically entails B, then B is an element in K. A belief set can be seen as a partial description of the world — partial because in general there are sentences A such that neither A nor $\neg A$ are in K.

Belief sets model the statics of epistemic states. I now turn to their *dynamics*. What we need are methods for updating belief sets. Three kinds of updates will be discussed here:

(i) Expansion: A new sentence together with its logical consequences is *added* to a belief set K. The belief set that results from expanding K by a sentence A will be denoted K^+_A.

(ii) Revision: A new sentence that is *inconsistent* with a belief set K is added, but in order for the resulting belief set to be consistent, some of the old sentences of K are deleted. The result of revising K by a sentence A will be denoted K^*_A.

(iii) Contraction: Some sentence in K is retracted without adding any new beliefs. In order for the resulting belief set to be closed under logical consequences, some other sentences from K must be given up. The result of contracting K with respect to A will be denoted K^-_A.

Expansions of belief sets can be handled comparatively easily. K^+_A can simply be defined as the logical consequences of K together with A :

(Def +) $K^+A = \{B: K \cup \{A\} \vdash B \}$

As is easily shown, K^+_A defined in this way is closed under logical consequences and will be consistent when A is consistent with K.

It is not possible to give a similar explicit definition of revisions and contractions in logical and set-theoretical notions only. To see the problem for revisions, consider a belief set K that contains the sentences A, B, A & B $\rightarrow$ C and their logical consequences (among which is C). Suppose that we want to revise K by adding $\neg C$. Of course, C must be deleted from K when forming $K^*_{\neg C}$, but at least one of the sentences A, B, or A & B $\rightarrow$ C must also be given up in order to maintain consistency. There is no purely *logical* reason for making one choice rather than the other, but we have to rely on additional information about these sentences. Thus, from a logical point of view, there are several ways of specifying the revision of a belief set. What is needed here is a method of determining the revision.

As should easily be seen, the contraction process faces parallel problems. In fact, the problems of revision and contraction are closely related, being two sides of the same coin. To establish this more explicitly, we note, firstly, that a revision can be seen as a composition of a contraction and an expansion. Formally, in order to construct the revision K^*_A, one first contracts K with respect to $\neg A$ and then expands $K^-_{\neg A}$ by A which amounts to the following definition:

(Def *) $K^*_A = (K^-_{\neg A})^+_A$

Conversely, contractions can be defined in terms of revisions. The idea is that a sentence B is accepted in the contraction K^-_A if and only if B is accepted in both K and K^*_A. Formally:

(Def -) $K^-_A = K \cap K^*_{\neg A}$

These definitions indicate that revisions and contractions are *interchangable* and a method for explicitly constructing one of the processes would automatically yield a construction of the other.

There are two methods of attacking the problem of specifying revision and contraction operations. One is to present *rationality postulates* for the processes. Such postulates are introduced in Gärdenfors (1984), Alchourrón, Gärdenfors and Makinson (1985) and discussed extensively in Gärdenfors (1988), and they will not be repeated here. A guiding idea for these postulates is that changes should be *minimal*, so that when changing beliefs in response to new evidence, one should continue to believe as many of the old beliefs as possible.

The second method of solving the problems of revision and contraction is to adopt a more *constructive* approach and build computationally oriented *models* of belief revision that can take a belief set (or some representation of such a set) together with a sentences to be added as input and which then gives a revised belief set as output. One idea in this area is that the sentences that are accepted in a given belief set K have different degrees of *epistemic entrenchment*. When determining which sentences to delete in the revision, the basic recipe is that one gives up those with the lowest degrees of epistemic entrenchment and retains those with the highest degree (Cf. Gärdenfors and Makinson (1988) for this approach).

In the theory of belief revision, the rationality postulates and the model approach are connected via *representation theorems* which say that all models in a certain class (for example using epistemic entrenchment to determine the revision function) satisfy a certain set of postulates (for example, the postulates (K*1) - (K*8) for revision as presented in Gärdenfors (1988)), and vice versa, any revision method satisfying these postulates can be identified with one of the models in the given class.

3.2 Belief revision models based on probability functions

Two central dogmas of Bayesianism are that states of belief can be represented by *probability functions* and that rational changes of belief can be represented by *conditionalization* whenever the information to be added is consistent with the given state of belief. However, there are other kinds of changes of belief that cannot easily be modelled by the conditionalization process. Sometimes we have to revise our beliefs in the light of some evidence that contradicts what we had earlier mistakenly accepted. And when $P(A) = 0$, where P represents the present state of belief and A is the new evidence to be accommodated, the conditionalization process is undefined.

And sometimes we give up some of our beliefs. This kind of change of belief is here, like above, called a *contraction*, and the goal of this subsection is to present a way of modelling this process for a probabilistic model of a state of belief.

In parallel with the situation for belief sets, one can distinguish three kinds of probabilistic belief changes:

(i) *Expansion*: where we start from $P(A) = \alpha$ for some sentence A and some α, $0 < \alpha < 1$, and where the expanded probability function P^+_A satisfies the criterion $P^+_A(A) = 1$.

(ii) *Contraction*: where we start from $P(A) = 1$ for some sentence A, and where the contracted function P^-_A satisfies the criterion $P^-_A(A) = \alpha$, for some α, $0 < \alpha < 1$.

(iii) *Revision*: where we start from $P(A) = 0$ for some sentence A, and where the revised probability function P^*_A satisfies the criterion $P^*_A(A) = 1$.

Expansions are normally modelled by conditionalization, i.e., $P^+_A(B) = P(B/A)$, for all B, but there is no similar explicit definition of the contraction and revision processes. However, in the same way as for belief sets, it is possible to define revisions of probability functions in terms of their contractions:

(Def P*) $P^*_A(B) = (P^-_{\neg A}(B))^+_A$

Or, using that expansion is modelled by conditionalization: $P^*_A(B) = P^-_{\neg A}(B/A)$ which is always well defined since $P^-_{\neg A}(A) > 0$ according to the postulate (P-1) below. This means that if we can give a satisfactory definition of probabilistic contraction, we have thereby also solved the problem of defining a probabilistic revision. So let us focus on probabilistic contractions.

When contracting a state of belief with respect to a belief A, it will be necessary to change the probability values of other beliefs as well in order to comply with the axioms of probability calculus. However, there are, in general several, ways of fulfilling these axioms. An important problem concerning contractions is how one determines which among the accepted beliefs, i.e., those A's where $P(A) = 1$, are to be retained and which are to be removed. One requirement for contractions of probability functions is that the *loss of information* should be kept as small as possible.

In Gärdenfors (1986) and (1988), I have formulated a number of postulates for contractions of probability functions. These postulates are based on the idea that the contraction P^-_A of a probability function P with respect to A should be *as small as possible* in order to minimize the number of beliefs that are retracted. In a sense that will be made more precise later, contractions can be viewed as 'backwards' conditionalizations. The postulates for probabilistic contractions can also be regarded as generalizations of a set of postulates for contractions in the case of belief sets.

I will here give a brief presentation of the postulates for probabilistic contraction. Formally, this process can be represented as a function from $\mathcal{P} \times \mathcal{L}$ to $\mathcal{P}$, where $\mathcal{P}$ is the set of all probability functions and $\mathcal{L}$ is the language that describes the space of events over which these functions are defined.[11] The value of such a contraction function, when applied to arguments P and A, will be called the contraction of P with respect to A, and it will be denoted P^-_A. Let us say that an event A is *accepted* in the state of belief represented by P iff $P(A) = 1$.

The first postulate is a requirement of 'success' simply requiring that A not be accepted in P^-_A, unless A is logically valid, in which case it can never be retracted:

(P-1) $P^-_A(A) < 1$ iff A is not logically valid.

It should be noted that this postulate does not say anything about the magnitude of $P^-_A(A)$. This leaves open a range of possibilities for an explicit construction of a contraction function. None of these possibilities will be ruled out by the remaining postulates. The value of $P^-_A(A)$ can be seen as a parameter in the construction of P^-_A.

The second postulate requires that the contraction P^-_A is only dependent on the content of A, not on its linguistic formulation:

(P-2) If A and B are logically equivalent, i.e., describe the same event, then $P^-_A = P^-_B$.

The following postulate is only needed to cover the trivial case when A is not already accepted in P:

(P-3) If $P(A) < 1$, then $P^-_A = P$.

So far, the postulates have only stated some mild regularity conditions. The next one is more interesting:

(P-4) If $P(A) = 1$, then $P^-_A(B/A) = P(B)$, for all B in $\mathcal{B}$.

This means that if A is first retracted from P and then added again (via conditionalization), then one is back in the original state of belief. This postulate, which will be called the *recovery* postulate, is one way of formulating the idea that the contraction of P with respect to A should be minimal in the sense that an unnecessary loss of information should be avoided. It also makes precise the sense in which contraction is 'backwards' conditionalization.

The final postulate is more complicated and concerns the connection between P^-_A and $P^-_{A\&B}$:

(P-5) If $P^-_{A\&B}(\neg A) > 0$, then $P^-_A(C/\neg A) = P^-_{A\&B}(C/\neg A)$, for all C.

In order to understand this postulate, we first present one of the arguments that has been proposed as a justification for conditionalization. Unlike all other changes of P to make A certain, conditionalization does not distort the probability ratios, equalities, and inequalities among sentences that imply A. In other words, the probability proportions among sentences that imply A are the same before and after conditionalization.

Now, if contraction may be regarded as 'backwards' conditionalization, then a similar argument should be applicable to this process as well. More precisely, when contracting P with respect to A, some sentences that imply $\neg A$ will receive non-zero probabilities, and when contracting P with respect to A&B some sentences that imply $\neg A$ or some sentences that imply $\neg B$ (or both) will receive non-zero probabilities. If, in the latter case, some sentences that imply $\neg A$ receive non-zero probabilities, i.e., if $P^-_{A\&B}(\neg A) > 0$, then the two contractions should give the same proportions of probabilities to the sentences implying $\neg A$, i.e., $P^-_A(C/\neg A)$ should be equal to $P^-_{A\&B}(C/\neg A)$, for all C. But this is exactly the content of (P-5).

This completes the set of postulates for probabilistic contraction functions. It should be noted that the postulates do not determine a unique contraction, but that they only introduce rationality constraints on such functions. Among other things, the value of $P^-_A(A)$ can be any number greater than 0 and smaller than 1. It is argued in Gärdenfors (1988) that rationality constraints are not enough to determine a unique contraction function (just as the probability axioms do not determine a unique rational probability function), but pragmatic factors must be added in order to single out the actual contraction. In the book, I introduce an ordering of 'epistemic entrenchment' among the beliefs to be used when determining which beliefs are to be given up when forming a particular contraction. The heuristic rule is that when we have to give up some of our beliefs, we try to retain those with the greatest epistemic entrenchment.

4. A Final Definition of Relevance

It is now time to return to the desideratum (I) formulated in the introductory section:

(I) If a belief state K is revised by a sentence A, then all sentences in K that are *irrelevant* to the validity of A should be retained in the revised state of belief.

We can now use the terminology introduced above to formulate this criterion more precisely:

(I*)   If $A \perp\!\!\!\perp C$, then $P^*_A(C) = P(C)$

When states of belief are modelled by belief sets, this criterion has as a special case:

(8)   If $A \perp\!\!\!\perp C$, then $C \in K^*_A$ iff $C \in K$

Parallel criteria should also be valid for contraction:

(I-)   If $A \perp\!\!\!\perp C$, then $P^-_A(C) = P(C)$

(9)   If $A \perp\!\!\!\perp C$, then $C \in K^-_A$ iff $C \in K$

By using (Def P*), it is easy to show that (I*) will follow from (I-). So what we want is a definition of irrelevance that will satisfy (I-) in addition to (R0) - (R4) and (R7).

To arrive at such a definition, first note that it follows from the postulate (P-4) that if $P(A) = 1$, then $P(C) = P^-_A(C/A)$, so the requirement that $P^-_A(C) = P(C)$ may as well be written $P^-_A(C) = P^-_A(C/A)$ in this case. And according to (P-3), we have in the case when $P(A) < 1$, i.e., the case when A is not known in the state of belief represented by P, that $P(C) = P^-_A(C)$, for all C. This means that the equality $P^-_A(C) = P^-_A(C/A)$ is a generalized version of the classical equality $P(C) = P(C/A)$ used in (D1) to define irrelevance. The more general equality also covers the case when $P(A) = 1$.

Using this equality in combination with the construction in (D3), we can now formulate the final definition of irrelevance:

(D4)   (a) $A \perp\!\!\!\perp C$ iff
$P(A) = 0$, or for all B such that $P^-_B(C) = P^-_B(C/B)$ and $P^-_A(A\&B) \neq 0$, it also holds that $P^-_A(C/A\&B) = P^-_A(C)$.

        (b) $A \mathbb{R} C$ iff not $A \perp\!\!\!\perp C$.

In the same way as for (D3') we can conclude, by letting B be a tautology, that $A \perp\!\!\!\perp C$ entails as a special case $P^-_A(C/A) = P^-_A(C)$.

Before we establish the properties of (D4), we need to reconsider one of the general postulates for revision. (D3) had the drawback that if $P(A) = 1$, then $A \perp\!\!\!\perp C$ for all C. This feature made it possible to satisfy (R2) without any exceptions in the limiting cases. However, one of the purposes of formulating (D4) is to eliminate this drawback, but this means that when $P(A) = 0$ so that $A \perp\!\!\!\perp C$, it is still possible to have $\neg A \mathbb{R} C$.

This violation of (R2) shows that it should be replaced by the following, slightly weaker version:

(R2') If $P(A) \neq 0$ and $P(A) \neq 1$, then $A \mathrel{R} C$ iff $\neg A \mathrel{R} C$.

The corresponding weakening of (1), which is equivalent to (R2') given (R1), is:

(1') If $P(A) \neq 0$ and $P(A) \neq 1$, then $A \mathrel{\amalg} C$ iff $\neg A \mathrel{\amalg} C$.

THEOREM 4: If the contraction function satisfies (P-1) - (P-5), then (D4) satisfies (R0), (R1), (R2'), (R3), (R4), (R7) and (I-).

Proof: (R0) and (R1) follow immediately from (D4). (R3) follows from (P-1) and (P-4) which entail that $P^-_{A \vee \neg A} = P$. To show (R2'), we proceed by verifying (1'). So assume that $P(A) \neq 0$, $P(A) \neq 1$, and $A \mathrel{\amalg} C$: We want to show that $\neg A \mathrel{\amalg} C$. By (P-3) we have $P^-_A = P = P^-_{\neg A}$. Suppose that for some B, $P^-_B(C) = P^-_B(C/B)$ and $P^-_{\neg A}(\neg A \& B) \neq 0$. We need to show that $P^-_{\neg A}(C/\neg A \& B) = P^-_{\neg A}(C)$, that is, $P(C/\neg A \& B) = P(C)$. If $P(A \& B) = 0$, then $P(C/\neg A \& B) = P(C/\neg A)$, since $P(C/A) = P(C)$, and we are done. So suppose that $P(A \& B) = P^-_A(A \& B) \neq 0$. Since $A \mathrel{\amalg} C$ we then know that $P(C/A \& B) = P(C)$. We need to distinguish two cases: (i) $P(B) < 1$. By (P-3) again, we then have $P^-_B = P$. Consider the following identities: $P(C) = P(C/B) = P(A/B) \cdot P(C/A \& B) + P(\neg A/B) \cdot P(C/\neg A \& B)$. Since $P(C/A \& B) = P(C)$, it follows from the fact that $P(A/B) + P(\neg A/B) = 1$ that $P(C/\neg A \& B) = P(C)$ which proves this case. (ii) $P(B) = 1$. Then $P(C/\neg A \& B) = P(C/\neg A)$. But, as established above $P(C/\neg A) = P(C)$ and we are done.

To prove (R4), it is sufficient to note that if C is contingent, then $P^-_C(C) < 1$ by (P-1), so $P^-_C(C) < P^-_C(C/C) = 1$ and hence $C \mathrel{R} C$.

The most difficult condition to verify is (R7). Assume that $A \mathrel{\amalg} C$ and $B \mathrel{\amalg} C$. The goal is to show $A \& B \mathrel{\amalg} C$. If $P(A \& B) = 0$, this follows immediately from (D4). So suppose that $P(A \& B) \neq 0$. It follows that $P(A) \neq 0$ and $P(B) \neq 0$. From the facts that $A \mathrel{\amalg} C$ and $B \mathrel{\amalg} C$, we know that $P^-_A(C) = P^-_A(C/A)$ and $P^-_B(C) = P^-_B(C/B)$, and consequently that $P^-_A(C/A \& B) = P^-_A(C)$ and $P^-_B(C/A \& B) = P^-_B(C)$. As a preliminary we show that $P^-_{A \& B}(C/A \& B) = P^-_{A \& B}(C)$.

Case 1: $P(A \& B) < 1$. It follows that either $P(A) < 1$ or $P(B) < 1$ and, from (P-3), that $P^-_{A \& B} = P$. If $P(A) < 1$, then also $P^-_A = P$, and since $P^-_A(C/A \& B) = P^-_A(C)$ it follows that $P^-_{A \& B}(C/A \& B) = P^-_{A \& B}(C)$. Similarly, if $P(B) < 1$, then $P^-_B = P$ and since $P^-_B(C/A \& B) = P^-_B(C)$, we know also in this case that $P^-_{A \& B}(C/A \& B) = P^-_{A \& B}(C)$. Hence, $P^-_{A \& B}(C/A \& B) = P^-_{A \& B}(C)$.

Case 2: $P(A \& B) = 1$. It follows that $P(A) = 1$ and $P(B) = 1$. By (P-4), this entails that $P^-_{A \& B}(C/A \& B) = P(C)$, so it suffices to show that $P^-_{A \& B}(C) = P(C)$. Consider the following identity: $P^-_{A \& B}(C) = P^-_{A \& B}(A) \cdot P^-_{A \& B}(C/A) + P^-_{A \& B}(\neg A) \cdot P^-_{A \& B}(C/\neg A)$. It follows from (P-4) that $P^-_{A \& B}(C/A) = P^-_B(C)$ and thus from the assumption that $P^-_B(C) = P^-_B(C/B)$ and (P-4) again that $P^-_{A \& B}(C/A) = P(C)$. If $P^-_{A \& B}(\neg A) = 0$, the identity thus reduces to $P^-_{A \& B}(C) = P(C)$ which is what we wanted to show. On the other hand, if $P^-_{A \& B}(\neg A) \neq 0$, we can apply (P-5) to conclude that $P^-_{A \& B}(C/\neg A) = P^-_A(C/\neg A)$. But from the assumption that $P^-_A(C) = P^-_A(C/A)$ it follows that $P^-_A(C/\neg A) = P^-_A(C/A)$, which by (P-4) reduces to $P^-_A(C/\neg A) = P(C)$. So also in this case, the identity reduces to $P^-_{A \& B}(C) = P(C)$ and we have shown that $P^-_{A \& B}(C/A \& B) = P^-_{A \& B}(C)$.

After this intermediate result, assume now that D is a sentence such that $P^-_{A \& B}(A \& B \& D) \neq 0$ and that $P^-_D(C) = P^-_D(C/D)$. Since $P^-_{A \& B}(C/A \& B) = P^-_{A \& B}(C)$

we can, by the same argument as above, conclude that $P^-_{A\&B\&D}(C/A\&B\&D) = P^-_{A\&B\&D}(C)$. We need to show that $P^-_{A\&B}(C/A\&B\&D) = P^-_{A\&B}(C)$ for all C. If $P(A\&B) = 1$, then this follows immediately from (P-4) and what was shown above. If $P(A\&B) < 1$, then $P(A\&B\&D) < 1$ and hence $P^-_{A\&B\&D} = P^-_{A\&B} = P$ and so $P^-_{A\&B}(C/A\&B\&D) = P^-_{A\&B\&D}(C/A\&B\&D) = P^-_{A\&B\&D}(C) = P^-_{A\&B}(C)$. This shows that (R7) is satisfied.

For (I-) finally, it is sufficient to note that if $P(A) = 1$, then $P^-_A(C) = P(C)$ follows from $P^-_A(C) = P^-_A(C/A)$ and (P-4), and if $P(A) < 1$, then $P^-_A(C) = P(C)$ is immediate from (P-3). This completes the proof of the theorem.

Apart from knowing that (D4) has the desired properties, we must also make sure that it does not lead to any triviality results. However, to prove this, one can use essentially the same example that was used in Gärdenfors (1978) to establish the non-triviality of (D3). The details are easy to verify but tedious so I will omit them.

5. Using Irrelevance in Constructions of Belief Revisions

The theorem proved in the preceeding section shows that if one starts from a probability contraction function, it is *possible* to define relations of relevance and irrelevance that satisfy the desired postulates. However, this procedure is like putting the cart in front of the horse, since it is more natural to take the irrelevance relation as *primitive* and then use this relation when *constructing* a belief revision function. In this section I will show how the notion of irrelevance can be exploited in such a construction. For simplicity, I shall only consider belief revision and contraction functions based on belief sets, but a similar approach can also be used for probabilistic revisions and contractions.

The key idea in the construction to follow is to use the irrelevance relation as a tool for *partitioning* the sentences in a belief set K:

(D≈) B *is relevance-equivalent to* C *in relation to* A, in symbols $B \approx_A C$, if and only if there is an E such that $E \perp\!\!\!\perp A$ and $E \to (B \leftrightarrow C)$.

The intuition behind this definition is that if the difference in the contents of B and C is irrelevant to A, then B and C should be treated equally when revising K with respect to A. We show that (D≈) indeed produces a partitioning of K:

THEOREM 5: If $\perp\!\!\!\perp$ satisfies (R0) - (R4) and (R7), then $\approx_A$ is an equivalence relation. In particular, if $B \perp\!\!\!\perp A$, then $B \approx_A T$. Furthermore, the mapping $B \to |B|_A$, where $|B|_A$ is the equivalence class of B under $\approx_A$, is a Boolean homomorphism.

Proof: It is trivial that $\approx_A$ is reflexive and symmetric. We need to show that the relation is transitive as well. So assume that $B \approx_A C$ and $C \approx_A D$, that is, there are sentences E and F such that $E \perp\!\!\!\perp A$, $E \to (B \leftrightarrow C)$ and $F \perp\!\!\!\perp A$ and $\vdash F \to (C \leftrightarrow D)$. By (R7) it follows that $E\&F \perp\!\!\!\perp A$ and by propositional calculus it is easy to derive $\vdash E\&F \to (B \leftrightarrow D)$. Thus $B \approx_A D$, and we have shown that $\approx_A$ is an equivalence relation. Since $\vdash B \to (B \leftrightarrow T)$, it follows that if $B \perp\!\!\!\perp A$, then $B \approx_A T$.

To show that the mapping $B \to |B|_A$ is a Boolean homomorphism, we need to show that equivalence classes are closed under negations and conjunctions.

If $B \approx_A C$, then there is an E such that $E \perp\!\!\!\perp A$, $\vdash E \to (B \leftrightarrow C)$, and consequently, by propositional logic, $\vdash E \to (\neg B \leftrightarrow \neg C)$, and so $\neg B \approx_A \neg C$. If $B \approx_A C$ and $B' \approx_A C'$, there are sentences E and F such that $E \perp\!\!\!\perp A$, $\vdash E \to (B \leftrightarrow C)$, and $\vdash F \perp\!\!\!\perp A$ and $F \to (B' \leftrightarrow C')$. By (R7) it follows that $E\&F \perp\!\!\!\perp A$ and by standard propositional calculus it is easy to derive $\vdash E\&F \to (B\&B' \leftrightarrow C\&C')$ and hence $B\&B' \approx_A C\&C'$. This completes the proof.

If we define the Boolean operations on the equivalence classes in the obvious way, i.e., $\neg |B|_A = |\neg B|_A$ and $|B|_A \& |C|_A = |B\&C|_A$ etc., it follows from the theorem above that the set |K| of equivalence classes will form a belief set, where all sentences that are irrelevant to A belong to the same equivalence class, i.e., $|T|_A$. This class will function as the tautology in |K|. If A is contingent, there are at least two distinct equivalence classes in |K| because we then have $|T|_A \neq |A|_A$.

If we assume that we have a belief contraction (or revision) function defined on |K|, then we can define a contraction (or revision) function on K in the following way

(D-) $B \in K^-_A$ iff $|B| \in |K|^-_A$.

The value of this construction is shown by the fact that if the contraction function on |K| satisfies (K-1) - (K-8) of Gärdenfors (1988), then the contraction function defined on K also satisfies (K-1) - (K-8) *as well as* (I-). The upshot is that if we have an irrelevance relation that satisfies the desired conditions defined on a belief set K, then we can, by the method presented here, construct a well-behaved contraction function for K that will satisfy (I-).

One important application area of belief revision theory is *updating logical databases*. From a computational point of view, the ultimate goal is to develop algorithms for computing appropriate revision and contraction functions for an arbitrary logical database (which models a state of belief). The proposed method will simplify the computations of belief revision functions since the number of equivalence classes, in all interesting cases, will be considerably smaller than the number of elements in the original belief set. Thus the amount of calculations required to compute the required belief revision will be reduced.

6. Conclusion

A general criterion for the theory of belief revision is that when we revise a state of belief by a sentence A, as much of the old information as possible should be retained in the revised state of belief. The motivating idea in this paper has been that if a belief B is irrelevant to A, then the general criterion entails that B should still be believed in the revised state. The problem was that the traditional definition of statistical relevance suffers from some serious shortcomings and cannot be used as a tool for defining belief revision processes. This led me to develop an amended notion of relevance that has the desired properties. In particular, the postulate (R7), is violated by the traditional definition. On the basis of the new definition, I have outlined how it can be used to simplify a construction of a belief revision method.

Notes

[1] Research for this article has been supported by the Swedish Council for Research in the Humanities and Social Sciences. I wish to thank Didier Dubois, David Miller, Henri Prade, Teddy Seidenfeld, and Wolfgang Spohn for helpful comments.

[2] Throughout this paper I will use probability measures defined on sentences. It is easy to translate the analysis presented here to probability measures defined on classes (or properties). 'Relevance' defined in terms of classes is a central concept in Salmon's and Greeno's theories of statistical explanation (cf. Salmon, 1971).

[3] Carnap (1950), p. 348.

[4] Keynes (1921), p.72.

[5] Keynes (1921), p. 55. I have changed Keynes's notation. Keynes gives no explicit definition of 'relevant' although (b) is obviously in line with his intentions.

[6] Carnap remarks in (1950), p. 420, that this definition is essentially the same as his definition of 'complete irrelevance' with the condition added that A not be logically false.

[7] Carnap(1950), p. 420.

[8] Keynes(1921), p. 55.

[9] In order to avoid contradicting (2), it must be assumed that A & B is consistent.

[10] Such an example can be found on pp. 362-63 in Gärdenfors (1978).

[11] I will still assume that $\mathcal{L}$ includes the standard propositional operators and is ruled by a logic including classical propositional logic.

References

Alchourrón, C.E., Gärdenfors P., and Makinson D. (1985), "On the logic of theory change: Partial meet contraction and revision functions", *The Journal of Symbolic Logic* 50: 510-30.

Carnap, R. (1950), *Logical Foundations of Probability*, Chicago: University of Chicago Press.

David, A.P. (1979), "Conditional independence in statistical theory", *Journal of the Royal Statistical Society, B*, 41, 1-31.

Geiger, D. (1990), "Graphoids: A qualitative framework for probabilistic inference", *Technical Report R-142*, Cognitive Systems Laboratory, UCLA.

Gärdenfors, P. (1978), "On the logic of relevance", *Synthese* 37: 351-67.

_ _ _ _ _ _ _ .. (1984), "Epistemic importance and minimal changes of belief", *Australasian Journal of Philosophy* 62, 136-57.

_ _ _ _ _ _ _ .. (1988), "The dynamics of belief: Contractions and revisions of probability functions", *Topoi* 5: 29-37.

_ _ _ _ _ _ _ . (1988), *Knowledge in Flux*, Cambridge, MA: MIT Press.

Gärdenfors, P. and D. Makinson (1988), "Revisions of knowledge systems using epistemic entrenchment", in *Proceedings of the Second Conference on Theoretical Aspects of Reasoning about Knowledge*, M. Vardi (ed.). Los Altos, CA: Morgan Kaufmann.

Harper, W. L. (1975), "Rational belief change, Popper functions and counterfactuals", *Synthese* 30: 221-62.

Keynes, J. M. (1921), *A Treatise on Probability*, London: Macmillan.

Pearl, J. (1988), *Probabilistic Reasoning in Intelligent Systems*, San Mateo, CA: Morgan Kaufmann.

Salmon, W. C. et al. (1971), *Statistical Explanation and Statistical Relevance*, Pittsburgh: Pittsburgh University Press.

Salmon, W. C. (1975), "Confirmation and relevance", in *Induction, Probability, and Confirmation*, Minnesota Studies in the Philosophy of Science, G. Maxwell and R.M. Anderson Jr. (eds.), Minneapolis: University of Minnesota Press, pp. 3-36.

Tversky, A. and Kahneman, D. (1982), "Causal schemas in judgments under uncertainty", in *Judgment under Uncertainty: Heuristics and Biases*, D. Kahneman, P. Slovic, and A. Tversky (eds.). New York: Cambridge University Press.

Part IX

MATHEMATICAL AND PHYSICAL OBJECTS

Between Mathematics and Physics[1]

Michael D. Resnik

University of North Carolina at Chapel Hill

The distinction between mathematical and physical objects has probably played a greater role shaping the philosophy of mathematics than the distinction between observable and theoretical entities has had in defining the philosophy of science. All the major movements in the philosophy of mathematics may be seen as attempts to free mathematics of an abstract ontology or to come to terms with it. The reasons are epistemic. Most philosophers of mathematics believe that the abstractness of mathematical objects introduces special difficulties in accounting for our ability to know them, to refer to them and even to entertain beliefs about them. These difficulties—supposedly absent even in the case of the most theoretical physical objects—make mathematical objects especially problematic and philosophically unattractive.

Few have questioned this epistemic thesis or the ontic distinction it presupposes. LaVerne Shelton (1980) challenged the abstract-concrete distinction some years ago in an unpublished APA address. Susan Hale continued this line in her dissertation and related articles (Hale 1988a, 1988b), and I raised doubts about Hartry Field's taking spacetime points and regions as a concrete, physical ontological foundation for his nominalization of physics (Field, Resnik 1985). Here I want to question the clarity of the mathematical-physical (and corresponding abstract-concrete) distinction by focussing on the ontology of theoretical physics.

According to most philosophers of mathematics, mathematical objects differ from physical objects in being causally inert and totally disconnected from spacetime. I will focus on these two ways of distinguishing mathematical and physical objects, for coupling either with the popular causal or informational theories of knowledge, reference and *de re* belief yields the dogma of the special epistemic inaccessibility of mathematical entities.

Of course, the inability of causal and informational epistemologies to account for our knowledge of mathematical objects counts as much against those epistemologies as it does against the belief in abstract, mathematical objects. But I do not plan to use this defense of mathematical objects here. Instead I will try to undermine the major epistemic critiques of mathematical objects by arguing that modern physics blurs the mathematical/physical distinction as drawn in spatiotemporal or causal-informational terms (section 1). In section 2, I will apply my conclusions concerning physics in questioning certain non-interference assumptions—claims to the effect that the physical world is metaphysically independent of mathematical existence and truth—that underlie recent attempts to free physical science from its dependence on mathematical objects.

1. Physics Blurs The Mathematical/Physical Distinction

Anti-realists in the philosophy of mathematics are invariably realists about physical objects. Indeed, pursuing an anti-realist program in the philosophy of mathematics seems to have little point if one is an instrumentalist in physics. An instrumentalist need only remark that the mathematical component of a physical theory is as much an instrument as the rest of it, and leave it at that. In contrast, we find mathematical anti-realists starting with a distinction between physical and abstract or mathematical entities and attempting either to combine an anti-realist interpretation of mathematics with scientific realism or to show that science can make do without mathematical objects. Thus it is appropriate to appeal to contemporary physics and cosmology to undercut attempts by mathematical anti-realists to distinguish between mathematical and physical objects.

1.1. Quantum Particles

Let me start then by considering some physical objects that upon further examination appear as mathematical as they do physical. I have in mind quantum particles. Due to the difficulties in interpreting to quantum theory, one might think quantum particles unsuitable candidates for physical objects. Relative to debates in the philosophy of physics this may be so; but from the standpoint of today's nominalist philosophers of mathematics, some of whom take spacetime points as physical objects, quantum particles count as both real and clearly physical.

The term "particle" brings to mind the image of a tiny object located in spacetime. But, on what seems to be the consensus view of the puzzling entities of quantum physics, this image will not do. Most quantum particles do not have definite locations, masses, velocities, spin or other physical properties most of the time. Quantum mechanics allows us to calculate the probability that a particle of a give type has a given "observable" property. But that does not even imply that if we, say, detect a photon in a given region of spacetime then the photon occupied that position prior to our attempts to detect it or that the photon would have been in that region even if we had not attempted to detect it. Prior to its detection a photon is typically in a state that is a superposition of definite (or pure) states, and quantum theory contains no explanation of how a photon or any other quantum system goes from a superposition into a definite state. Recent mathematical critiques of hidden variable theories indicate that this mysterious feature of quantum mechanics is virtually unavoidable.(Shimony) In the face of this, one might still say that quantum particles are tiny bits of matter with very weird properties—ones that are only partially analogous to classical physical properties. But there is a further problem.

When we have a system of several particles of the same kind, two photons, for example, there are no quantum mechanical means for tagging them at the beginning of an interaction and re-identifying them at the end of it. Suppose, for example, we run an experiment in which we detect two photons one above the other at one location and later detect two photons (intuitively, the same photons) again one above the other at a location to the right of the first. One might expect that there is a fact as to which photon is on top at the end of the experiment. But quantum theory recognizes no such fact; for such a fact must be described by reference to the trajectories the particles take in travelling between the two locations. Since their initial and final positions in spacetime are fixed, they have definite trajectories only if they have definite velocities, violating the uncertainty principle. The situation is even worse when we combine quantum theory with special relativity, for then there is no fact as to whether we have the same two photons. It might be, for example, that one photon splits into an electron-positron pair, whose members in turn annihilate one another producing the photon we detect. Moreover, between our two detections such processes might occur indefinitely many times. Actually, the same is true of experiments "tracking" a single photon. The one with which we start might be destroyed before we see the one with which we end.

Thus it is better to think of particles as features of spacetime—more like fields—rather than as bodies travelling through spacetime. This view is seconded in the following passage from a recent account of particle physics directed at other scientists

> In the most sophisticated form of quantum theory, all entities are described by fields. Just as the photon is most obviously a manifestation of the electromagnetic field, so too is an electron taken to be a manifestation of an electron field. ... Any one individual electron wavefront may be thought of as a particular frequency excitation of the field and may be localized to a greater or lesser extent dependent upon its interaction. (Dodd, p.27; See also Teller)

Now by thinking of the behavior of iron filings around a magnet we can get an intuitive grasp of what a magnetic field is and its intensity or direction at a given point. However, quantum fields are not distributions of physical entities, rather they are roughly distributions of probabilities. As the electron field varies its intensity over spacetime so does the probability of an interaction involving electrons.(Cf.Teller, pp.612-613.) And remember we cannot think of the electrons as definitely present prior to any electron interaction we observe.

How, then, are we to think of quantum particles and fields? My proposal is that we take the mathematics as descriptive rather than as "merely representational". Fields and particles are functions from spacetime points to probabilities. If that construes fields as mathematical or quasi-mathematical entities, so be it. One advantage to this approach is that it allows us to recognize the individual particles of a given type of particle field as mathematical components of that field. One disadvantage is that it rules out the possibility of explaining the connection between a quantum field and its observed manifestations in familiar physical terms. But quantum theory all but rules this out anyway.

Of course, this proposal blurs the distinction between mathematical and physical objects. Either physical objects become mathematical or mathematical objects acquire physical attributes. I have no difficulties with this. For adherents of the traditional view of physical objects, on the other hand, the challenge is to reduce quantum particles to an uncontroversially physical basis.

How might they respond? One approach would be to identify probabilities with propensities and quantum fields with distributions of propensities, and then argue that distributions and propensities, themselves, are physical entities. But it is unclear to me how one could make a compelling case that this is a genuine *physical* alternative to an openly mathematical account in terms of probabilities and functions.

An analog to Hartry Field's (1980) formulation of Newtonian gravitation would be much more convincing, if one could make it work. The idea would be to treat quantum particles and fields as structural features of spacetime specified in non-mathematical terms. Just as Field was able to avoid referring to quantities, such as temperature, by using physical predicates, such as 'warmer than', one might avoid referring to particles and fields by describing regions of spacetime as 'electronic', 'photonic' or 'electromagnetic'. I do not know whether it is possible to carry through this idea in any reasonably convincing way, and those far more expert than I have already expressed their doubts. (See, e.g., Hellman, p.116.) But let me add a worry of my own.

Suppose that we manage to define 'photonic', 'electronic', etc. in rich enough physical terms to characterize the structure of spacetime regions occupied by photons, electrons, etc. Now suppose that we want to describe the two photon case I discussed earlier. At the beginning and end of the experiment we have two photonic regions of spacetime. But there is no quantum mechanical fact as to whether the spacetime region connecting these events is just the sum of two photonic regions, or of photonic, electronic and positronic regions, and so on. Instead the region has a property corresponding to the superposition of infinitely many pure states. We can make sense of such superpositions using variables ranging over mathematically characterized pure states. In a Hartry Field style formulation, however, variables are restricted to spacetime points and regions. Thus we might be able to define spacetime predicates that allow us to reformulate talk of finitely many specific pure states and superpositions of them as talk of regions. But it is unclear to me how we can recover infinite superpositions without introducing questionable devices such as infinite disjunctions. Even then it may be that we cannot recover the full use of quantification over particles that physics requires. Unfortunately, I am too ignorant of quantum field theories to know how serious a problem this is.

1.2. Epistemic Breakdowns

On the traditional view, a major epistemic difference between physical and mathematical objects is that the former can and the latter cannot participate in a causal process that permits us to detect them. It is pertinent, then, that physicists now recognize physical entities suffering from the very epistemic disability ascribed to mathematical objects. Take, for example, the photon-electron-positron-photon transformations we discussed earlier. Quantum field theory posits processes of this type that happen so fast that they are in principle undetectable. Virtual processes, as they are called, need not even conserve energy or momentum so long as the total processes of which they are components do. Black holes in spacetime supply another example. According to physicist Clifford Will, "there is no way for any external observer to determine, for example, the total number of baryons... [inside a black hole] " and "there must exist [such] a singularity of spacetime at which the path or world line of an observer who hits it must terminate, and physics as we know it must break down". (Will, p.28)

I would expect someone to reply at this point that, unlike mathematical objects, black holes, virtual processes and other examples that one might cite are supposed to be part of the causal network and to have physical effects, and that, furthermore, the theories that posit them are empirical theories supported by empirical evidence. Yet doesn't this sort of response beg the question? True, given a mathematical/physical distinction, one can argue that mathematical objects don't participate in the causal network and mathematical theories remain silent concerning it. But in view of our earlier considerations, the point about the causal network is no longer evident. Indeed, we might now argue that mathematical objects, in the form of quantum fields, do participate in the causal network.

Furthermore, as I have argued elsewhere (Resnik 1989), even supposedly purely mathematical theories have observational consequences and are justified in part on the basis of observational evidence. For instance, we might appeal to recursion theory to predict the results of a computer run, or conversely, use the results of computer runs to convince ourselves of properties of a Turing algorithm or to support a step in a proof. Admittedly, these inferences depend upon taking certain auxiliary hypotheses for granted; and given a mathematical-physical distinction, one might argue that the real empirical content should accrue to those hypotheses rather than to the mathematical ones.

In any case, empirically confirmed theories are as much committed to the existence of numbers and other paradigmatic mathematical entities as they are to black holes or virtual processes, because these theories posit numbers and their ilk as the solutions to equations, dimensionless constants, and so on. Undercutting this point would require reformulating physics along the lines of Hartry Field's program. Given the heavy use that program so far made of spacetime regions and points, we would still need a clear mathematical-physical distinction to determine whether Hartry Field style physics succeeds in eliminating mathematical objects.

So far I have argued that 1) because quantum particles are probability fields or something akin to them, they will impede attempts to distinguish between mathematical and physical objects on spatio-temporal or causal grounds; and 2) that undetectable physical processes count against distinguishing mathematical and physical objects on the basis of physical detectability. Although the mathematical-physical distinction is central to the standard approach to the philosophy of mathematics, these claims underscore the need to clarify it and our current inability to do so.

2. Non-Interference Claims.

Hartry Field once argued along the following lines that good mathematics should be conservative though it need not be true: Physical theories are supposed to be non-conservative in the sense of having observational consequences, which do not follow from just our prior observational knowledge. Not so for mathematics; it is true in all possible worlds, a priori, and the realm it treats has no effect upon physical reality. Thus a mathematical theory with empirical consequences would have something wrong with it—it might even be inconsistent. (1980, pp.12-14.)

Field is right, of course, that good physics should have novel observational consequences. That is why physicists formulate their theories in terms that are prone to produce such consequences. For example, if they construct a theory introducing a new particle, they are very likely to give it a mass, spin, charge, etc., which are likely to allow them to derive observable consequences from their theory. To put the matter metaphorically, physicists typically equip their theories with at least girders extending towards the observable world even when they may not know how to build full bridges. Field is also right that mathematicians do not aim for observational consequences. Instead they try to describe structures that may or may not be realized in the physical world. Thus they do not assign "physical" properties to the objects they posit and find it far more important to relate them to other mathematical objects. This sociology may be responsible for the popular intuition that the mathematical realm does not interfere with the physical one, but plainly it is no argument for that intuition.

The intuition also surfaces in an argument of Terrence Horgan's defending a nominalization of science using counter-factual conditionals with mathematical antecedents Here is what he says:

Since sets are not supposed to be part of the world's spatio-temporal causal nexus, that nexus would be exactly as it is whether or sets existed or not; for sets would not causally influence the concreta in the spatio-temporal causal nexus, and the idea of their non-causally influencing those concreta is just unintelligible.(Horgan)

We find the intuition again in Geoffrey Hellman's recent book (Hellman), which contains an account of applied mathematics similar to Horgan's. Although Hellman translates statements of mathematicized science as second-order, strict implications, he paraphrases them as counterfactuals with antecedents such as "if X,f were any omega sequence". He notes that for his approach to work, we must assume that the presence of such omega sequences will not affect the way the world actually is. To see what is at stake here, consider the presumably true statement "The number of stars is finite". Suppose that Hellman rendered this as, roughly, "If X,f were any omega sequence, then some member of it would be greater than its proxy for the number of stars". Then an omega sequence of stars would falsify this translation. To avoid such irrelevant counter-examples, Hellman restricts this and similar translations of applied mathematical claims to omega sequences that do not interfere with the way the world actually is. Of course, such claims will be non-vacuously true only if it is logically possible for such non-interfering omega sequences to exist; and, accordingly Hellman recognizes the need for a postulating possibilities of this type.

At this point we should note an important difference between Hellman, on the one hand, and Horgan and Field, on the other. Hellman is a modal-structuralist. He believes that mathematics is about structures, but he does not posit them outright. Instead, he characterizes individual mathematical structures using modal operators and posits only that it is logically possible for them to be realized rather than that they are actually realized. If we may think of structures for a moment as universals, then Hellman's approach is Aristotelian—structures exist only as realized. By contrast, Field and Horgan think of mathematical objects under what is called the objects-platonist view; they see mathematical objects as full-fledged abstract objects that cannot be identified with anything physical.

Given the Field-Horgan view of mathematical objects, it is almost immediate that they do not interfere with the physical world, and, hence that it is logically possible for them to exist while the actual world remains as it now is. This presupposes, of course, that fields, waves, and particles are not mathematical—in particular, not functions or (impure) sets.

Hellman's position is much more subtle and his attempts to work it out are very interesting. He spends quite a bit of effort exploring ways in which we might formulate non-interference conditions. The first way is simply to take the phrase "does not interfere with the way the world actually is" as acceptable as it stands, and use this phrase in antecedents of counterfactual renderings of applied mathematical statements. This would change the statement about the stars to "if X,f were any omega sequence which would not interfere with the way the world actually is, then some member of it would be greater than its proxy for the number of stars". We would also use the phrase to state the following possibility postulate: it is possible that there is an omega sequence which does not interfere with the way the world actually is. Now, as Hellman emphasizes, making sense of these conditionals and this postulate presupposes that the terms in which they are formulated make sense. He takes this to mean that it makes sense to refer to the actual world (or the actual condition of a given physical system) apart from any relativization to a language or theory or conceptual framework, and so on (p.100). Of course, the world Hellman has in mind is supposed to contain no mathematical objects. So at this point he too employs the same presupposition used by Field and Horgan.

As it turns out, Hellman is not happy referring to the world as it actually is, because it presupposes a strong physical realism. Due to this, Hellman moves on to explore ways of formulating non-interference conditions in more precise and metaphysically neutral terms. His general strategy is to introduce non-mathematical predicates R and use them to formulate clauses of the form

> x,y,z,...,w stand in R if and only if x,y,z,...,w *actually* stand in R, for all *actual* objects x,y,z,...,w.

Given an appropriate set of predicates, we could use such clauses to state that in a given counterfactual situation all the actual objects have exactly the properties and relationships to each other that they actually have.

Can we find an appropriate set of predicates? Hellman divides this question in two: First, can we find a set of non-mathematical (or synthetic) predicates for picking out the set of actual objects under consideration in a given application? Second, can we be sure that through their extensions these predicates fix a physical world corresponding to our usual mathematical descriptions of the actual world? (Pp.129-139.) For example, it is not enough to have a predicate true of just stars, we must also insure that our clauses only have models in which the number of stars is whatever it actually is. Hellman devotes most of his effort to investigating the second question. In the end, he combines a cautious optimism that it has a positive answer with a lingering fondness for the strong realist approach (p.142).

Hellman's caution concerning the second question contrasts with the cavalier attitude he takes toward the first, ontological one. This, he says, "is relatively unproblematic, for the particular context of application usually involves a given domain of material objects to which a piece of mathematics is to be applied, and it suffices to cover this domain with synthetic predicates" (p.132). Alright, let us suppose that the context of application is quantum electrodynamics, QED. What are the synthetic predicates for picking out the relevant material objects? I guess Hellman would say something like 'electron', 'photon', 'positron', since those are the types of particles involved. But, in the light of our earlier attempt to understand what quantum particles are, it would be appropriate to object that we need a more basic set of synthetic predicates in order to reflect the apparent ontic complexity of electrons, photons and positrons. Since that might require Hellman to carry out more of a Hartry Field style nominalization of quantum physics than he wants, I can imagine him replying: "Look, although we may be unable to state what an electron is without using mathematical terms, electrons are whatever they are independently of our mathematical descriptions. And surely they are physical objects. Thus, since my predicate 'electron' is a primitive predicate true of just them, it is a suitable synthetic predicate for formulating non-interference conditions needed for applying mathematics in QED." Now I see two problems with this sort of reply. First, it takes exactly the strong realist stance towards electrons that at this point Hellman is trying to avoid taking towards the actual world. Second, it simply takes it for granted that electrons are physical objects. It seems unlikely to me that Hellman can carry out his program for applying mathematics without presupposing a problematic mathematical-physical distinction.

Setting this issue aside, let us ask how sure we can be that the physical world will permit Hellman's non-interference conditions to be met. Hellman's case is different from the Field-Horgan case, because, as far as I can tell, for Hellman, to add an omega sequence or a continuum or an iterative hierarchy to the world is not to add something abstract but rather something concrete. Adding new concreta to the world might change the behavior of the things already there. Of course, that depends upon what we mean by the actual world. Are we to count just its population and their properties and relations taken in extension? Or should we also include its laws? Adding a new planet to our solar system will not change the behavior of the other planets so long as the physics of the new world is appropriately different. Thus if we do not include its physics in our description of the actual world, then it is logically possible to add more stuff to it and leave the original stuff as it is. I do not see any place for the laws of physics in Hellman's sketches of how we use synthetic predicates to spell out non-interference conditions; so it may be that he would be willing to take the actual world *sine* physics approach to non-interference. But once we free our world of its physics, its unclear how and whether Hellman's position differs significantly from the Horgan-Field position, according to which the additional entities are abstracta.

In any case, other passages (pp.100-101) suggest that Hellman has the actual world *cum* physics in mind when he speaks of non-interference. So long as we restrict ourselves to applications of mathematics to small collections of physical objects viewed as "isolated systems", the possibility of there being additional physical objects outside such systems that realize various mathematical structures seems relatively unproblematic. But suppose we consider instead applications of mathematics to cosmology. As I understand it, one consequence of the Big Bang theory is that the amount of matter in the universe is finite. Now let us suppose that the Big Bang theory is part of a description of how the world actually is. Then it would not be logically possible for there to be an infinite amount of matter and for our world to remain as it actually is.

Here I am probably appealing to the world *cum* its physics and its initial conditions. We can avoid using "initial conditions" in this example, by going to even more speculative theories; for some cosmologists are trying to develop theories of the origins of the universe that dispense with initial conditions. Thus under the usual understanding of "physically possible" the actual universe would be the only physically possible one.

There is a way for Hellman to defuse these objections. By counting spacetime points or regions as already part of the physical world, the conditions for applying Hellman-style mathematics can often be met without invoking additional physical objects—perhaps even in the cosmology case. However, if we take the less controversial stand of not counting points and regions as part of the physical world, then even applying number theory forces one to confront non-interference problems. For excluding spacetime points and regions from the set of actual physical objects leaves a (presumably) finite set.

3. Conclusion

I will conclude with some observations from my own structuralist perspective. (See Resnik, 1981, 1988.) On this view, numbers, functions and sets (at least) are positions in structures, and their identity is determined by their relationships to other positions in the structure to which they belong. One consequence of this is that usually there is a fact as to whether the positions of one structure are identical to those of another only if the two structures are part of a containing structure. Taking spacetime and the natural number sequence, for instance, as separate structures deprives us of a context in which we can speak of there being a fact as to whether the natural numbers are identical to any omega sequence of spacetime points. In particular, it deprives us of a context in which it makes sense to say that while have admitted spacetime points we have admitted no numbers. From this structuralist point of view, nominalists should aim to avoid committing themselves to the *structures* exhibited by so-called mathematical objects and not merely to avoid committing themselves to mathematical objects *per se*. For there simply may be no fact as to whether they have met the latter aim. To illustrate this point, consider the next passage from Hellman.

> ...When it comes to mathematics, however, we need not regards its abstract structures as literally part of the actual world.... [it] is convenient (perhaps essential) in describing [a material] configuration in detail, but the configuration is "already there" prior to the mathematical description, and independently of any ... [mathematical structure] that may be invoked in such a description (pp.127-128).

But if the configurations (structures) are there, then, from the structural point of view, it is simply unclear as to what it is for the corresponding mathematical objects to fail to be there as well.

Returning to the theme with which I began this paper: rather than worrying about epistemic differences between mathematical and physical objects, it would be better to worry about epistemic differences between various structures—between, for example, infinite and finite ones, or between continuous and discrete ones. While we may find interesting epistemic differences between such structures, I see no reason for interesting epistemic differences to arise between the structures studied in physics and those studied in the traditional (non-global, non-foundational) branches of mathematics.

Note

[1] I would like to thank Susan Hale and Geoffrey Sayre-McCord for their helpful comments on earlier versions of this paper.

References

Dood, J.E. (1984), *The Ideas of Particle Physics: An Introduction for Scientists*. Cambridge: Cambridge University Press.

Field, H. (1980), *Science Without Numbers: A Defense of Nominalism*. Princeton: Princeton University Press.

Hale, S.C. (1988a), *Against the Abstract-Concrete Distinction*. Unpubished Ph.D. Dissertation, University of North Carolina.

_ _ _ _ _ . (1988b), "Spacetime and the Abstract/Concrete Distinction", *Philosophical Studies* 53:85-102.

Hellman, G. (1989), *Mathematics Without Numbers*. Oxford: Clarendon Press,

Horgan, T. (1987), "Science Nominalized Properly", *Philosophy of Science* 54:281-2.

Resnik, M.D. (1985), "How Nominalist is Hartry Field's Nominalism?", *Philosophical Studies* 47: 163-181.

———. (1981), "Mathematics as a Science of Patterns: Ontology and Reference", Nous 15: 529-50.

———. (1988), "Mathematics from the Structural Point of View", *Revue Internationale de Philosophie* 42: 400-24.

———. (1989), "Computation and Mathematical Empiricism", *Philosophical Topics* 17: 129-44.

Shelton, L. (1980), "The Abstract and the Concrete: How Much Difference Does This Distinction Make?" Unpublished paper delivered at the American Philosophical Association, Eastern Division Meetings.

Shimony, A. (1989), "Conceptual Foundations of Quantum Mechanics" in *The New Physics,* Paul Davies (ed.). Cambridge: Cambridge University Press.

Teller, P. (1990), "Prolegomenon to a Proper Interpretation of Quantum Field Theory", *Philosophy of Science* 57: 594-618.

Will, C. (1989), "The Renaissance of General Relativity" in *The New Physics*, Paul Davies (ed.). Cambridge: Cambridge University Press.

Elementarity and Anti-Matter in Contemporary Physics: Comments on Michael D. Resnik's "Between Mathematics and Physics"[1]

Susan C. Hale

California State University, Northridge

I believe that the nominalist-platonist debate which Professor Resnik tries to dissolve is even more of a non-starter than he appears to think, judging from his "Between Mathematics and Physics". I shall argue for this by showing, first, that conceptions of particles as mathematical, or quasi-mathematical, have a longer history than Resnik notices, and, second, that the current difficulties with thinking of elementary particles as tiny chunks of matters are more profound than he realizes. Finally, I hope to point toward a diagnosis of the confusion which leads nominalists to pursue ill-motivated debates and research programs.

As early as 1762 we find Roger Boscovich arguing against an atomistic or corpuscularian conception of particles. He used a conception of particles as simple, indivisible, and extentionless as one crucial feature in building a physical theory which he saw as superior to Newtonian mechanics in numerous respects, including both consideration of epistemic values, e.g., it is less *ad hoc* than Newtonian mechanics (pp. 13, 54-55), and greater agreement with empirical observations, e.g., his theory explains why the fixed stars do not "coalesce into one mass", whereas universal gravitation theory requires that the stars would collapse (pp. 16, 144-147). The most general motivation for his mathematical or quasi-mathematical conception of particles, though, came from his argument that a standard atomistic conception of particles leads to paradox in understanding collision. Boscovich argued that the standard view leads to any one of three unacceptable consequences: either a body involved in a collision has two distinct velocities at the same time, or there is compenetration of colliding bodies, or there are non-continuous changes in the velocites of colliding bodies (pp. 10, 24-25).

Boscovich conceived of points as "simply mathematical points surrounded by attractive and repulsive forces" (Gjertsen, p. 171). He distinguished between purely geometrical points and physical (or material) points. The *only* difference between the two is that any physical point "possesses the real properties of a force of inertia" and of other, attractive and repulsive, forces (p. 58). Perhaps, in order to preserve Boscovich's distinction, it would be best to label his points "quasi-mathematical", yet this label ought not obscure the fact that his points were not physical in the ontological terms used in the nominalist-platonist debate.

Although Boscovich's theory, including his quasi-mathematical conception of particles, did not widely influence influence physicists of his day, his conception of particles did prove fruitful later in the development of science, especially in Michael Faraday's theory of electrical conductivity, in which particles were merely centers of force, not tiny chunks of matter (Gjertsen, p. 172).

Of course, Einstein has superceded Newton and his opponents, and quantum electrodynamics has superceded Faraday: we are not tempted to base our conception of particles on those of Boscovich and Faraday, even had their conceptions been uncontroversial when they were writing.

The threat posed by contemporary physics to the ontological assumptions traditonal in the nominalist-platonist debate is deeper than Resnik realizes. Indeed, the conceptual presupposition of the distinction between mathematical and physical objects is endangered. This conceptual presuppositon, which Werner Heisenberg argues must be abandoned, is the picture of elementary particles combining in geomtrical or dynamical configurations to form larger material objects.

Heisenberg sees the challenge to this picture as motivated by a variety of different features of early quantum mechanics, but especially from P.A.M. Dirac's relativistic theory of the electron and from the discovery of the positron. Of this, Heisenberg writes:

> I think that this discovery of antimatter was perhaps the biggest change of all the big changes in physics in our century. It was a discovery of utmost importance, because it changed our whole picture of matter (pp. 31-32).

How does this conceptual change emerge from Dirac's work, according to Heisenberg? For brevity sake's, I only sketch Heisenberg's answer. Dirac's suggestion that positrons could be produced by pair-production, along with the experimental discovery of positrons, leads to abandonment of the law of conservation of particle number. As Heisenberg writes, "For instance, ...one could say that the hydrogen atom does not necessarily consist of proton and electron. It may temporarily consist of one proton, two electrons, and one positron" (p. 32). For any particle, with pair-production creating particles from energy, the number and configuration of particles of which it consists is indeterminate, so long as the total symmetry of the system is the same as the symmetry of the particle itself. This leads us to realize that "the elementary particle [is] not elementary anymore. It is actually a compound system, rather a complicated many-body system, and it has all the same complications which a molecule or any other such object really has" (p. 32). The properties of such systems can best be calculated from natural law, just as we would calculate the states of complicated molecules. Then we can conceive of these elementary, non-elementary particles as complicated states, perhaps as mathematically-described states of fields.

A natural response to this situation, given our atomistic preconceptions, is to think that we were simply wrong about which particles are elementary; those we thought to be elementary have turned out to be complex, so let's search elsewhere for our simple chunks of matter. Heisenberg thought that this heuristic motivated much of the search for quarks (pp. 16-17, 35). Yet this response betrays a fundamental misunderstanding, for it simply postpones the problem until we have identified the new family of particles to which we are tempted to ascribe elementarity. As Heisenberg points out:

> ...even if...quarks were to exist, we could not say that the proton consists of three quarks. We would have to say that it may temporarily consist of three

quarks, but may also temporarily consist of four quarks and one antiquark, of five quarks and two antiquarks, and so one. And all these configurations would be contained in the proton, and again one quark might be composed of two quarks and one antiquark and so on (pp. 35-36).

Heisenberg's point here is that the indeterminancy of quantum number goes all the way down — it occurs for all particles, so no family of particles can be elementary.

Another inviting response is to admit all this but to claim that it shows no more than that our concept of elementarity must be revised in the face of antimatter. Instead of ascribing elementarity to particles on the basis of simplicity, now we should ascribe elementarity on the grounds that a particle is not composed of particles of any different families (since any particle and its anti-particle are in the same family). Then, according to this response, protons are not elementary, since they are composed of quarks (and antiquarks), whereas quarks are elementary, since they can be composed only of quarks (and antiquarks).

But this response also misses Heisenberg's point: were any given quark composed of a particular number of quarks and antiquarks, this response would suffice, but the crucial difficulty which bothered Heisenberg is *numerical indeterminancy*. Although related to the numerical identity problem which Resnik notices (Ms., pp. 5-6), this is a different problem, for here an analogue of Benacerraf's multiple reduction problem, now for allegedly paradigm examples of physical objects, faces traditionalists in the nominalist-platonist debate.

Heisenberg's own response to this situation was to jettison the concept of an elementary particle and replace it with the concept of a fundamental symmetry group. This may seem like a simple expression of instrumentalist sympathies, and indeed Heisenberg's sympathies were instrumentalistic. But this response is not open to nominalists, at least not without a re-conception of their debate with platonists. Furthermore, this response misses the depth of Heisenberg's argument, for his considerations lead one to reject or reconceive the concept of matter even at the level of middle-sized dry goods; conceiving, e.g., a tin can containing artichoke hearts as a material object composed of smaller chunks of matter leads, ultimately, to the same indeterminancy that plagues quarks conceived as physical particles.

Heisenberg himself saw that the consequences of his argument were far-reaching. He recognized that his argument against elementary particles rests on undermining the distinction between elementary particles and compound systems (p. 59), which led him to understand particles as "stationary states of the physical system `matter'" (pp. 59, 66), and he uses scare quotes around the word `matter' in some instances. Particles are essentially described by their transformational properties under fundamental symmetry groups. But, then, as Heisenberg puts it, "theoretical understanding of particle physics can only mean an understanding of the spectrum of particles (p. 66). Particles, then, become, at their ontological (and realist) best, secondary structures, defined by combinng the dynamics of a system with the boundary conditions; the concept of a particle, or even of a spectrum of particle, has no role in the mathematical formulations of the dynamics of a system. So, for Heisenberg, dropping the concept of an elementary particle leads to downgrading the status of the particle concept itself; at best, the concept of a particle has a derivative conceptual status and particles are ontologically derivative as well, and the relations *divide* and *consist of* (obtaining between particles, or even larger chunks of matter) lose all literal meaning whatsoever. From here it is a short step to a re-conceptualization of matter itself. If the particle concept has been an inherent part of our concept of matter, as seems plausible, then undermining the particle concept is undermining the concept of matter. Fundamental symmetry groups provide a natural choice of replacement concepts (though there may be others), for it is by fundamental symmetry groups that we describe and understand what Heisenberg calls the "fundamental structures of nature" (p. 105), which we might think of as the physical system of matter. But this re-conceptualization leads us to believe that "these [fundamental] structures are much more abstract than we had hoped for" (p. 105) in the early days of quantum mechanics, prior to Dirac, to Pauli's and Fermi's work on beta-decay, and experimental detection of the positron.

If this picture is even along the right lines, the picture of physics with which we are left holds little consolation for traditionalists in the nominalist-platonist debate, for it undermines the legitimacy of the concepts used to draw the distinction between mathematical and physical entities. And, more importantly, this picture undermines the motivation for drawing the distinction in the first place: what is the significance of drawing an ontological distinction which is, at best, derivative from one side of the distinction alone (since clarification of the physical rests on clarification of (part of) the mathematical)?

Now, I believe, we have the conceptual tools needed to diagnose the confusion which leads nominalists to pursue ill-motivated, ill-conceived research programs. As Resnik points out, "Anti-realists in the philosophy of mathematics are invariably realits about physical objects" (Ms., p. 3). The challenge to the nominalist, as Resnik sees it, is to give a reduction of the mathematics allegedly necessary for physics to a physical ontology. So far, the nominalists' attitude toward physics looks to be an attitude which takes physics very seriously, giving it pride of place over other special disciplines (with the possible exception of epistemology). Yet nominalists change their attitude toward physics when confronted with its anti-atomistic themes; nominalists seem blithely to accept an ontological understanding of physics that, at a fundamental level, differs little from Newton's. Certainly nominalists accord strange quantum properties to little chunks of matter, properties which Newton never dreamt of; certainly nominalists do not believe that Newtonian laws accurately describe the behavior of these tiny chunks of matter. But still — there are those tiny chunks of matter — despite the early challenges of Boscovich and Faraday, and despite the much more radical challenge from contemporary "particle" physics. Thus, the nominalists' attitude toward physics is inconsistent or confused. If realism about physical entities and the primacy of physics are to be held, then the challenges to atomism coming from physics must be taken seriously. One cannot help oneself to the physics that is pleasing and dispose of the rest.

It seems that there is still a strong motivation for pursuing traditonalist problems: what is one to do otherwise? Rather than being tempted by the sway of tradition in this way, I would like to suggest that we should see the dissolution the nominalist-platonist debate as liberating, leaving us free to discover and pursue new problems. A cluster of new problems is suggested by Heisenberg's proposal that the particle concept be replaced by the concept of fundamental symmetry groups. Heisenberg, of course, did not know what physics would look like were this conceptual change to take place. For example, he did not know what questions would replace those which presuppose the concept of an elementary particle, such as "Can one divide quarks further, or are they indivisible?" (his own example uses `electrons', not `quarks'). He could not foresee the changes in directions of physical research to which his recommended re-conceptualization would lead. Nor could he foresee the overarching ontological consequences of such a change. Herein lie fresh, exciting conceptual questions for physicists and philosophers alike. Exploration into their answers may give us interesting directions to pursue in the epistmelogy of mathematics as well. The change Heisenberg proposes suggests that there is more continuity between the ontology of mathematics, the ontology of physics, and the ontology of the objects of our everyday experience than many philosophers have believed there to be, and we may find that a better understanding of this ontology suggests ways to better understand epistemology, including the epistemology of pure mathematics.

Note

[1] I would like to thank Michael D. Resnik, Mark Risjord, and Richard Zaffron for their helpful comments while I was initially preparing this commentary. Many participants at the PSA meeting raised good criticisms, but I would especially to thank Steven French and Hartry Field for their criticisms.

References

Boscovich, R.J. (1966), *A Theory of Natural Philosophy*, translated by J.M. Child. Cambridge, Massachusetts: M.I.T. Press.

Cartwright, N. (1987), "Max Born and the Reality of Quantum Probabilities", in *The Probabilistic Revolution*, Vol. II, L. Kruger, G. Gigerenzer, and M. Morgan (eds.). Cambridge, Massachusetts: M.I.T. Press.

Gjertsen, D. (1989), *Science and Philosophy: Past and present*. London: Penguin.

Heisenberg, W. (1983), *Encounters with Einstein*. Princeton: Princeton University Press.

Resnik, M.D. (1991), "Between Mathematics and Physics" this volume.

Problematic Objects between Mathematics and Mechanics

Emily R. Grosholz

The Pennsylvania State University

The relationship between the objects of mathematics and physics has been a recurrent source of philosophical debate. Rationalist philosophers can minimize the distance between mathematical and physical domains by appealing to transcendental categories, but then are left with the problem of where to locate those categories ontologically. Empiricists can locate their objects in the material realm, but then have difficulty explaining certain peculiar "transcendental" features of mathematics like the timelessness of its objects and the unfalsifiability of (at least some of) its truths. During the past twenty years, the relationship between mathematics and physics has come to seem particularly problematic, in part because of a strong interest in "naturalized epistemology" among American philosophers. The tendency to construe epistemological relations in causal and materialist terms seems to enforce a sharp distinction between mathematical and physical entities, and makes the former seem at best uncomfortably inaccessible and at worst irrelevant. Paul Benacerraf, for example, poses this situation as a conundrum. (Benacerraf 1965) And Hartry Field suggests that we might do best simply to scuttle any appeal to mathematical entities in a physico-philosophical description of the world. (Field 1980)

I would like to argue that what exists in mathematics are the items that figure in problems, especially the constellations of problems that we see as constituting its diverse domains. Mathematical entities only occur enmeshed in interesting problems; indeed, they come into view because they are both determinate and opaque and therefore problematic: one can formulate questions about them, and the questions don't have obvious answers. This way of talking about mathematical existence might seem psychological or historical, but my formulation points out its objectivity: once seen, a mathematical problem is not an accident of someone's subjectivity, nor a by-product of historical institutions.

What kind of epistemological picture is needed to explain how we can apprehend the items of mathematics as problematic? To see a triangle, for example, as a complex unity establishing a determinate but opaque relationship among its sides and interior angles, requires some transcendental as well as causal-material activity on our part. A causal-material account might be able to explain how we come to have an image of a triangle, or how we abstract the shape of a triangle from instances that we gradually learn to see as triangular, but it cannot explain why the resultant apprehension poses a problem.

Mathematics exists as a multiplicity of diverse domains. Problems and the items that figure in them form clusters whose discursive boundaries mark them off from other clusters. This also testifies to the objectivity of mathematical knowledge: certain kinds of problems, and not others, arise with respect to certain kinds of items. You can't say whatever you wish about a mathematical object, and you can't pose arbitrary questions about it.

But mathematical domains are not unrelated; they overlap at their boundaries, though the nature of the overlap always reveals opacities as it is determined. The partial unification of domains is not trivial because the domains really are distinct. Thus, establishing correspondences between domains is itself an interesting mathematical problem. But what items figure in such problems? In various studies published over the last decade, I have investigated mathematical research at the overlap of domains and found that one of its common features is the occurrence of "hybrids." The Cantor space at the intersection of logic and topology, and the Cartesian parabola at the intersection of algebra and geometry are examples of such hybrids. (Grosholz, 1985, 1990/91)

Hybrids are characteristically items that would not have arisen in one or the other domain investigated in isolation. Sometimes problems arise in a domain, but cannot be solved there. (The objectivity of items is attested not only by the problems in which they figure, but also by the intractability of those problems.) Then links with other domains must be forged so that other methods and perspectives can be brought to bear on the original problem, and in that process hybrids appear. They hover ambiguously, items not clearly belonging to one or the other domain, sometimes appearing almost contradictorily qualified as belonging to one *and* the other. And they may become the characteristic items of new domains arising at the intersection of old ones, as Leibniz' infinite-sided polygons which are also continuous curves (algebraic and transcendental) become the central focus of the emerging domain of the infinitesimal calculus. (Grosholz forthcoming)

In the seventeenth century, a new correlation was forged between the domains of mathematics and mechanics. This extended episode in the history of science ought to be closely examined, then, by philosophers of science concerned with the relationship between the objects of mathematics and physics. I have been especially interested to note that the interaction between mechanics and the infinitesimal calculus (arising itself during that period out of algebra, geometry and number theory) does not look appreciably different from the interaction between, say, algebra and geometry. Then perhaps the objects of physics are the items that intervene in the problems of physics; and a causal-materialist account of our knowledge of them falls short here just as it does in the case of mathematics. What makes the motion of a piece of string, a swinging pendulum, a planet, or the configuration of a hanging chain or a spinning globe, problematic?

1. Leibniz' Tractrix

I would like to discuss briefly one such example of the interaction of mechanics and mathematics in the seventeenth century, and point out certain of its important features, appealing explicitly to the model of the partial unification of mathematical domains just sketched. Typically, in this kind of unification problem-solving strategies from allied (but still distinct) domains are brought to bear on a family of problems that arise in one domain but cannot be solved there. This interaction changes the shape of the domains involved though they still maintain their autonomy, and tends to precipitate hybrids at the overlap of the domains.

The tractrix, Leibniz claims, is especially well suited to his new calculus. It was introduced to him in a drawing room in Paris when a mathematician dragged his watch across a table, describing a straight line with the free end of the fob, and asked the assembled guests what curve it traced. He added that because of the effect of friction, at any given point on the curve the direction of the motion of the watch is supposed to be along the fob, which is thus seen as the tangent to the curve. That's to say, he would never have asked his companions to examine the path of his watch on the table unless he had already learned to see its fob as a tangent, and its trajectory as an analyzable curve. (Bos 1988, pp. 9-12)

This parlor game owed its interest to the work of two generations of mathematicians on the problem of tangents. Leibniz retells the tale because his own work focussed on the inverse problem of tangents, and because mechanics had played a central role in his attempts to extend the new synthesis of geometry and algebra beyond the Cartesian program. Mechanics in the early 1670's was in part a practical exercise involving fountains, catapults and winches, and in part an emerging theory based on rediscovered texts by pseudo-Aristotle, Archimedes and Heron of Alexandria, as well as the writings of Descartes and Huygens. The use of "mechanical procedures" like this instance of the watch tracing a path on the table allowed Leibniz to locate, systematize and justify the introduction of transcendental curves and numbers, infinite series and reasonings involving infinitesimal magnitudes. And it was also one small step in the unification of practical and mathematical mechanics.

The mathematician asks for a description of motion under certain constraints, and the object of inquiry is the trajectory of the body in motion. Experience offers the watch on the table in Paris, whose motion is a problem; mathematics offers a curve which is already a hybrid because it exists as a geometrical shape, the solution of a differential equation in Leibniz's new calculus, an algebraic equation and a point-wise "mechanical" construction on the basis of other, constructing curves. A trajectory is a peculiar object to locate in experience. Its unity is not that of a physical object in its substantial oneness (a oneness which is indeterminate empirically but absolute transcendentally). Rather, its unity is that of a nexus of forces on the one hand, and on the other hand the unity of a mathematical curve, ambiguously defined as a hybrid.

Leibniz' tractrix can be embedded in two quite different kinds of diagrams. The first is a schematic picture of the drawing-room table and the watch. The x-axis is the base curve, along which the end of the watch-fob is pulled; if the cord-length is set equal to a, the initial position of the watch will be at a distance a up the y-axis and the end of the watch fob at the origin; call the line segment representing the watch and its fob PQ. Then as the end of the fob, Q, is pulled, P traces out the curve. This tracing specifies a differential equation, $dx = -(\sqrt{a^2 - y^2})/y \, dy$, which follows from the similarity of the characteristic (differential) triangle and the finite triangle with sides a, σ and y. (Diagram 1) In a sense it gives the tractrix in a nutshell: that curve in which the differences between the abscissae and the differences between the ordinates obey this relationship. But what curve is that? The differential equation sums up the situation, but it must be solved. (Bos 1988, pp. 21-22)

Integrating both sides (the variables are already separated) yields $ax = -a \int \sqrt{a^2 - y^2}/y \, dy$. In a second, geometrical rather than mechanical, diagram Leibniz shows how to construct the curve $z = -a(\sqrt{a^2 - y^2}/y)$. He draws it point by point, making use of certain lines and a circle quadrant to construct it as the auxiliary curve ZZA that he calls the *linea tangentium*; and then, assuming that the latter can be integrated, he constructs the tractrix by finding for every point y a point x whose distance from the y-axis is equal to the area under the auxiliary curve at y, divided by the constant a: XY = area YAZ / a. (Diagram 2) (Bos 1988, pp. 22-24)

The role of the tractrix in the two diagrams is very different. Its occurrence in both signals its hybrid nature, for the first is a schematic picture of its mechanical genesis, a watch being dragged across a table, and the second exhibits its relation as a point-wise construction to the circle and the *linea tangentium* in geometrical fashion. The first diagram augments geometry with a mechanical process; and it likewise imposes on mechanics a geometrical interpretation which in fact oversimplifies the dynamical situation. The diagram contains no representation of the variable time; the only vestige of the dynamical aspect of the real situation is the assumption of friction between the watch and the table (the fob itself is assumed frictionless). This can be read off the diagram in the assumption that the motion of the watch is always in the direction of the fob.

The first diagram is also as it were mathematically incomplete; it reveals the exact and determinate genesis of the curve and a few of its important properties, but in itself, like the differential equation correlated with it, does not show how to investigate the tractrix further or how to relate the tractrix to other curves. Thus the second diagram must supplement it. The tractrix can be identified by shape as the same entity in both diagrams, and the sameness of shape is fundamental, for it is what holds the variables associated with the curve together in an intelligible unity. But the bridge between the two diagrams that registers the distinction as well as the relation between the mechanical and geometrical contexts is the expression of the curve in terms of Leibniz' differential equation and its transformation into ordinary algebraic terms, that is, its solution. Thus the tractrix exists as a hybrid to which experience and indeed mechanics alone would never have drawn our attention, and which Cartesian geometry would never have generated without the Leibnizian extension to mechanical constructions.

2. Resnik's Structuralism

The writings of Michael Resnik represent an important counter-proposal to the causal-materialist account of mathematical knowledge. Resnik wishes to assimilate rather than separate the objects of mathematics and physics, and to buttress the ontological and epistemic status of mathematical objects. His structuralist account of the entities of mathematics characterizes them as positions in patterns and likens our acquaintance with them to our acquisition of knowledge about linguistic and musical patterns. This then leads him to explore the extent to which our knowledge of physical entities can also be understood as acquaintance with patterns. Claiming that the entities of quantum physics themselves look like patterns, he finds no reason to think that mathematical patterns could not be instantiated by them. His final pronouncement, however, is that what the mappings between mathematical and physical patterns look like must remain indeterminate, since there is no master pattern which includes them both. An important feature of his structuralist theory is that a fact of the matter about how structures correspond exists only when they are subpatterns of the same pattern. (Resnik 1981, 1982, 1990)

In this section, I want to criticize Resnik's account of mathematical entities as positions in patterns, because I think its logicist bias makes certain kinds of complex unities in mathematics difficult to see. Since they are just the kind of item which I have discussed as hybrids at the overlap of domains, their absence in his account also makes the application of mathematics to mechanics harder to understand. All the same, the central tenet of Resnik's structuralist theory that reference from structure to structure may be indeterminate is quite consonant with my own convictions. For our recurrent attempts to elaborate and revise inter-structural correspondences play an important role in the advancement of mathematical and scientific knowledge.

Resnik claims that mathematics does not present objects with an "internal" composition, given in isolation and with features independent of the structures in which they happen to occur. Rather, he writes, "The objects of mathematics, that is, the entities which our mathematical constants and quantifiers denote, are structureless points or positions in structures. As positions in structures, they have no identity or features outside of a structure." Nor, he adds, do they have any "internal structures." This rather polemical claim is immediately modified in what follows: clearly Resnik considers the objects of mathematical study to include structures themselves as well as points or positions.

Yet the suggestion persists that what mathematics is really about is points or positions, with stuctures intervening in a secondary way as relations among points or positions. For Resnik's two leading illustrations of what he means by positions or points are the natural numbers in sequence, and geometrical points in a (discrete and finite) spatial array. After all, mathematics has been regarded as the offspring of the Adam of numbers and the Eve of geometrical points. And the natural numbers considered as iterated units, and geometrical points, seem to have no independent presentability and no internal composition apart from the structures in which they occur.

Resnik defines a structure (or, to use the term he prefers, pattern) as "a complex entity consisting of one or more objects, which I call positions, standing in various relationships (and having various characteristics, distinguished positions and operations.)" (Resnik 1981, p. 530) Examples of structures are models of formal theories, like (N,S) the natural numbers with the successor function, and a finite, discrete pattern of geometrical points. In general, Resnik says, "patterns are specific models of theories (up to isomorphism)." (Resnik 1981, p. 536) So it seems that while patterns have an internal composition, or rather while they are a composition (whether internal or not is unclear), they cannot be given independent of the points or positions whose relations they are. There is nothing more to them than those relations.

Herein lies the difficulty. Neither points or positions, nor patterns, seem to be the kind of thing about which one could pose an interesting mathematical problem. What is there to say about a point, or the unit? If there is nothing to say, why should there be anything to say about relations among entities about which there is nothing to say? This puzzle is not an empty bit of sophistry. Rather, it indicates Resnik's failure to locate a middle ground of mathematical objects that exhibit an interesting, complex and problematic unity, objects that can indeed be given independently and which have internal composition.

In order to show the importance of this middle ground of mathematical objects, I need to exhibit the circularity in Resnik's exposition that rules them out ad hoc and to explain why he doesn't notice the circle. Take the case of geometry. Resnik sees geometry as a model of a formal theory couched in the language of predicate logic with some geometrical vocabulary added in the extralogical axioms. The quantifiers of the formal theory range over geometrical points, which then seem to be the true objects of geometry, and the model supplies relations among those objects. Geometrical points have no internal composition; and the logicization of geometry stipulates that points are the only object of geometry; so geometrical objects have no internal composition.

Resnik doesn't recognize this circularity (at least not in the papers cited), perhaps because of certain prejudices he shares with many other contemporary philosophers who look at mathematics through the lens of logic. The first is the assumption that mathematical domains are structured like logical theories. The second is Quine's dictum that the objects of a domain are what the quantifiers of such logical theories quantify over. The third is the logicist misunderstanding of analytic geometry, which views the domain of geometry as reduced to that of number, the continuous as reduced to the (infinitely iterated) discrete. Logicist reduction plays down the difficulty of establishing correlations between domains, and the opacities that remain in such correlations once they have been set up.

Descartes in the seventeenth century offered a similar misinterpretation of geometry. He wanted to present his mathematics as a relational structure instantiated by items that would be mere place-holders. These place-holders would have no internal structure of their own, and so would not impugn the generality of, or threaten to disrupt, the structures in which they stood. He chose straight line segments. This was in many ways a useful choice, but the place-holders were far from neutral. They changed both the geometry and the algebra of the problem-context in which Descartes worked, and excluded many important objects (areas, volumes, infinitesimals, curves) from his geometry of ratios and proportions, thus limiting in important ways the kinds of problems and solutions he could entertain. Moreover, his insistence that his relational structure was transparent to reason blocked his ability to see that his mathematics was in fact suspended between two nonequivalent kinds of structures (equations, and proportions), and that the way these structures worked was the product of historical debate, and certain decisions on the part of his mathematical antecedents. (Grosholz 1990/91, ch. 1 and 2)

Likewise, Resnik doesn't examine the consequences of the debt which his theory of patterns owes to twentieth century predicate logic and the set theory that arose alongside it. To suppose that any geometrical item is a logical-relational structure holding among points depends on two historical projects. The attempt to assimilate geometry to the realm of number inspired and continues to challenge transfinite set theory. The attempt to assimilate the realm of number to logic characterized the first decades of the development of predicate logic, leaving an indelible imprint. The former gives us the habit of thinking of the continuum as an infinite concatenation of discrete (number-like) points, and the latter of supposing that numbers behave like well-formed formulae.

Though in his discussion of congruences between patterns, Resnik rejects strict reductionism for the good reason that there is no master pattern in terms of which congruences might be established univocally, still he does not escape the influence of these venerable logicist projects of reduction. Predicate logic pretends to be a transparent structure, but actually logicians have chosen its characteristic place-holders, certain well-formed formulae, and that choice has consequences. It affects the shape of other structures to which predicate logic is applied both by amplification and suppression. Domains, objects and problems which do not lend themselves to the combinatorial, boolean shape of predicate logic tend to fade away in the eyes of logicians working under its insistent light. (Grosholz 1982)

I have argued, however, that the true objects of a mathematical domain are those entities about which problems arise, the foci of mathematical investigations. The objects that inspire mathematical research programs are profoundly interesting; their recalcitrance and mystery are as challenging as their revelations. No geometer would waste his or her time investigating dots. Moreover, the relations among mathematical domains like logic and geometry, geometry and number theory, or set theory and analysis, are not relations of reduction in any simple sense. The objects, problems and methods of these domains are too heterogeneous to be captured by a single morphism; indeed, part of the mystery that inspires mathematical research is the investigation of what happens at the indeterminate overlap of mathematical domains. Resnik himself makes this point, though expressed in terms of positions, patterns and the congruences that hold between patterns. He claims, "there is no fact of the matter whether an occurrence of a pattern is or is not the same as another except when they are both subpatterns of the same pattern," to explain the phenomena of multiple reductions between domains. (Resnik 1981, p. 546)

Resnik connects the indeterminacy of reference between patterns with the absence of any single master pattern, and of any true individuals in mathematics, with respect to which some absolute reference might be fixed. Both points are well taken. But I would observe that citing the absence of a master pattern also indicates the multiplicity of mathematical domains, though in a very abstract and general way. Also, the absence of true individuals (noninstantiables) (Gracia 1988) in mathematics is consistent with my claim that among mathematical instantiables are unities that have interesting internal complexity and can be given independently. A large part of mathematics is about these unities, which are purely formal and yet isolable and complex. I would add that such formal unities, because they are isolable and complex, can thus sometimes function especially well as schemata for individuals encountered in the natural world.

Euclid certainly did not take points as the basic objects of geometry. Euclidean geometry studies a variety of objects which, he is quite careful to point out, are very different from each other: points, lines (which are bounded by but not composed of points), and plane figures (which are bounded by but not composed of lines). (Heath 1956, pp. 153-232) In Euclid's view, the more complex entities could not be reduced to the simpler; indeed, their heterogeneity is to him so important that he bans the yoking together of points, lines and plane figures in ratios and proportions. His treatment of the objects of geometry reveals that the unity of points is trivial; the unity of lines is somewhat more interesting, since lines can be measured by lines; and the unity of plane figures, like triangles and circles, is so rich and various that it constitutes a research program of which the Pythagorean Theorem is the signpost and flag. (Heath 1956, pp. 349-368)

Euclidean plane figures like the right triangle are objects about which an important set of problems can be posed; these problems are not problems about points. They concern the endlessly interesting internal composition of the triangle, as a whole greater than the sum of its parts, the points or vertices that bound its sides, the lines that join its vertices, the angles that exist between those lines. That whole has the unity of shape that allows it to be given in a diagram, in isolation from all the other possible objects of geometrical study. (Susan Hale, in arguing for a distinction between the relational and intrinsic properties of geometrical entities, underscores my point here. Some mathematical entities such as curves, she concludes, do have properties which are not merely extrinsic and relational. (Hale forthcoming))

Thus, I claim that Resnik cannot explain the research program of Euclidean geometry simply by reference to structures as he defines them. Nor can his account of congruence between patterns explain the synthesis of geometry and algebra in the work of Descartes, of geometry and mechanics in Newton's Principia, or the synthesis of geometry, algebra, number theory and mechanics in the late seventeenth century writings of Leibniz, for those partial unifications were posed and elaborated in terms of isolable, internally complex mathematical objects. Descartes' Cartesian parabola,

Leibniz' tractrix and Newton's planetary ellipse were not treated as congruences in which the positions or points of one pattern are mapped onto or occur within the positions or points of another pattern. They were hypotheses about certain possible relations among domains made treatable as problems, that is, made into a research program, by the problematic unity of shape exhibited by higher algebraic and transcendental curves.

The realm of number, the hierarchy of sets, the formulae of logic will never give rise to the peculiar unity of geometrical shape. (Though one might argue that they have interesting, nontrivial unities of their own.) And points augmented by logically specified relations will never yield lines or figures. A triangle is not a position or a pattern in Resnik's sense, and it is not even a subpattern, since a subpattern is only a collection of points or positions united by logically specified relations. The founding of research programs in mathematics, and the partial unification of mathematical domains, indeed the very possibility of mathematical knowledge, requires the existence of objects that, while not individuals (noninstantiables) are yet isolable, internally articulated unities.

3. Structuralism and Space

In Euclid's geometry, space as a whole is not itself an object of study. One might say something about its properties: it has no boundaries, no regions, no parts, no separable components, no holes: it is isometric, isotropic, homogeneous, infinite, continuous, dense, non-separable. (But note how anachronistic it sounds to say most of that.) However, space as a whole does not intervene in problems: Euclid's problems have to do with points, lines and bounded plane figures. I have been urging that it makes sense to think of the objects of Euclidean geometry as heterogeneous, as many-sorted; now I want to urge the difference between the objects which articulate Euclidean space, and the space itself. The former are finite and bounded; they have parts and discernible regions apropos those parts. In the case of plane figures, they have the problematic unity of shape. Though we might want to say that they reveal the properties of space as a whole through the way they articulate that space, in almost every important respect they are unlike it.

Does space taken as a whole have a unity? As far as I know, no one in classical antiquity ever asked that question. We might say in retrospect that it has the unity of being without parts, separable components, boundaries, regions, etc. But that's a trivial unity, a unity without internal composition, a unity due to sheer indeterminateness or structurelessness. While it allows the objects of Euclidean geometry, and their congruences and similarities, to appear, in itself it has none of the features that makes them objects of study.

Perhaps I am being unduly enigmatic; I may make it seem like a mystery how determinate geometrical objects could "emerge" from indeterminate geometrical space. Historically, though, it was space as a whole that "emerged" as an object of study from the study of geometrical objects like points, lines and plane figures. And this process did not take place within geometry, but only after geometry had been combined with the domain of number in the seventeenth, eighteenth and nineteenth centuries, and then combined with set theory at the turn of the twentieth century. For once a continuum is seen in analogy with the discrete realm of number, and the discrete realm of number is reorganized to mimic the continuum, space itself becomes a hybrid, novel object which for certain purposes of problem-solving may be decomposed to point-numbers. And once set theory has been introduced, these infinitary collections of points can be seen as (geometric-numerical) sets, thus as bearers of internal structure and objects of study.

In twentieth century mathematics, we are used to talking about spaces as objects of study, not only Euclidean spaces, but non-Euclidean spaces and a variety of topological spaces whose properties are very different from Euclidean spaces (they may have separable components, holes, bumps, etc.), as well as highly infinitary function spaces whose points or positions are functions. Yet it is well to remember two things. First, these spaces are not objects of set theory per se, for set theory taken in itself would never have discovered them as objects, nor the problems in which they intervene. Rather, they are hybrids hovering ambiguously between analysis and set theory. To call them merely sets of points is to forget their origins.

Second, the difference between Euclidean space and for example Hilbert space brings to the fore an important tension in the modern framework between the demand for reduction and the demand for hierarchy. (Once again I would point out that while Resnik's structuralism is not reductionist, it includes and masks some reductionist assumptions.) We say that the points or positions in Hilbert space are functions; this make them look anomalous on Resnik's account, since points or positions aren't supposed to have any internal composition. The obvious response is that functions themselves are just sets of points. But then the difference between Euclidean space and Hilbert space is obscured, for they have both become just sets of points. Set theory alone (and, I would add, Resnik's structuralism) cannot reinstate the distinction, that is, the middle ground in which functions have a unity different from the unity of points and from the unity of space.

Resnik wants to talk about Hilbert spaces as structures with peculiar points or positions, Euclidean space as a structure of points, and the ordinary objects of Euclidean geometry as substructures of points. I have just argued that the kinds of unity in question are very different, as are the problems in which they figure, and that therefore something very important is lost in this account. But returning to Resnik's way of talking raises a further question: what kind of unity do structures have? For Resnik certainly thinks that we study structures as well as points or positions in mathematics; indeed, the fact that they intervene in structures is what makes points or positions susceptible of study. And if we study structures, they must exhibit some kind of unity or bring mathematical reality into some kind of unity, for nothing is knowable that isn't unified.

When Resnik talks about structures as models of formal theories in predicate logic, as he often does, the unity invoked is the unity of an axiomatized set of statements formulated in the language of predicate logic. For the pervasive, almost invisible logicism that runs through contemporary philosophy of mathematics imposes the attributes of logic on everything it touches. Now the unity of deductive systems is an extremely interesting kind of unity; even Euclid was interested in it. It has been studied in the twentieth century with the help of Boolean algebra and recursion theory. But it is not the kind of unity a geometric space has; and it is not the kind of unity a triangle or a function has.

Resnik's choice of the words "structure" or "pattern" reminds us that the objects of mathematics are not individuals (noninstantiables). All the objects of mathematics are instantiables and so look like universals; this explains in part their transcendence. But the objects of mathematics also exhibit kinds of formal unity which occupy a middle ground between the trivial unity of a point, unit or the empty set, and the imperfectly understood transfinitudes of set theory. Recognizing this aspect of mathematical reality helps to explain how mathematical domains are constituted as collections of related problems about certain kinds of objects; and how they are partially unified around hybrid objects, a synthesis that contributes to the growth of mathematical knowledge in important ways. It also helps to explain how mathematics can be instantiated by physical reality, in virtue of a process that gradually (though never completely) unifies mathematics and physics, despite the many ways in which physical objects differ from mathematical objects.

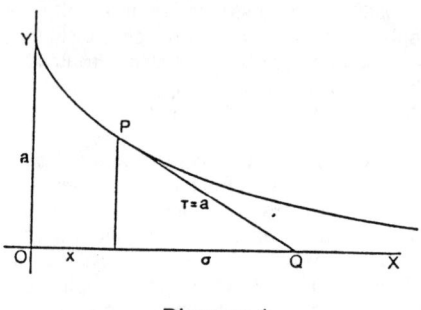

Diagram 1

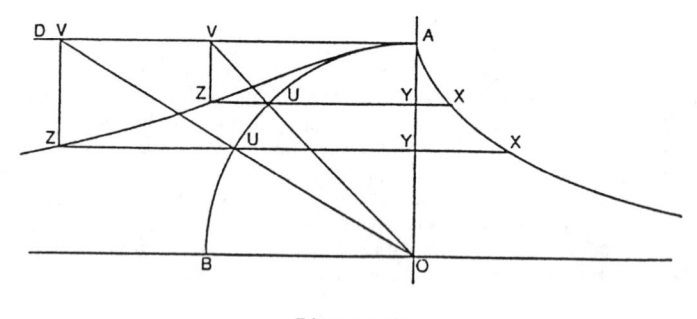

Diagram 2

References

Benacerraf, P. (1965), "What Numbers Could Not Be", *Philosophical Review* 74: 47-73.

Bos, H.J.M. (1988), "Tractional Motion and the Legitimation of Transcendental Curves", *Centaurus* 31: 9-62.

Field, H. (1980), *Science Without Numbers*. Princeton: Princeton University Press.

Gracia, J. (1988), *Individuality*. Albany: State University of New York Press.

Grosholz, E. (1985), "Two Episodes in the Unification of Logic and Topology", *British Journal for Philosophy of Science* 36: 147-57.

_____. (1990/91), *Cartesian Method and the Problem of Reduction*. Oxford: Oxford University Press.

_____. (forthcoming), "Was Leibniz a Mathematical Revolutionary?", in *Revolutions in Mathematics*, D. Gillies, ed. Oxford: Oxford University Press.

Hale, S. (forthcoming), "On Intrinsic and Relational Properties."

Heath, T.L. (1956), *Euclid's Elements*. Vol. I. New York: Dover.

Resnik, M. (1981), "Mathematics as a Science of Patterns: Ontology and Reference", *Nous* 15: 529-50.

_____. (1982), "Mathematics as a Science of Patterns: Epistemology", *Nous* 16: 95-105.

_____. (1990), "Between Mathematics and Physics", PSA 1990, Volume II, pp. 369-378.

Structuralism and Conceptual Change in Mathematics

Christopher Menzel

Texas A&M University

Professor Grosholz packs a lot into her interesting and suggestive paper "Formal Unities and Real Individuals" (Grosholz 1990b). In the limited space available I can comment briefly on its several parts, or direct more substantive comments at a single issue. I will opt for the latter; specifically, I want to address her critique of mathematical structuralism, as found especially in the writings of Michael Resnik.

I begin with a brief, hence necessarily caricatured, summary of Resnik's influential view. According to structuralism, the subject matter of a mathematical theory is a given *pattern*, or *structure*, and the objects of the theory are intrinsically unstructured points, or *positions*, within that pattern. Mathematical objects thus have no identity, and no intrinsic features, outside of the patterns in which they occur. Hence, they cannot be given in isolation but only in their role within an antecedently given pattern, and are distinguishable from one another only in virtue of the relations they bear to one another in the pattern (see, e.g., Resnik (1981)).

Grosholz's judgment on Resnik's structuralism is rather severe: "Neither points or positions, nor structures, seem to be the kind of thing about which one could pose an interesting mathematical problem." The reason for this, in the case of points, is simply that they have no intrinsic features; there simply is nothing to say about them *per se*. But since there is nothing to say about points, she continues, then because there is nothing more to structures than relations between points, there won't be anything interesting to say about structures, since these are relations among entities about which there is nothing to say.

The problem Resnik has run into, then, is that he has failed to locate a "middle ground" of genuine mathematical objects that can be given independently, and which are internally structured, internally complex. For, as any mathematician knows, "the true objects of a mathematical domain are those entities about which problems arise." And in contrast to Resnik's bland, unstructured points, these objects are "profoundly interesting," at once recalcitrant, mysterious, and revelatory; they are complex unities that are greater than the sums of their parts. A consequence of this, Grosholz notes, is that structuralism is woefully inadequate vis-a-vis the history of mathematics. For "the founding of research programs in mathematics, indeed the very possibility of mathematical knowledge, requires the existence of objects that are yet isolable, internally articulated unities."

Though I share some of Grosholz's concerns, I think her critique seriously misrepresents both the character and richness of Resnik's view. First of all, Grosholz reads too much into the centrality of points in Resnik's account. For I think the featurelessness of points leads her to conclude that, despite the importance of internal complexity exhibited by a structure in his theory, insofar as we try to objectify that structure, we must identify it with a point, and hence whatever structure we had was lost. Thus, she infers, Resnik finds himself hoist on his own petard: despite the obvious conceptual centrality of internal structure in mathematical research, it must be relegated to a secondary role in mathematical ontology, just out of Resnik's theoretical reach. Quite to the contrary, however, Resnik doesn't lose internal structure, at least nothing that is important in the notion; he just places it elsewhere.

Grosholz is correct to a certain extent: structures are in a precise sense secondary with respect to mathematical theories, if we say that the primary objects of a theory are those things it quantifies over, those objects in its intended domain. A structure is thus, so to say, what a theory is *about*, its subject matter, but not an object of the theory, as Peano Arithmetic is about, but does not quantify over, the natural number structure, and ZF is about, but does not quantify over, the iterative hierarchy of sets. In Wittgenstein-ese, a theory's subject matter is *shown*, but not spoken of. Grosholz's objection seems to be that structures thus remain forever shown, never spoken of, and hence, on Resnik's account, cannot play an appropriately primary role in mathematical research. However, this would be so only if the secondary character of structures were absolute. But in Resnik's account there is also a precise sense in which structures are nonetheless *primary* mathematical objects, hence structureless points, and yet with all the information Grosholz packs into the notion of internal complexity intact.

How so? It is a well known fact of model theory that what is shown at one level can be spoken of at another. That is, a structure S that constitutes the subject matter of one theory, and hence is not an object of that theory, can in fact become an object of a further theory capable of overtly expressing the relations between S and its points. But then what is the subject matter of *this* theory? It is a *new* structure that contains all the points of S, and a further point as well that plays the role of S itself qua mathematical object, in addition to new relations that depict the relations between S and its points. The internal complexity of mathematical objects does not disappear, but simply manifests itself as external complexity between further structureless points. The original structure itself becomes "positionalized," as Resnik has put it in a recent paper (Resnik 1988, p. 410).

Let's fix ideas with some examples. Consider the natural number structure (N,<), which we might represent pictorially as follows (taking the arrow to be transitive):

$$\bullet \longrightarrow \bullet \longrightarrow \bullet \longrightarrow ...$$

Now, Grosholz wants the structure itself to be an internally articulated mathematical object, which we might represent so:

$$[\bullet \longrightarrow \bullet \longrightarrow \bullet \longrightarrow ...]$$

Since there are no such internally structured objects for Resnik, this kind of representation is unavailable. But internal structure is just more relations. Thus, when we want explicitly to objectify the natural number structure, we consider it not in itself as an *isolated* point, but in relation to other points in a new pattern:

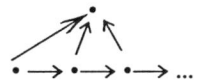

$$\bullet \rightarrow \bullet \rightarrow \bullet \rightarrow \ldots$$

Intrinsically, then, qua mathematical object, (N,<) is indeed a structureless point; but the information that Grosholz packs into the internal structure of her mathematical objects is nonetheless perfectly transparent within the larger pattern.

A second example: the right triangle. For Grosholz, a triangle has an

> endlessly interesting internal composition,...the points...that bound its sides, the lines that join its vertices, the angles that exist between those lines. That whole has the unity of shape that allows it to be given in a diagram, in isolation from all the other possible objects of geometrical study.

And again comes the charge that, because the objects of structuralism are featureless points, the internal composition of the right triangle that is exhibited so clearly in concrete diagrams is lost. But this attack against the structuralist is similarly misguided. *Qua* structure, the right triangle pattern has a rich and complex internal composition with vertex positions, line positions, angle positions, and their associated relations. Considered as part of a further pattern, *qua* object, the right triangle is indeed just a featureless point within the larger structure, but one which nonetheless bears all the appropriate structural relations to the vertex, line, and angle positions of the original structure—if the larger structure really does incorporate the right triangle, then it must exhibit all of the complexity of the triangle's structure as well.

From this perspective, there seems little that distinguishes Grosholz's picture from Resnik's in regard to the matter of internal complexity. The general point is that the complexity we associate with a mathematical object lies not in the object considered in isolation, but in *relations* that hold between several (perhaps many) associated objects. Given that we can objectify patterns, whether we choose to build this complexity directly into the object itself, and call this its internal structure, or reserve complexity for patterns alone, seems more a matter of taste than substance.

This failure to appreciate the richness and flexibility of Resnik's structuralism lies behind Grosholz's charges of explanatory inadequacy vis-a-vis the history of mathematics. If all mathematical objects are mere homogenous points, the argument seems to be, then we lose what is unique to each mathematical domain, and hence we cannot explain how distinct domains differ in a way that allows for mathematical progress and change. Her arguments here tie in with her own intriguing notion of a hybrid (Grosholz 1985, 1990a). According to Grosholz, a *hybrid* is a mathematical object that, as it were, straddles two or more distinct mathematical domains. Such objects typically arise when problems insoluble in one domain can be approached within another, overlapping domain. Thus, the tools of analytic geometry can solve problems relating to curves, e.g., parabolas, that are insoluble in classical Euclidean geometry. But the curves of analytic geometry are hybrids, neither wholly geometrical, nor wholly algebraic. The crucial point in this context is that such curves must be considered *autonomous and irreducible*, related to, but distinct from, Euclidean curves. To reduce curves generally to structures or positions in structures is to strip them of this ambiguous character that explains mathematical progress. Thus, Grosholz charges, Resnik cannot explain the synthesis of geometry and algebra in the work of Descartes,...or the synthesis of geometry, algebra, number theory and mechanics in the late seventeenth century writings of Leibniz, for those partial unifications were posed and elaborated in terms of isolable, internally complex mathematical objects.

But in light of the discussion above, the charge here will not stick; to the contrary, structuralism can provide an appealing account of mathematical change, in particular the specific sorts of change that lead Grosholz to her hybrids. The overlap of distinct domains, for the structuralist, is just co-occurrence of patterns.[1] The distinctive character of a co-occurring pattern in different contexts is explained by differences in the ways the pattern occurs (e.g., as a spatially located diagram or a point set), the tools that are used to study it in these different settings (ruler and compass vs. algebra or differential calculus), and the problems broached. Advances in tools especially can open up hitherto inexpressible avenues of research in the study of a given pattern or cluster of patterns. Mathematical progress generally thus consists in the acquisition of deeper insight into the nature of a certain pattern, or cluster of patterns; progress and change of the sort Grosholz seizes upon in particular occurs when such new insight is acquired by recognition of pattern co-occurrence across distinct domains, and the corresponding development of new and more powerful tools and systems of representation. For the structuralist there is therefore no need to postulate a special hybrid object "hovering ambiguously" between distinct domains that share similar structure. One need not go beyond the structure *simpliciter*. This is not to say that the notion of a hybrid has no significant philosophical role to play. It is especially noteworthy as a device for characterizing the above sort of conceptual advance in mathematics. Grosholz errs, I think, in investing it with ontological, rather than simply conceptual, significance.

Now, of course, to analyze the advances of past mathematical pioneers in structuralist terms is not to say that they understood what they were doing along overtly structuralist lines. However, Grosholz appears to saddle the structuralist with such a thesis when she points out that

> Descartes' Cartesian parabola [and] Leibniz' tractrix...were not treated as congruences in which the positions or points of one pattern are mapped onto or occur within the positions or points of another pattern.

Structuralism is a *metaphysical* thesis, a thesis about what mathematical objects *are*, not a historical thesis. Though we should certainly be able to understand the historical development of mathematics in structuralist terms—e.g., along the lines sketched in the previous paragraph—nothing at all follows about the views of Descartes, Leibniz, and their fellows in regard to the nature of mathematical objects. How Descartes and the other pioneers did in fact conceive the denizens of the mathematical universe is no doubt crucial for understanding the actual historical development of mathematics.[2] All the structuralist needs to argue is that these conceptions, however prominent in the minds of the pioneers, might diverge from the metaphysical facts of the matter.

Finally, it should be mentioned that the structuralist—Resnik in particular—can do more than simply defend against Grosholz's charges. For in addition, structuralism promises—if still imperfectly—some significant advantages over Grosholz's position. First, it brings an appealing unity to mathematics: all mathematics is ultimately grounded in patterns. Its diversity, as argued, springs from the great multiplicity of patterns there are, the way they cluster into separate domains, the great variety of ways in which they can be represented, and the great variety of tools by means of which they can be studied. Second, perhaps more important, structuralism potentially yields considerable epistemological dividends, since knowledge of patterns generally can be grounded in simple perceptual situations. Given the ontological unity of mathematics according to structuralism, this in turn promises an epistemological ground for all of mathematics. Resnik thus purports to tell us what mathematical objects are, and to do so in a way that does justice to the richness of mathematics while yet making them epistemologically tractable. Grosholz, by contrast, seems to be subject to the usual problems that plague her—to all appearances—rather traditional variety of platonism.

Resnik's structuralism is not without its difficulties; but it appears to be undiminished by Grosholz's attacks.

Notes

[1] Grosholz (forthcoming) seems to admit as much herself: of Leibniz' application of the calculus to the sphere of mechanics she writes, "[t]hus another domain is brought into alignment with number theory, algebra and geometry, *sharing structure with them*...but maintaining a partial independence as well" (my emphasis).

[2] Indeed, the notion of a hybrid might be useful here as well. In treating a curve as an infinite-sided polygon, Leibniz arguably wasn't conceiving of it *simply* as a Euclidean plane figure, or an object of analytic geometry, since those domains do not permit such a representation. By the same token, it would be anachronistic to characterize him as conceiving of his discovery as consisting in deeper knowledge of a single pattern. It might thus be appropriate to characterize his own conception as directed toward a new object—a hybrid—closely related to its fellows, but nonetheless distinct, with a similar but distinct internal articulation.

References

Grosholz. E. (1990a), *Cartesian Method and the Problem of Reduction*. Oxford: Oxford University Press.

———. (1990b), "Formal Unities and Real Individuals," in *PSA 1990*, Volume Two, Fine, A., Forbes, M., and Wessels, L. (eds.). East Lansing: Philosophy of Science Association.

———. (1985), "Two Episodes in the Unification of Logic and Topology, *British Journal for the Philosophy of Science* 36: 147-57.

———. (forthcoming), "Was Leibniz a Mathematical Revolutionary?"

Resnik, M. (1981), "Mathematics as a Science of Patterns: Ontology and Reference," *Nous* 15: 529-550.

———. (1988), "Mathematics from the Structural Point of View," *Revue Internationale de Philosophie* 42: 400-424.

Part X

RUDOLF CARNAP CENTENNIAL

The Unimportance of Semantics[1]

Richard Creath

Arizona State University

Our deepest commitments about history are reflected in how we break it down into periods. (Cf. Galison 1988) By drawing a break at a certain point we emphasize the novelty and importance of a new development. It is also how we contain and dismiss certain work as no longer relevant. Thus, in the history of physics we break the story with Newton, both to emphasize his roles in bringing previous developments to a close and in initiating new lines of work, and also to suggest that the ongoing practice of physics thereafter can appropriately in large measure ignore what preceeds Newton. Periodizing history is essential to understanding it, including when we periodize the work of a given writer. In philosophy, anyone who did not see a gulf between Kant's early work and his critical philosophy or between Wittgenstein's *Tractatus* and his *Philosophical Investigations* would be missing something enormously important. But periodization can also be dangerous in blinding us to the continuities between periods and in erroneously suggesting that it is safe to ignore what has come before. Nowhere is this more true than in standard treatments of the work of Rudolf Carnap.

Carnap's work is typically divided into four periods: There is the Carnap of the *Aufbau* (1928) which lasts until the early 1930s; there is the Carnap of syntax which lasts till the ink is dry on *The Logical Syntax of Language* (1934), say 1935; there is the semantical period which lasts till the mid- or late-1940s; and finally there is a period devoted to confirmation and probability theory and other broadly pragmatic matters in the philosophy of science. Some would add a Kantian or pre-critical period on the front; others would draw the lines in slightly different ways. But usually Carnap's work is thus drawn—and quartered.

Even Carnap encouraged this periodization; in fact, he appealed to it to avoid annoying questions about his earlier work. For example, in the late 1940s Carnap visited Minnesota. Wilfrid Sellars was at the time teaching a seminar on the *Aufbau* and so peppered Carnap with questions about it. Carnap couldn't escape; they were riding together in a car from the airport. When Carnap could contain his exasperation no further, he cut off discussion of the *Aufbau* by saying "But that book was written by my grandfather!" (Sellars 1975, p. 277) On the present occasion I have no wish to argue against periodization per se; it is essential to the historian's task. Moreover, I have no wish to argue (my title notwithstanding) either that semantics is unimportant to us or that Carnap's work in his so-called semantical period is uninteresting or unimportant. Instead, what I shall urge is that there is no great cleavage either between Carnap's syntactical and semantical work or between his semantical and later work. Of course, there are differences, but they are not as important as the continuities. And we will mislead ourselves about all of it if we are guided by the standard periodization.

The first thing I want to do, however, is to review, perhaps a bit cynically, what I take to be the reasons or causes behind the standard periodization. Thereafter, I can review what is new and imporatant in the *Logical Syntax* that binds the rest of Carnap's work together and then use that to argue against the standard periodization more specifically.

It is easy to see why someone now would want there to be a wall after the *Logical Syntax*. After all, it says that all logic (and for that matter all philosophy) is syntax, and we know this to be false. In addition, it rejects a truth predicate and with it reference and designation, all of which we know to be central to our enterprise. Besides, it is so *hard*! It seems to have hundreds of pages of incredibly complicated English-free notation, all in defense of some thesis that we know to be wrong anyway.

By contrast, *Meaning and Necessity* (1947), which is emblematic of semantics, is so readable, so easy, and so accessible to anyone who has had standard contemporary graduate training in philosophy. We may quarrel with this or that claim in the book, but its view is still in the mainstream.

The argument for a cleavage between the semantics and the later work is similar in form. In the later period Carnap's work becomes wholly absorbed with philosophy of science and especially with probability and confirmation theory, a concern which seems absent in the semantical period. In this metaphysical age, worrying about the epistemology of science seems peripheral to the mainstream of philosophers. It is rather an odd taste to be thus charitable for two reasons: First, by looking at what we thought was wrong we might learn something. We might learn that it wasn't so wrong, or we might learn that scattered among the falsehoods were a lot of important truths. Second, we might come to understand quite differently and perhaps a bit better what already seemed familiar and right.

All in all, then, the reasons for the standard periodization hinge on a determination to focus on that which seems accessible, relevant, and largely right to *us*. Insofar as this exemplifies a principle of charity it is not at all disreputable. I think, however, we should resist the urge to be thus charitable for two reasons: First, by looking at what we thought was wrong we might learn something. We might learn that it wasn't so wrong, or we might learn that scattered among the falsehoods were a lot of important truths. Second, we might come to understand quite differently and perhaps a bit better what already seemed familiar and right.

To make my case for a different periodization, or at least for less of a wall on either side of semantics, I need now to go back to the *Syntax* to see what was interesting and new about that. In fact I will spend a good deal of time on the development of the *Syntax*, but if we understand the importance of that clearly enough, the rest of my argument will fall neatly and fairly quickly into place.

To find out what was new in the *Syntax* we have to ask what Carnap was committed to just before he wrote that book. Well, his deepest commitment was to a thoroughgoing empiricism, which consisted of two parts. The first of these was a rejection of intuition. This supposed transempirical mode of knowing independent matters of fact was much loved by Platonic and Cartesian rationalists and apparently by Frege and Russell as well. The rejection of intuition is what the elimination of metaphysics is all about. The second part of Carnap's empiricism is the conviction that the meaning of a claim is somehow the mode of its verification. This link between meaning and knowing or justification derives more or less straightforwardly from Hume, but it was not dependent on any particular form of verification or testing. He could change his mind about how to test at the drop of a counterexample without giving up the crucial linkage to meaning. Whatever is not appropriately linked to experience is unintelligible. Metaphysics, thus, is not false, not unjustified, but utterly without cognitive meaning. Even so, the slogan that meaning is the mode of verification is at best only a strategy for a theory of meaningfulness and only hazily suggests a full fledged theory of meaning that would include synonymy, implication, and confirmation.

Now the severest problem of a traditional empiricism, such as Carnap's, is the question of what to do about mathematics and logic. (We might add epistemology to that list.) This is where logicism, which Carnap appropriated from Frege and Russell, comes in. If mathematics can be reduced to logic as logicism says, then the severest problem for empiricism is narrowed to what to do about logic. For this Carnap accepted Wittgenstein's doctrine in the *Tractatus* that logic is no news at all. Thus, empiricism is saved.

Carnap had a variety of other commitments before *Syntax* which should be briefly mentioned. He had always defended a kind of conventionalism, but on closer inspection this comes to nothing more than a Duhemian underdetermination thesis. He was also committed to something that might be called the possibility of alternative *Aufbau*s. Even before that book was published Carnap agreed that one could set up a constructional system in various ways. One could do it on an autopsychological basis, as he in fact did following the empiricist tradition and especially Russell. Or one could do it on a physical basis as Neurath had insisted. Now Carnap also seemed to think that some constructional systems were more "correct" than others, but what the possibility or the unequal correctness of the alternatives came to was not very well worked out.

A third early commitment beyond empiricism was to a philosophy of geometry according to which alternative mathematical geometries are really differing implicit definitions of the terms they contain. Thus, they do not disagree. This very early belief (It is expressed in Carnap's disertation.) undoubtedly derives from Hilbert, but it was also reinforced by Carnap's association with Schlick. If Quine is right that there is no fathoming the subdoctoral mind (Quine in Dreben 1990), then there is no hope of finding out what prompted Carnap to accept this in the first place. Finally, Carnap is committed to a full scale fallibilism. Not only are theories uncertain, but so too is the observational basis on which they rest. This idea is not yet there in the *Aufbau*, but it pre-dates the *Syntax*.

So much for background. When Carnap sat down to write what was to become *The Logical Syntax of Language* he most certainly did not have the whole actual outcome in mind. In fact, all he intended to say was that one could indeed describe the logical form of a language and do so within that language itself. That one can talk about logical form is directly contrary to Wittgenstein's claim that logical form could be shown but not said, and Carnap proposed to accomplish it via Godel's technique of arithmetization. That one could talk about logical form within that language was very important to Carnap, von Neumann, Neurath, and others apparently because it was confusedly connected with the question of the unity of the language of science. (Contrary to Carnap's first impression, separating the object and metalanguages would appear to be no more insidious than type theory.) In order for a language to be able to describe itself that language has to be fairly weak. Stronger languages, including those in which classical mathematics could be expressed, must be ruled out. So there was to be no discussion or defense of these stronger languages, no principle of tolerance, and no discussion of general syntax. If Carnap had left Syntax at this point it would have been rather tame stuff.

In the process of writing *Syntax*, however, Carnap came to see that the forms of a variety of stronger languages could be described, albeit in a metalanguage stronger than the object languages themselves. Even more importantly he saw that by transfinite means (essentially by allowing an omega rule) the incompleteness established by Gödel of any language sufficiently strong to express classical mathematics could in important respects be repaired. Given the utility of these stronger languages and the metamathematical means to deal with them, such stronger languages could no longer be dismissed. This discovery was as electrifying to Carnap as the discovery that there are non-Euclidian systems was to geometers in the 19th century. (And for similar reasons.)

In order to cope with the discovery, Carnap made use of two old commitments from the pre-*Syntax* days. First, he used the idea that philosophic issues or doctrines can be absorbed into matters of logical form. The claim was originally Wittgenstein's, but he was concerned with a single language. The point still applies when we have a variety of languages. Importantly, among the philosophic commitments embedded in a language are the epistemic ones. I shall return to this later in greater detail. Given that an epistemology is embedded in each of the languages, there can be even in principle no non-question-begging epistemic grounds for preferring one language or logic or philosophy over another. Second among the pre-Syntax commitments to be used, Hilbert's implicit definition approach to geometry was waiting in the wings. Now, Carnap could generalize it. Alternative sets of logical rules could be thought of as implicit definitions of the philosophic terms they contain. Alternative philosophic presuppositions are not in conflict; they are just different ways of speaking. Like any theory of implicit definition, Carnap's theory of meaning here (insofar as we can call it that) is both *functionalist* and *holist*. It is functionalist because it specifies a system of relations that the words must bear to one another without saying more concretely what will do the job. It is holist because it is the totality of rules that defines each term. If one were going to use the word 'meaning', the meaning of a term would be its function within the whole system of rules.

Because the alternative sets of logical rules are thought of as definitions we get a thoroughgoing conventionalism with respect to philosophy and certain basic parts of science. This is encapsulated in the Principle of Tolerance. And the conventionalism we get is vastly more powerful than the Duhemian underdetermination thesis that had gone before. Underdetermination, remember, imagines that questions of what counts as evidence and what logical relation theory must bear to evidence are *settled* in advance. It then notices that, given plausible answers to those questions, for any amount of evidence, more than one theory will bear the appropriate relation to it. Carnap's new conventionalism, by contrast, ultimately says that what counts as evidence and what the appropriate logical relations are (even what the logical consequence relation itself is) *are all up for grabs too*. Carnap's is a very radical view.

By making it a *linguistic* conventionalism Carnap avoids the result that anything goes. What is conventional is meaning. The word 'unicorn' considered merely as a noise has no meaning. But we can go on to make it meaningful by specifying linguistic conventions in any way we choose. In these conventions are embedded epistemic standards. If the conventions are chosen in one way it will be contingently correct to say 'There are unicorns' and chosen in another way (i.e., in a way that gives 'unicorn' a different meaning) it will be contingently correct to say 'There are no unicorns'. Once the meanings of the various words have been fixed, however, the logical and epistemological standards have likewise been fixed by being embedded in the conventions that gave the words meaning. What we get is not "ways of world making" but "ways of word making".

Carnap's conventionalism is tempered in another way, too, by his pragmatism. There may be no non-question-begging epistemic grounds for choosing among the alternative languages (logics, philosophies), but some languages are more convenient than others. Inconsistent languages are pragmatic disasters, and so are languages without inductive rules. It is not necessary to establish that a language is maximally or even minimally convenient before using it, but philosophic discussion (where it is not wholly misguided) must be pragmatic. Qua pure logicians our job is merely to trace out the consequences of this or that convention. This is an engineering conception of philosophy. Of course, we may make proposals and defend them as useful, but here we go beyond pure philosophy and enter the empirical. Carnap's pragmatism is encouraged but not contained in his association with the Bauhaus school of artists and architects. This association has recently been illuminatingly explored by Peter Galison. (1990) Carnap's pragmatism was likewise encouraged but not contained in Neurath's attempts at social engineering and by political and social changes within Europe more generally. The contrast with the Platonic ideal of philosophy, or for that matter with Kant's, couldn't be greater. The development of Carnap's radical conventionalism and pragmatism really is a watershed and really does begin a new period in his philosophy.

Before proceeding to his later work, however, I want to explore the *Syntax* still further. The first thing to do it to look at his pre-*Syntax* views to see how much they changed, and among these the place to begin is with logicism. Under the new conventionalism one no longer has to reduce mathematics to logic in order to resolve the epistemic problems of empiricism. Mathematical axioms can be treated as implicit definitions quite directly. Mathematics would then be true in virtue of meaning and hence by convention even without any reduction to logic. Such a reduction is still possible and desirable. Any reduction is an economy and thus pragmatically attractive. This is like reducing the number of axioms in propositional logic; it is nice but hardly earthshattering. In the (very weak) sense of thinking that the reduction is both possible and desriable Carnap is still a logicist, but this is logicism virtually in name only. In fact it had been absorbed into formalism. The reduction no longer carries with it any special epistemic benefits. Both logic and mathematics can be conventional or analytic with or without the reduction.

So what happens to the rest of Carnap's pre-*Syntax* views? As expected, he remains an empiricist, but perhaps surprisingly, empiricism itself becomes a convention, albeit one for which there are powerful pragmatic reasons. (Carnap 1936-37, p. 33) His anti-metaphysical stance, that is his rejection of intuition, is retained and systematized. The most basic ontological commitments are matters of convention. Carnap also gives the appearance of being a nominalist, especially in the last section of the book. But this section seems to have been written early, probably before his conventionalist breakthrough. In any case, whatever the appearances, Carnap is a neutralist, even a noncognitivist, about basic ontological claims. Outside the language there is nothing to say; inside a language those questions have been trivially settled. The nominalist appearance is given by the fact that the language that Carnap prefers and would propose is a nominalist one, but that is hardly ontological commitment. Even Carnap's verificationism (better called confirmationism) is retained and deepened. He now has more nearly a full theory of meaning, and by identifying the epistemic rules as implicit definitions of the terms they contain, he is able to give some substance to what had been only a slogan: that meaning is the mode of verification.

A variety of other pre-*Syntax* views should be mentioned as well. His Duhemian conventionalism (the underdetermination thesis) is retained, but it pales beside his radical new conventionalism. There is still the possibility of alternative *Aufbau*s, but now this can be given a systematic account, not a mere suggestion. Gone, however, is the idea that one of these constructional systems is more correct, though some may be more convenient. The early Hilbertian philosophy of pure geometry is obviously generalized and moved to center stage. Finally, the fallibilism is still there, but it, too, is now a convention. Like empiricism itself, though, it is a convention for which there are powerful pragmatic reasons.

The actual text of *The Logical Syntax* proceeds first by example. It takes a weak language and then a stronger language and shows how to describe their logical forms, presumably on the grounds that showing is easier than saying how to do this. The last part of the book is a discussion of the philosophic import of syntax, but between the examples and the finale is a discussion of general syntax which is the real heart of the book. Here Carnap's primitive terms are 'is a sentence' and 'is a direct consequence of'. The first of these exhausts what we call syntax. Plainly, Carnap means to go beyond that. Not only is the logical consequence relation itself semantical, as we use the term, but so are truth tables, interpretation, and analyticity, all of which Carnap discusses. The treatment especially of the last of these is surprisingly close to a full semantical account. No doubt Carnap is still wrong in thinking that all philosophy is syntax, even in his broader sense, but he is not nearly as wrong as we might have thought given our use of the word.

All of this raises the question of why Carnap rejected the concept of truth. There is a rumor afoot, propagated by Hartry Field, that it was because truth did not admit of physicalist defintion. (Field 1972) This cannot be right about Carnap. Field need not worry, though, because the historical question does not bear directly on Field's own program. Carnap does discuss truth in *Logical Syntax*, and what he says is very odd. Most of the discussion concerns the semantical paradoxes, as though he were worried that any concept of truth were unavoidably inconsistent. That would be respectable but still a mistake. But Carnap goes on to show just how to avoid the contradiction, namely by formulating the truth predicate in a metalanguage distinct from the language to which that predicate applies. So why doesn't Carnap accept this? Well, it might have been because he still wanted to do the logic within the language it described. But he doesn't say this. And it would be a rather weak reason. And more importantly, he has plainly renounced this want in the rest of the book. What he does say is that this approach would not yield a genuine syntax: "*For truth and falsehood are not proper syntactical properties*; whether a sentence is true or false cannot generally be seen by its design, that is to say, by the kinds and serial order of its symbols." (Carnap, 1937, p.216) What kind of reason is this?! Of course truth is not syntactical in this sense, but the question is why philosophy should be restricted to syntax in so narrow a way. Coffa interprets the argument as verificationism, but it is not even that. In technical terms this is just plain 'goofy'. It is as though 'and' is not a logical term on the ground that it is not a purely logical matter whether birds sing *and* Caesar marched. Or that 'two' is not properly definable in logic because it is not a purely logical matter whether there are two toads in Transylvania.

The argument against truth is so bad that it is plausible to assume that Carnap was antecedently prejudiced against the concept of truth. If so, he might have thought that his cavalier attitude about it in *Syntax* was harmless. Certainly Carnap was so predisposed, for under the pernicious influence of Neurath truth would have been called "metaphysical" and "absolutist". It was metaphysical because its acceptance somehow committed one to a domain of Russellian facts or Kantian things-in-themselves. It was absolutist because it was somehow confused with certainty, which in non-analytic cases Carnap's and Neurath's fallibilism forbids. This hypothesis about the underlying causes of the rejection of truth is confirmed when Carnap finally accepts the notion soon after *Syntax*. He does not say: now I see that truth is physicalistically definable. Instead, he says in effect: now I see that truth and confirmation must be distinguished and also that truth is not metaphysically loaded. (Carnap 1936)

What is sad about the whole episode is not only that truth is in fact entirely compatible with the conventionalism and pragmatism of *Logical Syntax*. Rather, it is that the background prejudices against truth actually *fly in the face* of the central lessons of that book, namely, its epistemic conventionalism and it ontological neutrality. Obviously, these are hard lessons to learn and ones which even Carnap had not fully absorbed.

It may be a blunder to identify truth with certainty, but it is no mistake at all to recognize that the enterprise of *Logical Syntax* is open to an epistemic interpretation. Carnap did not care for the word 'epistemology' because he thought it to have been preempted by psychologists and by philosophers practicing the work that Frege dismissed as psychologism. But that need not mislead us. What Carnap is investigating is the pure structure of epistemology. His most basic relation, that of direct consequence, is after all the relation that reasons bear to that which they immediately justify. If you want to find out what some foreign speaker means by some term, go find out what arguments the speaker accepts. (Carnap 1950b, p.37) Something is conventional just in case it could have been otherwise, and there is no epistemic reason for choosing among the alternatives. Carnap emphasizes the centrality of analyticity, but something is analytic just in case its justification can be traced back to conventions and nothing more. Thus, to be analytic is to have a special epistemic status. Analyticity is no mere substitute for truth within logic and mathematics, a substitute which is needed when one lacks a concept of truth but which is obsolete when one has a real truth concept. In effect, the analytic is the epistemically conventional. By seeing meaning as provided by implicit definitions and these as exhibiting the epistemic structure of the language, Carnap's verificationism is built in at the very foundation. The epistemic dimension of all this never changed. Later, when he addressed the issue of what we could believe about theoretical entities, it boiled down for him to the question of whether we should adopt by convention a rule of inference that would justify theoretical claims on observational evidence. He himself was willing to adopt such a convention, but instrumentalists were not. (Realists on the other hand often spoke as though the inference rule had been ordered by God, and there was nothing left for human convention to decide.) In any case, my point is not about scientific realism, but rather merely to illustrate that Carnap's notion of logic, from *Syntax* onward, is broad enough to include what we call epistemology. Certainly Carnap makes quite a point of calling his enterprise the logic of science.

If we count *The Logical Syntax of Language* as fundamentally epistemic, as we should, then it may seem that there are two overwhelming ommissions in the book. There is only the most rudimentary discussion of induction and confirmation, and there is almost nothing said of observation. Perhaps these omissions can be forgiven in a work devoted primarily to classical logic and mathematics. But still we should want more. As far as induction is concerned, it seems that Carnap did not know what to say at this point.

What he does say is sadly deductivist. What is required is that his most basic relation, that of direct consequence, be generalized in a natural way to include partial implication. This would then be an account of inductive inference, of confirmation, and of logical probability. Indeed, all of Carnap's discussions of probability and later philosophy of science can be thought of as attempts to carry out this very program. Whatever the success of Carnap's theory of logical probability it must be seen as continuous with the program of Syntax and in fact as trying to develop a proposal for a workable convention, perhaps one that would be an explication of our usual conventions.

As far as observation is concerned the story is a bit more complicated. Carnap did talk about this in "On Protocol Sentences" (1932), published while he was writing *Syntax* and in "Testability and Meaning" (1936-37) published just after. What he had to say was important, perhaps even on the right track. But not only do we recognize it as inadequate, Carnap did too. Unfortunately, in the nearly forty years remaining to him he did not see how to develop these ideas further in a satisfying way. In any case from the early 1930s Carnap is a conventionalist about observation as well. The protocol sentences and reports are not themselves conventions, but rather what we take as a protocol is conventional. Empirical psychology will help with the pragmatic question of how to revise or formulate the observation language, but there is no convention independent fact of the matter about whether protocols will concern sensory experiences, or physical objects, or both. Carnap hoped that the observable features of things could be marked by a special vocabulary and hence that observationally justified beliefs could be picked out syntactically. Then the fact that protocols need not be justified by inferring them from other beliefs might somehow be marked via the direct consequence relation. We know that this won't work and that observation is a vastly more complicated affair than just a special vocabulary. Carnap knew it too, but he did not know just how to improve the account. Moreover, he was more interested in saving both observation and theory while damning metaphysics than in exploring the limits or nature of observation. Though he hadn't yet figured out how he would carry it off, Carnap is quite explicit about being a fallibilist even about observation reports. Now I think that something along the lines that Carnap wanted (but never found) can be devised. This, however, is neither the time nor the place. What we can say now is that the conventionalist aspects of Carnap's approach to observation have for the most part been neither appreciated nor explored. We can also say, despite the omissions on observation and induction, that Carnap is giving us in *Syntax* the structure of an epistemology, or at least trying to do so.

Now it may seem odd in a paper nominally devoted to semantics that so far I have talked almost exclusively about the so-called syntax period. Of course, my thesis is that by misunderstanding *Syntax* we are led to exaggerate the differences between semantics and the work before or after and also to misunderstand Carnap's semantics itself. A correct understanding here will show Carnap's last four decades are, if not a seamless fabric, then at least a more or less continuous development. Carnap is not lurching from one misguided enthusiasm to another. Rather he has one broadly consistent leading idea, which demands our attention and which is too little understood.

With a clearer picture of *Syntax* in mind we can now address the supposed break between *Syntax* and semantics. As we saw there is a lot of what we call semantics in *Syntax*. It is in the consequence relation, in the truth tables, in the work on interpretation, and especially clearly in the discussion of analyticity. Of course, a concept of truth is officially rejected in *Syntax*, but when Carnap finally does accept truth, this is fully consistent with his conventionalist program. The confusions that had prevented the acceptance of truth are avoided not by abandoning that program but by carrying it through. (The conventionalism by itself will guarantee that the language is ontologically neutral. It will also guarantee that the protocols can have any desired degree of justification; if, as Carnap thinks, it is not handy to let them be certain, then don't. If truth can be defined in the way that *Syntax* outlines or that Tarski shows, then truth is a completely separate matter.) The move to semantics does not change the content of Carnap's conventionalism, but it does change the form. Now Carnap speaks of truth, reference, designation, and the like, and shows how such notions are interdefinable. The theory of truth and reference amounts to an implicit definition in the metalanguage not to a physicalist reduction. That Carnap now talks cheerfully "of" propositions, properties, meanings, etc. is taken by some to show that he has given up his old nominalism and has become a Platonist. This is twice wrong. He wasn't a nominalist before, and he isn't a Platonist now. Instead he remains a metaphysical non-cognitivist, and this is founded on his ongoing conventionalism. He doesn't have to repeat himself. After all, he thinks he is in the friendly and thoroughly pragmatic U.S. and that all this can be taken for granted.

To see that his conventionalist epistemological theme is sustained, let us very quickly review the major writings of the semantical period. In "Truth and Confirmation" (1936) Carnap admits, indeed insists, that the confusion of truth and certainty is just a mistake. There is certainly no repudiation of his conventionalism. The first book where he is working out his semantical views is *Foundations of Logic and Mathematics*. (1939) It title notwithstanding this is largely about the interpretation of science. He reiterates and defends his conventionalism for logic (no doubt in response to Quine's "Truth by Convention" (Quine 1936)) and reaffirms that his logic is a logic of science. The books *Introduction to Semantics* (1940) and *Formalization of Logic* (1943) can be considered as a pair. *Formalization* was written first and contains a result sufficiently dramatic that Carnap decided to preceed it with *Introduction to Semantics* both to prepare the reader for *Formalization* and as a textbook on semantics. What is so surprising? It is that logic too is open to non-standard interpretations. This arises out of his functionalist approach to definition; if logic proceeded by latching onto Platonic (convention independent) entities it is unlikely that this problem would ever have come up. Even *Introduction to Semantics* is at great pains to stress its continuity with what has gone before.

Meaning and Necessity is the book that we now tend to think of when we think of Carnap's semantics, no doubt because it is so delightfully readable. On this topic it is interesting to note that Church complained that the book was mere informal prolegomena and that publication should await the development of a strictly formal system. (In other words, don't publish until it has been made totally unreadable.) In any case the book arose out of the Quine-Carnap correspondence, that is, it was prompted by Quine's worries about meaning and analyticity. I have already argued that analyticity is at bottom an epistemic notion. At this point in the debate, however, Quine had not yet brought forward his alternative theory of knowledge, and Carnap was still reluctant to use the word 'epistemology'. So Carnap did not emphasize this aspect, even though it is there. Instead he concentrated on presenting clarification and technical improvement. His method of intension and extension, as Carnap was quick to agree, is really a method of intension. (The intensions do all the work.) But on closer inspection it turns out that expressions have the same intension if and only if they are alike in point of epistemic functioning. There are also non-essential changes in the presentation of analyticity; the definitions are now given in terms of state descriptions. These changes are the direct products of his emerging theory of inductive confirmation, work on which is occuring simultaneously. Finally, even the theory of modality that is given in the book is designed to make it empiricistically acceptable by making it a linguistic, i.e., conventional, affair.

The last major publication of the so-called semantical period is "Empiricism, Semantics, and Ontology" (1950a). This is basically a reiteration of his epistemic conventionalism and pragmatism and the ontological non-cognitivism that follows from it. The tone of the paper is very much: I cannot believe I need to repeat this stuff; I've been saying it all along. And indeed he had.

We are now in a position to evaluate the periodization of Carnap's work that puts a sharp break between *Syntax* and semantics. The case for the break is not very convincing. The radical conventionalism and pragmatism that was the main message of *Syntax* is still there. It may be altered in form, but not in substance. The ontological neutrality is still there, and the epistemic interpretation which makes sense of *Syntax* is still every bit as illuminating as before. In the end, Carnap's logic is still a logic of science. If our concerns are no longer epistemological, then so much the worse for us. As historians we should avoid a periodization which imposes our limitations on Carnap. I do not deny that there are changes over time in Carnaps's work. Certainly his program broadened. Certainly he emphasized, as every scholar does, what was new rather than old. Certainly the move from Prague to Chicago, and with it the change of context and critics, altered how he chose to present his ideas. But just as certainly there is no break in Carnap's fundamental conventionalism.

As for the supposed break between semantics and the work on probability and philosophy of science, very little needs to be said that has not already been covered. The probability work, though it is presented as describing a purely logical relation, is of obvious epistemic import. It is essentially an attempt to carry out the generalization of the direct consequence relation that we earlier described as the unfinished business of *Syntax*. Even the accounts of meaning now offered are overtly epistemological. Although I haven't the room to demonstrate this here, when we come to see the probability as an attempt to formulate conventions, the whole set of which give the meaning for the terms of the language, then many of the criticisms of *Probability* (1950b) can be blunted. If the epistemic reading of *Syntax* is acceptable then *Syntax* and the late work on probability and philosophy of science are of a piece. If the program of *Syntax* is continued in both semantics and probability, then it is unlikely that there is a break between the latter two. It is even hard to specify just when the break is supposed to have occurred. The probability work began well before *Meaning and Necessity* was published, and indeed it is reflected in the text of that work as well. In fact, Carnap lists *Logical Foundations of Probability* (1950b) as Volume Four of his Studies in Semantics. I think he knew what he was doing.

What then shall we make of all of this? Certainly, we are entitled to conclude that the standard periodization of Carnap's work with which we began is at best an exaggeration. It seems to me, however, that the moral goes far deeper than this. The legend that Carnap's thought made a radical shift in the mid-1930s is all that has licensed us in isolating and ignoring *The Logical Syntax of Language*. But what is really radical and well worth studying in Carnap's work is precisely the epistemic conventionalism and pragmatism that he announced there. It is also what is constant throughout the rest of his work. Once we see this, we have a whole new way of approaching his semantics and especially the topic of analyticity. Analyticity becomes an epistemic notion, and this fact can be used to respond to the criticisms of Quine among others. The work on probability will likewise have to be reevaluated, keeping in mind that conventionalism and a functionalist approach to meaning are being presupposed. If this is done, it seems likely that at least some of the objections to Carnap's probability theory can be answered.

How we periodize our history (and Carnap's) matters. But getting the historical record straight is only secondary. What is primarily at stake here is how we can learn from the past and how we can set that to work in our ongoing philosophy. Carnap's conventionalism and pragmatism is a rich and powerful and largely neglected tool to be used by us—now. It is the future that matters most about the past. And by reexploring our past we can hope to shape that which is to come.

Notes

[1] I would like to thank colleagues Jane Maienschein and Michael White and fellow symposiasts Burton Dreben and Michael Friedman for comments on an earlier version of this paper. I would also like to thank the College of Liberal Arts and Sciences of Arizona State University for a research travel grant.

References

Carnap, R. (1928), *Der Logische Aufbau der Welt*. Berlin-Schlachtensee: Weltkreis-Verlag.

——. (1932), "Über Protokollsätze", *Erkenntnis*, III: 215-228. (Translated as Carnap 1987.)

——. (1934), *Logische Syntax der Sprache*. Wien: Julius Springer. (Translated as Carnap 1937.)

——. (1936), "Wahrheit und Bewährung", *Actes du Congrès international de philosophie scientifique*, Sorbonne, Paris, 1935. 4. *Induction et probabilité*. *Actualités scientifique et industrielles*, 391, Paris: Hermann & Cie. 18-23. (Translated as Carnap 1949.)

——. (1936-37), "Testability and Meaning", *Philosophy of Science* 3: 419-471, 4: 1-40.

——. (1937), *The Logical Syntax of Language*. London: Kegan Paul Trench, Trubner & Co. (Translation of Carnap 1934.)

——. (1939), *Foundations of Logic and Mathematics. International Encyclopedia of Unified Science* I,3. Chicago: University of Chicago Press.

——. (1940), *Introduction to Semantics*. Cambridge, MA: Harvard University Press.

——. (1943), *Formalization of Logic*. Cambridge, MA: Harvard University Press.

——. (1947), *Meaning and Necessity: A Study in Semantics and Modal Logic*. Chicago: University of Chicago Press.

_____. (1949), "Truth and Confirmation", in *Readings in Philosophical Analysis*, Herbert Feigl and Wilfrid Sellars (eds.), New York: Appleton-Century-Crofts, 119-127. (Translation of Carnap 1936.)

_____. (1950a), "Empiricism, Semantics, and Ontology", *Revue internationale de philosophie*. IV,11: 20-40.

_____. (1950b), *Logical Foundations of Probability*, Chicago: University of Chicago Press.

_____. (1987), "On Protocol Sentences", *Nous*, XXI: 457-470. (Translation of Carnap 1932.)

Dreben, Burton, (1990), "Quine", in *Perspectives on Quine*, Robert B. Barrett and Roger F. Gibson (eds.), Cambridge, MA: Basil Blackwell, 81-95.

Field, Hartry, (1972), "Tarski's Theory of Truth", *Journal of Philosophy*, LXIX: 347-375.

Galison, Peter, (1988), "History, Philosophy, and the Central Metaphor", *Science in Context*, II: 197-212.

_____,(1990), "*Aufbau*/Bauhaus: Logical Positivism and Architectural Modernism", *Critical Inquiry*, XVI: 709-752.

Quine, W.V., (1936), "Truth by Convention", *Philosophical Essays for A.N. Whitehead*, O.H. Lee, ed., New York: Longmans, 90-124.

Sellars, Wilfrid, (1975) "Autobiographical Reflections" in *Action, Knowledge and Reality: Critical Studies in Honor of Wilfrid Sellars*, Hector-Neri Castaneda (ed.), Indianapolis, IN: The Bobbs-Merrill Co., 277-293.

Part XI

IMPLICATIONS OF THE COGNITIVE SCIENCES FOR PHILOSOPHY OF SCIENCE

Implications of the Cognitive Sciences for the Philosophy of Science[1]

Ronald N. Giere

University of Minnesota

1. Introduction.

Does recent work in the cognitive sciences have any implications for theories or methods employed within the philosophy of science itself? The answer to this question depends first on one's conception of the philosophy of science and then on the nature of work being done in the various different fields comprising the cognitive sciences. For example, one might think of the philosophy of science as being an autonomous discipline that is both logically and epistemologically prior to any empirical inquiry. If the cognitive sciences are empirical sciences, then research in the cognitive sciences could not have any significant implications for the philosophy of science. And that would be the end of the matter. Logical Empiricism is now typically understood as having exemplified this point of view.

More specifically, Logical Empiricism took it for granted (i) that scientific knowledge should be understood as ideally having the structure of a formal logical calculus, and (ii) that the empirical warrant for scientific claims is given by directly observed data together with formal rules which determine the weight of the evidence for or against the particular claims in question. In a Logical Empiricist framework, therefore, scientific knowledge is structured linguistically and the epistemological relationship between evidence and theory is a linguistic relationship. Actual human cognition, as investigated by the cognitive sciences, is irrelevant to such relationships. This was part of the legacy of Frege, who preached forcefully against the sin of psychologism. One of the main messages of this paper is that it is finally time to put this legacy in its proper perspective, which is, that it is not particularly useful for understanding the nature of science.

The situation is not much different if one adopts the picture of the philosophy of science developed by philosophical critics of Logical Empiricism such as Lakatos (1970) or Laudan (1977, 1984). Both agree with Logical Empiricists that theories are linguistic structures. They disagree only that there is any quasi-logical epistemological relationship between data and individual theories. For Lakatos, Laudan, and their followers, appraisal applies only to a series of theories. The question for both is whether the whole series of theories is "progressive" or not, where progress is measured either in terms of predicted novel empirical content or problem solving effectiveness. The traditional connections between rationality and truth (or probability) are broken. Rationality is reduced to a kind of progress, but the proposed kinds of progress have little to do with actual human cognition. So here again there can be no substantial implications for the philosophy of science from the direction of the cognitive sciences.

Things appear quite different from the viewpoint developed by Kuhn thirty years ago in *The Structure of Scientific Revolutions* (1962) and in subsequent philosophical writings (1977). For Kuhn, scientific knowledge is not adequately captured by laws, theories, or other linguistic structures. Nor is there any epistemic relationship between linguistically formulated theories and descriptions of evidence. Rather, scientific knowledge is embodied in a scientific community which shares a family of "exemplars" — specific examples which are judged to exhibit solutions to important problems. There is no epistemic relationship which abstracts from the actual judgments of a scientific community about the exemplary status of a family of specific examples, or about the applicability of certain of these exemplars to a new problem.

Here there is ample room for input from the cognitive sciences. Relevant inquiry into the nature of exemplars and of judgments about exemplars is the kind of thing one could, at least in principle, expect to find being studied by cognitive scientists. Kuhn himself clearly recognized this possibility with his original appeal to gestalt psychology and subsequent appeals to cognitive psychology and psycholinguistics (Kuhn 1977).

With these examples in mind, one can make a stab at formulating some conditions on any conception of the philosophy of science which could accommodate the relevance of research in the cognitive sciences. (i) The philosophy of science must be *naturalistic* in the sense that its claims are subject to test by empirical data in the same way that ordinary scientific claims are so subject. This is hardly sufficient, however, since Laudan (1977) claims his theory of science to be naturalistic in that it is to be authenticated by appeal to the history of science. So we must add either (ii) understanding the nature of scientific knowledge, its laws, theories, or whatever, requires understanding the cognitive capacities and activities of scientists; or (iii) judging the epistemic merit of scientific knowledge claims requires appeal to the judgmental capacities and activities of scientists; or both (ii) and (iii).

Now what about the cognitive sciences? What implications have they for a "cognitive" philosophy of science satisfying the above conditions? Within the cognitive sciences, it is common to distinguish three overlapping disciplinary clusters which tend to be thought of as providing three different *levels* of analysis: (i) cognitive neuroscience, (ii) cognitive psychology, and (iii) artificial intelligence (AI). The standard view is that the functional units get larger and more abstract as one moves up the hierarchy from neuroscience to AI. The implications for the philosophy of science are different, and sometimes conflicting, depending on which cluster of disciplines one examines. I shall consider all three in order of increasing abstractness.

2. Implications from Neuroscience.

The person who has done most to exploit recent developments in neuroscience in the service of the philosophy of science is Paul Churchland (1989). For Churchland, the brain is a network consisting of layers of neuron-like units. At each moment, every unit exhibits an activation level which it transmits to units in the next "higher" layer. More importantly, each pathway from one unit to units in the next higher layer is characterized by a "weight," which may be positive or negative, that regulates the relative strength of the signal transmitted along that particular pathway. Thus, the strength of the signal coming into any higher level unit is a weighted sum of the activation levels of the units which feed into that unit. If we assume that the activation levels of units in the initial layer are determined by sensory inputs, then the activation levels of units at all higher layers are thereafter determined by the set of weights which regulate the strength of signals transmitted from unit to unit through the various levels. The output would be the set of activation levels at some higher layer of units.

An exemplar of this kind of network is a simple three-layered network designed to distinguish rocks from mines in the bottom of a harbor (1989, pp. 164-69). The input layer consists of 13 units each representing the average energy in a range of the frequency spectrum of a sonar signal. Each of these units is connected, with a given relative weight, to each of seven units in the second, so-called "hidden layer." Similarly, each of these seven units is connected, again with given weights, to both of two output units, representing "rock" and "mine " respectively. A resultant activation level of near one for the rock unit and near zero for the mine unit means that the network has processed the input as being a signal from a rock rather than a mine.

This network is "trained up" by being fed examples of sonar spectra from known objects. At first the network gives indeterminate answers. But after each example the weights are adjusted slightly so as to improve its answer. After a few thousand such training exercises, the network not only gives unambiguously correct responses for most of the original examples, but also for new examples. It has succeeded in isolating general features of mine and rock signals that permit reliable discrimination between the two.

If we say that the network has developed a *representation* of mine and rock signals, then this representation has some interesting features. First, the representation is not fundamentally propositional in nature. The network does not function by performing anything like logical operations on statements. There are no encoded statements on which to operate. Second, the representation is not localized anywhere in the network, but *distributed* throughout the network as a whole. Churchland argues that the representation is best characterized in terms of the total set of *weights* that regulate the propagation of activation strengths from the layer recording the sensory inputs to the output layer.

Neural networks function primarily as pattern recognizers. Thus, as Churchland points out, the human brain is particularly well-suited to implement a theory of science which takes Kuhnian exemplars as primary. The normal science activity of recognizing a new problem as being similar to an older, exemplary problem is just the kind of thing for which, on Churchland's view, scientists' brains ought to be particularly well adapted. This is an exciting, and, I think, fundamentally correct, insight.

Churchland goes on to suggest that we *identify* a scientist's theory with the weight vector that characterizes that scientist's neural network. This could apply to a scientist's whole neural network, a truly global theory, or just to some part, which would be a more local theory. In either case, it follows immediately that perception is necessarily theory laden. For perception just is the process of propagation up the network that yields a perceptual judgment, and that process is characterized by a particular weight vector.

Here I think that Churchland's reductionist proclivities have gotten the best of him. His enthusiasm for connectionist models has temporarily blinded him to the obvious fact that any adequate cognitive theory of science will require consideration of cognitive structures at higher, and more abstract, levels of functional organization. This general point can be brought home with a few commonplace examples.

There is a whole subculture built around the activities of creating, producing, performing, marketing, and listening to music. Neuroscience no doubt has something to contribute to a theoretical understanding of the culture of music. But focusing solely on the synaptic weights of the brains of composers, performers, and listeners seems a hopeless way to pursue this subject. We need to be able to talk about individual musical works, types of music, styles of performance, and so on, at a higher level of organization.

Similarly, suppose that physicalism is true. That is, the whole world and everything in it, including all living things, is nothing more than one big quantum mechanical system. Nevertheless, attempting to pursue evolutionary biology at the quantum mechanical level would be a hopeless scientific enterprise. We cannot so easily dispense with talk of things like genotypes and founder populations. So also with an understanding of the culture and evolution of science as a cognitive activity.

One source of our disagreement may be that Churchland wants scientific theories to be solely in the brains of scientists. But this supposition ignores the obvious fact that scientists use a wide variety of "external" representational devices such as diagrams, graphs, and, of course, written words and equations. Maybe we should invoke an even more radical notion of distributed representations than that provided by neural networks. Even Kuhn's exemplars may have to be thought of not as being localized in the brains of individual scientists, but as distributed both among the brains of many scientists, and also among their many external representational devices.

It is understandable that Churchland should wish to solve all the major problems of the philosophy of science, like the theory ladenness of observation, at the level of neuroscience. Current connectionist models of the brain are impressive. But there is also a more modest role for neuroscience. Churchland himself gives a clear statement of this more modest role for epistemology in general:

> Making acceptable contact with neurophysiological theory is a long-term constraint on any epistemology: a scheme of representation and computation that cannot be *implemented* in the machinery of the human brain cannot be an adequate account of human cognitive activities. (1989, p. 156, italics added)

Implementation does not require the *identification* of higher-level functional entities with features of neuronal entities. It only requires that whatever is done with the higher-level entities be something than can actually be accomplished by flesh and blood scientists with flesh and blood brains. But Churchland's main point is unassailable. Being compatible with established results in neuroscience is a necessary condition for any adequate cognitive theory of science as a human activity.

3. Implications from Cognitive Psychology.

Among several good examples of research in the cognitive sciences being used to approach problems in the philosophy of science is Nancy Nersessian's study of the development of electrodynamics from Faraday to Einstein (1984, 1988, 1991). Unlike Lakatos (1970) or Laudan (1977), Nersessian is not concerned to show that conceptual change is rational in some special sense. And contrary to both classical Logical Empiricism and the original Kuhn, she argues that conceptual development in science is continuous but not cumulative. To develop an adequate theory of how such development might occur, she invokes ideas originating within cognitive psychology. These include theories of "mental models," like that of Johnson-Laird (1983), and of both analogical and imagistic thinking. Here I would like to highlight several general features of her account — features that I think should be part of any cognitive theory of science.

Nersessian argues that one cannot construct an adequate theory of conceptual change by focusing only on abstract structures — be they "concepts," "mental models," or whatever. The reason, I think, is that such structures possess no internal dynamical forces which could make changes of one type rather than another. They may contain structural possibilities for change, but they contain nothing to make changes happen.

The alternative, as Nersessian insists, is to think of conceptual change as something that is accomplished by cognitive agents. So a theory of conceptual change is necessarily a theory of the capacities and activities of cognitive agents. In my own terms (Giere 1989a), the basic units of analysis for a cognitive theory of science should be individual scientists — not concepts, theories, etc. So the problem of conceptual change becomes the problem of how scientists develop new conceptual structures.

Taking the individual scientist as the basic unit of analysis provides a natural bridge between the philosophy of science and both the history of science and the sociology of science. In spite of much recent historical research focusing on the social and institutional aspects of science, the individual scientist remains at the center of historical narratives. Similarly, recent sociology of science emphasizes the role of various types of human interests and human interactions in the development of particular sciences. But interests and human interactions can only influence the development of science in so far as they influence the thoughts and actions of individual scientists. Putting the scientist at the center of a cognitive theory of science thus makes it possible causally to connect a variety of interests and interactions with the actual historical course of science.

There are several possible ways of understanding projects like Nersessian's. One way is as an attempt to show that beginning with (i) a particular model (theory, or whatever) and (ii) the natural cognitive abilities of scientists, one can explain the development of a successor model. For example, to explain how Maxwell developed his field theory, one would need only his starting point with Faraday's theory plus an understanding of the relevant cognitive mechanisms at Maxwell's disposal.

This clearly is not Nersessian's project. I mention it only for comparison. In any actual scientific development, the later model would be dramatically underdetermined by the earlier model plus cognitive mechanisms. One would have to add at least some other models used by the scientist in question, as Maxwell used various mathematical models unknown to Faraday.

A second possible project requires creating a new, cognitive version of the internal-external distinction by claiming that only other "scientific" models need be considered, and not, for example, religious or political models. My own view would be that one cannot possibly legislate a priori what sorts of models one might have to consider in order to explain how a particular scientist got from the initial to the final set of scientific models. Any kind of model might play an important role. Robert Richards' (1987) study which suggests the influence of religious models on Darwin's thinking provides a recent example of this third sort of project.

A still more inclusive project would be one that includes not only other models, but motivations and interests as well. The recent sociological literature contains many cases for which it is claimed that one cannot explain how the new models came to be developed or accepted without invoking a variety of interests (Shapin 1982). If one understands the realm of the cognitive as excluding motivation and interest, then this fourth possible project would deny that there are always "cognitive" explanations of the development of new models. I would prefer to classify motivational factors as cognitive, so that then all explanations of new models might be appropriately cognitive. In any case, I suspect that this fourth project is more inclusive than many, including Nersessian, would prefer, but I doubt that anything less can do justice to the historical facts.

Given the origin of their concerns in the two decades following Kuhn's analysis of science, Nersessian and others have quite naturally focused on conceptual *change*. What is emerging from this work, I think, is an increasing realization that we lack a clear, widely shared theory of the nature of conceptual structures themselves. We are in the position of trying to develop a theory of conceptual change without having a good theory of concepts — the things that are supposed to change.

That we should be in this situation is not all that surprising. In their reaction to both Logical Empiricism and Kuhn, Lakatos and Laudan maintained the Logical Empiricist view that theories are fundamentally sets of statements. They sought to replace the *epistemological* doctrine of Logical Empiricism that there can be empirical support for individual theories with the doctrine that there can only be progress in a historical series of theories.

Most other attempts to talk about conceptual change likewise assumed a Logical Empiricist account of the nature of theories. The problem was to understand how one got from one theory, so characterized, to another. What Nersessian and others are coming to realize is that one cannot construct an adequate account of the process of theoretical change using a Logical Empiricist account of the nature of theories. One needs a better account.

It is not just that the Logical Empiricist account presupposes that it is adequate to think of a scientific theory as an axiomatic system formulated in first order logic. Rather it is that the Logical Empiricist account takes the basic representational relationship to be that of the *truth* of an individual *statement*. Nor is the later, more holistic, Quineian conception of theories much better. It merely moves us from individual statements to sets of statements. Truth remains the basic goal for representational success.

A "semantic", or "model theoretic", account of theories (Giere 1988; Suppe 1989; van Fraassen 1980, 1989) takes us a step in the right direction, but only, I now think, a small step. On this view, a model has the logical function of a predicate. Thus, rather than talking about statements being true or false, we talk about predicates being *true of* the world (or not). This is still too narrowly linguistic. We need a representational notion that is broad enough to encompass graphical, imagistic, and pictorial representations as well.

Here we find, I think, what is right now the most promising problem area in the philosophy of science for which the cognitive sciences might provide interesting solutions. Nor is it the case that the cognitive sciences lack candidates for the replacing of linguistic, or linguistic-like, structures as the basic representational devices of the sciences. Rather, there are too many candidate devices. It seems that every research area in the cognitive sciences employs a different notion. Thus the "mental models" of cognitive psychologists like Neisser (1976) and Johnson-Laird (1983) are different.

And these are different from the "mental models" of more developmentally oriented psychologists like Carey (1985) or Gentner (Gentner and Stevens 1983). And these are different again from the "mental models" of cognitive linguists like Lakoff (1987). The job for a cognitive philosopher of science, therefore, is not simply to find an adequate notion of representation to import from cognitive psychology. Rather, the task is to forge from these diverse conceptions a conception of representation in science that is adequate for constructing a good theory of scientific development. But the first step is finally to abandon the legacy of Frege and Russell and to realize that statements, or sets of statements, are just one type of representational device - and maybe even not the most important type employed in the actual practice of science.

Abandoning this tradition is a lot easier said than done. In addition to new theories of representation, we may need what used to be called a new metaphysics. In an age of naturalism this would be called a new "theoretical perspective." The old perspective was obtained by reading the structure of language into the world. So the world was thought of as consisting of states of affairs which mirror the structure of statements. A better perspective may be to begin with the world as something with many levels of complexity, so that it might be pictured from many angles. Different pictures may capture different aspects of the complexity, or different levels of the complexity. Simple questions of truth or falsity become irrelevant. But that does not mean that one is not genuinely representing the world.

I am inclined to take the metaphor of pictures quite seriously. Rather than taking representation by statements as fundamental, we should take the way in which pictures represent the world as fundamental. So there may be something to a picture theory of meaning after all, except that it is not statements themselves that picture the world. Rather, statements are just one type of device that may be used in constructing a picture, or model, of the world. It is the model that pictures the world. The problem, then, is to understand that relationship.

4. Implications from Artificial Intelligence.

More than most other disciplines within the cognitive sciences, artificial intelligence has been split by connectionism. It is here that the idea of a "second cognitive revolution" is most applicable. As is already clear from my earlier discussion of neuroscience, connectionism provides a way of linking the top and the bottom of the cognitive sciences hierarchy. Here I will confine my remarks to good old-fashioned, rule-based AI.

There are two sorts of traditional AI activities which have potential implications for a properly naturalized philosophy of science. One is the development of programs that can perform a variety of scientific tasks which go well beyond number crunching. These include: (i) Discovery programs, inspired by Herbert Simon and implemented by Pat Langley and others, which, using fairly general heuristics, can uncover significant regularities in various types of data. Programs such as BACON (Langley, et al 1987) and KAKEDA (Kulkarni and Simon 1988) provide exemplars of such programs. (ii) Programs which generate and evaluate causal models in the social sciences. Here the prototypes are the TETRAD programs developed by Clark Glymour and associates (1987). (iii) Programs which aid in the classification and resolution of anomalies arising in the course of theorizing and experimentation. Lindley Darden (1991) is currently developing such programs with particular reference to genetics.

Programs of this nature are potentially of great scientific utility. That potential is already clear enough to keep good people working on developing them further

(Shrager and Langley 1990). The implications of these sorts of programs for a cognitive philosophy of science are mainly *indirect*. The fact that they perform as well as they do can tell us something about the structure of the domains in which they are applied and about possible strategies for theorizing in those domains. In general, the role of these programs in science is more something to be explained by a cognitive philosophy of science than a resource to be deployed in developing such explanations.

Other philosophers of science, by contrast, have advocated philosophically much more ambitious projects for applying standard AI techniques to the philosophy of science. Paul Thagard (1988, 1989, 1991) provides a prominent example. Thagard explicitly seeks to distance himself from Logical Empiricism. (i) He regards his philosophy of science as a species of naturalized philosophy of science . (ii) He has a much more liberal view of the nature of scientific theories. For Thagard, theories are conceptual or propositional networks rather than axiomatic systems. (iii) Going beyond standard deductive and inductive logics, his model of scientific validation is based on a notion of explanatory coherence which includes such things as analogy.

Nevertheless, it still seems to me that the differences between Thagard's project and that of Logical Empiricism are at the level of *implementation* rather than of fundamental principles. For example: (i) Thagard maintains the fundamental view that scientific knowledge has a structure that can be adequately captured by some sort of propositional system. (ii) He maintains that the reasoning required in choosing one theory over another can be analyzed solely in terms of relationships among propositions. (iii) He advances a project for getting from the descriptive claims of a naturalistic philosophy of science to normative rules, such as inference to the best explanation. And he clearly intends that the normative force behind such rules be more than a matter of empirically based, means-end reasoning (1988, ch. 7).

Thagard would also like to think that his propositional structures are embodied in actual human thinking, thereby bridging the divide between AI and cognitive psychology. But so far there is little evidence that this is even possible, let alone actual. The interesting question is what Thagard would do in the face of empirical evidence that actual humans either can not or do not utilize anything like explanatory coherence in deciding among rival theories. But clear evidence one way or the other is likely to be a long time coming. We are left at the moment with a propositional structure whose normative claim on our allegiance remains obscure, and the hope that the minds of scientists embody such structures.

Thagard's writings contain two different approaches to the fundamental question of the nature of representation in science. His earlier account (1988) portrays scientific theories as a type of *production system*. In light of the well-known logical equivalence between production systems and axiomatic systems, Thagard argues that the equivalence pertains only to "expressive" power and not to "procedural" operations. That is, production systems are computationally more tractable and perhaps even psychologically more realizable than axiomatic systems. This useful distinction, under the labels "informational equivalence" and "computational equivalence" has long been employed by Herbert Simon (1978) to distinguish, for example, the difference between linguistic and pictorial representations (Larkin and Simon 1987).

In his more recent writings (1989, 1991), Thagard distinguishes conceptual systems from theories as sets of propositions utilizing the concepts from the corresponding conceptual system. He portrays conceptual systems as networks of localized concepts linked by part and kind relationships. For example, in the Ptolemaic conceptual system, the Sun and Mars are both planets, which are a kind of star, that is, wandering stars. In the Copernican conceptual system, by contrast, the Sun is a star, while the Earth and Mars are planets, which are now regarded not as stars but as satellites of the sun. Both systems deal with the same set of objects, but they are categorized differently.

The conceptual networks exhibited by Thagard and others seem to me to provide only a relatively superficial description of the endpoints of a scientific revolution. At most they provide ways of cataloguing scientific revolutions according to how conceptual networks get restructured. These networks provide no account whatsoever of the *dynamics* of conceptual development. Unlike Nersessian, Thagard does not even attempt to develop an account of how scientists, beginning with one conceptual system, construct a later conceptual system. For this project it would be better to begin with the *pictures* constructed by members of the rival camps. For example, a Ptolemaic picture would show the Earth at the center of a set of concentric circles, one of which represents the orbit of the Sun. A Copernican picture would have the Sun at the center with the orbit of the Earth represented by one of the concentric circles. The structure of Thagard's conceptual networks is embodied in these pictures, but the pictures included much more. And we know such pictures played a role in the actual thinking of participants at the time.

On Thagard's view (1989, 1991) the new conceptual system replaces the old because the new theory exhibits greater explanatory coherence relative to data shared by both theories. The overall explanatory coherence of any theory is a function of binary coherence relationships among statements of the theory together with statements of evidence. Thagard has developed a program, ECHO, which performs the requisite calculations of relative explanatory coherence for rival theories. The theoretical importance of his account, however, lies in the theory of explanatory coherence, which I cannot reproduce here. For those who wish to pursue the matter further, I offer the following caveats.

In the examples Thagard presents, such as the Copernican Revolution (Thagard 1991; Nowak and Thagard 1991), the respective theories are reconstructed by Thagard and his associates from classical texts. It is difficult to assess whether the process of reconstruction itself has introduced biases in favor of the known historical outcome. Even if it has not, the texts used were typically produced in a historical context in which the objectives were as much rhetorical as scientific. So Thagard may simply be analyzing the rhetorical structure of a text, not the reasoning of any of the participants (Giere 1989b). In any case, Thagard provides no independent evidence either that the participants made the choices they did because they perceived the greater explanatory coherence of the Copernican theory, or that their minds just naturally worked according to his principles.

A determination of relative explanatory coherence begins with two existing rival conceptual systems together with their corresponding theories. This account, therefore, assumes a version of the Logical Empiricist distinction between discovery and justification. But even assuming such a distinction, an account of the choice of one theory over a rival theory would have to include much more of the actual historical context, including the motivations and interests of the participants, than is embodied in the sorts of relationships among propositions countenanced in Thagard's theory of explanatory coherence. And it would have to include an understanding of experimentation as more than merely a means of producing evidence statements.

The differences between Thagard's approach and my own may be summarized by considering the difference in the labels we employ: *computational* philosophy of science versus *cognitive* philosophy of science. As I see it, a computational theory of science is one species of cognitive theory of science, one drawing primarily on the resources of traditional, rule-based AI. Those resources may be useful, but they seem to me far too limited for the task.

5. Conclusion.

In the wake of Kuhn's work, many philosophers of science managed to hold on to the basic framework of Logical Empiricism by marginalizing and assimilating Kuhn's insights as being concerned solely with the temporal development of scientific theories, a topic which simply was not on the agenda of Logical Empiricism. One finds now a similar reaction to the idea of importing theories and methods from the cognitive sciences into the philosophy of science. There is a tendency to marginalize and assimilate such studies as being concerned only with conceptual change, scientific discovery, or creativity.

My view is that, as was the case regarding Kuhn's historical critique, this is a defensive attempt to avoid facing the challenges to the fundamental assumptions of Logical Empiricism, particularly the assumption that the basic representational device in science is a statement, a set of statements, or some similar linguistic structure. The message coming from at least some of the cognitive sciences is that this simply is not so. The implication of these cognitive sciences for the philosophy of science is that the philosophy of science needs to be rethought from the ground up.

Note

[1] The author gratefully acknowledges the support of the National Science Foundation and the hospitality of the Wissenschaftskolleg zu Berlin.

References

Carey, S. (1985), *Conceptual Change in Childhood*. Cambridge: MIT Press.

Churchland, P.M. (1989), *A Neurocomputational Perspective*. Cambridge: MIT Press.

Darden, L. (1991), "Strategies for Anomaly Resolution". In *Cognitive Models of Science*, ed. R.N. Giere, Minnesota Studies in the Philosophy of Science, vol. 15. Minneapolis: Univ. of Minnesota Press.

Gentner, D. and Stevens, A.L. (1983), *Mental Models*. Hillsdale, NJ: Erlbaum.

Giere, R.N. (1988), *Explaining Science: A Cognitive Approach*. Chicago: University of Chicago Press.

_ _ _ _ _ _. (1989a), "The Units of Analysis in Science Studies". In *The Cognitive Turn: Sociological and Psychological Perspectives on Science*, ed. S. Fuller, M. DeMey, T. Shinn, and S. Woolgar, Sociology of the Sciences Yearbook. Dordrecht: D. Reidel.

_ _ _ _ _ _. (1989b), "What Does Explanatory Coherence Explain?" *Behavioral and Brain Sciences* 12 (1989), 475-76.

Glymour, C. et al. (1987), *Discovering Causal Structure: Artificial Intelligence, Philosophy of Science, and Statistical Modelling*. Orlando, FL: Academic Press.

Johnson-Laird, P.N. (1983), *Mental Models*. Cambridge: Harvard Univ. Press.

Kuhn, T.S. (1962), *The Structure of Scientific Revolutions*. Chicago: Univ. of Chicago Press (2nd ed. 1970).

_____. (1977), *The Essential Tension*. Chicago: Univ. of Chicago Press.

Kulkarni, D., and Simon. H. (1988), "The Processes of Scientific Discovery: The Strategy of Experimentation". *Cognitive Science*, 12:139-175.

Lakatos, I. (1970), "Falsification and the Methodology of Scientific Research Programmes". In *Criticism and the Growth of Knowledge*, ed. I. Lakatos and A. Musgrave. Cambridge: Cambridge Univ. Press.

Lakoff, G. (1987), *Women, Fire, and Dangerous Things: What Categories Reveal About the Mind*. Chicago: Univ. of Chicago Press.

Langley, P., Simon, H.A., Bradshaw, G.L., and Zytkow, J.M. (1987), *Scientific Discovery*. Cambridge: MIT Press.

Larkin, J.H., and Simon, H.A. (1987), "Why a Diagram is (Sometimes) Worth a Thousand Words". *Cognitive Science*, 11: 65-99.

Laudan, L. (1977), *Progress and Its Problems*. Berkeley: Univ. of California Press.

_____. (1984), *Science and Values*. Berkeley: Univ. of California Press.

Neisser, U. (1976), *Cognition and Reality*. New York: Freeman.

Nersessian, N.J. (1984), *Faraday to Einstein: Constructing Meaning in Scientific Theories*. Dordrecht: Nijhoff.

_____. (1988), "Reasoning from Imagery and Analogy in Scientific Concept Formation". In *PSA 1988*, vol. 2, ed. A. Fine and J. Leplin, 41-47. East Lansing, MI: The Philosophy of Science Association.

_____. (1991), "How do Scientists Think? Capturing the Dynamics of Conceptual Change in Science". In *Cognitive Models of Science*, ed. R.N. Giere, Minnesota Studies in the Philosophy of Science, vol. 15. Minneapolis: Univ. of Minnesota Press.

Nowak, G. and Thagard, P. (1991), "Copernicus, Ptolemy, and Explanatory Coherence". In *Cognitive Models of Science*, ed. R.N. Giere, Minnesota Studies in the Philosophy of Science, vol. 15. Minneapolis: Univ. of Minnesota Press.

Richards, R. (1987), *Darwin and the Emergence of Evolutionary Theories of Mind and Behavior*. Chicago: Univ. of Chicago Press.

Shapin, S. (1982), "History of Science and its Sociological Reconstructions". *History of Science* 20:157-211.

Shrager, J. and Langley, P. (1990), *Computational Models of Discovery and Theory Formation*. Palo Alto, CA: Morgan Kaufmann Publishers, Inc.

Simon, H.A. (1978), "On the Forms of Mental Representation". In *Perception and Cognition: Issues in the Foundations of Psychology*, ed. C.W. Savage, 3-18, Minnesota Studies in the Philosophy of Science, vol. 9. Minneapolis: Univ. of Minnesota Press.

Suppe, F. (1989), *The Semantic Conception of Theories and Scientific Realism*. Urbana, IL: Univ. of Illinois Press.

Thagard, P. (1988), *Computational Philosophy of Science*. Cambridge: MIT Press.

_____. (1989), "Explanatory Coherence". *Behavioral and Brain Sciences.* 12: 435-467.

_____. (1991), *Conceptual Revolutions*. Princeton: Princeton Univ. Press.

van Fraassen, B.C. (1980), *The Scientific Image*. Oxford: Oxford Univ. Press.

_____. (1989), *Laws and Symmetry*. Oxford: Oxford Univ. Press.

Paradigms and Barriers[1]

Howard Margolis

University of Chicago

> Having for thirty years believed and taught the doctrine of phlogiston... I for a long time felt inimical to the new system, which represented as absurd that which I hitherto regarded as sound doctrine; but this enmity... springs only from force of habit... [Black to Lavoisier, 1791]

1. Introduction

This paper is abstracted from a forthcoming book which defends a particular answer to the question of just what it is that shifts when a paradigm shifts. The claim is that what shifts are habits of mind. And in particular the claim is that the most striking cases of paradigm shift will characteristically turn on a shift in some single, uniquely critical, habit of mind: the barrier. An account of a radical discovery — discovery that prompts the Kuhnian symptoms of incommensurability, so that intuitions that seem irresistible to some seem perverse to their rivals — then characteristically turns on how some individual got past the barrier (escaped the critical habit of mind), while at least for a while others could not do so.

The study spells out the argument in detail and supports and illustrates it with a series of historical cases. But since the basic claims strike first hearers more readily as obviously wrong than as probably right, I can't hope to persuade many readers with the brief summary here. However, I will sketch out the main lines of the argument, and indicate the sort of historical applications developed in the full study: perhaps enough to make the idea seem at least possibly right.

The argument is an extension of the account of persuasion and belief in my Patterns, Thinking & Cognition (1988), where I try to show that a remarkable range of cognitive features (and in principle all cognition) can be accounted for in terms of sequences of pattern-recognition. The new work provides a stronger version of the argument about paradigm shifts worked out in that earlier book. There the argument is applied (in particular) to give an account of how the heliocentric idea could lay on the table, waiting to be noticed, for 14 centuries before someone finally was able to see it. For as Neugebauer (1968) and others have pointed out, everything needed for the heliocentric argument is in Ptolemy's *Almagest*. In the new work summarized here and in other work in press, I elaborate on the Copernican account, and add new accounts of two further memorable episodes: the overthrow of phlogiston and the emergence of probability.

2. Habits of mind

The central notion of habits of mind refers to scenarios or templates or patterns that guide intuitions in the automatic, non-conscious, hard-to-change way characteristic of physical habits. I had a good deal to say about that in *Patterns*. The new study begins with detailed discussions of why (in terms of their neural character) habits of mind can reasonably be taken to be essentially identical to physical habits, and hence why it makes sense to suppose that the way habits of mind operate, their resistance to change, and the conditions under which they nevertheless sometimes do change closely parallels what we know about physical habits. I must treat the notion of habits of mind with properties akin to those of physical habits as something the reader is already familiar with, which in a rough sense at least is surely true.

Our special concern is with what binds a community in the special way that yields the Kuhnian symptoms of incommensurability across the beliefs of rival communities. Suppose for a moment that the relation between habits of mind and paradigms were actually something like the relation between physical habits (and as a reviewer has suggested, also habits that characterize a style of play) and games. On that view, we could think of people operating within a paradigm as developing certain characteristic habits of mind that fit with and ordinarily facilitate their work. The situation (on this kind of view) is like that of someone who plays tennis or squash, and accordingly develops habits that facilitate play in those games. Theories and descriptions of equipment and procedures in a science would be the analog of rules and descriptions of the equipment and layouts for the games.

Changes in habits of mind (if this view were sound) would go along with a paradigm shift, as some habits would change if a person switches from squash to tennis. But for paradigms as for sports a change in habits of itself could not change the activity. Nor would it make sense (if the parallel with games were sound) to say that habits of mind are constitutive of a paradigm. That would be like saying that the habits characteristic of people expert in squash or tennis are constitutive of the games themselves.

In contrast to all that, on the view here habits in fact are constitutive of paradigms. To put the point in the most extreme way, on this view shared habits of mind are the only *essential* constituents tying together a community in the way that makes talk of sharing a paradigm fruitful. Talk of a paradigm without particular habits of mind (I want to argue) is like talk of a square without a perimeter.

Entrenchment in certain habits grows out of intense, specialized experience with certain kinds of activity, as entrenchment in the physical habits that make an expert tennis player or violinist grow out of the practice of those activities. Further, as is stressed in the full study though I will not have space to develop the point here, what prove to be particularly potent barriers turn out to be *entangled* in several different ways in the thinking and practice of members of the expert community. See the detailed discussion of the "nested spheres" sense of the world in Chapters 11 and 12 of *Patterns*. Hence on the view here the central puzzle for understanding what bound together a certain community (made communication easy within the community and made it hard to communicate with a rival community when that appeared) would be identification of habits of mind which tacitly guide various critical intuitions.

A paradigm shift then *is* special sort of shift in habits of mind. In particular, we are interested in just those cases where the shift is in some way essential for emergence of a new idea in science. We expect to find that the more striking the Kuhnian symptoms of incommensurability, the more effective an account focussed on some shift in habits of mind will prove to be.

So sometimes new ideas appear which create a marked sense of cognitive shock in the first audience, and which eventually yield (when successful) the Kuhnian sense of conversion among those convinced (apparent, for example, in Black's remark to Lavoisier quoted at the outset), and of being swept aside by something that does not really make sense among those who are not. What marks such cases, on the argument here, is that a shift — ordinarily entirely tacit and unconscious — in some critical habit of mind is required to see the new idea in a way that makes it look insightful rather than illusory.

The argument here might seem more congenial if I spoke of a paradigm as defined by a "point of view" rather than habits of mind. But there is good reason not to do that. A person is ordinarily conscious of a point of view. But unless specifically and effectively prompted, a person is ordinarily unconscious of the operation of habits, and indeed is to a large extent unaware of the existence of habits. A person ordinarily can try a different point of view, and certainly understands what is being asked when someone proposes a look at things from another point of view. But we can't try out a different habit, and the claim that a person has a habit is something that usually needs to be backed up by much more than an appeal to introspection.

So while the habits of mind claim pressed here overlaps some of what is meant by the more familiar comment that people within a paradigm share a point of view, it is not merely another way of stating such a claim. A person can be highly expert in a formal theory, able to give an elegant presentation of how observations can be interpreted from that point of view, but not be operating in the paradigm at all. For example, modern writers like Neugebauer 1968 and Price 1959 can take a Ptolemaic point of view, but no modern writer can be entrenched in Ptolemaic habits of mind — we all inescapably see the Ptolemaic models in relation to Copernican beliefs. This easily yields striking divergences from the intuitions of an actual Ptolemaic astronomer. From the time of Ptolemy to the time of Copernicus, no *astronomer* left even a hint of seeing the heliocentric possibility. But on Neugebauer's or Price's account that was easy to do, hence their unflattering judgments on the Copernican achievement. And indeed, logically, the transformation from Ptolemaic to Copernican astronomy is easy. But cognitively it was enormously difficult, as discussed in detail in the full study.

Since habits of mind characteristically are unnoticed, how do we ferret them out? Certainly an attempt to specify the complete set of habits of mind characteristic of a paradigm is not plausible. But we never need undertake that impossible task. We only need to identify those habits of minds which are critical for distinguishing a paradigm from some historically conspicuous rival. Pragmatically, habits can only be made visible by setting them against the background of some alternative. We can become aware of our own habits of walking by comparing our some other gait we observe, as we can become aware of our own habits of speech by comparing our habits of speech to those of some other community which shares the language but with differences of accent and lingo. In our context, if we can identify an at least implicit alternative, we could try to say something about habits that would facilitate one way of doing things (for habits of mind, some sorts of intuitions) and that would act as a barrier to the alternative. But we need never undertake the impractical task of providing some total account of habits of mind characteristic of a paradigm. There is not some in-principle unbounded scope to the enterprise.

A subtle but essential point here concerns the distinction between a rule or axiom (such as that heavenly motions must be circular, or compounded of circles) and the habit of mind that makes that rule seem beyond questioning. The latter, I try to show, is always linked to experience in the world, allowing that experience in the world includes work with the computations, diagrams, and other intellectual and physical apparatus peculiar to the theory at issue. It is important to distinguish between the intuitive sense that grows out of experience and the merely verbal (however useful and unavoidable) approximation of that habitual sense of things.

But relative to physical habits, habits of mind are particularly likely to go unnoticed. What we can observe are only reports of intuitions, which ordinarily do not directly reveal the habits of mind that prompt them. The socially-shared habits of mind with which we are particularly concerned here may be even harder to discover. To people within the community, what will prove to be critical for the barrier analysis may be intuitions that seem too obviously right to prompt discussion. They are in the realm of what "everybody knows" and takes for granted. The important role that thought experiments have played in the history of science — always proposed by individuals who wish to challenge prevailing views — is as a device for prompting people to notice that something totally taken for granted could be wrong (Kuhn 1977).

But this shows that the critical habits of mind are unlikely to be superficially obvious, hence routinely discussed in accounts of the events. But it is not hard to see how to proceed: namely by close examination of what was being taken for granted, what persistently ignored prior to the paradigm shift but not so afterward: in particular, where did good arguments or evidence seem to be missed or dismissed; where did bad arguments seemed to go unchallenged?

Summing up: Entrenchment in particular habits of mind shared across an expert community is (on the argument here) exactly what defines operating within some paradigm, tacitly guiding key intuitions within the community, facilitating communication and many aspects of constructive work, but also constraining what can be seen as making sense.

3. Gap versus barrier views

The barrier argument (as against what is at least implicit in more familiar views) suggests a parallel with views of adaptation before and after Darwin. The most natural intuition is that what looks like a set of arrangements to serve some function implies some agent or pressure which intends or forces that set of arrangements. Before Darwin, no naturalist had ever really escaped that sense of things, usually made explicit in the intuition that good design implies a designer. But on the Darwinian account good design can emerge by a process of natural selection in which there is no design. Design can appear as the result of a process which itself intends nothing.

The parallel I want to suggest for paradigm shifts takes the extremely natural view that the cause of difficulty in understanding a new position, or difficulty in reaching it, is some logical distance between the two positions: what I will call the gap. On that view, we could think of a revolutionary paradigm shift as a shift requiring an intrinsically large logical discontinuity between the theories or practices characteristic of the new and old paradigms.

Alternatively, though, we can consider the quasi-Darwinian possibility that paradigm changes which provoke the symptoms of a Kuhnian revolutionary episode are just those where there happens to be an important cognitive *barrier* between the new and old paradigms. If that is the right story, then we can get a cognitive crisis even if there is no intrinsically difficult logical distance between the two. Contrariwise (on this view) even when there is a substantial logical distance, the transition can go smoothly (the discovery occurs soon after the knowledge needed is available, and its contagion across the relevant community is rapid) because there happens to be no habit of mind sufficiently robust in the context to constitute a strong barrier. We get what can be articulated as a "normal revolution", in contrast to the cognitively stressful characteristics of a Kuhnian revolution.

In the full study, this point is particularly evident in an analysis of the overthrow of phlogiston. Logically Lavoisier's initiative can be seen as merely an episode in the larger story of the rise of pneumatic chemistry. The latter required that what had always been a substance (air) now be seen as a category (today, gases), and that the varied substances within this category play just as essential a role in chemical processes as until then been allowed only for solids and liquids. In contrast, as has often been pointed out, Lavoisier's new theory of combustion can be assimilated to Stahl's phlogiston theory by simply defining phlogiston to be negative oxygen. Logically, consequently, the former transition involves a conceptual shift far more drastic than the latter. But in the view of not only Lavoisier but of Priestley and other adversaries, it was the overthrow of phlogiston that was revolutionary, while the radical conceptual shifts of that came with the rise of pneumatic chemistry were absorbed with remarkably little controversy.

This distinction between "normal" and "Kuhnian" revolutions (those which reveal the symptoms of a cognitive crisis vs. those which do not) relates to the more general distinction I am drawing of *gap* vs. *barrier* views of paradigm shifts. In terms of the gap view, the question of whether a scientific development is worth describing as a paradigm shift depends on the extent of the conceptual change required for its assimilation. But on the barrier view, major conceptual change is only a correlate, not a defining or necessary feature, of the cognitively stressful developments Kuhn has taught us to think of as involving paradigm shifts.

On the gap view, the logical distance separating a prevailing view and some conflicting new proposal may be comfortably narrow, so that a person can just step across it. But wide gaps require a leap. Very wide gaps require an exceptional good jumper and exceptionally favorable conditions for the leap to be feasible. Wider gaps still are beyond the reach of any plausible human leap. If the new idea is reachable at all it will only be later in the history of the science, when a paradigm has developed which does not require so forbidding a leap. On this view, once we know the prevailing theories before and after a paradigm shift, a judgment about the size of the logical gap will explain why a dramatic transition (a leap) was required to get across in some cases, but only a normal science evolution from earlier ideas in others. Although simplified, I think this is close to the implicit picture usually held. Thagard (1990) provides a particularly explicit 'gap' account of the overthrow of phlogiston as a remapping of networks of rules, taxonomies, and so on. This turns on a very different (and more familiar) view of how the overthrow proceeded than I am led to on the barrier view. And the record, I try to show, supports the barrier view, not the "usual story" of secondary accounts.

In a 'gap' account, habits of mind play no essential role. Whether a particular transition is revolutionary or not depends on how different the rival theories are, how extensive a remapping is required to go from one formal network to another, and so on. From the 'gap' view, there may not be much in Kuhn's notion of paradigms beyond a single word for the complex of theories, equipment, and procedures as described in a textbook. Talk of incommensurability is likely to seem just puzzling.

But now consider an alternative which more congenial to the habits of mind claim. Here the critical problem for a revolutionary paradigm shift is not some necessarily difficult logical gap that needs to be leaped (though that might be present), but the robustness of the most severe cognitive barrier that blocks the path to discovery. Ordinarily, on this view, anyone who can get over the barrier can get across the gap. Hence the critical focus is not on a comparison of theories ex ante vs. ex post, and from that to judge the size of the logical gap. Rather, the critical focus is within the gap, where the primary question concerns identifying the critical barrier within that gap. Of course the bigger the gap, the more room for a really forbidding barrier somewhere along the discovery path. But here the size of the gap is only a correlate of the cognitive difficulty of the transition, not the cause of it, nor necessarily even a well-marked correlate. The psychological character of the barrier turns essentially on the difficulty of breaking an entrenched habit of mind.

Of course a person might also have difficulty related to habits because he is missing some critical facilitating habit as well as by entrenchment in an incompatible habit. Correspondingly, therefore, a paradigm shift might turn on the emergence of a new habit of mind rather than on the breaking of a deeply entrenched one. So although the argument here is framed in terms of the latter case, it implies the converse case as well. I may be unable to hit a certain kind of tennis shot, or to see the point of a particular sort of argument, for either reason. Hence the barrier might take the inverse form of a missing facilitating habit. I give a number of concrete applications in the full study, but the main one for a positive barrier is the new account of the overthrow of phlogiston, the main illustration of a missing facilitating habit concerns the emergence of probability.

In either case when we see a shift which shows the Kuhnian symptoms (delays in seeing a possibility that is logically available, incomprehension or indifference among its first audience to arguments that eventually come to seem hard to resist) we expect to be able to identify a particularly important habit of mind, and to be able to give an account of how it came to arise and how it came to be overcome. We are prepared to find that even when the symptoms of delay or incomprehension are striking that nevertheless the steps necessary to get from one the old theory to the new might be logically easy. That was true for the transition from Ptolemaic to Copernican mathematical astronomy, as Neugebauer and Price have argued, and as I try to show in a more transparent way in the study. The same holds even more transparently for the two other major cases in the new study (concerning phlogiston and probability).

Hence, summing up the barrier claim: There may exist a habit of mind (or several such) which yield intuitions incompatible with some novel idea, where the novel idea is one that once grasped, turns out to be very powerful. Call that a strategically located habit of mind. And occasionally, for some subset of strategically located habits, the relevant habit could not only be strategic but also highly robust, so that escaping it is difficult even after some striking anomaly challenging that habit had come on the scene.

A habit of mind that meets these conditions for a particular case (that is: a habit of mind that is both robust and strategically located) I will call a barrier for that case. Of course habits of mind far more often block ideas which would not only look worthless or perverse to someone entrenched in those habits, but would in fact be worthless or perverse. Once again, as with physical habits, ordinarily habits are the key to fluent and efective expert performance. It requires peculiar circumstances for the reverse to hold. The whole point of the Darwinian parallel is to notice that the reverse cases, like favorable mutations, though rare, must occasionally occur. And like favorable mutations, habits of mind that are at once both adversely stubborn and strategic, even though rare, will play a major role in how things develop.

4. The uniqueness argument

The stronger claim I will finally introduce is that characteristically, the barrier is *unique*, which readily prompts objections along these lines:

Even if habits of mind in fact play the constitutive role urged here, why should it follow that some particular habit of mind will ordinarily be critical for a particular paradigm shift? For it must be a network of interacting habits of mind that characterizes a paradigm, hence a network of habits that must change. So why not suppose that the empirical anomalies or whatever else serves to challenge what had been taken for granted might develop at a number of different points in the network, or in several more or less at once, with no particular habit of mind playing a uniquely critical role?

And even if we are only concerned with what happened in some particular historical case (not with abstract might-have-beens), why not even then multiple points on which the discovery proceeded? And yet again, even if there was a single point of initial escape from the commitments of the prior paradigm, might that not be something merely idiosyncratic for the particular discoverer in particular circumstances: hence something which even if identifiable might be of no real significance for understanding the development of science, or for understanding more generally the emergence of radically novel ideas?

Summing up these objections using an analogy from *Patterns*: If discovery is like finding a path to the top of a mountain, why suppose that any feasible route would encounter a certain barrier, or that one particular point of difficulty on the path actually taken (even if that was somehow the only path that might have been taken) was uniquely the barrier that was critical?

As usual in such a situation, the complimentary components of an answer are a conceptual argument for the plausibility of the claim, then an attempt to show successful application of the argument. The full study, of course, seeks to provide both. But here I can give only a shortened version of the conceptual argument.

Any anomaly must weaken a theory, and make further challenges to the theory easier, and perhaps make previously invincible intuitions visibly problematical. But in the language I used in *Patterns*, ordinarily anomalies are tamed and after a period are no longer capable of prompting serious doubts about the basic theory. Consequently, even what logically seems to be a striking anomaly will still have only a limited window of opportunity in which it might be fruitfully exploited. It is not hard to find striking illustrations of this propensity in the historical record, as with various tamed anomalies for Ptolemaic astronomy mentioned in *Patterns,* p. 263, or as I show holds for the increase of weight of calxes in the forthcoming Lavoisier analysis. It is that powerful propensity to tame anomalies that makes it plausible to suppose there is one identifiable habit of mind newly challenged (or in the converse case, some facilitating habit of mind which emerges just prior to this discovery) which is ordinarily critical for putting a radical novelty within reach.

For phlogiston — a particularly easy case for exposition, since the relevant habit of mind is rooted in everyday experience (not expert knowledge) — we are all familiar with seeing flames emerge from the burning fuel, and with the gradual disappearance of the fuel as this process continues. If we look to the history of ideas about combustion (as in Gregory 1934) it is clear that this "phlogistic" intuition — under other names of course — long predates Stahl's elaboration of that idea, and indeed can be traced back to the earliest recorded speculations about the nature of fire. Somehow fire is congealed in combustible substances, and the flames we see leaping up during combustion reveal the escape of this congealed matter of fire.

In other cases (for example, the Copernican case, as developed in detail in *Patterns*), the critical habit of mind is not so obvious. Nevertheless, examination of the history can be expected to reveal a plausible candidate. For if such a barrier exists, it must leave evidence of its presence in a puzzling inability of those involved to see possibilities that are logically easily available, and which (after the transition) no one seemed to be capable of missing.

But when the tendency to tame anomalies and the possibility of strategic and stubborn habits of mind are considered together, the conjecture that a barrier will ordinarily be unique becomes hard to avoid. Since habits are rarely either quickly made or quickly broken, much repetition is characteristically required to displace an entrenched habit. But anomalies eventually lose their bite. If beyond a momentary escape from the barrier habit of mind, there is only a further barrier — not a striking novelty that motivates the repeated effort required to break a deeply entrenched habit — then the opportunity to go further will ordinarily be lost. The anomaly will be tamed, rather than the habit broken. But given that point, it becomes almost tautological that, if the barrier argument in general is correct, then the stronger claim will also be true: that a characteristic of revolutionary discovery is that there will ordinarily be some unique critical barrier.

Putting this argument in another way: Just what we mean when we say a habit of mind was a barrier in some context is that if given a chance (if not blocked by that barrier) some idea incompatible with that habit of mind would work in strikingly effective ways. An anomaly creates a window of opportunity for established ideas to be challenged. But over time anomalies are ordinarily tamed. And habits are not broken abruptly, any more than they are made abruptly: repeated experience is required to break a habit even after some transient circumstances has allowed a momentary departure from the habitual intuition. Hence unless an escape from some habit of mind quickly reveals further ideas or observations sufficiently remarkable to assure that the individual will be highly motivated to repeat the experience (to try to glimpse whatever was striking again, ponder its significance, think about how it might make sense), then the episode is likely to be transient, so that nothing deep happens until some new anomaly arises, or new circumstances bring renewed salience to what had become a tamed anomaly.

So suppose we have a case in which we can point to more than one candidate for barrier: the Copernican case treated in detail in *Patterns* provides a splendid example. For here the obvious candidate for barrier — the profound intuitive sense of the Earth as fixed and stable — does not in fact turn out to be the barrier. Instead the critical habitual sense of things turns on a way of seeing the structure of the world that could be acquired in the deeply entrenched and entangled way we need only through the practice of a thoroughly expert Ptolemaic astronomer: what I called in *Patterns* the "nested-spheres" sense of the structure of the world. But when multiple potential barriers are involved on the way to some discovery, they will always be to some extent interdependent: if one were broken or even weakened the other would become more vulnerable. Hence (for two salient candidates for barrier) the main possibilities following a breach of the first barrier become these:

If there are breakthrough effects, the second apparent barrier *may* now have lost much of its potential to block the new idea. On the account worked out in *Patterns,* that occurs in the Copernican case. Or if there are no breakthrough effects, the anomaly will be tamed. Then both barriers would remain in place. Or (the final main possibility) if there are breakthrough effects but the second barrier remains strong, then the breakthrough effects are themselves remembered as a revolutionary discovery, with the eventual fall of the second barrier seen as a second revolutionary episode. In Kuhnian language, we would have a striking paradigm shift with the first breakthrough, and another later on with the second — as with the fall of Ptolemaic astronomy after Copernicus, then decades later, the replacement of the Ptolemaic technical apparatus by Kepler's ellipses.

But for all three possibilities, although multiple habits of mind may be critical for some idea to fully emerge, we still have, for each episode of cognitively radical discovery, one critical barrier

Summing up in Lakatosian language, but with a meaning substantially different from the rational reconstructionist sense Lakatos had in mind: The normal response to anomalies is some change in the protective belt not a disruption of the hard core. For anomalies are fragile, gradually losing their power to shock, hence to incite interest in radical alternatives to what has become the accepted view of things. The dynamics of the process consequently almost assure that to the extent that an episode is well-marked by Kuhnian symptoms of incommensurability, then some particular habit of mind is likely to be uniquely crucial for that episode. The extension of the Copernican analysis and the additional case material developed in the new study, I try to show, demonstrates a striking capacity of this line of argument to yield fresh insight into some of the most intensely studied episodes in the history of science.

Note

[1] Support from the Program in History and Philosophy of Science, National Science Foundation is gratefully acknowledged. I am indebted to Tom Nickles for providing the occasion to write this summary, and (without implying agreement with the result) I am indebted to Nickles, David Hull and Thomas Kuhn for valuable comments.

References

Giere, R. (1988), *Understanding Science.* Chicago: University of Chicago Press.

Gregory, J.C. (1934), *Combustion from Heracleitos to Lavoisier.* London: Arnold.

Kuhn, T. (1962), *The Structure of Scientific Revolutions.* Chicago: University of Chicago Press.

_____. (1977), "A Function for Thought Experiments, in *The Essential Tension.*" Chicago: University of Chicago Press.

Margolis, H. (1988), *Patterns, Thinking & Cognition.* Chicago: University of Chicago Press.

_____. (1991). "Tycho's System and Galileo's Dialogue", *Studies in History and Philosophy of Science* 22: 259-275..

_____. (forthcoming). *Paradigms and Barriers.* chicago: University of Chicago Press.

Neugebauer, O. (1968), "On the Planetary Theory of Copernicus", V*istas in Astronomy* 10: 89-103.

Thagard, P. (1990), "The Conceptual Structure of the Chemical Revolution". *Philosophy of Science* 57:183-209.

Barriers and Models: Comments on Margolis and Giere

Nancy J. Nersessian

Princeton University

Giere's paper calls for nothing less than a "paradigm change" in philosophy of science and provides a overview of what developments in cognitive science might contribute to a new philosophical theory of the growth, development, and change of scientific knowledge. Specifically, he challenges the position - still dominant in postpositivistic philosophy - that the basic representational device of the scientist is a linguistic structure. Rather, he proposes that the basic device is a "model" or "picture" by which I take him to mean some form of structural analog to the events, entities, or processes under investigation. I am in basic agreement with his analysis and with his assessment of current cognitive approaches to analyzing science; especially with his contention that, at least for the present, cognitive psychology has the most to offer a cognitive philosopher trying to understand how models are constructed and function within a scientific community.

As Giere points out, I have for some time been investigating in what ways joining philosophical analyses with those in the cognitive sciences might contribute to a more refined understanding of the nature and processes of scientific development and change. As part of this investigation, I have proposed a new interdisciplinary method - "cognitive-historical analysis" - as an approach to examining the actual practices of scientists who have created and changed representations of nature. The premises that form the basis of that proposal are: (1) the temporal perspective of history is needed to understand development and change in science and (2) the representational and problem-solving practices of scientists are outgrowths of ordinary cognitive resources. I see Margolis' analysis as fitting into the cognitive-historical genre, and thus want to focus my commentary on it. While his analysis is thought provoking, I believe it is fundamentally mistaken as an account of scientific change. And - more importantly for this session - it provides the wrong model for how a cognitive philosopher of science should investigate the nature and processes of scientific change.

Margolis' analysis is mistaken primarily because it gives no role to the scientist as the problem solver who brings about change. It provides the wrong model for a cognitive philosopher because it selectively and uncritically adopts data and speculative hypotheses from the cognitive sciences, gives more weight to these findings and interpretations than the state of these investigations warrant, and tries to fit cases of scientific change into the cognitive accounts. What is wrong with this approach is that it fails to take into account that the cognitive sciences are themselves in an embryonic state. There are no fully-developed theories one can simply take off the shelf and apply to science. In fact, at present, cognitive theories are largely uninformed by sophisticated scientific practices, making the fit between them and cognitive theories something that still needs to be determined. As Giere points out, a major task facing cognitive philosophers is to forge an interpretative framework adequate for understanding scientific change from cognitive analyses coupled with examinations of scientific practices. Margolis' project seems, rather, to be that of fitting the historical data to speculative hypotheses developed from work in the cognitive sciences.

Margolis' purpose is to provide a general model for how paradigm shifts occur in science. He takes as premise the claim of his recent work that all cognition reduces to pattern recognition. This claim emerges from: (1) an argument for how the mechanism of pattern recognition could have evolved in the brain and (2) evidence from psychological studies that humans are bad at drawing logical inferences from propositional representations and quite good at such things as perceptual inferences from diagrams and reasoning via mental models (non-propositional structural analogs of entities, events, and processes). Although I am not evaluating that work, the evidence he cites does not support the reductionist claim: all the empirical studies show is that pattern recognition is an important aspect of human cognition. I point this out here because Margolis' tendency is to generalize beyond what even cognitive scientists would see as the implications of their work. From the premise that all cognition reduces to pattern recognition he develops the thesis that scientific discovery can be explained in terms of the breaching of a cognitive "barrier" that allows recognition of a new pattern.

The "barrier" mechanism of scientific change works as follows. Margolis maintains that operating within a paradigm is exercising a "habit of the mind". Here he is drawing from work in cognitive science that treats the question of how knowledge is represented in the mind. Some theorists have argued that knowledge is represented in organized units variously called "schemata", "scripts", or "frames". The differences among these need not concern us here. The main claim is that specific organizational units are activated when we encounter or think about a situation or problem. Margolis' "habits" are reflex-like ways of understanding that arise from the automatic employment of the knowledge units available to a scientist working in specific paradigm. Additionally, he claims that among these "habits" there exists a unique barrier that prevents us from seeing another available and better pattern that is waiting to be discovered once the barrier is "breached". In his central example, he maintains that the Copernican hypothesis "lay on the table with no one able to see it" for 1400 years. "All the necessary information had been available" to anyone who would have been able to breach the nested spheres barrier of the Ptolemaic framework.

It is important to keep in mind that Margolis' intention is to use findings in the cognitive sciences to give more explicit formulation to some basic Kuhnian insights. Specifically, Kuhn's claim that doing normal science requires working under the exemplars it provides becomes, in Margolis' analysis, the claim that normal science involves exercising habits, i.e., automatically employing the specific schemata arising from the paradigm. These habits condition the expert's deepest intuitions employed in understanding and making judgments and present an obstacle to change. The incommensurability - or failure of communication - Kuhn maintains occurs in the initial stages of a scientific revolution - is, here, due to the psychological "fact" that those deeply entrenched in existing habits find it difficult to breach the barrier and are thus unable to recognize the new pattern even when it is pointed out to them. Kuhn himself offers the metaphor of the "gestalt switch" as a way of understanding these phenomena. It is difficult to see what Margolis offers in his analysis that goes beyond the "gestalt switch" metaphor in characterizing revolutionary change. Just as with that metaphor, Margolis' account makes recognizing a new pattern appear to be something that simply happens to scientists rather than the outcome of an extended period of problem solving and construction by scientists.

This brings us to the heart of the problem. Margolis promises new insights into "the evolution of radically new ideas"; i.e., into the mechanisms of representational change in science. But his mechanism - breaching a cognitive barrier - gives a relatively minor role to the actual activities of the cognitive agents: the scientists. While I agree with him that a radical social constructivist approach cannot explain how new representations arise and become viable competitors to accepted views, the proper response is not to take all construction out of scientific change. There is much historical research to support the view that new scientific representations are constructed in a problem-solving process and that the scientists involved in this process employ a wide range of cognitive strategies for solving problems. Some things Margolis says indicate that he might agree with this, but his central thesis that there is a unique cognitive barrier in every paradigm and that the new patterns are out there waiting in the wings to be seen once this is breached - available, e.g., in the Copernican case for 1400 years - gives no role to the specific problem situation of the scientist in creating the new paradigm.

That seeing a new pattern lies at the heart of major scientific change is obvious. To say scientific change involves seeing new patterns neither explains anything nor gives any insights into the evolution of new ideas. We still want to know why, e.g., Copernicus saw a pattern that Ptolemy failed to recognize. Historically it is simply wrong that nothing had changed in 1400 years and that Copernicus had nothing available to him that was not available to Ptolemy. Among other factors, numerous technical problems had been encountered in those 1400 years and, significantly, the equant that Ptolemy had introduced as the center of motion of the system was seen by Copernicus as conflicting with the neo-platonic desideratum of uniform motion about the geometrical center of a planet's orbit. As a result, Copernicus' framing of the problem situation differed significantly from that of Ptolemy. Contrary to Margolis' analysis, a new pattern only becomes available when there is a set of problems to which it seems to offer a solution and whatever might constitute a cognitive barrier in a situation is relative to a specific framing of the problem situation. The new pattern is constructed as a solution to the new framing of the problem and it is the structure of this framing that enables specific patterns to emerge as possible solutions.

Since Margolis aims to provide a general account of scientific change I will underscore this point by way of different historical case. What can Margolis' analysis provide in the way of illumination about why Einstein saw the pattern of the special theory of relativity that many of his contemporaries came so close to but failed to recognize? Margolis' response should be that he was able to breach the aether barrier. But this response does not explain why Einstein was able to do this and how the new pattern emerged. My answer would be: Einstein framed the problem situation in a fundamentally different way. The new framing came from a process of reflection on and reassessment of the foundations upon which classical mechanics and electromagnetism rested. His framing of the problem situation included: (1) the failure of attempts to reduce electromagnetism to mechanics and vice versa; (2) Planck's analysis of black-body radiation, which indicated to him that both electromagnetism and mechanics are inadequate in regions small enough for fluctuation phenomena to count; and (3) specific problems with Lorentz' electron theory. Relative to this framing of the problem situation, Einstein claimed the solution was to put electromagnetism and mechanics on an equal footing. In this case the aether becomes "superfluous", i.e., the barrier, and the special theory of relativity, the new pattern.

There are several historical puzzles Margolis' framework would need to explain relating to this case. One puzzle is that Faraday - with a quite different framing of the problem situation - recognized and breached the aether barrier. Yet, this did not lead him to the pattern of the special theory of relativity. Faraday's field theory is not that of Einstein. Another is that Lorentz recognized the aether barrier (and understood the special theory) but refused to breach it. Among other reasons, for him, a satisfactory explanation of the contraction of moving rods had to include a causal explanation for the phenomenon and this involved understanding the interaction of the aether with the molecular forces in the rod. Historical data such as these cannot be accommodated by Margolis' model of scientific change.

In sum, I find the model of scientific change presented by Margolis offers little insight into the processes of representational change. This being said, I do think the cognitive sciences provide a valuable resource for philosophy and history of science. In particular, the cognitive sciences can contribute to inquires into the dynamic forces through which a vague speculation gets articulated into a new scientific theory, gets communicated to other scientists, and comes to replace existing representations of a domain. As Giere points out, ongoing investigations in cognitive psychology are contributing new insights into the agency of individual scientists in constructing new representations of nature. These will aid philosophers in their attempts to construct new theories of development, evaluation, and choice of scientific theories. They will also enrich historical analyses of specific instances of scientific creativity, whether the innovation led down a dead end, to "winning" science, to the same point via different routes, etc.

Putting emphasis on the individual scientist, though, by no means requires a "new-fashioned" internalist analysis, as critics such as Fuller have claimed. Nature may endow the individual with specific cognitive mechanisms, but nurture influences acquisition, development, and use of cognitive strategies and conceptual resources. Investigations are underway as to how "situated" cognition is and also to what role interests and motivations play in cognition. Historical and sociological research over the last twenty years has shown us that there is no denying the importance of the facts that scientists are situated within scientific communities and wider social contexts and that they acquire cognitive strategies and conceptual resources from both of these contexts. Indeed, what makes cognitive-historical analysis so attractive is its potential to provide a much-needed synthesis of the cognitive and the social dimensions of knowledge construction.

Some Twists in the Cognitive Turn

Steve Fuller

Virginia Polytechnic Institute and State University

1. The Rhetoric of Cognitivism

Contrary to what the name suggests, the recent work of Howard Margolis (1987) and Ronald Giere (1988) demonstrates that the "cognitive turn" in the philosophy of science is *not* simply the application of cognitive science to the study of science. For one thing, neither one is what Jerry Fodor (1981) has called a "methodological solipsist," that is, someone who wants to account for thought processes without presupposing an account of the world of which those thoughts are about. Margolis, in fact, comes perilously close to Fodor's anti-cognitivist foe, J. J. Gibson (1979), whose "ecological" perspective requires that an organism's thought processes be specified in terms of structures in the environment, "affordances," capable of satisfying the organism's desire to know. The Margolian focus on overcoming "barriers" to alternative "habits of mind," as the psychic basis of Kuhn's paradigm shifts, only serves to highlight this broadly "functional" side of thinking that typically has no place in cognitive science. Although more methodologically electic than Margolis, Giere is also more the solipsist in suggesting that the interesting story to be told about how conceptual change occurs— one which requires the resources of the neuroscientist, psychologist, and computer simulator—transpires entirely within the head of the individual scientist. Consequently, Giere says virtually nothing about the character of the environment (natural or social) such that conceptual change proceeds as it does. The big difference, of course, is that whereas Fodor and some cognitive scientists still suppose that solipsism implies apriorism, or at least innatism, Giere does not.

In what sense, then, are Margolis, Giere, and many of the people they cite taking a *cognitive* turn? I would say that it is in a quite conservative sense, insofar as their turn is biased toward findings and interpretations that support the image of the scientist as a competent, largely self-sufficient human agent. Consequently, they downplay research pertaining to the cognitive limitations of individuals, especially the failure of individuals to appreciate the context-dependence, and hence global inconsistency, of their thought and action. Moreover, our cognitivists underestimate the cognitive power that is gained via group communication and technological prostheses. But in the course of displaying these biases (cf. Fuller 1989, for the opposing biases), the cognitive turn has brought to light important philosophical issues—"metaphysical" ones, as Giere himself now suggests—that previously eluded philosophers of science (cf. Fuller forthcoming). They pertain to the bearers of scientific properties: Where in the empirical world do we find knowledge, theories, rationality, concepts?—to name just four philosophical abstractions hitherto left in ontological limbo. Margolis and Giere are clear about arguing for the individual scientist as the relevant locus. Their focus is "cognitive" in the familiar sense of being concerned more with the individual's thought *processes* than with the *products* of her thought. That is probably because neither challenges the idea that these processes produce the sorts of things that more traditional philosophers of science would regard as having "cognitive content," such as theories. As a result, while the cognitive turn tends to give us a full-blooded sense of what theorizing is like (e.g. a pattern of neural activation), we are still left with a rather pale, abstract sense of what theoretical output is like. For example, I suspect that different "styles" of theorizing radically underdetermine the types of theoretical texts that are written, yet it was those texts that initially led philosophers and historians to believe that there was something cognitively special about science.

2. Some Mistaken Identities

My own perspective on the cognitive turn in the philosophy of science is very much like Marx's on the capitalist turn in the history of political economy. In capitalism, relations among people are mistaken for properties of things. What Marxists mean by this claim is that goods do not have an inherent value, or natural price, but only an exchange value that is determined by the social relations among the capitalist, worker, and consumer. Likewise, I believe that, in its attempt to locate abstractions in the empirical world, cognitivism mistakenly identifies (1) rational reconstructions for actual history, (2) properties of groups for those of individuals, (3) properties of language for those of the mind, and (4) properties of society for those of nature. I will consider each in turn.

(1) Like Piaget's genetic epistemology, the Margolian account of paradigm shifts as the overcoming of cognitive barriers is more pedagogy than history of science. In other words, teachers could use Margolis to get students to see beyond the shortcomings of their current framework to a more comprehensive one—but only once that next stage of comprehensiveness has already been achieved by the scientific community. His is a method for meeting standards rather than setting them. Margolis' confusion here probably stems from his insensitivity to the normative dimension of Kuhn's account of scientific revolutions. In particular, unlike the way it is used in politics, where it makes sense to speak of "failed revolutions," all of Kuhn's revolutions are success stories. That is, the only cognitive changes that he recognizes as "scientific revolutions" and "paradigm shifts" are the ones that moved scientists closer to our current paradigms. Beyond that, Kuhn has little to say about how such revolutions occur, for that would involve accounting for a variety of individuals, most with interests quite distinct from those of the original revolutionary, who nevertheless found that person's work of some use for their own. Thus, Margolis mistakes reconstructed history for the real thing because he typifies in one individual a process that is better seen as distributed across a wide range of individuals.

(2) This last point is worth emphasizing, as it brings into focus the simplistic sociology that often informs the cognitive turn. Kuhn is more to blame here than either Margolis or Giere— especially his tendency to characterize scientists as having a common mindset or worldview, which, in turn, makes it seem as though, for a given paradigm, once you've seen one scientist, you've seen them all. Sociologists regard this typification of the group in the individual as a methodological fallacy, the "oversocialized conception of man" (Wrong 1961). The problem with the conception is

that, in attempting to account for the social dimension of thought, it actually renders the social superfluous by ignoring how the interdependence of functionally differentiated individuals makes it possible for a group to do certain things that would be undoable by any given individual. Philosophers are prone to an oversocialized conception of people because of bad metaphysics. They tend to treat a part-whole relation as if it were a type-token one: to wit, society is an entity that emerges from the arrangement of distinct individuals, not a universal that exists through repeated instantiations. Indeed, I am inclined to think that the signature products of cognitive life—knowledge, theories, rationality, concepts—are quintessentially social in that they exist only in the whole, and not in the parts at all. For example, it is common for cognitive psychologists to treat conceptual exemplars, or "prototypes," as templates stored in the heads of all the members of a culture, when in fact they may be better seen as concrete objects that function as public standards in terms of which the identities of particular items are negotiated (cf. Lakoff 1987). It may well be that each party to such a negotiation has something entirely different running through her mind, but their behaviors are coordinated so as to facilitate a mutually agreeable outcome.

(3) Continuing in the spirit of the last remark, if one is looking for an account of the brain that starts with minimal common capacities and then builds up quite different neural networks, depending on an individual's experience, one need look no further than the promising array of parallel distributed processing (PDP) models. However, contrary to what Giere seems to think, I believe that the extreme context-sensitivity of PDP models implies that whatever sustained uniformity one finds among members of a scientific community *cannot* be due to any uniformity in their thought patterns, but rather to some uniformity in the public character of their behavior, especially the language in which members of that community transact business. (In fact, that might be the *point* of scientific language.) For, if PDPers are correct about the variety of neural paths that can lead people to say, do, and see roughly the same things, then I take that to be an argument for the nervous system *not* providing any particular insight into the *distinctiveness* of science as a knowledge producing activity. (Of course, PDP would still say a lot about "how we know the world" in the looser sense of successfully adapting to the environment.) But even if one were to find this conclusion outlandish, it remains to be seen whether the cognitivists have a story to tell about scientific communication, the means by which findings are ultimately judged to be normal, revolutionary, or simply beside the point. From works such as Nersessian (1984), which Giere cites approvingly, it would seem that communication is the process by which a later scientist reproduces an earlier scientist's thought processes in order to continue a common line of research. However, if thought is as context-sensitive as PDPers suggest, then it is unlikely that this story could be literally true—especially if the relevant thought processes are defined in terms of what we now, only in retrospect, regard as a "common line of research." And even if a later scientist wanted to pursue an earlier scientist's work, it is not clear that either her means or her motives would involve the reproduction of that work (cf. Wicklund 1989). My guess is that the "concept maps" and other heuristics that cognitivists elicit from scientific texts are more formal analyses of scientific rhetoric that conveyed the soundness of the scientist's work than representations of "original" scientific reasoning that readers followed step-by-step in their own minds. This is by no means to demean the accomplishment, but simply to put it in perspective.

(4) Finally, perhaps the grossest sociological simplification behind the cognitive turn may be termed its "visually biased" social ontology: to wit, social factors operate only when other people are within viewing distance of the individual; if no one is in the vicinity, then the individual is confronting nature armed only with her conceptual wiles. The solitary laboratory subject working on psychological tasks—the

source of much of Giere's evidence—certainly reinforces this image, but Margolis is the bigger offender in failing to see that cognitive patterns are memories of socially framed experiences, which are resistable and replaceable only in socially permissible ways. The project of altering one's point of view, not merely for the sake of entertaining the alternative, but for making the alternative the basis of one's subsequent research, involves the simultaneous calculation of what philosophers have traditionally called "pragmatic" and "epistemic" factors. This serves to bind "the social" and "the natural" in one cognitive package that cannot be neatly unraveled into, respectively, impeded and unimpeded thought processes. Relevant to this point is the *Machiavellian Intelligence Thesis*, recently proposed by two Scottish animal psychologists (Byrne & Whiten 1987). They argue that cognitive complexity is a function of sociological complexity, such that the organisms which respond to environmental changes in a less discriminating fashion tend to be the ones with a less structured social existence. One conclusion that Byrne and Whiten draw is that the complexity of nature distinctively uncovered by science may be little more than a reflection of the combination of people who must be pleased, appeased, or otherwise incorporated before a claim is legitimated in a scientific forum. A more simply organized science would, then, perhaps reveal a simpler world.

References

Byrne, R. and Whiten, A. (eds.) (1987), *Machiavellian Intelligence*. Oxford: Oxford University Press.

Fodor, J. (1981), *Representations*. Cambridge MA: MIT Press.

Fuller, S. (1989), *Philosophy of Science and Its Discontents*. Boulder: Westview Press.

Fuller, S. (forthcoming), "Naturalized Epistemology Sublimated," *Studies in History and Philosophy of Science* 22.

Gibson, J. J. (1979), *The Ecological Approach to Visual Perception*. Boston: Houghton Mifflin.

Giere, R. (1988), *Explaining Science*. Chicago: University of Chicago Press.

Lakoff, R. (1987), *Women, Fire, and Dangerous Things*. Chicago: University of Chicago Press.

Margolis, H. (1987), *Patterns, Thinking, and Cognition*. Chicago: University of Chicago Press.

Nersessian, N. (1984), *Faraday to Einstein*. Dordrecht: Nijhoff.

Wicklund, R. (1989), "The Appropriation of Ideas," in *Psychology of Group Influence*, P. Paulus (ed.). Hillsdale NJ: Lawrence Erlbaum, pp. 393-424.

Wrong, D. (1961), "The Oversocialized Conception of Man," *American Sociological Review* 26: 184-193.

Part XII

THREE VIEWS OF EXPERIMENT

Allan Franklin, Right or Wrong

Robert Ackermann

University of Massachusetts, Amherst

I regret to inform you that Allan Franklin is unable to be here because of the consequences of his collision with a truck in Boulder, Colorado several weeks ago. The three of us have decided to proceed with the symposium in his honor, even though it is now missing its fulcrum. The original point of the symposium was to have an informed discussion of two versions of atomic parity-violation experiments, versions that embody opposed philosophical conceptions of the experiments. The first conception is embodied in Andy Pickering's account in his *Constructing Quarks,* an account that is explicitly criticized by Allan Franklin's more recent discussion in his *Experiment, right or wrong* (Pickering 1984; Franklin 1990). The symposium would have brought this confrontation into focus, with Franklin's presentation of his critique followed by Pickering's rejoinder at this symposium, both of them in sufficient command of the detailed history of the atomic parity-violation experiments to allow for the possibility of a useful exchange of differences of opinion. In Franklin's absence, Pickering commands the field of relevant scientific detail here by default, and we are reduced out of courtesy to the missing position to more general issues concerning normative and constructionist accounts of experimentation. After we have spoken briefly in turn about these more general issues, we will have whatever discussion may be provoked from the floor before ending the session.

In Allan Franklin's absence, I have taken on the task, not just of chairing the symposium, but of presenting his views sufficiently so that the papers of Pickering and Lynch have a semblance of live context. Franklin's new book is a continuation of the historical studies he began in his earlier *The Neglect of Experiment* (Franklin 1986). The point of this work (in conjunction with well known studied by Latour and Woolgar, Pickering, Lynch, Hacking, Knorr-Cetina, and others) is to rescue the notion of experimentation that seems so crucial to an understanding of science from the disembodied form that it took in older empiricisms where experimentation was regarded as a mechanism for producing data regarded as factual assertions that could be used to test the truth claims of theory. As these recent and quite varied detailed studies of experimentation show, experimentation is a complicated concrete process culminating in data that are often subject to different interpretations, that is, data that are not as factlike as the older empiricisms had assumed. Franklin is probably to be singled out on this terrain, not only for his background expertise as an experimental physicist, but for his normative attitude that experiments can be divided (in principle) into the good and the bad in sufficient time to effect valid discriminations between rival scientific theories. In short, Franklin's work, in the full context of these studies, has to be located in the area of that position which sees experiment as providing an environment of settled fact that selects among theoretical mutants, rather than seeing experiment as one element that is articulated or negotiated with approximately equal weight against other elements in a sort of open scientific dialogue.

It's pretty easy to see Franklin's general attitude encapsulated in the title of his new book, *Experiment, right or wrong*, and even more clearly in his Preface. The Preface recounts Richard Feynman's efforts to visit Tannu Tuva, a small Asian country well-known for its postage stamp issues, "in the right way," that is, without taking advantage of his prominence as a scientist. After several years of inquiry and preparation, Feynman died two weeks before permission arrived. Franklin's moral to this narrative is that science should proceed like Feynman did in seeking admission to Tannu Tuva, that is, "in the right way." The assumption that there is a right way, based on the epistemology of experiment to be developed in the book, is transparent. But, of course, other morals can be drawn from Feynman's quest, perhaps the most obvious being that if you search for something in the right way, it will always arrive too late. This seems to me to be the nub of any discussion of Franklin's work. Can the right way of experimentation, whatever that is, be identified sufficiently early on that the advance of science can be plotted along rational paths, in some clear sense of rationality? Whatever that sense of rationality, it apparently must contain a normative component that identifies a right way for rational scientists, against which individual deviations have to be seen, not as variants increasing the social gene pool of scientific possibilities, but as errors.

Without going into the details of the atomic parity-violation experiments, the issue between Franklin and Pickering comes down to an agreement that the scientists involved *chose to accept* certain experiments and their interpreted results, but to a disagreement as to what it means to say that these choices were reasonable. In Franklin's imagery, Pickering regards the experiments that were not accepted as mutants that were slain by a decision not to let them live, whereas Franklin regards them as mutants that died of natural causes, i.e., mutants that died because they were bad experiments, and could not be nourished in an appropriate field of data.

Let me quote Franklin quoting Pickering:

> We saw in the preceding section that in 1977 many physicists were prepared to accept the null results of the Washington and Oxford experiments and to construct new electroweak models to explain them. We also saw that by 1979 attitudes had hardened. In the wake of experiment E122, the Washington-Oxford results had come to be regarded as unreliable. In analysing this sequence, it is important to recognize that between 1977 and 1979 there had been no *intrinsic* change in the status of the Washington-Oxford experiments. No data were withdrawn, and no fatal flaws in the experimental practice of either group had been proposed. What had changed was the *context* within which the data were assessed. Crucial to this change of context were the results of experiment E122 at SLAC. In its own way E122 was just as innovatory as the Washington-Oxford experiments and its findings were, in principle, just as open to challenge. But particle physicists *chose* to accept the results of the SLAC experiment, *chose* to interpret them in terms of the standard model (rather than some alternative which might reconcile them with the atomic physics results) and therefore *chose* to regard the Washington-Oxford experiments as somehow defective in performance or interpretation (1990, p. 174).

Commenting on this passage, Franklin says:

> Though I do not dispute Pickering's contention that choice was involved in the decision to accept the Weinberg-Salam model, I disagree with him about the reasons for that choice. In Pickering's view, "The standard electroweak model unified not only the weak and electromagnetic interactions: it served also to unify practice within other diverse traditions of HEP (high-energy physics) theory and experiment ... Matched against the mighty traditions of HEP, the handful of atomic physicists at Washington and Oxford stood little chance" (Pickering, *Constructing Quarks*, pp. 301-2). In my view, the choice was a reasonable one based on convincing, if not overwhelming, experimental evidence (1990, p. 174).

Later, summarizing an intervening exposition and discussion of the experiments, Franklin says:

> My interpretation of this episode differs drastically from Pickering's. The physics community chose to accept an extremely carefully done and carefully checked experimental result that confirmed the Weinberg-Salam theory. This view is supported by Bouchiat's 1979 summary. After hearing a detailed account of the SLAC experiment by Prescott, he stated "To our opinion, this experiment gave the first truly convincing evidence for parity violation in neutral current processes ... In addition, the most plausible alternative to the W-S model, that could reconcile the original atomic physics results with the electron scattering data, was tested and found wanting. There certainly was a choice, but, as the "scientist's account" or evidence model suggests, it was made on the basis of experimental evidence. The mutants died of natural causes (1990, pp. 180-181).

So, what is under dispute comes down to this question: Does a reasonable choice of an interpreted experiment as supporting some theoretical conjecture rather than another come down to accepting that the experiment provides (normatively) reliable experimental evidence, where reliability can in some sense be objectively calculated, or does it mean that to be reasonable is to agree to abide by a consensus arising out of open negotiation concerning all of the aspects involved in some area of scientific investigation. In other words, can truth be grounded in science in any stronger way than appealing to the limits of an (admittedly fallible) scientific consensus? Can rationality be discerned in the decisions of individuals, or is it a property of a group process in scientific investigation? It's fairly obvious that this revives an old epistemological dispute about the nature of truth that recurs in the history of philosophy, and just as obvious, I think, that the problem may result from supposing that there are just two basic positions to be considered.

Franklin is determined to use Bayesian theory to develop a notion of reliable evidence that would lay down formal tracks along which a rational discussion of evidence would have to move. It's a crucial question whether his Bayesian representation proves anything, or whether it is simply a formal representation of his already existent assumptions about the relative weight of evidence. I suspect the latter. For example, a basic assumption that gets coded into Franklin's Bayesian analysis is that the validity of an experimental result is increased by independent confirmation, that is, by the same result obtained from two different experiments that are regarded as somehow equivalent. I wouldn't quarrel with the concrete examples that are given by Franklin (and Hacking) of this phenomenon; but the question is whether a general characterization of this phenomenon can be represented in Bayesian symbolism. Franklin's informal representation of the Bayesian principle goes like this:

> Thus, if we wish to know the correct time, it is better if we compare watches than if either of us looks at our own watch twice (1990, pp. 107-108).

This is hardly self-evident. For example, if I have a watch known to be very accurate, and you have an erratic watch, we may agree that two looks at my watch (to verify that it is running) can give a better estimate of the time than averaging the time shown on your watch with the time shown on my watch, particularly if they show a large divergence. Something like this happens in the relevant scientific examples, since not just any two experiments increase a consensus about validity in practice. A new experiment improving an old design completed by a scientist known to be a good experimenter may indicate an accuracy of data far superior to that offered by comparing these data to the data of an arbitrarily chosen other experimenter. It's quite obvious that scientific gossip and folklore correctly influences experimental interpretation and estimates of which experiments should be compared, but it's far from obvious that this fact can be captured in any Bayesian representation of evidential relationships between experimental results before the negotiation of gossip and discussion has reduced the field of all data to the data that count and some feeling for relationships of reliability between data derived from different experimental sources, at which point the Bayesian representation can model well enough what might have been the thinking of those who have turned out to be correct. In short, Bayesian formalism, like other formalisms, depends upon reasonable background assumptions in its use; otherwise, it can lead us far from intended goals. Curiously, Franklin's informal gloss captures this fact, congenial to constructionism, in its hypothesis that *we* wish to know the correct time (jointly), otherwise it has no application. The point I'm making does not necessarily tell in favor of constructionism. After all, the anthropological or sociological investigations of working scientific laboratories that we have all undertaken by invasions of laboratories well known to the scientists under observation, so that they cannot be regarded as free from well canvassed distortions that can be found in such investigations. There simply is no way of objectively telling whether laboratory conversations, no matter how apparently informal, but recorded by investigators under circumstances where the investigation can't be concealed, represent what would occur in the absence of investigation, or represent instead a version of what the investigated population thinks should be its representation. Even if such investigations are enormously helpful in correcting certain kinds of idealist and normative misconceptions about the practice of science, there should be worry that these investigations may misrepresent what occurs in laboratories because the style of constructivist investigation overemphasizes the goal of cooperation in reaching agreement *within* laboratories, tends to assume that the cessation of overt disagreement means that consensus has been reached, assumes that everything can be (in principle) questioned in laboratory discussions because of an already existing view that science is going to be descriptively rational, and so forth. The problem is that laboratory talk, like all human talk, is based partly on silences that have to be *interpreted* by an investigator. Merely recording them, but not discussing them, may amount to supposing that they are not important, which is very likely not the case.

Lynch's work is much closer to constructionism as I've described it than Pickering's, but it's not at all clear that any form of constructionism needs to conflict with Franklin's form of experimental realism, provided that it would make sense to think of levels of scientific description for different purposes, in which the apparent relative disorder of high resolution analyses gives way to apparent order on lower resolution, just as relative molecular chaos may give way to regular cellular processes or individual confrontations of many kinds between individual soldiers may give way to the loss of the right flank in the analysis of a battle. Franklin's argument that the scientific community should be expected to occasionally go against the weight of evidence if the constructionist account were true is nugatory if this is taken into account. Constructionist accounts trace "weight of evidence" as it is under negotiation, in real time, until what the weight of evidence is can be subjected to consensual agreement. After that, Franklin has to be right, but his notion of weight of evidence may always come too late to catch the constructionist account opening a window on scientific irrationality. It should also be noted that the practice of following a series of related experiments on a single topic, followed by all of the approaches under consideration, may lose sight of side paths switching in and out of these sequences. Such switching can provide comparisons to changing weight of evidence in related fields that are not explicitly noted in the sequences under study, either in publications or inside scientific conversations, since one scientist affected by results in another field may have no knowledgeable audience for relevant observations. The community of conversation may only be able to communicate (without costly learning episodes) on the focus of community investigation.

Pickering's modulating position between purely constructionist description of the micro-practices of laboratories and Franklin's normative coercions can be seen in the conceptual apparatus that he brings to his discussion of experimentation. Let's turn to his interesting observation that the "scientific articulation of the real is the product of a pragmatically achieved, three-way reinforcement between material practice, instrumental modelling of the practice and modelling of the phenomenal world" (1987). This view is developed from the study of a particular sequence of experiments that Pickering analyses as though these three factors were all plastic resources equally open to adjustment until a satisfactory resonance between all three could be achieved. This is very clever, if only because by positing three equally plastic resources, the indeterminate 3-body collision problem is modelled, and the path that investigation will follow cannot be predicted. Pickering doesn't say that all three resources are equally plastic, but by calling them all "plastic resources," and noting that any of the three can be adjusted at any point, Pickering strongly suggests this reading, and the suggestion would be supported by all of those laboratory studies that tend to suggest that everything is, in principle, open to question and negotiation.

I think it would be worth exploring the existence of possible asymmetries in the plasticity of Pickering's resources, since such asymmetries might provide a clue as to how constructionism, if it's an accurate description of scientific practice, can also capture the failure of relativism that ought to turn up in an accurate description of scientific practice, without defeating relativism in advance by a notion that terms in scientific discourse correspond to items in the real world, so that there is an intrinsic metric of truth in scientific discourse. What are the three plastic resources? One (A) is the material resource of the apparatus or material experimental set-up, another (B) is the conceptual resource that explains the working of A, and the third (C) is a theoretical model or set of such models. (A, B, and C may denote appropriate sequences.) In some sense, A and B together yield data that are relevant to an evaluation of C, while C stands in turn as an evaluator of A and B, since data relevant to C must be produced. The traditional point of comparison is that between the data and C, but since the data are produced from A and B, the conceptual variability inherent in B prevents any naive experimental realism with respect to the data. This is an elegant representation of ideas to be found in Hacking (especially) and other sources, here extended by Pickering to provide a conceptual resource for discussing the development of a series of physical experiments until such a time as the problem initiating the experiments might be regarded as settled.

At the start of a sequence of experimentation, we can assume that A, B, and C are distinct resources. When such a sequence ends, typically, A and B have collapsed,

and the reliable data now being generated ends the experimental sequence with certain consequences for C. That A and B typically collapse is reflected in the fact that scientific papers find it sufficient to state B and the data, that is, to present an account of how the finally successful apparatus works. Typically, the data and C do not collapse, so that data and theory (or at least one of the theoretical models in C) remain in a tension sufficient to fuel further scientific development of a new triad (A', B', C') can be generated by a plausible or interesting variation in one of the endpoints of the original sequence. Thus Pickering's account involving three plastic resources and standard accounts are pretty much equivalent at the end of an experimental sequence. It's in the interval, as A and B are brought into resonance, that we need to concentrate our attention. Let's assume that A and B are not in synch at the start of some sequence, necessitating a change in A or B or both. Is there anything we can say about changing the apparatus as opposed to changing the theory of the apparatus? Change in either A or B can result from change in the other. But there is a difference. Some changes in A can be seen as improvements in terms of a valuation that is not sensitive to variations in B or in C. Getting an apparatus to run more smoothly or more quickly, for example, can be an obvious improvement that may (or may not) necessitate a change in B. If a change in B is required due to the relationship of A and B, it usually can be accomplished. Perhaps it has always been accomplished. On the other hand, when changes in B can be seen to be improvements, it is not always possible to change A to fit B because of some some material consideration, and sometimes changes in A that result show that the theoretical improvement was illusory.
Although this does not begin to initiate a detailed analysis, there is a sense in which A appears to be a less plastic resource than B or C. To repeat, changes in A can often be seen (in real time, without waiting for accomodation by B) as *improvements*, whereas "improvements" in B don't begin to count unless A is actually altered and realizes the improvements conjectured. It's conceivable that this small asymmetry can account, ultimately, for large scale directions of scientific progress and for the objectivity and rationality of those directions.

Why isn't this possibility more widely recognized? I think the answer is that writing about experimentation automatically privileges B and C, that is, talk about experimentation, since that's what can be written down. There is the further fact that grounding rational lines of inquiry in lucky discoveries of improvement in apparatus seems embarrassing to experimenters, who might like to be granted powers of thought, and who might also crave an image of scientific rationality. Therefore, it is not all that frequent that an experimental paper freely admits that a breakthrough occurred when someone tried some "sticky tape," "waste plastic material that happened to be on hand" or "a new kind of oil" to doctor a balky piece of equipment, but such incidents do occur. So, there's a bias against sticky tape in the original accounts, and then again in philosophical reflection. In my opinion, we have to work against this bias, and against the temptation to produce smooth symmetric theories of experimentation. Let me come back to Allan Franklin for a moment. I pointed out to him in a review (and in conversation) that the only real representation of experiment (A, as opposed to B) in his first book is the glorious photo of a mess of a laboratory on the dust jacket. The photo on the dust jacket of his second book is that of someone's laboratory notes and data. This is precisely a wrong direction, I think, in order to get a grasp on A, or real apparatus. Philosophers still need to get sticky tape on their fingers. In short, ladies and gentlemen, we need to get down and get dirty before we will have an appropriate understanding of experimentation.

References

Franklin, A. (1986), *The Neglect of Experiment*, Cambridge: Camridge University Press

_____. (1990), *Experiment, right or wrong*, Cambridge: Cambridge University Press.

Pickering, A. (1984), *Constructing Quarks*, Chicago: The University of Chicago Press.

_____. (1987), "Against Correspondence: A Constructionist View of Experiment and the Real," in *PSA 1986*, Volume Two, Fine, A. and Machamer, P. (eds.). East Lansing: Philosophy of Science Association. pp. 196-206.

Reason Enough?
More on Parity-Violation Experiments and Electroweak Gauge Theory

Andy Pickering

University of Illinois at Urbana-Champaign

In recent years a unified strategy in dealing with constructivism has been emerging in the writings of historians and philosophers of science. In my own experience, the strategy is exemplified in the long critiques of all or parts of my book, *Constructing Quarks* (*CQ*), set out by Paul Roth, Peter Galison and Allan Franklin. These critiques have two common features. First, the substance of constructivist claims is more or less ignored, in favour a fictional version that simply asserts the opposite of what the critic wants to affirm, which is, second, that the evolution of science should be grasped in terms of some relatively simple and unsituated concept of 'reason' (or 'logic' or 'persuasive argument').[1] Allan Franklin's discussion of the history of parity-violation experiments in atomic and high-energy physics exemplifies both of these features.[2] Concerning the first, the position he attributes to *CQ* is summarised as a pure negative: Pickering, he says, 'obviously doubts that science is a reasonable enterprise based on valid experimental or observational evidence' (165).[3] This negative is set up to lead into its inverse, which is Franklin's own position: that science is a reasonable enterprise based on valid etc etc. I resent Franklin's gloss of my analysis of science, as I resent Roth's and Galison's, but I largely let that pass here. My main concern is with Franklin's conception of 'reason' and what that concept can accomplish in helping us understand science. I think it is important to set out my position on this at least once in my life, because I believe that ideas like Franklin's are actually the main impediment to understanding what constructivism amounts to, what it claims, and what problems it addresses. Until the weaknesses of positions like his are exposed, there is, I think, little hope of useful dialogue between constructivists and large sections of the history and philosophy of science community.

The thrust of my argument is simple. Whatever the image that Franklin's rhetoric might conjure up in the minds of his audience, it is untrue that I deny that science is a reasonable enterprise, or that evidence has a constitutive role to play in the production of scientific knowledge. The problem I see in Franklin's understanding of science is rather that there are *too many reasons* to be found in science, and that these reasons point in all sorts of directions. They cannot therefore be understood as unproblematic explanations of why science proceeds historically in one direction rather than another. Rather than arguing this assertion in the abstract, I want to discuss the specific passage in the history of particle physics that Franklin takes to establish his own position, but first I need to clarify one aspect of that position.[4] Franklin campaigns under the banner of what he calls the 'evidence model', but there are actually two threads to his argument. He wants to suggest, first, that there is some especially 'reasonable' and theory-independent way of extracting 'reliable' empirical conclusions from a confused field of evidence. And, once such an extraction has been made, he wants to suggest that certain implications follow for theory-choice — implications that are caught up in his 'evidence model'. I take these threads in turn.[5]

1. What is the evidence?

The case in dispute concerns the discovery of parity-violating effects in electron-hadron interactions. It is not disputed that conflicting evidence on this topic was offered by experimentalists in the period 1976 to 1981, evidence that came from a range of bench-top atomic-physics experiments and one high-energy physics experiment, experiment E122, performed at the Stanford Linear Accelerator Center (SLAC). The question is, how did the scientific community make sense of this field of evidence? Franklin presents his own way of making sense of it, which he regards as especially reasonable. My feeling is, though, that there were any number of reasonable ways of making sense of the data, none of which can be especially singled out as better than the others. Since my strategy is simply to make a non-exhaustive list of reasonable alternatives, I rehearse Franklin's way first and number it '1'.

1) Franklin notes that the atomic-physics experiments in question were very difficult, were plagued with interpretative and systematic errors and were incapable of agreeing with one another: some reported the expected parity-violating effects predicted by the Weinberg-Salam (WS) model, others reported that any such effects were much smaller than the model predicted, if they existed at all. Franklin's suggestion is that the most reasonable stance towards these experiments as a class is to forget about them, to regard the results pro and con the WS model as, in effect, neutralising one another. He then argues that the remaining high-energy physics experiment, E122 at SLAC, was, instead, trustworthy, and therefore it was reasonable to accept the findings of E122, namely that parity violation in electron-hadron interactions does indeed occur at the rate predicted by the WS model. As he puts it, parodying a passage from *CQ*: 'Scientists *chose*, on the basis of reliable experimental evidence provided by the SLAC E122 experiment, to accept the Weinberg-Salam theory. They *chose* to leave an apparent, but also quite uncertain, anomaly in the atomic parity violation experiments for future investigation' (192).

My first comment on this phase of Franklin's essay is that it is hard to see that anything especially philosophical is at stake. Franklin, it seems to me, simply offers us an artful reading of the scientific publications designed to lead up to the conclusion he wants to reach. If there is some system behind his commentary, he does not make it explicit. However, my intention is not to argue with him about such niceties, so I move on to my second comment. It is that Franklin's way of evaluating the evidence in question seems to me quite reasonable. I cannot, in the abstract, see anything wrong with reasoning about this confused field of data as he does. What I dispute about Franklin's reasonableness, though, is its uniqueness. To make this point, I continue with a list of some other ways of thinking about the same data that also appear reasonable.

2) Franklin asserts, in the passage quoted above, that the physics community in fact reasoned as he does about the evidence on parity violation, but this is not quite correct.[6] He is right to say that E122 decided the issue for most physicists, but he is wrong if he imagines that this was accompanied by a reasoned judgement that the

atomic-physics experiments neutralised one another. Rather, the presumption within the physics community was that something was wrong with just those atomic-physics experiments that failed to detect the parity violation that E122 had found. Thus, in what I think was the first major review talk to follow the announcement of E122's results, the reviewer discounted the negative findings on parity-violation (from groups working at the Universities of Oxford and Washington) while including the positive findings (from groups at Novosibirsk and Berkeley) in his calculations of the phenomenological parameters describing the electroweak interaction. Oddly enough, Franklin discusses this review in his essay (quoting from *CQ*), as he does another authoritative review talk from a workshop on neutral-current interactions in atoms held two months later. There, as Franklin notes, the reviewer concluded: '*As a conclusion on this bismuth session, one can say that parity violation has been observed roughly with the magnitude predicted by the Weinberg-Salam model*' (173). Just to be clear, let me state that the bismuth experiments in question were the atomic-physics experiments performed at Oxford, Washington and Novosibirsk, and that the first two arrived at null results; only the Novosibirsk group had data consistent with the predictions of the WS model.[7]

So, it seems reasonable to say that the physics community did not reason quite as Franklin does, reasonable though his way of reasoning is. Where Franklin feels that the atomic-physics experiments neutralised one another, the physicists themselves excluded from their calculations those experiments that produced evidence against the WS model, while including the experiments that went along with the model. Should we therefore say that the physics community reasoned unreasonably? I don't think so. What I do think is that to see that their reasoning was reasonable one has to recognise that a kind of dichotomous logic was at work: either the WS model was right about parity violation or it was wrong. If it was right, as indicated by E122, then those atomic-physics experiments that failed to find parity violation had somehow to be in error and should not be taken into account. Again, this reasoning seems quite defensible to me. What seems less defensible is the artful way in which Franklin in his essay glosses over the gap, already evident there, between his own reasoning and that of the physics community. He has to do this, though, because pointing to the dichotomous logic at work in the actual reasoning of the physics community draws attention to the importance of theoretical context in this instance of scientists' reasoning about evidence. Back to this in a moment; first, some more reasons.

3) A lot hinges upon E122 in Franklin's reasoning, as it did in the physics community's. No-one disputes that E122's findings just about settled matters. Does that mean that the performance and interpretation of E122 was itself beyond dispute? As far as Franklin is concerned, the answer is yes, and he gives his reasons for thinking so. He recites at great length the checks that the SLAC experimenters gave as reasons for believing that their parity-violating signal was genuine and not an artefact of their apparatus (176-80). I agree with Franklin that such checking is important for scientists to do, and for science-studies to think about. I agree that it has a constitutive role in scientists persuasion of themselves and others of the reliability of their findings. It is quite reasonable to feel the force of arguments based upon such checks. But it is at this point that I start to feel that Franklin's rhetoric is somewhat disingenuous. I note, for example, that the atomic-physics experimenters, even those who reported that parity-violation did not exist, did checks too — though Franklin makes no mention of them in his essay.[8] If checking is all that is at stake, I wonder whether there is any special reason for trusting E122 and putting it in a special category all by itself. Why wouldn't it be reasonable to lump E122 together with the atomic-physics experiments and let them all neutralise one another? Actually, I think it would be reasonable to proceed thus, and hence to conclude that there was no reliable data to be had on parity-violation in the

period under consideration. And to reinforce this conclusion — to make it seem even more reasonable—I can point out just how anomalous an experiment E122 was in the history of particle physics and in the history of science in general.

Experiment E122 was performed just once and then disassembled. No experiment like it has been performed since, and no experiment like it seems likely ever to be performed. Now, twelve years later, the findings of E122 stand as the sole record of parity-violation in high-energy electron scattering. As I pointed out in *CQ*, this is a situation as far removed as can be from the standard philosophical paradigm of intersubjectively replicated evidence. If I suggested that this in itself were sufficient grounds for reasonable doubt about the findings of E122, would I just be playing a philosopher's game with no relevance to the real world? Here I can borrow some more rhetoric from Franklin's paper. He concludes with some moral tales concerning theoretical presuppositions which insinuated themselves into famous experiments. Fortunately, he says, 'the importance of the experiments led to many repetitions and to the correction of these early results' (1988, p. 28). Where does that leave E122? If I were a physicist who in 1978 had said that I preferred to wait for E122 to be replicated before making up my mind up parity violation, would I have been acting unreasonably? I think not.[9]

4) Suppose that despite qualms concerning the replicability of the findings of E122 one were inclined to accept them. And suppose further one were sufficiently impressed by the competence of Patrick Sandars, the leader of the Oxford atomic-physics collaboration, to accept their null-result at face value. One might, for example, have been so intimidated by Sandars in the undergraduate teaching laboratory that one decided to become a theoretical physicist — as I was and did. Having become such a theorist, would it then be unreasonable to try to find some variant of electroweak gauge theory that could reconcile the null-result of the Oxford and Washington atomic-physics experiments with the positive findings of E122? Again, I think this course of action would be reasonable, and could be made to seem even more so by, for example, noting the very large error bars on the pro-WS Berkeley data, and by undermining the pro-WS Novosibirsk data by pointing to the dubious track record of Soviet physicists in bench-top experiments that attempt to address topics of interest in particle physics (on the rest masses of neutrinos, for example). Franklin writes as if the possibility of devising such alternative models had been ruled out once and for all by experiment E122 (180). But though it is true that E122 analysed their data in a way that displayed the improbability of a particular class of variant gauge theories, the so-called 'hybrid models', I do not believe that it would have been impossible to devise yet more variants.[10]

So, I have listed four quite different ways of reasoning about the evidence available on electron-hadron parity violation in the late 1970s: Franklin's way, that discounts all the findings of the atomic-physics experiments; the physicists' way, that discounted only those atomic-physics experiments that disagreed with the WS model; a line that challenges E122 on the grounds of replicability; and a line that credits both E122 and the Washington-Oxford atomic-physics experiments and tries to reconcile them. As I said before, this list is not exhaustive — it is easy enough to think of yet more ways of reasoning about the data — but it is enough, I think, to call into question Franklin's suggestion that there is some uniquely reasonable way of figuring out what Nature was trying to tell physicists in this instance. And thus, as I see it, an explanatory problem remains open — that of understanding why, from my indefinite list, physicists in fact took the second option, crediting the pro-WS data as they did. One can exclaim, 'but it's reasonable!' until one is blue in the face, but the problem remains.[11]

One last point in this connection. It requires no great powers of the imagination to construct the list of reasonable readings of the data that I have just given. One question that arises in my mind is therefore that of why Franklin and the other critics of constructivism are so obsessed with the first option. The answer is, I think, a moral one. The second and fourth items on my list (though not the third) implicate theory in a straightforward way in the assessment of evidence; and Franklin *et al* seem to have a moral conviction residing at a level *beyond reason* that such implication of theory is a bad thing. Evidently I do not share their morality, but I would make two comments concerning it. First, it would perhaps be possible to devise a system of institutional arrangements in which the moral purity of the empirical base of science could be maintained as Franklin desires. The products of that system would not, however, be the same as those of what we presently call science. Mainstream quantitative US sociology might emerge pretty much unscathed, but a lot of modern physics would be ruled unscientific. Second, in accordance with the inversion strategy I mentioned at the beginning, Franklin and his fellow moralists portray the constructivist enemy as their own opposite, the veritable Anti-Christ of epistemology. If morality requires the purity of the empirical base, then constructivists are accused of insisting on its desecration. If we suggest that evidence is not everything, then it must be nothing; all evidence is just an imposition of theoretical prejudice. We are the dreaded 'theory-firsters' in Peter Galison's forgettable phrase. Against this I remark that my list of reasonable ways of proceeding includes items 1 and 3 as well as 2 and 4. If scientists behave like 'theory-firsters', as particle physicists increasingly did in the period covered by *CQ*, don't shoot the messenger.

2. The evidence model?

The first thread of Franklin's argument was intended to establish a pure realm of evidence for science, in preparation for the second thread concerning theory-choice and the 'evidence model'. I have already suggested that the clean split between assessing evidence and choosing theories is hard to maintain in the instance under dispute. As I argued in *CQ*, it seems clear that the physics community evaluated evidence and made their choice of theory both at the same time, quite reasonably, as described under the second item on my list. Thus, I suspect that parity violation is not a very perspicuous choice of example for advancing the Franklinian cause. But I want to make things difficult for myself in what follows, by pretending that the experiments straightforwardly supported the predictions of the Weinberg-Salam model. My reason for doing so is to highlight some further shortcomings of Franklin's 'evidence model'.

What is the 'evidence model'? 'The evidence model', says Franklin, 'explains adherence to scientific beliefs in terms of their relationship to valid experimental evidence' (162). Thus, in the present instance, and forgetting about the problem of deciding what the 'valid' evidence was, it was reasonable to adhere to belief in the WS model, since the predictions of the model stood in a relation of agreement to the evidence. Now, as usual, I am happy to assent to this assertion, it seems reasonable enough to me. But still I feel that something fishy is going on here, which I can best summarise by saying that Franklin's 'evidence model' *affects to explain much more than it actually does*. The key question to consider is: what follows from the evidence model? what are the implications for future practice of reasonable belief grounded in evidence? In the case in point, after E122 physicists took the WS model for granted as a trustworthy description of the world of electroweak phenomena and largely ceased to explore alternative possibilities. So perhaps this is what Franklin understands as the reasonable behaviour that follows from the 'evidence model'. It sounds reasonable to me. But it is important to recognise that as a general prescription for scientific practice, this understanding of the 'evidence model' is a recipe for vicious conservatism.

To emphasise this point, I turn to what appears to be the knockout punch at the end of Franklin's paper. 'If the social constructivist view were correct', he says, 'then one would expect to find at least one episode in which the decision of the scientific community went against the weight of the experimental evidence. No such episode has been provided' (this volume).[12] I think there is something desperately wrong with this formulation. I can think of many examples that run counter to it. Consider the two great theoretical conjectures that run through the history of what I call the 'new physics' of elementary particles — the quark model and gauge theory. The quark model postulated the existence of particles carrying third-integral electric charges which were known not to exist from half-a-century's worth of experimentation, and the electroweak gauge theory as laid out by Salam and Weinberg in 1967 immediately predicted the existence of weak neutral currents which were known not to exist from many observations of K-decays and neutrino interactions. It was crucial to the history of modern particle physics that both of these models were initially elaborated in the face of, not with the support of, the available evidence.

How could Franklin respond to this observation? He could start, I suppose, by defending the letter of the passage quoted above and insisting that 'the scientific community' did not accept the quark or Weinberg-Salam models in their early years; only certain subsections of the community jumped on the relevant bandwagons. But then, what might one say about those subsections? Here Franklin's closing remark — 'that scientists, being human, are fallible and do not always behave as they ought to, should surprise no one' (this volume) — might come into play. The founders of the theoretical wing of the new physics — including heavyweights like Nobel laureates Gell-Mann, Weinberg and Salam — were indeed human, and in this instance they were exercising their prerogative and acting unreasonably. But I doubt whether the most rabid empiricist would want to go that far. Franklin's proper response to these important episodes would be, I imagine, that these laureates and their followers did have reasons for elaborating theories that appeared to be false. But what might these reasons be? Presumably either that the evidence that made the theories false was itself suspect — in which case we are back where we started, with my observation that there is more than one reasonable way of reasoning about a field of evidence — or that there is a further category of reasons that bear upon theoretical practice that just escapes the 'evidence model'. Either way, my point is established: there are just too many reasons around for reason to stand as an explanation of the development of science. My conclusion concerning Franklin's 'evidence model' is therefore either that it implies the kind of conservatism that would have ruled out the development of the quark model and gauge theory — and thus, incidentally, would have made the historical episode under discussion unthinkable — or that it is toothless as explanation: it explains utterances of belief in very special circumstances and nothing else.[13]

3. The trouble with reason

So far, I have been querying Franklin's way of thinking about science on its own terms. This is an exercise that is necessary from time to time. But I want to close by stepping outside Franklin's chosen frame. I offer three brief statements to indicate what I see as general problems of Franklin-type analyses of reason.

- Franklin's argument about the episode under discussion has enough plausibility to make it worth disputing. That would not have been the case if the topic had been, say, the development of QCD, the gauge theory of the strong interaction. There the relation between theory and experiment was so different from the theory-testing paradigm of traditional philosophy that Franklin's evidence model could have gained no purchase whatsoever (Pickering 1984, pp. 309-46). From this one could conclude that the development of QCD was itself unreasonable, or, as I do, that the theory-testing paradigm is in general a misleading starting point for thinking about science.

- Franklin's model of the scientist is that of a static reasoner, a weigher of evidence and a comparer of evidence and predictions. I do not deny that scientists engage in such practices, but I do deny that such a model can take us very far in understanding science. The model is just too thin. Most importantly it conceals the temporal dimension of scientific practice, the fact that scientists live not just in the present but in the past and future as well — in a field of goals and histories. This was what I tried to grasp and analyse in *CQ* in my model of what I called the dynamics of practice — although you would never guess it from Franklin's critique (or Galison's). I do not claim to have said the last word on this subject then or since, but I do continue to claim that attention to the temporality of practice is necessary if one wants to understand why, in the midst of the proliferation of reasons, science develops as it does.[14]

- An exclusive focus on scientific reasoning contributes to the strange blindness of traditional philosophy to the material dimension of science. In *CQ* I remarked upon the gross shift in the material practices of experimenters that was part and parcel of the establishment of QCD (1984, 347-82), and the more subtle shifts that accompanied the development of electroweak gauge theory, including those surrounding the establishment of parity violation in electron-hadron interactions.[15] I concluded that scientists had to learn proper ways of conducting themselves in the material world at the same time as they learned how to think about it. I continue to find this a striking and important observation, though, of course, it can find no expression in a philosophical discourse organised just around reason. For me, this indicates that we need a new philosophy (Pickering 1990).

Notes

[1] Roth and Barrett (1990), Galison (1987); for my replies see Pickering (1990, forthcoming a). For a similar critique of constructivism more generally, see Giere (1988) and for my critique of the critique, Pickering (forthcoming b).

[2] To avoid confusion, I should explain that in 1988 Allan Franklin invited me to participate in a 1990 PSA symposium organised around his account of the parity-violation experiments (Franklin 1988). That account subsequently appeared as Chapter 8 of his book (1990) though with certain changes (none of any substance as far as my arguments are concerned). A highly abridged version of Franklin's argument appears in this volume. My essay was written as a response to Franklin (1988) but, except where indicated, page number citations in what follows are to the book (1990). Pickering (1984, pp. 290-302) is my account of the historical developments presently under discussion.

[3] This formulation appears at the end of a sequence of glosses and translations which begins with quotations from Trevor Pinch and continues via references to 'interests' (162-3) — an analytical concept that I have self-consciously abstained from using since around 1980.

[4] Two further lines of argumentation also support my conclusion that there are too many reasons in science. The first relates to my own experience in exploring the history of particle physics. In the course of that research I have met and often interviewed and collected documentation from many 'deviant' scientists — scientists who reject, say, all or parts of the present gauge-theory orthodoxy. These deviants prove to have better worked out substantive and philosophical reasons for their heretical attitudes (which differ widely from one to the other) than the proponents of the orthodoxy. Secondly, there is now a pretty extensive literature on the analysis of scientists' discourse and argumentation. The image of the scientific actor that emerges from this literature is that of a person artfully constructing reasons for particular purposes, and not of a Franklinian automaton ruled by reasons beyond her control (Gilbert and Mulkay 1984). It seems to me that both Franklin's present essay and the atomic-physics publications discussed below would be extremely fruitful sites for the study and documentation of situated reason-giving.

[5] This is an appropriate point to confess that Franklin has found a substantive error in *CQ*'s account of the experiments in question: he is right that the 1977 publications from Washington and Oxford made no reference to hybrid unified electroweak models, though, as he concedes, such models were discussed in the literature of the time (169). On another putative error, I think the mistake is Franklin's. When I stated in *CQ* that 'The details of the [Novosibirsk atomic-physics experiment] were not known to Western physicists', it is clear from the context that I was referring to the period up to and including the review talk by F. Dydak given at the European Physical Society Conference, 27 June-4 July 1979. My statement is thus not refuted, as Franklin supposes (171), by the presence of the Soviet physicists at the meeting held in Cargèse, 10-14 September 1979. My other supposed errors are dealt with in the text and notes below.

[6] Also: 'The physics community chose to await further developments in the atomic parity violating experiments, which, as I have shown, were uncertain' (180).

[7] Franklin also mentions a third review talk — Commins and Bucksbaum (1980) — and states that 'They [Commins and Bucksbaum] regarded the situation with regard to the bismuth results as unresolved' (p. 173, note 16). In fact, that review treated the Washington and Oxford experiments as having no data at all, while including the positive findings from Novosibirsk and Berkeley in its calculations of the parameters of the electroweak interaction and concluding that these parameters were in 'very satisfactory agreement' with the WS model (Commins and Bucksbaum, pp. 38-41, 48-51, quotation at 51).

[8] Thus, for example, Franklin dwells on the systematic errors reported in the 1977 account of the Oxford atomic-physics experiment (169), but makes no mention of the passage immediately following where the experimenters discuss their procedure for getting round these errors: they randomly interspersed their measurements on bismuth vapour with measurements on a dummy tube not containing bismuth (Baird *et al.* 1977, 800). The rhetorical potential of the artful pruning of quotations is nicely brought out in Ashmore (1988, 138-9).

[9] As Franklin puts, 'Unlike statistical errors, which can be calculated precisely, systematic errors are both extremely difficult to detect and to estimate' (191). How fortunate there were none of the latter in E122. As a speculation, I offer the thought that the uncritical reception of E122 may have depended somewhat on the 'politics of experiment'. Parity violation was the last of a series of major discoveries made at

SLAC during the 1970s; to challenge E122 would have been to challenge the credibility of the laboratory itself (and, of course, to challenge both gratuitously, since there was no other accelerator at which comparable measurements could be made). Possibly connected with this speculation is the fact that the early reports of the findings in E122 in the scientific press credit the experiment to Richard Taylor, while the leader and moving force of the E122 collaboration was actually Charles Prescott (see, for example, *Times* (1978), *New Scientist* (1978), Walgate (1978) and *Physics Bulletin* (1978); on Prescott's role, see Pickering (1984, 298-9). Taylor was the experimenter who had led the collaboration responsible for the discovery of scaling at SLAC in the late 1960s (Pickering 1984, 127-31).

[10] Open-ended recipes for the construction of variations on the electroweak theme had first been written down in 1972: see Pickering (1984, 181).

[11] At various points in his essay, Franklin writes as if the issue between us concerns 'evidential weight' rather than 'reason'. Thus, 'The issue seems to turn on the relative evidential weight one assigns to the original Oxford and Washington atomic physics results and to the SLAC E122 experiment ... Pickering seems to regard them as having equal weight. I do not' (174), 'Pickering claims that the decision of theory choice excluded evidence, of equal weight, that argued against that choice' (1988, p. 27), 'I believe he [Pickering] made an incorrect judgement on the relative evidential weight of the two different experiments' (1988, p. 29, note 3). It is understandable that a Bayesian might want to speak of 'evidential weight' as if it could be read off the surface of a published text (see Franklin 1986, Ch. 4), but my remarks here on 'reason' can serve equally well to demonstrate the problematic nature of 'evidential weight'.

[12] In this connection Franklin mentions studies of the discovery of the weak neutral current by myself and Peter Galison, before stating without argument that he believes Galison's account 'supports the evidence model' and is 'more persuasive' (164). The interested reader might consult the works cited by myself and Galison, as well as Pickering (1989).

[13] Note that this conclusion applies to all static articulations of 'reason' (see below), including Bayesianism, and not just to the particular version of the 'evidence model' that is proposed in Franklin's present essay.

[14] For more recent discussions of my understanding of the dynamics of scientific practice, see Pickering (1990, forthcoming a). Pickering (1990) contains a discussion of how the static dimension of scientific reasoning can integrated with an understanding of the dynamics of practice.

[15] In this connection, certain developments concerning the Washington atomic-physics experiment might repay detailed attention, though I failed to mention them in *CQ*. As Franklin notes, the Washington group published new data in 1981 that agreed with the predictions of the WS model. There they included a table listing the different experimental runs that they had performed since their earliest experiments on bismuth vapour (Hollister *et al.* 1981, p. 645, Table I). It is interesting to note that these runs were categorised by the kind of laser used. The earliest run had used a parametric oscillator (this resulted in the 1976 publication reporting a null-result) as had the second run (which resulted in an even tighter upper limit on the extent of parity violation, as reported in their 1977 publication). The group then switched to using a gallium-aluminium-arsenide laser diode: they first used a transverse-junction-stripe diode, which led to measurements confirming the null-results of the 1977 publication and which were reported in an unpublished PhD thesis; then they worked with two other stripe

diodes and a channeled-substrate-planar diode and obtained results consistent with the WS model. The results from the positive experiments (with the second and third stripe diodes and with the planar diode) were averaged to give the stated result of the 1981 paper. There is, then, a *prima facie* case for thinking of the history of this experiment as being that of the tuning of experimental techniques to the production of credible phenomena: the experimenters were finding out what kind of laser to use, in a very particular sense, in the course of their material and interpretative practice.

Although not strictly relevant, I cannot resist two further remarks concerning the Washington experiment. First, I note that gallium arsenide was also the source of polarised electrons for experiment E122. Second, though the Washington group never remarked upon it, the measured positive effect reported in the 1981 publication was actually within the quoted experimental error of their 1976 null-result (they quote values of a quantity $10^8 R$ stated to be -10.4 ± 1.7 and -8 ± 3 respectively; their Table I, which I am reading here, is a masterpiece of obfuscation). What turned a null-result into a confirmation of theory was that calculations of the expected effect had decreased by a factor of two between 1976 and 1981.

References

Ashmore, M. (1988), "The Life and Opinions of a Replication Claim: Reflexivity and Symmetry in the Sociology of Scientific Knowledge", in *Knowledge and Reflexivity: New Frontiers in the Sociology of Knowledge*, S. Woolgar (ed.). Beverly Hills and London: Sage, pp. 125-54.

Baird, P.E.G. *et al.* (1977), "Search for Parity-Nonconserving Optical Rotation in Atomic Bismuth", *Physical Review Letters* 39: 798-801.

Commins, E. and Bucksbaum, P. (1980), "The Parity Non-Conserving Electron-Nucleon Interaction", *Annual Reviews of Nuclear and Particle Science* 30: 1-52.

Franklin, A. (1986), *The Neglect of Experiment*. Cambridge: Cambridge University Press.

_____. (1988), "The Way Mutants Meet Their Deaths: The Case of Atomic Parity Violation Experiments". Unpublished draft, University of Colorado, Boulder, dated 11 Feb. 1988.

_____. (1990), *Experiment, Right or Wrong*. Cambridge: Cambridge University Press.

_____. (this volume), "Do Mutants Have to be Slain, or Do They Die of Natural Causes? The Case of Atomic Parity Violation Experiments".

Galison, P. (1987), *How Experiments End*. Chicago: University of Chicago Press.

Giere, R.N. (1988), *Explaining Science: A Cognitive Approach*. Chicago: University of Chicago Press.

Gilbert, G.N. and Mulkay, M. (1984), *Opening Pandora's Box: A Sociological Analysis of Scientists' Discourse*. Cambridge: Cambridge University Press.

Hollister, J.H. *et al.* (1981), "Measurement of Parity Nonconservation in Atomic Bismuth", *Physical Review Letters* 46: 643-6.

New Scientist (1978), "Major Boost for Unified Theory", 22 June, p. 824.

Physics Bulletin (1978), "Left at Last?", 29: 396.

Pickering, A. (1984), *Constructing Quarks: A Sociological History of Particle Physics*. Chicago and Edinburgh: University of Chicago Press/Edinburgh University Press.

_____. (1989), "Editing and Epistemology: Three Accounts of the Discovery of the Weak Neutral Current", in *Knowledge and Society: Studies in the Sociology of Science, Past and Present*, Vol. 8, L. Hargens, R. A. Jones and A. Pickering (eds). Greenwich, CT: JAI Press, pp. 217-232.

_____. (1990), "Knowledge, Practice and Mere Construction", *Social Studies of Science* 20: 682-729.

_____. (forthcoming a), "Beyond Constraint: The Temporality of Practice and the Historicity of Knowledge", to appear in *Philosophical and Historiographic Problems about Small-Scale Experiments*, J. Buchwald (ed.).

_____. (forthcoming b), "Philosophy Naturalized a Bit", to appear in *Social Studies of Science* 21(3).

Roth, P.A. and Barrett, R.B. (1990), "Deconstructing Quarks: Rethinking Sociological Constructions of Science", *Social Studies of Science* 20: 579-632.

Times (1978), "Physics: Confirmation of Unified Theory", 16 June.

Walgate, R. (1978), "Success for Unified Field Theory", *Nature* 273: 584.

Allan Franklin's Transcendental Physics

Michael Lynch

Boston University

Does Allan Franklin's study of atomic parity-violation experiments provide convincing evidence against social constructivism? According to Franklin (1990a, p. 2), "when questions of theory choice, confirmation, or refutation are raised they are answered on the basis of valid experimental evidence . . . [and] there are good reasons for belief in the validity of that evidence." Franklin asserts that social constructivists take the opposite position: "They would say that it is not the experimental results, but rather the social and/or cognitive interests of the scientists, that must be used in the explanation." Having set up the contrasting positions, he then asks the reader, "which of us is telling the more plausible story?" (Franklin 1990b, p. 163). Consistent with his evidence model, he proposes to discriminate between the two opposed positions by consulting historical evidence. He describes the results of two different sets of experiments on atomic parity-violations, and assesses the extent to which the experimental data match predictions based on the Weinberg-Salam unified theory of electroweak interactions. The earlier experiments, performed in the mid-1970s at Oxford and Washington did not support the W-S theory, whereas the later experiments, performed in 1979 in the Soviet Union and at Berkeley and SLAC, supported that theory. Franklin emphasizes that the later experiments used different arrays of equipment, as well as a variety of procedures for checking results, eliminating backgrounds, and ruling-out possible sources of artifact.

Franklin argues that the physics community's eventual preference for the later (SLAC, etc.) experiments was "reasonable" (i.e., justified by evidence and procedural rationality). He argues that this preference was not based on absolute grounds, but that it was justified by the superior "weight" of the evidence. As he reconstructs the situation in the 1970's particle physics community, physicists recognized at the time that the Oxford-Washington experiments were problematic for several reasons. Not only did the experimental results conflict with predictions based on an accepted theory, the calculations of the theoretically predicted effects of passing polarized light through bismuth vapor were uncertain, and the experimental techniques were untried. "These were extremely difficult experiments, beset with systematic errors of approximately the same size as the predicted effects" (Franklin 1990b, p. 176). In contrast, the later experiments were more convincing, not only because they seemed to support the W-S theory, but because they employed more reliable procedures which produced less systematic error in the experimental results. Franklin goes on to give a nauseatingly detailed account for those of us who are not trained in physics, and if we take his word for it (and I have no reason not to) particle physicists had good reason to accept the USSR-Berkeley-SLAC experiments, even though they found no "fatal flaws" in the results of the Oxford-Washington experiments. He draws two related conclusions from this: (1) The evidence model accounts for the historical episode, and (2) the evidence model is supported by the historical evidence to a greater extent than is the social constructivist view. I can imagine that both of these claims are disputable, although I will only take issue with the second claim. I am not going to try to support the converse of Franklin's argument, i.e., by claiming that social constructivism accounts for the evidence better than does the evidence model. Instead, I shall question the way Franklin initially sets up the opposition between his evidence model and a social constructivist position. In my view, Franklin's attempt to use an historical case study to settle the realist-constructivist debate is symptomatic of his more general inattention to the difference between epistemological argumentation and situated practical reasoning. However diligently and competently he describes the atomic parity-violation experiments and their results, he miscasts the position he says he is arguing against and the choice between his position and social constructivism is undecidable on that basis.

1. The Evidence for the Evidence Model

Franklin argues (1990b, p. 163) that his "evidence model applies to both science and the study of science." In a footnote (n. 3, p. 163) he acknowledges that "Some readers may worry that I am using the evidence model to decide whether or not an evidence model applies to science." This is not a serious problem, he says, because there are no guarantees "that the view that scientists use such a model will be supported by the evidence." As far as he is concerned, the relationship between an epistemological position and an historical case study is analogous to that between predictions based upon a physical theory and relevant experimental evidence. This analogy is a fairly "thin" one, especially if we accept what Franklin has to say about the "epistemology of experiment." According to him, experimental instruments and techniques incorporate checks, triangulation procedures, and other strategies for establishing the validity of results. However, it is not clear to me how his schematic reconstructions of experiments could themselves be comparable to the material practices and assessments of evidence they describe. Franklin approvingly cites Peter Galison's (1987) argument to the effect that the modern particle physics community houses separate material cultures in which theorists, instrument makers, and experimentalists hone their skills and develop their collective interests. This situation, Galison argues, offers practical constraints on the testing of theories, since the experimentalists and instrument makers do not simply follow after the demands of the theorists, but act in accordance with their own distinctive traditions. However, there is no comparable independence between Franklin's articulation of his evidence model and the narrative descriptions he uses as evidence for it. He crafts both of them from within the same literary space. Nor is his general epistemological model precisely constrained by the sorts of material, institutional, and practical conditions that Galison identifies in the particle physics community. In brief, while Franklin's evidence model may be well argued and convincingly documented, his documentary methods do not incorporate anything like the practical and social constraints that Galison identifies for physics experiments.

Franklin tells us that particular scientists had "good reasons" for acting as they did in the cases he reconstructs, and that they accepted evidence because it was "valid". He employs a Bayesian approach to reconstruct scientists' probability judgments about the relationship between hypotheses and evidence, but he admits (Franklin 1990a, p. 100) that this procedure does not reflect the actual judgments scientists made at the time. And, I would add, in many instances it is questionable whether it makes sense even to use rough probability estimates to reconstruct their judgments. A great deal turns upon what exactly might be meant by such terms as "good reasons" and "valid evidence," as both expressions permit a wide range of applications to particular cases. Franklin's case description supports his initial claim, but his story is motivated by and organized around the epistemological lesson he uses it to elaborate, and I fully expect that his constructivist interlocutors would be able to recite a different story of the "same" case supporting their claims. Even if Franklin and Galison are correct when they say that theories and experimental results in physics are not "so plastic that they can always be brought into agreement with each other" (Franklin 1990a, pp. 158-9, n. 25), I would not say this about the relationship between general epistemological claims and historical case descriptions. So, I have considerable doubt about the way Franklin sets up his study as a "test" of what he seems to think are two mutually exclusive epistemological theories.

2. Sociology of Scientific Knowledge Explanations

Franklin presents a strict opposition between his evidence model and the social constructivist "view" (note that he does not call it a "model"). The evidence model states that scientists' choices are based on "good reasons" and "valid evidence," whereas he leads us to think that social constructivists believe the opposite, that such choices are based on poor reasons, or no reasons at all, and that scientists disregard experimental evidence. Consider two of his characterizations: (a) "Obviously I do not agree with the social constructivists that all pictures of the world are equally good" (Franklin 1990b, p. 163); and (b) "[Pickering (1984)] obviously doubts that science is a reasonable enterprise based on valid experimental or observational evidence" (p. 165).

Franklin does not always characterize social constructivism so starkly, but the way he phrases these two characterizations maximizes the contrast between the explanations given by his evidence model and and those given by social constructivists. In his view, the social constructivists entirely discount the role of experimental evidence, and they treat theory-laden "interests" as the sole basis for the construction and interpretation of experimental results. The contrast between the two positions, as Franklin presents them, can be concisely stated as follows:

EVIDENCE MODEL: theory choice, confirmation, and refutation are made on the basis of valid experimental evidence.

INTEREST MODEL: theory choice, confirmation, and refutation are *not* made on the basis of valid experimental evidence; instead, they are based on social interests.

This way of setting up the comparison implies that actions based on social interests are incompatible with evidential justification. Although Franklin's set up may facilitate a clear choice between the two positions, it gives a very misleading picture of what the social constructionists have argued. Social constructivism is not a single position, but for the most part Franklin focuses only on two loosely organized "programmes" in British sociology of science: the Edinburgh School's "strong programme" in the sociology of knowledge (Barnes 1974; Bloor 1976), and a related "empirical relativist" case-study approach (Collins 1985; Pinch 1986). In David Bloor's (1976, p. 1) terms, the programmatic aim is to investigate, and sometimes to explain, "the very content and nature of scientific knowledge."

The strong programme in the sociology scientific knowledge built upon Karl Mannheim's (1936) *Wissensoziologie*, by proposing to "strengthen" its domain of application. Mannheim was concerned with the question of how to demonstrate the relationship between historical systems of knowledge and their "existential" conditions. In a famous passage, he suggested how this could be done:

> The existential determination of thought may be regarded as a demonstrated fact in those realms of thought in which we can show (a) that the process of knowing does not actually develop historically in accordance with immanent laws, that it does not follow only from the "nature of things" or from "pure logical possibilities", and that it is not driven by an "inner dialectic". On the contrary, the emergence and crystallization of actual thought is influenced in many decisive points by extra-theoretical factors of the most diverse sort. These may be called, in contradistinction to purely theoretical factors, existential factors. This existential determination of thought will also have to be regarded as a fact (b) if the influence of these existential factors on the concrete content of knowledge is of more than mere peripheral importance, if they are relevant not only to the genesis of ideas, but penetrate into their forms and content and if, furthermore, they decisively determine its scope and the intensity of our experience and observation, i.e. that which we formerly referred to as the "perspective" of the subject. (Mannheim 1936, pp. 239-40)

Mannheim acknowledged that not all knowledge is equally amenable to this mode of explanation, since some of the propositions in mathematics and the "exact" sciences seem to be immanently accountable. The historical stability and consensual use of a statement like "two times two equals four" makes it impossible to show how the content of the statement reflects the particular social position of its users (Mannheim 1936, p. 272).[1] The form of the statement "gives no clue as to when, where, and by whom it was formulated," unlike an artistic work whose composition can give art historians many clues for assigning it to a particular artist or genre of art, associating it with historically relative stylistic conventions, and explicating the relevant artistic community's presuppositions about the nature of the artistic subject. Similarly, a text or argument in the social sciences can easily be traced to a "school" or "perspective" like Marxism, functionalism, or rational-choice theory.

Bloor and other social constructivists take issue with Mannheim's apparent "exemption" of mathematics and the exact sciences from the purview of the sociology of knowledge, and they argue that this exemption is a consequence of his belief in the transcendent reality of mathematical objects and natural laws. As I understand Mannheim's position, however, he is not subscribing to an absolutist position on the "nature of things" or "pure logical possibilities," any more than he is endorsing a Hegelian conception of the "inner dialectic" of ideas. Instead, he is discussing the requirements for *demonstrating* the "existential determination of thought" against the claims of various absolutist and transcendental philosophies. Naturalism, logical determinism, and dialectics challenge the sociology of knowledge with obstinate arguments generated from within, or on behalf of, rival philosophical commitments. Such arguments are not easily displaced, and Mannheim recommends a methodical procedure for accomplishing their displacement in particular cases. This procedure has two basic steps:

(a) Employing historical comparisons to show that an "immanent theory" cannot entirely explain the contents and historical development of the system of knowledge in which it is situated. This procedure is used to demonstrate that such a theory cannot unequivocally and exhaustively attribute the state of its knowledge at any given time to "the nature of things," "pure logical possibilities," or an "inner dialectic".

(b) Specifying the social conditions (the local historical milieux, class interests and group 'mentalities', rhetorical strategies, etc.) that influenced the development and content of the given state of knowledge.

Mannheim strongly opposed transcendental and absolutist philosophies, so it might seem that he would dismiss the very possibility that knowledge *could ever* "develop historically in accordance with immanent laws." Nevertheless, he was unable to find a way to demonstrate that an expression like "2 x 2 = 4" could be explained by extra-theoretical "existential" factors. Bloor, Barnes, Collins and other contemporary sociologists of knowledge addressed this problem by drawing upon a variety of sources to supplement and broaden Mannheim's explanatory program. The "strong programme" in the sociology of knowledge retained Mannheim's basic two-step form of demonstration, while modifying it to cover science and mathematics. With appropriate modifications of Mannheim's terms, adherents to the strong programme sought to show that:

(a) While scientists and mathematicians may act in accordance with the immanent logic of theory, their actions are not unequivocally determined by "nature of things" or "pure logical possibilities." On the contrary, the extension of a mathematical rule or scientific theory is determined by socialized judgments and practical interests that limit the field of acceptable applications.

(b) The influence of social factors on the concrete content of scientific and mathematical knowledge is of more than peripheral importance. Intra- and extra-scientific factors influence the acceptance of theories and the interpretation of experimental evidence.

Sociologists of scientific knowledge who adhere to the strong programme often accomplish the first step with the aid of arguments from philosophy of science about the underdetermination of theories by facts and the theory-ladenness of observation, and they make use of more general skeptical arguments about the relationship between signs and meanings.[2] Radicalizing Kuhn, they tend to view historical controversies as particularly illuminating phenomena. Their descriptions of controversies demonstrate that consensus is essentially fragile, that controversies end without being definitively settled, and that stable scientific fields often include disgruntled members who ascribe the consensus in their fields to 'mere' conformity. Historical and ethnographic documentation provides the necessary leverage for contesting the unequivocal determinacy of the "nature of things" or "pure logical possibilities," and for demonstrating the contingent nature of consensus within particular disciplines. The second step is accomplished by using diverse sources from sociology, anthropology and the philosophy of language. Bloor (1976), for instance, calls upon Durkheim's basic method for linking the symbolic content of religious ritual and magical belief to the structural divisions within the tribe. He and Barnes (1983) update Durkheim's second-hand anthropology by making use of Mary Douglas' cognitive anthropology, and particularly her "grid-group" scheme for linking the properties of a group to the cognitive style of its members' beliefs and arguments. Barnes, Bloor, and Collins also make use of Mary Hesse's (1974) "network" approach to the organization and entrenchment of culturally-specific classificatory schemes. This approach enables a demonstration of non-arbitrary (i.e., relational) variations between the configurations of similar semantic domains in different knowledge communities.

Franklin's main target is Andy Pickering's (1984) *Constructing Quarks*, an exhaustively detailed and innovative exemplar of the strong programme's research policies. As its title suggest, the study reviews the series of theoretical and experimental developments since the 1960's that culminated in the establishment of what Pickering calls the "quark/gauge theory world-view." This world-view was populated by new theoretical entities, including "quarks," which were said to be fundamental constituents of protons and neutrons. Gauge theory provided incentive for particle physicists to pursue funding for increasingly massive and powerful instruments to "penetrate" more deeply into the inner structure of matter. In line with the strong programme's two-step method of demonstration, Pickering (1984, p. 6) contests what he calls "the scientist's version" of the immanent development of a series of experiments supporting the new theories on the composition of matter. He cites the familiar philosophical arguments on the underdetermination of theories by experimental facts, and he also argues that the "facts" themselves are "deeply problematic". This, he says, is because the factual status of experimental data depends upon fallible judgments about whether equipment was functioning properly, effective controls were made, and relevant signals were correctly distinguished from noisy backgrounds. Moreover, he argues, the "factual" sense and meaning of the data were developed through the use of models, analogies, and simulations which aligned those data with theoretical pre-conceptions. Pickering proposes that the relation between theory and experimental data is one of "tuning" or "symbiosis" rather than independent verification of theory by fact. His historical account demonstrates the "potential for legitimate dissent" on questions of experimental procedure and theoretical interpretation of data.[3] He describes the debates between different research groups, and uses their discrepant accounts as a basis for demonstrating the multiplicity of possible interpretations of the relevant experimental events and their theoretical implications. To explain how scientists manage to accomplish experimental interpretations and theory-choices he introduces a concept of "opportunism in context," a way of describing how physicists pursue the particular experimental-interpretive pathways that enable them to exercise their professional skills and make use of the most "interesting" of the available theoretical developments.

Pickering's study is distinguished by its close attention to experimental practices and instrumentation. He discusses the available designs for bubble-chamber apparatus, methods for interpreting traces of sub-atomic particles, and computer simulation procedures used in experiments on "weak neutral currents". His pragmatic focus is consistent with a recent trend in social studies of science toward descriptions of experimental instrumentation, technique, and analysis (Shapin and Schaffer 1985; Gooding 1988). The more abstract, theory-based conception of knowledge familiar from earlier socio-historical studies is gradually turning into a more particularistic conception of the material sites, artifacts, and techniques of 'knowledge production'. The focus is more intensive and "internal" (in the non-rationalist sense), as the aim is to identify the pragmatic strategies and informal judgments made at the worksite when researchers sort through "messy" arrays of data and decide whether equipment is working properly.

As I understand it, Franklin's characterization of the strong programme's thesis is more than a little overdrawn, and in my view his historiographic "test" does not actually discriminate between mutually exclusive accounts of the relationship between theory and experiment. Social constructivists do *not* say that experimental evidence is irrelevant to theory choice, confirmation, and refutation. Nor do they argue that there are no good reasons for belief in the validity of evidence. Instead, they argue that experimental evidence does not *compel* acceptance of a single theory, or, in Pickering's (1984, p. 5) terms, "experiment cannot *oblige* scientists to make a particular choice of theories." Franklin apparently does not disagree with Pickering on this point. Franklin (1990a, p. 147) agrees, at least partially, with the Duhem-Quine underdetermination thesis, which states, as he puts it, that "no finite set of confirming instances can entail a universal statement," and he gives a homely example: "No matter how many white swans one sees it does not entail that 'all swans are white.' A single instance can, however, refute a universal statement. Observation of a single black swan refutes 'all swans are white'." (I am not so sure about the latter part of this statement, that a single black swan would refute the statement "all swans are white." Unless the anomalous instance were determined to be representative of a coherent population, variety or species of swan, it would most likely be viewed as a freak, mutant, or victim of an oil spill.) Franklin then says that, for him, "compel" means "having good reasons for belief," and he characterizes these good reasons in terms of pragmatic strategies and plausibility judgments. It is not clear to me whether this version of experimental "compulsion" is incompatible with the constructivist position, since constructivist explanations only require that, as mentioned earlier, scientific developments "do not follow only from the 'nature of things' or from 'pure logical possibilities'." The important words here are "only" and "pure". The explanatory program does not prohibit the possibility that scientists use the ruling-out and reality-testing strategies that Franklin describes, it only requires that these not be treated as absolutely compelling or exclusive grounds for belief.

3. "Good Reasons" and Valid Evidence

What is at stake in the debate between social constructivism and realism can perhaps be clarified by focusing upon two related questions: (a) whether "good reasons" for accepting evidence imply validity, and (b) whether sociologists and historians of science should take a partisan position on the scientific arguments they study.

(a) The first question concerns whether the fact that theory choice, confirmation, and refutation are made on the basis of experimental evidence justifies the conclusion that such evidence is "valid". From a sociological point of view, the uncontroversial fact that scientists typically give reasons for their choices — reasons that they hope will be accepted as good reasons for the validity of the evidence — does not justify treating such reasons as causes for consensual belief in the validity of the evidence. In my reading, "valid" is a gratuitous term in Franklin's assertion that theory choices are settled "on the basis of valid experimental evidence." Even if we grant that scientists base their theoretical judgments on experimental evidence, and that retrospective analyses of the evidence show systematic patterns consistent with such judgments, are we compelled to conclude that the evidence was "valid"? As Hacking (1983, p. 54) puts it, "To add 'and photons are real', after Einstein has finished, is to add nothing to the understanding.... If the explainer protests, saying that Einstein himself asserted the existence of photons, then he is begging the question. For the debate between realist and anti-realist is whether the adequacy of Einstein's theory of the photon does require that the photons be real."

Sociologists and historians of science have described numerous cases where members of scientific communities come to agree that particular experimental or observational results are valid. So, for instance, I take it that in the case of atomic parity violation experiments, Pickering and Franklin both agree that as Franklin (1990b, p. 165) puts it, "By 1979 the Weinberg-Salam theory was regarded by the high-energy physics community as established." But where Franklin wants to explain the establishment of this theory by citing the validity of experimental evidence for it, Pickering aims to describe the historical process without initially making assumptions about which evidences were or were not valid.

(b) Instead of opposing or rejecting evidential accounts of theory choices, sociologists of knowledge try to remain uncommitted to the extant 'beliefs' in the communities they study. This policy dates back at least to Mannheim's (1936) attempt to distinguish the sociology of knowledge from an epistemologically "relativist" position by saying that relativism retains an absolutist standard of evaluation when it confuses the insight that 'all knowledge is relative to the knower's situation' with the conclusion that 'all knowledge-claims must be doubted.' Presuming to doubt *all* knowledge is no less absolutist than presuming that there must be a ground for all true knowledge. So, instead of advocating relativism, Mannheim argued for a "relationist" concept of knowledge. Rather than opting for a radically individualistic conception of knowledge, he proposed that particular ideas are situated within historical and social circumstances. Such ideas might not be justifiable, according to absolutist standards of rationality, but this should not discount their adequacy in terms of the relevant epistemic community's categorical judgments and validity claims. For Mannheim, "relational" knowledge — knowledge cultivated within a living community of understandings — could be dynamic without necessarily being arbitrary, and he argued that a "non-evaluative general total conception of ideology" could be attained.

> The non-evaluative general total conception of ideology is to be found primarily in those historical investigations, where, provisionally and for the sake of the simplification of the problem, no judgments are pronounced as to the correctness of the ideas to be treated. ... The task of a study of ideology, which tries to be free from value-judgments is to understand the narrowness of each individual point of view and the interplay between these distinctive attitudes in the total social process (Mannheim 1936, p. 80).

Although contemporary sociologists of science are critical of many of Mannheim's views, they share his aim to "step back" from the systems of knowledge studied without discounting the cultural and pragmatic 'validity' of that knowledge. Franklin (1990b, p. 162) quotes a line from Trevor Pinch (1986, p. 8), saying about scientific "beliefs" that "*many pictures* can be painted, and furthermore, ... the sociologist of science cannot say that any picture is a better representation of Nature than any other." Pinch is not saying that "all pictures of the world are equally good" (Franklin, 1990b, p. 163) nor is he denying that normative appraisals of the evidence are part of the picture. Instead, he is making a point about the relationship between the sociology of knowledge and the scientific fields it studies. The important phrase in his remark is "the sociologist of science cannot say . . . ." Pickering makes a similar point about history of physics. He argues that the historian's descriptive task is different from the naturalistic endeavors of the scientists studied. Historians write about human actions, whereas physicists attempt to "discover the structure of nature" (Pickering 1984, p. 8). Like other constructivist sociologists of science, Pinch and Pickering recommend that sociologists should attempt to study the contemporaneous actions of scientists, while remaining detached from the scientists' naturalistic commitments.

Franklin has fewer qualms about retrospective descriptions based on currently accepted physics, in part because he accepts the distinction between context of discovery (or, as he prefers, the context of "pursuit") and context of justification. He says he is interested mainly in justification of physicists' choices, and he comfortably makes assertions like the following: "During the 1960s events now attributed to weak-neutral currents were seen but were attributed to neutron background. At the time there was no theoretical prediction of such currents" (Franklin 1990b, p. 164). From his point of view, what physicists later took to be the case can be used to define what physicists had "seen" in the 1960's. Social constructivists have a different aim entirely, as they try to describe the operative conditions under which scientists perform their collective activities. From their point of view it is absurd to say that physicists had "seen" evidences of weak-neutral currents before they had the relevant concept. Instead, it would be more appropriate to say that they *saw* fluctuations in the neutron background. The descriptive task would then be to follow the series of theoretical innovations, experiments, negotiations, arguments, controversies, and the like, from which the 'discovery of weak neutral currents' eventually emerged. Such a history would not reproduce the physicists' historicized account of the discovery, it would attempt to recover the series of scientists' actions together with their historicized achievement (Garfinkel, Lynch, and Livingston 1981).

One can certainly question whether historians and sociologists can indeed detach themselves from retrospective understandings, and one can surely doubt that it is possible to describe scientists' actions without making judgments about the correctness of those actions and the theoretical entities they implicate. But it misses the point to read Pinch or Pickering to be saying that evidence is irrelevant to theory choice or that all theories are equally acceptable. Franklin's reconstructions of experiments on parity-violation experiments do not refute their claims, since they do not make the claims he refutes. In a way, he is speaking as one of the "natives" Pinch and Pickering study when he insists that evidences are "valid" and that there are "good reasons" for the choices scientists make.

4. Transcendental Physics

To return to the question with which I began this paper: Does Franklin's evidence compel us to favor his evidence model instead of Pickering's constructivist model of "opportunism in context"? My answer is "No," since Franklin's evidence is not independent of the articulation of his model, and the position he tries to persuade us to reject is a caricature of a constructivist argument. Short of awarding Franklin with a decisive victory, we could perhaps consider giving him more modest credit. But to do this, we need to locate where exactly his account differs from those of his constructivist interlocutors. Franklin's major challenge to constructivism seems to be that the evidence provided by the Washington-Oxford experiments was doubtful from the beginning, and that this was recognized at the time by members of the physics community. According to him, the uncertainties in the calculations and in the experimental procedure were such that physicists did not accept the results with a great deal of confidence. When the results from the 1979 experiments were presented, according to Franklin, physicists had clear procedural and evidential grounds for preferring them to the Washington-Oxford results. He argues that researchers forged ahead on the basis of relative (but not absolute) assurance that the evidence supporting the W-S theory was stronger than the evidence against it, and later experiments further justified their judgments. So, according to Franklin they had "good reasons" *at the time* for favoring the experiments supportive of the W-S theory. His analytic procedures mute this claim somewhat, since his retrospective assessment of the evidence confuses the issue of what physicists at the time made of the relevant experimental results.

At this point, to assess (or contest) Franklin's claim seems to require historical research about just how physicists understood the Washington-Oxford experiments in the late 1970s. Franklin does supply some testimony about this, but he confuses the issue with his overriding concern about whether the physicist's choices were, in the end, justified. From his account, we do not get any sense of there being a diversity of views in the mid-1970's particle physics community, nor does he give us a dynamic picture of the various arguments that may have circulated within that community and of the temporal "careers" of those arguments. His narration of experimental practices is rather static, as Robert Ackermann pointed out in a review of Franklin's (1986) earlier book:

> An irony of Franklin's book is that an actual experimental set-up is only portrayed once in its fully contingent form, and that in the glorious confusion of apparatus in the dust jacket photograph. Inside, as in all 'histories' of experimentation, experimental set-ups are given in schematic diagrams that portray the theory of how apparatus could work so as to produce meaningful data, and observational data are represented in the smoothed form gathered from properly working apparatus ... (Ackerman 1989, p. 188)

I think the same could be said for Franklin's more recent reconstruction of the atomic parity violation experiments. Although I am not skeptical about Franklin's claim that scientists orient to evidence and have good reasons for acting as they do, I think it is worth distinguishing the reasons particular scientists give for their judgments from an account of how an idealized reasoner would assess reconstructed arrays of experimental data. This does not come down to a difference between contexts of discovery and justification, since provisional justifications are constructed when scientists progressively work through the contingencies in a novel experimental situation. Franklin's inventories of experimental checks, calibrations of equipment, and so forth, suggest some of the ways in which experimentalists construct justifications, but again he bases his "epistemology of experiment" on abstracted experimental designs stated in written reports. To gain an appreciation of the contingent production of "live" experimentation, perhaps an example will help. The example I will use is not drawn from particle physics, nor is it an example of an experiment. The example is taken from an account of a discovery in astronomy, and I use it here because it provides a simple and dramatic demonstration of a progression of actions unfolding in time in a "scientific" research situation.

5. An Excursion Into Astronomy

Below, I have reproduced portions of a transcript of a conversation that was recorded aboard NASA's Kuiper Airborne Observatory, while it flew over the Indian Ocean in March 1977. The transcripts were presented in an article published that same year by three of the members of the team (Elliot, Dunham, and Millis 1977, pp. 414-15). According to the researchers, their expedition was designed to record high-quality photo-electric light curves of a star (SAO 158687) as it was eclipsed by the planet Uranus. They believed that the data would enable them to find how the temperature and other properties of the planet's atmosphere changed with height above its surface, and by coordinating their observations with those from a few ground-based observatories, they hoped to be able to get precise measures of the diameter and oblateness of the planet. On the appointed night, they flew along a path calculated to be in the shadow of the eclipse, and they set the telescope on the star and begin recording its light curve on a chart recorder. About a half-hour before the predicted eclipse, the following conversation ensued:

(The main speakers in the transcript include Jim Elliot, principal investigator; Ted Dunham, data recorder; Jim McClenahan, NASA mission director; Al Meyer, telescope operator; and Pete Kuhn, meteorologist. Fourteen other members of the NASA team and flight crew were also aboard.)

Dunham:	What was that? What was that?
Elliot:	What?
Dunham:	This!
Elliot:	I dunno. Was there a tracker glitch?
Meyer:	Nothing here.

Dunham:	Uh-oh. No, I don't think it's anything here, it's clearly duplicated in both channels.
Elliot:	Yeah, I mean, clouds, or . . .?
Meyer:	Ask Pete.
Dunham:	Pete, what's your water vapor?
Kuhn:	Eight point nine.
Dunham:	Well, that's pretty low.
McClenahan:	What happened?
Elliot:	Well, we got a dip in the signal here, which was either due to a loss or momentary glitch in the tracker, or a cloud whipping through.
Dunham:	Okay, I think somebody should have the responsibility of always watching the focal plane there. I suppose that a lot of people are.
Elliot:	But no one caught that one.
McClenahan:	Nobody caught that one. . . .

The transcript continues after a break of a minute or two:

Dunham:	OK, I got a deep short spike here.
Elliot:	I wonder if we're getting any clouds?
Dunham:	No, Pete said we had . . . microns of water.
Kuhn:	There's no clouds; I mean, truthfully, there's nothing up here.
Elliot:	Well, maybe this is a D ring. ["This comment, which causes general laughter, was prompted by a team joke: If we didn't observe an occultation, we could use the data to put an upper limit on the optical depth of a hypothetical ring around Uranus!" — From Elliot *et al.*, (1977) p. 414.]
Dunham:	With a normal optical depth of three, right? Another one.
Elliot:	Yeah, those are real — I guess.
Oishi:	Boy, that was a deep one.
Kuhn:	Yep. There's no indication of any fog at all.
McClenahan:	Doesn't seem to be any bore-sight shifting.
Elliot:	Yeah, that's good. I think we're getting real — could be small bodies — the satellite plane is face on, or it could be just small bodies like thin rings.
. . .	
Elliot:	Maybe it's something to do with Uranus, because they seem to be about the same amplitude on that scale. Nominal occultation in twenty minutes.
Mink:	Right.
Dunham:	Another one!
Elliot:	It's definitely the star being occulted somehow.
. . .	
Dunham:	There's another one!

The full eclipse occurred on schedule, and afterwards the researchers found out that their recording of the 'dips' was corroborated by Perth Observatory. Two days later, one of the researchers noticed that the dips in the light curve before and after the eclipse matched up almost perfectly, and this seemed to indicate that they were very thin rings, and not satellites (moons) as they had previously thought. Although Herschel claimed to see Uranus' rings when he discovered the planet in the late eighteenth century, this later was dismissed as an impossibility.

Ostensively, this excerpt from a "live" sequence of observations supports Franklin's thesis. The succession of phenomenal "dips" and the observing team's intervening checks on water vapor and instrument tracking can be cited as examples of an "epistemology of experiment." The transcript enables us to follow, in rapid succession, how a surprising and singular anomaly becomes progressively "attached to nature" as the team deploys its specialized skills and monitors the equipment to eliminate the possibility of a tracking error or cloud interference. The succession of dips provides a kind of naturally occurring basis for ruling out possible interpretations of the prior dips and honing in on a more restricted set of phenomenal possibilities. The astronomers themselves seem to subscribe to Franklin's thesis. In their published article, Elliot, Dunham, and Millis (1977, p. 414) give a 'Franklinian' interpretation of the transcript. Alongside the transcript they present commentaries on what they later determined they were seeing at the time. For instance, just before Dunham exclaims "What was that?" the article tells us that "First secondary occultation appears on the chart record, but is not noticed for almost a minute." They also identify "Delta ring occultation" at the point in the transcript where Dunham remarks, "OK, I got a deep short spike here." This ex-post facto identification of what the researchers were "really" looking at enables readers to gain an ironic appreciation of the transcripted fact that Elliot's remark "Well, maybe this is a D ring" at the time draws laughter from his colleagues. A determination made two days later — that the "dips" were evidence of occultations of the star by planetary rings — is used in the article to define what the researchers saw and failed to see at the time.

This way of conceptualizing the "actual events" in the sequence may seem plausible, natural, and even irresistible. Nevertheless, it is not an accurate historical description, if by "historical description" is meant an account that identifies the significance historical agents "attach to" the events in their life-world at a particular time. Instead, what we might call a "transcendental" vantage point equips the reader with a fore-knowledge of what was determined afterwards; a fore-knowledge that consequently acts as a backdrop for defining what the speakers in the transcript were "really" seeing. Franklin's account of parity-violation experiments is a slightly refracted version of such a transcendental history, as he does not hesitate to use a practically and historically-established account to identify how specific historical agents managed to achieve it. Moreover, Franklin tends to treat observation and reasoning as monological phenomena, whereas the transcript allows us to appreciate that observation and reasoning were accomplished by what might be called an observing assemblage; a multi-receptive, multi-bodied, and internally communicating socio-technical unit. The voices in the transcript testify to various coordinative and communicative actions that occur as the observing assemblage adjusts and reshapes itself in light of the latest in the series of "dips". Actions occur simultaneously on several fronts, through a flexible distribution of specialists and readable technologies coordinated by supervisory remarks and commands. What is especially impressive about the observing assemblage's organic maneuvers is that they are inextricable from the practical situation; a situation that includes the airborne observatory, along with its trajectory, its equipment, its staff, and its particular mission. The "epistemological" strategies are deeply embedded in that practical situation.

6. Conclusion

I have argued that Franklin's dispute with constructivism is miscast before he even begins to talk physics. Constructivists do not deny the role of practice, materials, and evidence. Nor do they say that all evidence is equally good or that scientists do not act without having good reasons. As Wittgenstein (1969, §105) asserts,

> All testing, all confirmation and disconfirmation of a hypothesis takes place already within a system. And this system is not a more or less arbitrary and doubt-

ful point of departure for all our arguments: no, it belongs to the essence of what we call an argument. The system is not so much the point of departure, as the element in which arguments have their life.

When uncertainties emerge they are resolved (when they are resolved) in reference to an unquestioned background. The background for scientific experiments includes accepted theories, concepts, procedures, and organizational arrangements, as well as entities and forces that are assumed to exist, laws that are held to be invariant, general and particular conceptions of how the experimental instruments are designed to operate and how they are operating now, assessments of the competency of the staff, and trust of previous runs of the experiment. Some of these hold fast as unquestioned bases for noticing and resolving particular uncertainties. On the surface, this seems to support Franklin's argument, but upon further examination it does not. Wittgenstein resists making any suggestion that those things that we hold fast, or that we use as practical "tests" for assessing more tenuous matters, are therefore "real" or "certain" in some metaphysical sense. Franklin seems to want to upgrade the *praxeological validity* (Garfinkel et al., 1988, p. 22) of experimental practices into "epistemological strategies," and while this may seem warranted by the case materials he presents, it has the effect of detaching these strategies from the local equipmental and technical environments that enable and at the same time frustrate scientists' attempts to make experiments work.

Notes

[1] Also see Mannheim (1936, p. 79). Stephen Turner (1981, p. 231, n. 3) observes that, contrary to what is assumed in many criticisms, Mannheim's exemption of the truths of arithmetic from sociological explanation was not made "on the ground of a criterion of 'rationality'."

[2] For a concise account of the use of the underdetermination and theory-ladenness theses in sociology of scientific knowledge, see Knorr-Cetina and Mulkay (1983).

[3] Pickering's training as a physicist is indispensable for this procedure, since it enables him to make claimably "legitimate" counterfactual assessments on what the experimenters he examines *could have* concluded. His competency enables him to avoid engaging in the sort of armchair relativism where general philosophical arguments assure the possibility of interpretative alternatives to those considered by the actual participants in an historical case. So, in a sense, his account is also a "scientist's version" although one that expresses perhaps an unusual set of theoretical and methodological commitments.

References

Ackerman, R. (1989), "The New Experimentalism," *British Journal for the Philosophy of Science*, 40: 185-90.

Barnes, B. (1974), *Scientific Knowledge and Sociological Theory*. London: Routledge and Kegan Paul.

Barnes, B. (1983), "On the conventional character of knowledge and cognition," in *Science Observed: Perspectives on the Social Study of Science*, K. Knorr-Cetina and M. Mulkay (eds.). London: Sage, pp. 19-51.

Bloor, D. (1976), *Knowledge and Social Imagery*. London: Routledge and Kegan Paul.

Collins, H. (1985), *Changing Order: Replication and Induction in Scientific Practice*. London: Sage.

Elliot, J., Dunham, E., and Millis, R. (1977), "Discovering the Rings of Uranus," *Sky and Telescope* 53(6): 412-16.

Franklin, A. (1986), *The Neglect of Experiment*. Cambridge, UK: Cambridge University Press.

_____. (1990a), *Experiment Right or Wrong*. Cambridge, UK: Cambridge University Press.

_____. (1990b), "Do mutants have to be slain, or do they die of natural causes? The case of atomic parity-violation experiments," Chapter Eight of *Experiment Right or Wrong*. Cambridge, UK: Cambridge University Press, pp. 162-192. (An abbreviated version of this chapter appears under the same title in *PSA 1990*, Volume 2.)

Garfinkel, H., Lynch, M., and Livingston, E. (1981), "The work of a discovering science construed with materials from the optically discovered pulsar," *Philosophy of the Social Sciences* 11: 131-58.

_____., Livingston, E., Lynch, M., MacBeth, D., and Robillard, A. (1988), "Respecifying the natural sciences as discovering sciences of practical action, I & II: Doing so ethnographically by administering a schedule of contingencies in discussions with laboratory scientists and by hanging around their laboratories," unpublished paper, Department of Sociology, UCLA.

Galison, P. (1987), *How Experiments End*. Chicago: University of Chicago Press.

Gooding, D. (1988), "How do scientists reach agreement about novel observations?" *Studies in History and Philosophy of Science* 17: 205-30.

Hacking, I. (1983), *Representing and Intervening*. Cambridge, UK: Cambridge University Press.

Hesse, M. (1974), *The Structure of Scientific Inference*. London: Macmillan.

Knorr-Cetina, K. and Mulkay, M. (1983), "Introduction: Emerging principles in social studies of science," in *Science Observed: Perspectives on the Social Study of Science*, K. Knorr-Cetina and M. Mulkay (eds.). London: Sage, pp. 1-18.

Mannheim, K. (1936), *Ideology and Utopia*. London: Routledge & Kegan Paul.

Pickering, A. (1984), *Constructing Quarks*. Chicago: University of Chicago Press.

Pinch, T. (1986), *Confronting Nature*. Dordrecht: D. Reidel.

Shapin, S., and Schaffer, S. (1985), *Leviathan and the Air Pump*. Princeton: Princeton University Press.

Turner, S. (1981), "Interpretive charity, Durkheim, and the 'strong programme' in the sociology of science," *Philosophy of the Social Sciences* 11: 231-44.

Wittgenstein, L. (1969), *On Certainty*, G.E.M. Anscombe and G.H. von Wright (eds.). Oxford: Basil Blackwell.

Do Mutants Have to be Slain, or Do They Die of Natural Causes?: The Case of Atomic Parity Violation Experiments[1]

Allan Franklin

University of Colorado

In *Constructing Quarks* (1984) Andrew Pickering discussed the early experiments on atomic parity violation performed at Oxford University and at the University of Washington and published in 1976 and 1977. The results disagreed with the predictions of the Weinberg-Salam (W-S) theory of unified electroweak interactions. Another experiment, performed at the Stanford Linear Accelerator Center in 1978, on the scattering of polarized electrons from deuterons confirmed the theory. Pickering regards the Oxford and Washington experiments as mutants, slain by the SLAC experiment.

By 1979 the W-S theory was regarded by the high-energy physics community as established, despite the fact that as Pickering recounts, "there had been no *intrinsic* change [emphasis in original] in the status of the Washington-Oxford experiments." (Pickering 1984, p. 301). In Pickering's view "particle physicists *chose* [emphasis in original] to accept the results of the SLAC experiment, *chose* to interpret them in terms of the standard model (rather than some alternative which might reconcile them with the atomic physics results) and therefore *chose* to regard the Washington-Oxford experiments as somehow defective in performance or interpretation." (Pickering 1984, p. 301). The implication seems to be that these choices were made solely so that the experimental evidence would be consistent with the W-S theory, and that there weren't good, independent reasons for them.

In this paper I will reexamine the history of this episode, presenting both Pickering's interpretation and an alternative explanation of my own, arguing that there were good reasons for the decision of the physics community. It will also explore some of the differences between my view of science and that proposed by the "strong programme" or social constructivist view in the sociology of science. A central feature of this view is that change in the content of scientific knowledge is to be explained or understood in terms of the social and/or cognitive interests of the scientists involved. The evidence model I advocate explains adherence to scientific beliefs in terms of their relationship to valid experimental evidence. This model also argues that there are good reasons for belief in that experimental evidence.

1. The Experiments

It had been experimentally demonstrated in 1957 that parity, or left-right symmetry, was violated in the weak interactions. This feature of the weak interactions had been incorporated into the Weinberg-Salam theory. The theory predicted that one would see weak neutral-current effects in the interactions of electrons with the strongly interacting particles. The effect would be quite small when compared to the dominant electromagnetic interaction, but could be distinguished from it by the fact that it violated parity conservation. A demonstration of such a parity-violating effect and a measurement of its magnitude would test the W-S theory. One such predicted effect was the rotation of the plane of polarization of polarized light when it passed through bismuth vapor. This was the experiment performed by the Oxford and Washington groups. They jointly reported preliminary values for R, the parity violating parameter, of $R = (-8 \pm 3) \times 10^{-8}$ (Washington) and $R = (+10 \pm 8) \times 10^{-8}$ (Oxford), in disagreement with the W-S predictions of $(-3 \text{ to } -4) \times 10^{-7}$ (Baird 1976).

Pickering offers the following interpretation. "Bismuth had been chosen for the experiment because relatively large effects were expected for heavy atoms, but when the effect failed to materialise a drawback of the choice became apparent. To go from the calculation of the primitive neutral-current interaction of electrons with nucleons to predictions of optical rotation in a real atomic system it was necessary to know the electron wave- functions, and in a multi-electron atom like bismuth these could only be calculated approximately....Thus in interpreting their results as a contradiction of the Weinberg-Salam model the experimenters were going out on a limb of *atomic* theory." (Pickering 1984, pp. 295-6). Pickering attributes all of the uncertainty in the comparison between experiment and theory to the theoretical calculations and none to the experimental results themselves.

The comparison was even more uncertain than Pickering implies and included such uncertainties. The experimenters reported systematic effects which they believed did not exceed $\pm 10 \times 10^{-8}$, and which were not yet fully understood. Thus, there were possible systematic experimental uncertainties of the same order of magnitude as the expected effect. The novelty of the experiments also tended to make the validity of the experimental results uncertain. The theoretical calculations of the expected effect were also uncertain, with the largest and smallest results differing by a factor of approximately two.

In 1977, both groups published more detailed accounts of their experiments with somewhat revised results (Lewis 1977 and Baird 1977). Both groups reported results in substantial disagreement with the predictions of the W-S theory. The Washington group reported a value of $R = (-0.7 \pm 3.2) \times 10^{-8}$, which was in disagreement with the new prediction of approximately -2.5×10^{-7}. This value was also inconsistent with their earlier result. This inconsistency, although not discussed by the experimenters, was discussed within the atomic physics community and lessened the credibility of the result. The Oxford result was $R = (+2.7 \pm 4.7) \times 10^{-8}$, again in disagreement with the W-S prediction. They noted, however, that there was a systematic effect in their apparatus of order 2×10^{-7} radians, which certainly cast doubt on the result. Pickering reported that the papers also "described two 'hybrid' unified electroweak models, which used neutral heavy leptons to accommodate the divergence with the findings of high energy neutrino scattering." (1984, p. 297). Although such models were discussed in the literature at the time, there is no mention of such speculation in these two experimental papers.

How were these results viewed by the physics community? Frank Close (1976) summarized the situation by noting that as the atomic physics results stood, they appeared to be inconsistent with the predictions of the Weinberg-Salam model supplemented by atomic physics calculations. Pickering states that, "if one accepted the Washington-Oxford result, the obvious conclusion was that neutral current effects violated parity conservation in neutrino interactions and conserved parity in electron interactions." (1984, p. 296). Close discussed this possibility along with another alternative that allowed the high energy neutrino experiments to show parity nonconservation while the low energy atomic physics experiments would not. "Whether such a possibility could be incorporated into the unification ideas is not clear. It also isn't clear, yet, if we have to worry. However, the clear blue sky of summer now has a cloud in it. We wait to see if it heralds a storm." (Close 1976, p. 506).

In Pickering's view, the 1977 publication of the Oxford and the Washington results indicated that, "the storm that Frank Close had glimpsed had materialised..." (Pickering 1984, p. 298), and this is supported in a summary paper by David Miller (1977). Nevertheless, I believe that the uncertainty in these experimental results made the disagreement with the W-S theory only a worrisome situation and not a crisis. In any event, the monopoly of Washington and Oxford was soon broken.

The experimental situation changed in 1978 when Barkov and Zolotorev (1978a,b, 1979, 1980b), two Soviet scientists from Novosibirsk, reported a series of measurements on the same transition in bismuth as the Oxford group. Their results agreed with the predictions of the W-S model. According to Pickering, "the details of the Soviet experiment were not known to Western physicists, making a considered evaluation of its result problematic." (1984, p. 299). This is simply not correct. During September, 1979 an international workshop devoted to neutral current interactions in atoms was held in Cargese (Williams 1980). This workshop was attended by representatives of virtually all of the groups actively working in the field, including Oxford, Washington, and Novosibirsk. At that workshop not only did the Novosibirsk group present a very detailed account of their experiment (Barkov and Zolotorev, 1980a), "but also answered many questions concerning possible systematic errors." (Bouchiat 1980, p. 364)

In early 1979, a Berkeley group reported an atomic physics result for thallium that agreed with the predictions of the W-S model (Conti 1979). Although these were not definitive results - they were only two standard deviations from zero - they did agree with the model in both sign and magnitude.

It seems fair to say that in mid-1979 the atomic physics results concerning the Weinberg-Salam theory were inconclusive. The Oxford and Washington groups had originally reported a discrepancy, but their more recent results, although preliminary, showed the presence of the predicted parity nonconserving effects. The Soviet and Berkeley results agreed with the model. Dydak (1979) summarized the situation in a talk at a 1979 conference. "It is difficult to choose between the conflicting results in order to determine the *eq* [electron-quark] coupling constants. Tentatively, we go along with the positive results from Novosibirsk and Berkeley groups and hope that future development will justify this step (*it cannot be justified at present, on clearcut experimental grounds* [Emphasis added].)" (1979, p. 35).

Pickering states that, "Having decided not to take into account the Washington-Oxford results, Dydak concluded that parity violation in atomic physics was as predicted in the standard model." (1984, p.300). I find little justification for Pickering's conclusion. Dydak was attempting to determine the best values for the parameters describing neutral-current electron scattering and had tentatively adopted the results in agreement with the W-S model. He concluded nothing about the validity of the standard model. Bouchiat was more positive. His summary paper concluded *"that parity violation has been observed roughly with the magnitude predicted by the Weinberg-Salam theory* (emphasis in original)." (Bouchiat 1980, p. 365). In another summary, Commins and Bucksbaum (1980) regarded the situation with regard to bismuth as unresolved.

The situation was made even more complex when a group at the Stanford Linear Accelerator Center (SLAC) reported a result on the scattering of polarized electrons from deuterium that agreed with the W-S model. (Prescott, 1978, 1979). Pickering concludes that scientists chose to accept the SLAC results, and the W-S theory, and chose to reject the early Oxford-Wahington results. In his view this was to unify practice within high energy physics. While I do not dispute Pickering's contention that choice was involved in the decision to accept the Weinberg-Salam model, I disagree with him about the reasons for that choice. In my view, the choice was a reasonable one based on convincing, if not overwhelming, experimental evidence

The issue seems to turn on the relative evidential weight one assigns to the original Oxford and Washington atomic physics results and to the SLAC E122 experiment on the scattering of polarized electrons. Pickering seems to regard them as having equal weight. I do not. I argued earlier that the situation with respect to atomic parity violation was very uncertain, both experimentally and theoretically

The moral of the story is clear. The early atomic physics experiments were extremely difficult, beset with systematic errors of approximately the same size as the predicted effects. There is no reason to give priority to the earliest measurements, as Pickering does. One might suggest, rather, that these earlier results were perhaps less reliable because not all of the systematic errors were known.

I will now examine the arguments presented by the SLAC group in favor of the validity and reliability of their measurement. I agree with Pickering that, "In its own way E122 was just as innovatory as the Washington-Oxford experiments and its findings were, in principle, just as open to challenge." (Pickering 1984, p. 301). For this reason, the SLAC group presented a very detailed discussion of their experimental apparatus, results, and the experimental checks they had done.

The experiment depended on a new high intensity source of longitudinally polarized electrons. The polarization of the electron beam was changed randomly to minimize the effects of drifts in the experiment. It had already been demonstrated that polarized electrons could be accelerated with negligible depolarization. In addition, both the sign and magnitude of the beam polarization were measured periodically by observing the known asymmetry in elastic electron-electron scattering from a magnetized iron foil. The experimenters also checked whether or not the apparatus produced spurious asymmetries. They measured the scattering using the unpolarized beam from the regular SLAC electron gun, for which the asymmetry should be zero, and found no effect at the level of 10^{-5}. Changes in beam polarization and the effect of beam energy were measured, for two different detection systems, and the results agreed both with each other and with theoretical calculations.

A serious source of potential error came from small systematic differences in the beam parameters for the two helicities. These could, conceivably, have caused apparent, but spurious, parity violating asymmetries. These quantities were carefully monitored and a feedback system used to stabilize them. The most significant imbalance was less than one part per million in the beam energy, which contributed -0.26×10^{-5}

to A/Q^2. This is to be compared to their final result of $A/Q^2 = (-9.5 \pm 1.6) \times 10^{-5}$ GeV/c^{-2}. This result was regarded by the physics community as a reliable and convincing result.

Contrary to Pickering's claim, hybrid models were both considered and tested by E122. The results are shown in Figure 1 and the superiority of the W-S model is obvious. For W-S they obtained a fit to the experimental data with a χ^2 probability of 40%. The hybrid model had a χ^2 probability of 6×10^{-4}, "which appears to rule out this model." My interpretation of this episode differs drastically from Pickering's. The physics community chose to accept an extremely carefully done and carefully checked experimental result that confirmed the Weinberg-Salam theory. The physics community chose to await further developments in the atomic parity violating experiments, which, as I have argued, were uncertain. The subsequent history of these experiments during the 1980s shows that although other reliable atomic physics experiments confirm the W-S theory, the bismuth results, although generally in agreement with the predictions, are still somewhat uncertain. In addition, the most plausible alternative to the W-S model, that could reconcile the original atomic physics results with the electron scattering data, was tested and found wanting. There certainly was a choice made, but, as the 'scientist's account' or evidence model suggests, it was made on the basis of experimental evidence. The mutants died of natural causes.

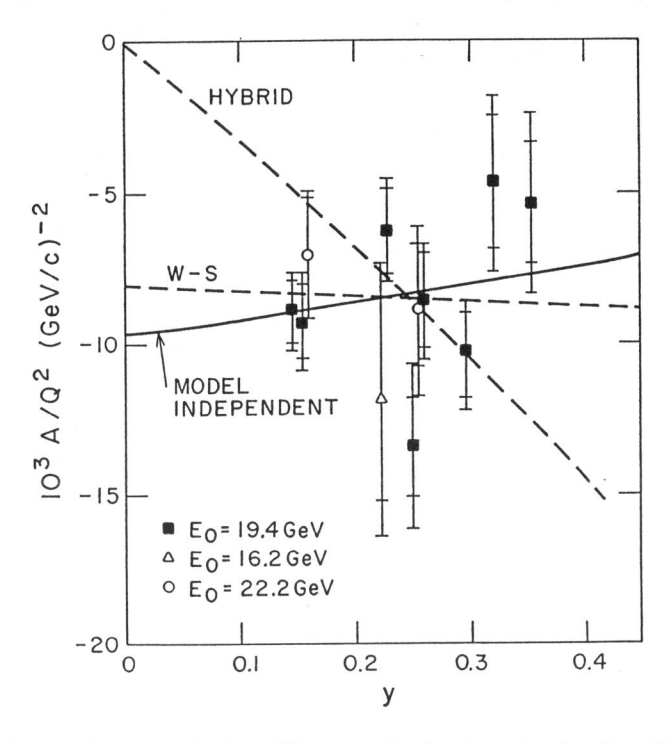

Figure 1. Asymmetries measured at three different energies plotted as a function of $y = (E_o - E')/E_o$. The predictions of the hybrid model, the Weinberg-Salam theory, and a model independent calculation are shown. "The Weinberg-Salam model is an acceptable fit to the data; the hybrid model appears to be ruled out" (Prescott 1979).

2. Discussion

There are several points worth making about this episode of atomic parity violation experiments. Perhaps most important is that the comparison between experiment and theory can often be extremely difficult. This is particularly true when, as in this episode, one is at the limit of what one can calculate confidently and what one can measure reliably.

Pickering remarks that by 1979, and presumably to this day, there had been no *intrinisic* change in the early Washington and Oxford results. In the sense that no one knows with certainty why those early results were wrong, he is correct. Nevertheless, since those early experiments, physicists have found new sources of systematic error, not dealt with in the early experiments. The redesign of the apparatus has, in many cases, precluded testing whether or not these effects were significant in the older apparatus. While one cannot claim, with certainty, that these effects account for the earlier, presumably incorrect, results, one does have reasonable grounds for believing that the later results are more accurate. The consistency of the 1980s measurements enhances that belief.

It seems clear that the evidence model fits this episode better than Pickering's model. Scientists *chose*, on the basis of reliable experimental evidence provided by the SLAC E122 experiment, to accept the Weinberg-Salam theory. They *chose* to leave an apparent, but also quite uncertain, anomaly in the atomic parity violation experiments for future investigation.

This episode also demonstrates that scientists make judgements about the reliability of experimental results that coincide with what one would decide on epistemological grounds. The SLAC group argued for the validity of their experimental result using strategies that coincide with an epistemology of experiment. (See Franklin 1990, ch. 6). As we have seen, the scientific community accepted their arguments. If the social constructivist view were correct then one would expect to find at least one episode in which the decision of the scientific community went against the weight of experimental evidence. No such episode has been provided. Nevertheless, I believe that the social constructivists are correct in insisting that in some cases the cognitive interests of the scientists do play a role in establishing experimental results, and thus in theory choice. (See Galison 1987, ch. 2 and Franklin 1986, ch. 5 for examples).

Suppose, however, that someone did present a case in which the decision went against the experimental evidence. Would that destroy the evidence model? I think not. I believe that the evidence model describes not only what scientists should do, but also what they, in fact, do most, if not all, of the time. That scientists, being human, are fallible and do not always behave as they ought to, should surprise no one.

Note

[1] This paper is a very abbreviated version of the arguments presented in Chapter 8 of Franklin (1990).

References

Baird, P.E.G. et al. (1976), "Search for Parity Nonconserving Optical Rotation in Atomic Bismuth", *Nature* 264: 528-9.

_ _ _ _ _ _ _. et al. (1977), "Search for Parity-Nonconserving Optical Rotation in Atomic Bismuth", *Physical Review Letters* 39: 798-801.

Barkov, L.M. and Zolotorev, M.S. (1978a), "Observations of Parity Nonconservation in Atomic Transitions", *JETP Letters* 27: 357-61.

_ _ _ _ _ _ _ _ _ _ _ _ _ _ _ _ _. (1978b), "Measurement of Optical Activity of Bismuth Vapor", *JETP Letters* 28: 503-6.

_ _ _ _ _ _ _ _ _ _ _ _ _ _ _ _ _ _. (1979), "Parity Violation in Atomic Bismuth", *Physics Letters* 85B: 308-13.

_ _ _ _ _ _ _ _ _ _ _ _ _ _ _ _ _ _. (1980a), "Parity Violation in Bismuth: Experiment", in Williams (1980): 52-76.

_ _ _ _ _ _ _ _ _ _ _ _ _ _ _ _ _ _. (1980b), "Parity Nonconservation in Bismuth Atoms and Neutral Weak-Interaction Currents", *Zhurnal Eksperimental' noi i Teoreticheskoi Fiziki (JETP)* 52: 360-9.

Bouchiat, C. (1980), "Neutral Current Interactions in Atoms", in Williams (1980): 357-69.

Close, F.E. (1976), "Parity Violation in Atoms?", *Nature* 264: 505-6.

Commins, E. and Bucksbaum, P. (1980), "The Parity Non-Conserving Electron-Nucleon Interaction", *Annual Reviews of Nuclear and Particle Science* 30: 1-52.

Conti, R. et al. (1979), "Preliminary Observation of Parity Nonconservation in Atomic Thallium", *Physical Review Letters* 42: 343-6.

Dydak, F. (1979), "Neutral Currents," in *Proceedings of the International Conference on High Energy Physics,* Geneva, 27 June-4 July, 1979. Geneva: CERN, pp. 25-49.

Franklin, A. (1986), *The Neglect of Experiment*. Cambridge: Cambridge University Press.

_ _ _ _ _ _. (1990), *Experiment, Right or Wrong*. Cambridge: Cambridge University Press.

Galison, P. (1987), *How Experiments End*. Chicago: University of Chicago Press.

Hollister, J.H. et al. (1981), "Measurement of Parity Nonconservation in Atomic Bismuth", *Physical Review Letters* 46: 643-6.

Lewis, L.L. et al. (1977), "Upper Limit on Parity-Nonconserving Optical Rotation in Atomic Bismuth", *Physical Review Letters* 39: 795-8.

Miller, D.J. (1977), "Elementary Particles—A Rich Harvest", *Nature* 269: 286-8.

Pickering, A. (1984), *Constructing Quarks*. Chicago: University of Chicago Press.

Prescott, C.Y. et al. (1978), "Parity Non-Conservation in Inelastic Electron Scattering", *Physics Letters* 77B: 347-52.

_____. (1979), "Further Measurements of Parity Non- Conservation in Inelastic Electron Scattering", *Physics Letters* 84B: 524-8.

Williams, W.L. (ed.) (1980), *Proceedings, International Workshop on Neutral Current Interactions in Atoms*, Cargese, 10-14 September, 1979. Washington: National Science Foundation.

Part XIII

COMPUTER SIMULATIONS IN THE PHYSICAL SCIENCES

Computer Simulations[1]

Paul Humphreys

University of Virginia

1. Introduction

A great deal of attention has been paid by philosophers to the use of computers in the modelling of human cognitive capacities and in the construction of intelligent artifacts. This emphasis has tended to obscure the fact that most of the high-level computing power in science is deployed in what appears to be a much less exciting activity: solving equations. This apparently mundane set of applications reflects the historical origins of modern computing, in the sense that most of the early computers in Britain and the U.S. were devices built to numerically attack mathematical problems that were hard, if not impossible, to solve non-numerically, especially in the areas of ballistics and fluid dynamics. The latter area was especially important for the development of atomic weapons at Los Alamos, and it is still true that a large portion of the supercomputing capacity of the United States is concentrated at weapons development laboratories such as Los Alamos and Lawrence Livermore.

Computer simulations now play a central role in the development of many physical sciences. In astronomy, in physics, in quantum chemistry, in meteorology, in geophysics, in oceanography, in crash analysis of automobiles, in the design of computer chips, in the planning of the next generation of supercomputers, in the discovery of synthetic pharmaceutical drugs, and in many other areas, simulations have become a standard part of scientific practice. My aim in the present paper is simply to provide a general picture of what computer simulations are, to explain why they have become an essential part of contemporary scientific methodology, and to argue that their use requires a new conception of the relation between theoretical models and their applications.[2]

Why should philosophers of science be interested in this new tool? Mostly, I think, because the way that simulations are developed and implemented forces us to reexamine a lot of what we tend to take as the right way to characterize parts of mathematically-oriented methodology and theorizing. Where this reexamination takes us will become clear as we go along, but before I discuss computer simulations specifically, I want to make some general points about the role of mathematical models in physical science. Let's begin with a claim that ought to be uncontroversial, but is not given enough emphasis in philosophy of science. The claim is: *One of the primary features that drives scientific progress is the development of tractable mathematics.* Whenever you have a sudden increase in useable mathematics, there will be a concomitant sudden increase in scientific progress in the area affected. This should not really need to be pointed out, but so much emphasis is placed on conceptual changes in science that powerful instrumental changes tend to be downplayed. This kind of sudden increase in mathematical power happened with the invention of the differential and integral calculus in the middle of the seventeenth century; it happened with the sudden explosion of statistical methods at the end of the nineteenth century, and I claim that the ability to implement numerical methods on computers is, in the late twentieth century, as significant a development as those earlier inventions. But what kind of development is it? Has it introduced a distinctively different kind of method into science, as Rohrlich (1991), for example, claims, or is it simply a technologically enhanced extension of methods that have long existed? If computer simulation methods are simply numerical methods, but greatly broadened in scope by fast digital computation devices with large memory capacity, then the second 'just much more of the same' view would be correct, and the situation would be similar to that in mathematics, where the introduction of computer-assisted proofs, such as were used to execute the massive combinatorial drudgery involved in the proof of the four colour theorem, is often regarded as not having changed the fundamental conception of what counts as a proof. My own view is that the situation is more complex than this simple dichotomy represents, because the introduction of computer simulation methods is not a single innovation but a multifaceted development. Let's begin with a couple of simple examples to show why mathematical intractability is an important constraint on scientific models.

2. Practical and Theoretical Unsolvability of Models

Take what is arguably the most famous law of all, Newton's Second Law. This can be stated in a variety of ways, but its standard characterization is that of a second order ordinary differential equation:

$$F = md^2y/dt^2 \qquad <1>$$

To employ this we need to specify a particular force function. In the first instance, take

$$F = GMm/R^2 \qquad <2>$$

as the gravitational force acting on a body near the Earth's surface (M is the mass of the Earth, R its radius). Then

$$GMm/R^2 = md^2y/dt^2 \qquad <3>$$

is easily solved. But the idealizations that underlie this simple mathematical model make it hopelessly unrealistic. So let's make it a little more realistic by representing the gravitational force as $GMm/(R + y)^2$, where y is the distance of the body from the Earth's surface, and by introducing a velocity-dependent drag force due to air resistance. We obtain

$$GMm/(R + y)^2 - c\rho s(dy/dt)^2 - md^2y/dt^2 \qquad <4>$$

Suppose we want to make a prediction of the position of this body at a given time, supposing zero initial velocity and initial position $y = y_0$. To get that prediction you have to solve <4>. But <4> has no known analytic solution — the move from <3> to <4> has converted a second-order, linear, homogeneous ODE into a second-order, non-linear, homogeneous ODE, and the move from linearity to non-linearity turns simple mathematics into intractable mathematics. Exactly similar problems arise in quantum mechanics from the use of Schrodinger's equation, where different specifications for the Hamiltonian in the schema

$$H\Psi = E\Psi$$

lead to wide variations in the degree of solvability of the equation. For example, the calculations needed to make quantum mechanical, rather than classical, predictions in chemistry about even very simple reactions, such as the formation of hydrogen molecules when spin and vibration variables are included, are extremely difficult and have only recently been carried out. (An explicit discussion of the differences between ab initio and semi-empirical methods in quantum chemistry is given below.)

You might say that this feature of unsolvability is a merely practical matter, and that as philosophers we should be concerned with what is possible in principle, not with what can be done in practice. But recent investigations into decision problems for differential equations have demonstrated that for many algebraic differential equations [ADE's] (i.e. those of the form

$$P(x, y_1, ..., y_n, y_1^{(1)}, ..., y_m^{(1)}, ..., y_1^{(n)}, ..., y_m^{(n)}) = 0$$

where P is a polynomial in all its variables with rational coefficients) it is undecidable whether they have solutions. For example, Jaskowski (1954) showed that there is no algorithm for determining whether a system of ADE's in several dependent variables has a solution in [0,1]. Denef and Lipshitz (1984) show that it is undecidable whether there exist analytic solutions for such ADE's in several dependent variables around a local value of x. (Further results along these lines, with references, can be found in Denef and Lipshitz (1989)). Obviously, we cannot take decidability as a necessary condition for a theory to count as scientifically useful, otherwise we would lose most of our useful fragments of mathematics, but these results do show that there are in principle, as well as practical, restrictions on what we can know to be solvable in physical theories.[3]

There is a methodological point here that needs emphasis. While much of philosophy of science is concerned with what can be done in principle, for the issue of scientific progress what is important is what can be done in practice at any given stage of scientific development. That is, because scientific progress involves a temporally ordered sequence of stages, one of the things that influences that progress is that what is possible in practice at one stage was not possible in practice at an earlier stage. If one focusses on what is possible in principle (i.e. possible in principle according to some absolute standard, rather than relative to constraints that are themselves temporally dependent) this difference cannot be represented, because the possibility-in-principle exists at both stages of development. So although what is computable in principle is important for, say, the issue of whether computational theories of the mind are too limited a representation of mental processes, what is computable in practice is the principal feature of interest for the methodologies we are considering here.

This inability to obtain specific predictions from mathematical models is a very common phenomenon, because most non-linear ODE's and almost all PDE's have no known analytic solution. In population biology, for example, consider the Lotka-Volterra equations (first formulated in 1925)

$$dx/dt = ax + bxy$$

$$dy/dt = cy + dxy$$

where x = population of prey, y = population of predators, a (>o) is the difference between natural birth and death rates for the prey, b(<o), d(>o) are constants related to chance encounters between prey and predator, c(<o) gives the natural decline in predators when no prey are available. With initial conditions x(o) = e, y(o) = f, there is no known analytic solution to the equation set.

These examples could be multiplied indefinitely, but I hope the point is clear: clean, abstract, presentations of theoretical schemas disguise the fact that the vast majority of those schemas are practically inapplicable in any direct way to even quite simple physical systems. This is not the point that models are never applicable to real systems: the point here is that even with radical idealizations, the problem of intractability is often inescapable, i.e. in order to arrive at an analytically treatable model of the system, the idealizations required would often destroy the structural features that make the model a model of that system type. This problem is widespread, and cuts across both sciences and subfields of those sciences, although it is more prevalent in some fields than in others.

These problems put severe limits on the applicability in practice of the standard, syntactically formulated method of hypothetico-deductivism, for most of the equations that represent the fundamental or derived theories of physics, chemistry, and so on cannot be used in practice to make precise deductive predictions from those representations together with the appropriate initial or boundary conditions. I should say here that I want to remain neutral as far as possible about the relative merits of the syntactic and semantic (or structuralist) reconstructions of theories. Although the semantic approach has definite advantages, both accounts are logical reconstructions of scientific practice. Because we are concerned here to stay as close as possible to considerations that present immediate problems to actual scientific practice, the debate over the merits of these reconstructions has only an indirect relevance to our interests. It is worth noting, however, that the issue of practical unsolvability means that the formulation of a theoretical model in some specific mathematical representation, rather than as a set of metamathematical structures, is an inescapable concern, and that whereas the semantic approach generally considers different linguistic formulations as mere linguistic variants of an underlying common structure, linguistic reformulations frequently have a direct impact on the ease of solvability of a mathematical representation, and hence this level cannot be ignored completely. In particular, I want to urge that what is of primary interest here is the mathematical form of equation types and not their logical form. To be specific: one could reformulate <1>, <2>, <3> and <4> in a standard logical language by using variable-binding operators, thus forcing them into the standard quantified conditional form that serves as the representation of laws in the traditional syntactic approaches, but to do this would be to distort what is crucial to issues of solvability, which is the original mathematical form.

It is this predominance of mathematically intractable models that is the primary reason why computational physics (and similar methods in other sciences), which provides a practical means of implementing non-analytic methods, constitutes a significant and, I think, a permanent, addition to the mathematical methodology of science.

3. Definitions of Computer Simulation

Here, taken more or less at random, are some suggestions that have been made for characterizing computer simulations: "Simulation is the technique by which understanding the behaviour of a physical system is obtained by making measurements or observations of the behaviour of a model representing that system." (Ord-Smith

(1975), p.3) "This is what simulation is all about, i.e. experimenting with models" (ibid, p.3)

> A precise definition of simulation is difficult to obtain...the term simulation will be used to describe the process of formulating a suitable mathematical model of a system, the development of a computer program to solve the equations of the model and operation of the computer to determine values for system variables . (Bennet (1974), p.2)

> The mathematical/logical models which are not easily amenable to conventional analytic or numeric solutions form a subset of models generally known as simulation models. A given problem defined by a mathematical/logical model can have a feasible solution, satisfactory solution, optimum solution or no solution at all. Computer modelling and simulation studies are primarily directed towards finding satisfactory solutions to practical problems. (Neelamkavil (1987), p.1).

> Simulation is a tool that is used to study the behaviour of complex systems which are mathematically intractable. (Reddy (1987), p.162)

Because of the variety of uses to which the term 'simulation' has been put, I am reluctant to try to formulate a general definition. It would be more profitable at this stage to simply explore the methods that are used under categories 1), 2), and 3) in section 4 below. We can, however, formulate a working definition based on the last characterization, which needs to be modified in three ways. First, simulation is a set of techniques, rather than a single tool. As the other quotations indicate, it would be hard to make a case for the view that there is an underlying unity to the set, at least at the present state of development of the field. Second, the systems that are the subject of simulations need not be complex either in structure or behaviour. As we also saw earlier, mathematical intractability can affect differential or integral equations having a quite simple mathematical structure, as in the case of the motion of the body falling under the influence of gravity, subject to a velocity-dependent drag force. The behaviour of this system is not unduly complex, merely hard to predict quantitatively without numerical techniques. Third, many computer simulations turn analytically intractable problems into ones that are computationally tractable, and we do not want to exclude numerical methods as a part of mathematics.

We thus arrive at the following working definition which captures what is common to almost all the simulations with which I am familiar.

Working Definition. A computer simulation is any computer-implemented method for exploring the properties of mathematical models where analytic methods are unavailable.

Some further remarks may be helpful. Although the everyday use of the term 'simulation' has connotations of deception, so that a simulation has elements of falsity, this has to be taken in a particular way for computer simulations. Inasmuch as the simulation has abstracted from the material content of the system being simulated, has employed various simplifications in the model, and uses only the mathematical form, it obviously and trivially differs from the 'real thing', but in this respect, there is no difference between simulations and any other kind of mathematical model, and it is primarily when computer simulations are used in place of empirical experiments that this element of falsity is important. But if the underlying mathematical model can be realistically construed (i.e. it is not a mere heuristic device) and is well-confirmed, then the simulation will be as 'realistic' as any theoretical representation is. Of course, approximations and idealizations are often used in the simulation that are additional to those used in the underlying model, but this is a difference in degree rather than in kind.

Next, in order for something to be a computer simulation, the whole process between data input and output must be run on a computer, whereas computational physics can involve only some stages in that process, with the others being done 'by hand'. Third, because computer simulations are usually oriented towards approximate solutions rather than exact solutions, they can be viewed as optimization devices that sometimes involve satisficing criteria. This approach underlies the variational method mentioned earlier, it underlies the simulated annealing method frequently used in connectionist models of perception and problem solution (see McClelland and Rumelhart (1986), especially Chapter 6), and it underlies many other intuitive 'good enough' criteria used in other areas.

4. Can Computer Simulation Be Identified With Numerical Methods?

What is computer simulation? The terminology is so widely used that it is hard to find a core meaning, but here are some central uses:

1) To provide solution methods for mathematical models where analytical methods are presently unavailable.
2) To provide numerical experiments in situations where natural experimentation is inappropriate (for practical reasons) or unattainable (for physical reasons). Under the former lie experiments that are too costly, too uncertain in their outcome, or too time consuming. Under the latter lie such experiments as the rotation of angle of sight of galaxies, the formation of thin disks around black holes, and so forth.
3) To generate and explore theoretical models of natural phenomena.

It may seem that use 1) is simply the use of numerical methods for solution purposes. To examine this claim, we need some definitions. *Numerical mathematics* is concerned with obtaining numerical values of the solutions to a given mathematical problem. *Numerical methods* is the part of numerical mathematics concerned with finding an approximate, feasible, solution. *Numerical analysis* has as its principal task the theoretical analysis of numerical methods and the computed solutions, with particular emphasis on the error between the computed solution and the exact solution.

Can we identify numerical methods with computer simulations? Not directly, because there are at least two additional features that a numerical method must have if it is to count as a computer simulation. First, the numerical method must be applied to a specific scientific problem in order to be part of a computational simulation. Second, the method must be computable in real time and be actually implemented on a concrete machine.

Beyond this, there is an important potential distinction between uses 1) and 3). In 1), the development of the model is made along traditional lines: some more or less fundamental theory is brought to bear on the phenomenon, theory which at least in its abstract, general, form is well understood and confirmed. Deductive consequences are drawn out from this theory to bring the general theory into contact with the specific area under investigation, and then the computational implementation of these consequences constitutes the simulation of the system. In contrast, in use 3), the development of the models is partly empirical, partly theoretical, and partly heuristic, with exploration and feedback from the simulation playing an important role in this development.

This distinction is not clearcut, and especially in use 3), elements from uses 1) and 2) often play a significant role. The difference is similar to a distinction that is often drawn in quantum chemistry between *ab initio* methods and semi-empirical methods (see e.g. R. McWeeny and B.T.Sutcliffe (1969), Chapter 9). Three kinds of treatments can be used to predict the energy levels of molecular orbitals. *Ab initio* methods use the actual Hamiltonian for the system in Schrodinger's equation. Idealizations are made, such as a fixed nucleus, only electrostatic interactions between particles, and non-relativistic calculations, and these idealizations are often drastic, but the goal is to represent as many of the important features of the molecules as possible. Then using a 'trial function' it calculates the solution 'exactly'. Semiempirical methods estimate some parameters in the orbital states that are difficult to calculate directly by empirical data or by numerical approximation, and then proceed as in the ab initio case. Model level methods use a Hamiltonian that deliberately omits some important influences on the energy levels, such as inter-electron interactions.

The distinction here between *ab initio* methods and model level methods seems to me to be quite arbitrary, since both use idealizations, and the interesting difference is that between *ab initio* methods and semi-empirical methods, and this is an appropriate place to discuss the differences between fundamental and phenomenological models. This distinction reflects the 'bottom up' and 'top down' methods familar from other areas of methodology, and there is a significant divergence of views about whether models should be constructed on the basis of some underlying general theoretical considerations, or whether instrumentally successful but theoretically ungrounded models should be used when the theoretical approach is infeasible. Both kinds are used in simulations and I see no reason to deny the appropriateness of either. I choose to focus on fundamental models here, primarily for two reasons. The first is one of expertise, or lack of it. Phenomenological models are usually highly specific devices constructed for the purpose of representing some specific phenomenon. A great deal of physical, chemical, or biological knowledge goes into their assessment, justification, and use (this is one area where 'physical intuition' is clearly an important consideration) and for this reason, such simulations can be assessed only by those actively working with them. The second reason for emphasizing fundamental models here is that this makes a comparison of simulation methodology with traditional philosophical views on theory structure and application much easier, for the latter is oriented almost exclusively towards fundamental theory. This said, a few remarks about the relation between the two approaches in the case of *ab initio* and semi-empirical models might be appropriate. (Semi-empirical models are not the same as phenomenological models, in that the former are still guided to a considerable extent by theory, but for the first reason just mentioned, I am not in a position to address phenomenological models in any detail.)

One important result of the availability of large-scale computational power is that whereas many idealizations in models, or the use of semi-empirical methods, were once forced upon chemists because the model had to result in tractable analytic mathematics, the idealizations made in *ab initio* methods now need not be determined primarily by that constraint, but are set by (a) limits on computational power available, (b) the ability to mathematically represent in the Hamiltonian complex influences on the energy levels (c) the availablity of numerical methods to approximate the representations in (b). This is a clear example of computational chemistry: the use of computers to allow one to treat models that could not be used without them. Indeed, these methods illustrate an interesting trade-off: These numerical methods allow one to deal with more realistic theories, and the increased use of approximations in the mathematics allows a decreased use of idealizations in the physics. This still leaves the treatment of most molecules currently outside the scope of *ab initio* methods, and given the restrictions due to (a) that are discussed below, no purely *ab initio* method will ever be fully computationally feasible, but the important point is that more and more systems that were once untreatable by fundamental approaches can now have theoretically justified quantum mechanical methods brought to bear on them. Compare this with methods that rely on the variation theorem. (See Eyring et al (1944) for a development of this theorem). The theorem states "If a normalized trial function S satisfies the relevant boundary conditions but is otherwise arbitrary, then $<S/H/S> \geq E_0$, where E_0 is the lowest eigenvalue, the equality applying when S is an exact solution." (Further applications of this procedure can be used to find approximations to other eigenfunctions.) Then the best wave function is obtained by varying the parameters in a trial function until the lowest energy is obtained. Here, the computational methods allow exploratory investigations that would not be possible without computers, and these are different from theoretically based methods in that although theory may be used as a guide to which parametric family of functions to explore (Gaussian or Slater orbitals are usually used, however, making the contribution of theory minimal), the final result is a matter of computational trial and error rather than explicit theoretical derivation. Moreover "A wave function that gives a good [estimate of the] energy does not necessarily give a particularly good value for another quantity, for example the dipole moment, whose expectation value may arise principally from somewhat different regions of space." (McWeeny & Sutcliffe, op. cit., p.235). I also find these figures from McWeeny and Sutcliffe, op.cit, p.239 revealing: Abstract developments of quantum mechanics require an infinite set of basis vectors to represent states. For the finite basis sets that actual applications need, suppose that m atomic orbitals are used (in the linear combination of atomic orbitals representation of molecular orbitals — the LCAO method). Then one needs $p = m(m+1)/2$ distinct integrals to calculate one-electron Hamiltonians, and $q = p(p+1)/2$ distinct integrals to calculate electron interaction terms. This gives

$$m = 4 \quad 10 \quad 20 \quad 40$$

$$q = 55 \quad 1540 \quad 22155 \quad 336610$$

This is a clear case where computational constraints, which are extra-theoretical and here involve primarily memory capacity, place severe limitations on what can be done at any given stage of technological development. This is different in principle, I think, from the constraints that the older analytic methods put on model development, because there new mathematical techniques had to be developed to allow more complex models, whereas in many cases in computational science, the mathematics stays the same, and it is technology that has to develop. The use of trial orbitals that I mentioned earlier in connection with the variational method seems to show that a very crude model can give an apparently realistic representation of the system. That is, in deciding upon the appropriate potential energy function to use in the Hamiltonian, suppose we choose one corresponding to a Slater atomic orbital of the form

$$V(r) = -fn/r + [n(n-1) - l(l+1)]/2r^2$$

where f is a parameter representing the effective field affecting the electron. Then using these atomic orbitals as the finite basis, we have to decide where the expansion of the state function will be truncated. Then, given various trials R, and trial orbitals S for another electron, we have to minimize the energy

$$E = 2 <R/h/R> + (2<RS/g/RS> - <RS/g/SR>)$$

Although there is a great deal of computation involved here, and certainly trial and error 'experimentation', there is also a good deal of theory lying in the background to justify the method, and even though the atomic orbitals used are pretty crude approximations, they are still guided by a physical model that has some theoretical justification.

I am thus going to treat each of uses 1), 2) and 3) above as part of computational physics (chemistry, etc.), and to consider computer simulation as a subset of the methods of computational science. Much more needs to be said about what is special to simulations, but I hope that the example just discussed shows that the interplay between theory, experiment, and computation in computational science entails that it is not to be identified with numerical methods, and a fortiori, neither should computer simulations.

Notes

[1] Research for this paper was supported by NSF grant DIR-8911393. I should like to thank Fritz Rohrlich for helpful discussions in connection with the PSA symposium.

[2] When examining this activity, we must be wary of one thing, which is that the field of computer simulation methods is relatively new and as such is rapidly evolving. Techniques that are widely used now may well be of minor interest twenty years hence, as developments in computer architecture, numerical methods, and software routines take place. The specific details of different kinds of simulation methods, such as finite-difference methods and Monte Carlo methods will be explored in a future paper, and some examples of currently used simulations are given in the following paper by Rohrlich.

[3] A further source of difficulty, at least in classical mechanics, involves the imposition of nonholomorphic constraints (i.e. constraints on the motion that cannot be represented in the form

$$f(r_1,...,r_n,t) = 0$$

where $\{r_i\}$ are the spatial coordinates of the particles comprising the system). For a discussion of these constraints, see Goldstein (1980), pp.11-14.

References

Bennett, A.W. (1974), *Introduction to Computer Simulation*. St. Paul, West Publishing Co.

Denef, J. and Lipshitz, L. (1984), "Power Series Solutions of Algebraic Differential Equations", *Mathematische Annalen*, 267:213-238.

_____. (1989), "Decision Problems for Differential Equations" *J. Symbolic Logic* 54:941-950.

Eyring, H., Walter, J., and Kimball, G. (1944), *Quantum Chemistry*. New York, J. Wiley and Sons.

Goldstein, H. (1980), *Classical Mechanics*, (2nd Edition). Reading, Mass, Addison-Wesley Publishing Co.

Jaskowski, S. (1954), "Example of a Class of Systems of Ordinary Differential Equations Having No Decision Method for Existence Problems" *Bulletin de l' Academie Polonaise des Sciences*, Classes III, vol. 2, pp. 155-157.

McClelland, J. and Rumelhart, D. (1986), *Parallel Distributed Processing*, Vols 1,2. Cambridge, The MIT Press.

McWeeny, R. and Sutcliffe, B.T. (1969), *Methods of Molecular Quantum Mechanics*, New York, Academic Press.

Neelamkavil, F., (1987), *Computer Simulation and Modelling*, New York, J. Wiley.

Ord-Smith, R.J. and Stephenson, J. (1975), *Computer Simulation of Continuous Systems*. New York, Cambridge University Press.

Reddy, R. (1987), "Epistemology of Knowledge-Based Systems", *Simulation*, 48:161-170.

Rohrlich, F. (1991), "Computer Simulation in the Physical Sciences", in PSA 1990, Volume 2, A. Fine, M. Forbes, and L. Wessels (eds). East Lansing: Philosophy of Science Association, pp. 497-506.

Computer Simulation in the Physical Sciences

Fritz Rohrlich

Syracuse University

1. Introduction

The central claim of this paper is that computer simulation provides (though not exclusively) a qualitatively new and different methodology for the physical sciences, and that this methodology lies somewhere intermediate between traditional theoretical physical science and its empirical methods of experimentation and observation. In many cases it involves a *new syntax* which gradually replaces the old, and it involves *theoretical model experimentation* in a qualitatively new and interesting way. Scientific activity has thus reached a new milestone somewhat comparable to the milestones that started the empirical approach (Galileo) and the deterministic mathematical approach to dynamics (the old syntax of Newton and Laplace). Computer simulation is consequently of considerable philosophical interest. In view of further technical developments in the near future, computer experts suggest that we are at present only at the very beginning of this new era.

The above claim is made here for the physical sciences only, but the suggestion is strong that computer simulation provides qualitatively new methodologies also for the biological and for the social sciences.

After an introduction into the notions of complex systems (Section 2) and phenomenological theories (Section 3), I review some of the changes brought about in the old relationship between determinism and predictability by the existence of deterministic chaos (Section 4). Then I shall present the dilemma of the theoretical model when there is a discrepancy between the theoretical model and empirical data (Section 5). Two specific examples of computer simulation are given (Sections 6 and 7) one involving the old syntax and one the new. The last three Sections are devoted to a discussion of the qualitatively new features of "doing science" by computer simulation using the two examples and providing further arguments in support of the above claim.

2. The complexity of physical systems

"Complexity" can be measured in various ways. One obvious way is the distinction of composite systems by the number of their components often called "objects", "bodies", or "particles": the three-body Helium atom (one nucleus and two electrons), the ten-body naive solar system (ignoring moons, asteroids, interplanetary matter, etc.), the DNA molecule consisting of thousands of nucleotides, or our galaxy containing hundreds of billions of stars.

The two-body problem in Newtonian dynamics can be reduced to a one-body problem; it contains one interaction between the two bodies. The three-body system cannot be reduced and contains three interactions between the three pairs of bodies. In general, the N-body problem involves $N(N-1)/2$ interactions between pairs. Thus, if one uses the number of interactions as a measure of complexity, that number increases like N^2.

Another measure of complexity for the purpose of theoretical analysis is the number of degrees of freedom, f. $f = 3$ for a single free particle since it can move in three independent directions; $f = 3^N$ for a gas of N particles.

As one considers systems of greater and greater complexity, the mathematical task of describing them becomes increasingly forbidding. Computers therefore seem to be a great boon in dealing with complexity. But one must not be led to *the naive view of scientific computing*: the computer simply provides much needed speed in carrying out a large number of mathematical operations, and it helps in keeping track of many variables. As will be demonstrated, the computer does a great deal more than that.

3. Phenomenological theories

The most fundamental physical theory involves exactly three empirically determined entities corresponding to the three "dimensions" of length, time, and mass. All other empirical quantities can be expressed as dimensionless ratios which the theory must be able to predict if it is to be fundamental. That theory is of course the goal of convergent realism.

Our actual theories are (still?) very far from that. At present even our best theory of elementary particles (which completely ignores all gravitational phenomena), the so-called 'standard theory', requires at least 17 parameters (if one assumes that all three neutrinos are massless). All these parameters must be determined by fitting theory to experiment.

On the other extreme are very "phenomenological" theories. They involve a large amount of empirical input and usually are competent only over a relatively narrow domain of physical phenomena. They are typically almost devoid of "unobservable" constructs and thus appeal to the instrumentalist. Well-known examples include geometrical optics and thermodynamics.

Almost all actual theories are somewhere between these two extremes; they are "phenomenological" (read non-fundamental) in one or more respects. As an example, Maxwellian electrodynamics is phenomenological in that it requires as empirical input the functional relations between the fields E and D and between the fields B and H.

A classification similar to the one suggested here was given by Bunge (1964). He distinguishes "black box theories" (similar to our phenomenological theories) and "transparent box theories" (similar to our fundamental theories). But he does not like the designation "fundamental" because even the above "standard theory" of elementary particles is to some extent phenomenological. I do not find this to be a strong argument against the juxtaposition of "fundamental" and "phenomenological".

4. Predictability collides with mathematical facts

Paul Humphreys (preceding paper) has already elaborated on the difficulties encountered in solving nonlinear differential equations which occur in physical theories. I shall deal with this subject from a slightly different perspective.

Mature physical theories are typically quantitative. They contain a mathematical structure which acts as the "syntax" of the theory. On the most advanced level of research, progress is often made by exploring different such structures: can one achieve empirical adequacy by a suitable interpretation of one of them? This poses an often very serious practical problem. How is one to determine the implications of a given mathematical structure if its application poses great technical mathematical difficulties?

Consider the subject of mechanics. Since Laplace, physicists have become accustomed to a standard procedure: the equations of motion of an object are second order differential equations whose solutions are uniquely determined by two initial conditions; these therefore determine the trajectory of the object uniquely for all times in the future, viz. the position as a function of time, $x = f(t)$. We have Laplacian determinism, and perfect predictability seems possible: "just give me pencil and paper and I shall construct the world's future from its present state."

Laplace was borne out by spectacular successes. A hitherto unknown planet (Neptune) was discovered in 1846 by a prediction based on the judicious applications of the equations of motion of Newtonian gravitation theory and mechanics; and in 1930 Pluto was found in a similar way. What a triumph for determinism and predictability!

Closer examination of these calculations reveals that the predictions were *not* based on exact solutions of the equations of motion but on perturbation theory (an approximation method of solution.) That approximation method was necessitated by the complexity of the system (the interaction of several planets with one another as well as with the sun had to be considered). In fact, it was proven more than one hundred years ago by Poincaré that even if there are only three bodies involved, the equations of motions are *not integrable*, i.e. there is no analytic solution that specifies the trajectories of the objects; *there is no function f* so that $x = f(t)$. Thus, while the differential equation yields a deterministic motion, that motion cannot be predicted because there is no function f. Only step by step integration can yield the solution. *Determinism has been divorced from predictability* in non-integrable systems.

The fact that perturbation theory nevertheless gives good results is misleading: it is based on an expansion which is only semiconvergent! It diverges asymptotically yielding no solution even though its first few terms converge and give good approximations.[1]

Since Poincaré, who had no digital computer at his disposal but drew his conclusions from qualitative mathematical reasoning, we know that the equations of mechanics permit a neat separation into integrable and non-integrable motion. We have been misguided since Laplace by the invalid assumption that all mechanical systems are integrable and that therefore determinism implies predictability. This is simply not the case; mathematical facts here contradict the philosophical claim of predictability as a necessary consequence of determinism.[2]

The motion specified by a differential equation can always be determined from given initial conditions by a step by step calculation taking small time intervals one at a time. In the large majority of cases (the non-integrable ones), this step by step procedure yields no regularity (such as periodicity or other known functional form f for x = f(t)). And it is here where the computer can be of tremendous help.

The important characteristic that distinguishes integrable from non-integrable motion is *the sensitivity of the motion to the initial conditions*. Two integrable motions that differ only very slightly in their initial conditions will be very close for a long time; the trajectories will diverge from one another only linearly. But two non-integrable motions whose initial conditions differ only very slightly will yield trajectories that diverge very rapidly from one another; the divergence is exponential. Even after a relatively short time will they be completely different; the farther one tries to predict, the more does that prediction depend on the precision of the intial conditions.[3]

These facts have far-reaching consequences. They imply that *integrable motion is predictable and non-integrable motion is unpredictable*. In the latter case one speaks of a *deterministic chaos*. Chaotic motion can be repeated only with initial conditions that are *exactly* the same. This is never possible in practice: no experiment can be repeated with mathematically identical initial conditions. This loss of predictability in non-integrable system is the real impact on physical science of the statement made by Paul Humphreys (previous paper) that "most nonlinear ordinary differential equations and almost all partial differential equations have no known analytic solutions".

The above description of the revolution in predictability presents a seemingly more desperate picture than is actually the case. The following must be kept in mind.

(1) Perturbation methods (semiconvergent series) around a nearby analytic solution are often possible and have been successfully used for predictions as is well-known from astronomy (see the above example of the predictions of Neptune and Pluto). The most successful example in this respect is relativistic quantum electrodynamics where an asymptotic expansion (an actually only semiconvergent perturbation theory) yields numerical predictions that in some cases exceed ten significant figures of accuracy.

(2) Phenomenological models are available or can be constructed which provide predictions to the desired accuracy. Such model are often of a probabilistic or stochastic nature and I shall present one in Section 7.

(3) Step be step integration yields usable results even where the trajectory is chaotic provided the trajectory is computed only for a sufficiently short time interval after the initial conditions. Thus, in weather forecasting, two predictions for tomorrow rather than for next week that are based on slightly different initial conditions today may not diverge uselessly far from one another.

5. The theoretical model dilemma

A theoretical model[4] is in a sense intermediate between a scientific theory and an actual physical system. As an idealized description it attributes to the latter properties which may or may not be faithful to its actual nature. Which properties are more faithfully portrayed depends on the purpose of the model. The "degree of truth" in the model - object correspondence has been called *verisimilitude* and has been defined at length by Popper (1962). Its evaluation involves a comparison with empirical evidence.

There are no rules for constructing theoretical models; they are based on educated guesses and highly trained intuition. They belong to the "art" of doing science.

In any case, how good or bad a particular model is from the point of view of physical theory can only be ascertained by relating the model to the scientific theory that is competent for it. Here it is essential to know how well the theoretical model *approximates* the theory; otherwise its empirical failure has unknown bearings on the theory. The approximation may be conceptual, mathematical, or both. Mathematical approximations are often necessitated by technical difficulties.

This situation results in the following serious *dilemma*: any disagreement between the model and the empirical evidence, i.e. any lack of verisimilitude, can either be due to the model qua approximation to the theory, or it can be due to the theory of which the model is an approximation. In the former case, the model is simply a poor approximation to the theory; in the latter case, the underlying theory itself is at fault. We shall see later how this dilemma can be resolved by computer simulations.

6. Simulation of atomistic mechanisms

"...large scale molecular dynamics computer simulations, which are in a sense *computer experiments*[5], where the evolution of a system of interacting particles is simulated with high spatial and temporal resolution by means of a direct integration of the particles' equation of motion, have greatly enhanced our understanding of a broad range of material phenomena."

This is a quote from the introduction to a recent paper by Landman et. al. (1990) in which the following computer simulation is presented.

The tip of a pin (Ni) is lowered to the plane surface of a piece of metal (Au); after contact with the surface it is raised again. What exactly is taking place on the atomic level? This is the question to which their computer simulation is addressed.

The simulation assumes 5000 atoms in dynamical interaction, 8 layers of 450 atoms constituting the Au metal, and 1400 atoms the Ni tip. The latter are arranged in 6 layers of 200 atoms followed by one layer of 128 and one of 72 atoms forming the tip. Additional layers of static atoms are assumed both below the Au and above the Ni; their separation provides a measure of the distance during lowering and raising of the Ni tip. The interatomic interactions are modeled quantitatively by means of previously established techniques and are appropriate for a temperature of 300 K. To this end the *differential equations of motion* were integrated by numerical integration in time steps of 3.05×10^{-15} sec.

The tip is lowered at a rate of $\frac{1}{4}$ Å per 500 time steps. This involves recomputing the state of the system for each of the consecutive positions of the tip. One can give a graphic presentation of each of these states of the system so that the whole process can be shown as a short film. Five states of the system are shown in the five attached figures; atoms belonging to different atomic layer have been computer-colored differently for easier identification; the present black-white reproduction does not do justice to the original.

There is clearly a tendency to forget that these figures are the result of a computer simulation, of a calculation; they are not photographs of a material physical model. But one certainly seems to be participating in an *experiment* rather than in a purely theoretical study.

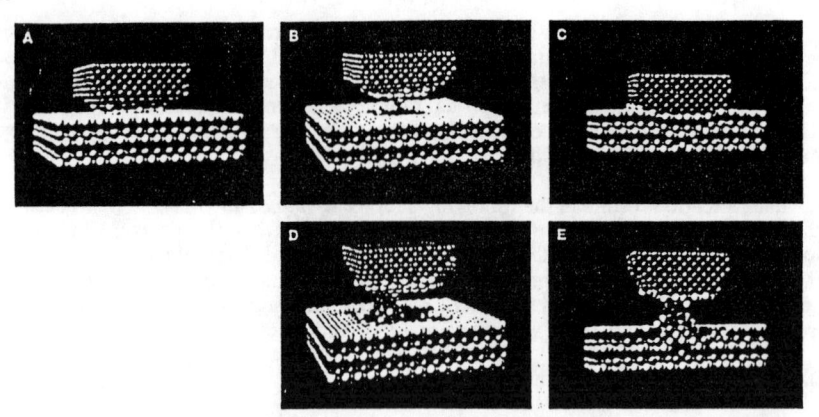

Fig. A. As the tip is lowered to smaller and smaller distance from the surface, an instability suddenly occurs: the tip "jumps down" to make contact with the surface which "jumps up". This "jump to contact" is due to an attractive force between the Ni and the Au atoms that is felt when the tip and the surface get close enough; it leads to rapid contact. The rise of the Au surface toward the tip can be clearly seen in this figure. **Fig. B.** As the tip is raised after contact, it takes with it part of the top atomic layer of the Au metal surface as well as a few atoms of its second layer. **Fig. C.** When the tip is pressed into the surface causing an indentation, the surface is seen to give way and a deformation arises that reaches all the way down through six layers. **Fig. D.** Separation after indentation produces a neck of Au atoms involving atoms from several of the top layers. The Ni tip is "wetted" by the Au atoms. Note also that several Ni tip atoms remain in the Au surface. **Fig. E.** A cut through Fig. D shows that the formation of the neck induces a deformation of the layers beneath the surface.

Can this model at all be tested by a comparison with actual experiments? It can. The calculations provide a value for the vertical force between the tip and the surface as a function of time. And this force can actually be measured. Such a measurement has not been possible until very recently since the force involved is so extremely small. The basic tool for such a measurement is the recently invented scanning tunneling microscope together with suitable additions (Binnig et. al. 1986). The measurements were done and reported in the same paper by Landman et. al. (1990). They are in satisfactory qualitatively agreement with the simulation. However, the quantitative agreement is poor for several reasons: there is a difference in the size of the Ni tip and there are long-range interactions that have been neglected in the simulation model, and there are other differences. Future work will attempt to improve on these matters.

7. Simulation of the evolution of a spiral galaxy

Astronomy has the great disadvantage relative to physics that it does not allow controlled experiments. And the natural evolution of stars and galaxies is much too slow to be observed by mankind. Luckily there are plenty of samples out there to fill in the observation of putative evolutionary stages.

An approach by means of fundamental theory to the problem of galaxy formation would of course involve Newtonian gravitation theory. But a typical galaxy has a mass of 10^8 to 10^{12} times the mass of the sun and has a corresponding number of stars. An account of the dynamics of that large a system of bodies is clearly impossible. Its complexity is too great.

In such a case, a phenomenological model is indicated. Such a model has been developed over a period of years by several investigators. It is a *statistical* model. The most recent summary of that work is contained in a paper by Seiden and Schulman (1990). But the earlier paper by Seiden and Gerola (1982) can also be consulted with profit. The short film galaxy evolution I showed in my presentation was prepared by Seiden.

Stars form by condensation from interstellar gas, live for about a billion years (give or take a factor of ten or so depending on their size) and then die. Their death occurs usually by an explosion that provides gaseous matter which becomes the material for further star formation. That process is repeated over and over again within a galaxy. Furthermore, spiral galaxies also rotate and must therefore have originated in a rotating gas. The condensation of gas in one region is triggered by a shock wave that most likely originated in a star explosion in some nearby region. This suggests that stars form in a self-propagating way and that this process is of a stochastic nature. The model advocated by the authors is indeed called "stochastic self-propagating star formation" (SSPSF).

The paradigm of this model is *percolation* known to us from making coffee. Water finds its way through the coffee grinds until it percolates all the way through. Percolation is at the base of very many phenomena: the penetration of oil through sand, the spreading of a fire in a forest, the progress of the epidemic of a contagious desease, etc. It involves a probability p of propagation from one location to a neighboring one. If p is too small the process will stop. But beyond a certain critical value p_c it will continue to the outermost reaches of the medium.

Since spiral galaxies have the shape of a thin circular disk the problem can be assumed to be two-dimensional. Such a disk can be divided into cells (sectors) as shown below. Space is thus treated in *discrete cells rather than as a continuum*.

If a star is born in one particular cell (black dot in the figure), there is a probability p that after a certain time interval another star will be born in one of the neighboring cells (indicated by circles). Since the whole disk consists of circular bands of cells, each band can be made to rotate at its own speed simulating the observed rotation (200 km/sec, say) as SSPSF is taking place. The time steps are taken to be 0.01 billion years and the radius of the disk is taken to be that of a typical galaxy.

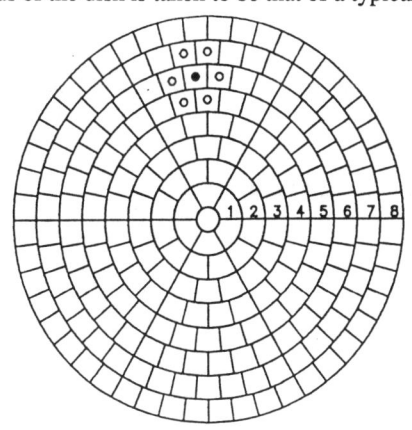

This is the galaxy evolution model and it is now just a matter of carrying it through on the computer and comparing the final result with what is seen in the sky. There are various parameters to be taken from observation: the rotational speed, the radius of the galaxy, etc. There are also adjustable parameters such as the probability for star formation within a given time period. The question is whether trial and error will yield a final appearance of the simulated galaxy that is close enough to what is observed. The figure (Seiden and Schulman 1990) shows the comparison between the final state of the simulation on the right and the photographs of the two corresponding galaxies, NGC 7793 (top) and NGC 628 (bottom) on the left. The agreement is indeed quite satisfactory (and better than the poor reproduction of the photographs seems to indicate).

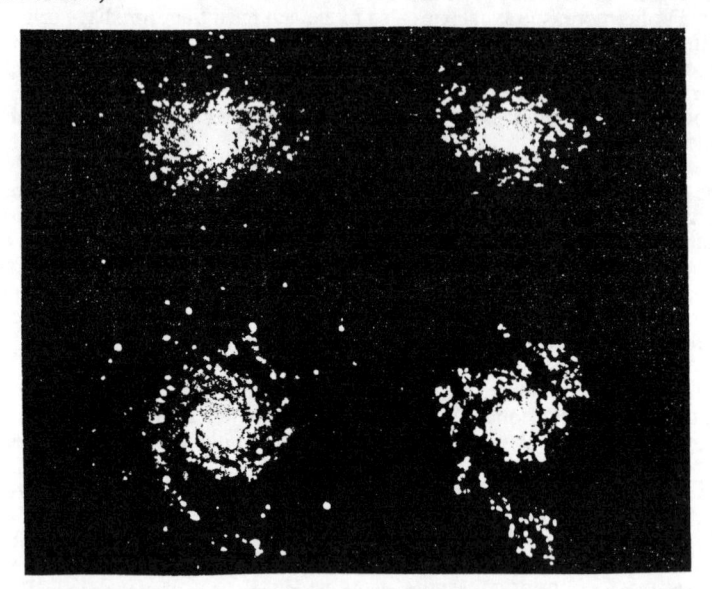

8. Model Experiments

T.H. Naylor (1966) offers the following definition. He says that computer simulation

" ... is a numerical technique for *conducting experiments*[5] on a digital computer which involves certain types of mathematical and logical models that describe the behavior of .... systems over extended periods of time."

Indeed, the first essential characteristic of simulation is modeling. And specifically for modeling on a digital computer the second essential characteristic is the ability of these *theoretical* models to allow one to conduct experiments. This is emphasized in the above definition as well as in the description given by Landman (1990) quoted at the beginning of Section 6.

In the atomistic mechanics example the Ni tip could be lowered so that it either just touches the Au surface or indents it; and the result of that action and of lifting it again can be seen in *atomic detail* in a practically continuous way. There is at present no way to carry out an actual experiment that would show that much detail over so long a time evolution.

In the galactic evolution model one can *see* the increasing number of stars and their formation into spiral arms during rotation. And one can change the attitude of the picture to conform to what is actually observed in the sky.

Simulations thus permit *theoretical model experiments*. These can be expressed by graphics that are *dynamically* "anschaulich" (a difficult to translate German word that means literally "visualizable" and that is best translated as "perspicuously clear", "vivid", and "graphic" but connotates an intuitive perception). Such "pictureableness" is psychologically and intuitively of tremendous value to the working scientist. It confirms (or corrects) his preconceived ideas and shows directions for improvements, for better models, or even for better theories. This has been elaborated in historical cases by Miller (1984).

In this respect computer simulations have something in common with *thought experiments* (Kuhn 1964/77 and Humphreys 1990). While the thought experiments have become historically famous in the context of resolving difficulties or "crises" in Kuhn's terminology, they are also in daily use (in a less dramatic way) in the imagination of every innovator whether he or she invents a new "widget" or designs a new experiment. Both thought and model experiments can be invaluable aids to perception.

9. A New Syntax

But there is yet another component to computer simulation that has so far not been touched upon. It is implicit in the second example (the galactic evolution simulation) but not in the first one. It is the use of *cellular automata* (CA) instead of differential equations as the syntax of a physical theory. Cellular automata were first suggested by von Neumann (Burks 1966) and Ulam (1960).

Recall that in the galactic evolution model one does not only treat time as discrete (making small steps of fixed time intervals) but also space; the SSPSF model of the galaxy consists of cells (rather than points) which are the primitive elements of the space over which the galaxy extends. In general, a CA is a mode of description in which (1) space and time are treated as discrete variables, (2) every cell can take on only one of a finite set of values, and (3) those values are determined synchronously (i.e. for all cells at the same instant of time) by a fixed rule that involves only the neighborhood of the cell in question. Thus, only *logical* steps are involved rather than *mathematical* ones as in the case of differential equations.

In the first example (the atomistic dynamics) the application of the theory involves the integration of differential equations which are an essential part of the mathematical structure of the applicable theory (quantum mechanics). In the second example (the galactic evolution) the CA method is used which is appropriate for a percolation theory. The syntax of the first theory is mathematical, that of the second logical. And this difference is of extreme importance.

(1) The general mathematical tool for the study of the time development of a system until now has been the differential equation. As discussed earlier (Section 4) most equations are not integrable so that analytic solutions are not available in most cases. Recourse to computers becomes a necessity. But on the computer this equation has to be integrated numerically which necessarily entails round-off errors causing the computing errors to increase with increasing time.

The new method of CA has none of these problems. There are no round-off errors so that the evolution of a CA model can be followed for an arbitrary length of time without any loss of accuracy whatever: the results are always *exact* because of the nature of the CA (Toffoli 1984).

(2) The CA method is ideally suited for digital computers (especially those with parallel architecture.) A relatively simple algorithm is involved.

(3) The CA method can easily be applied to models of complex systems involving many degrees of freedom. Differential equations are limited to a small number of degrees of freedom and become unmanageable very quickly as the complexity increases.

(4) The CA method resolves the model dilemma presented in Section 4. Since there is no approximation involved in applying the theory (all calculations are exact), any disagreement between the model and the empirical data can be blamed directly on the theory which the model realizes.

(5) The problem of predictability takes on a very different form. It is no longer the unpredictable deterministic chaos which is the dominant mode of solution. Rather, one distinguishes four classes of systems amenable to the CA method each with a different type of predictability (Wolfram 1984). Class I: predictability is trivial; class II: predictability depends only on local initial conditions; class III: predictability requires an increasingly large number of initial data as time increases (more and more initial cells are involved); class IV: the system is unpredictable.

(6) Class III systems show self-organizing behavior: in time new structures arise from structureless initial conditions. This is a fascinating new feature. It demonstrates that in complex systems not only can chaos arise from order but also conversely, order can arise from chaos. A number of examples of the latter situation are known in physical systems (Gaponov 1990) and provide yet another instance of the qualitative difference of models with a CA syntax from those based on differential equations.

These features more than justify taking CA seriously for computer simulation models of physical systems (Vishniak 1984).

10. Conclusion

I have pointed out that the syntax of theories for the evolution of physical systems which traditionally has been a mathematical structure based on differential equations can also be based on the syntax of cellular automata. These show qualitative differences of a far-reaching nature relative to differential equations. Physically, CA type models are necessarily of a phenomenological nature rather than of a fundamental one. But they are ideally suited for complex systems which make up the bulk of all systems we encounter in nature and which have so far largely escaped analysis. One could define computer simulation in a narrow sense as a method devoted entirely to the modeling of complex systems by means of the CA syntax.

Secondly, if we consider computer simulation in the wider sense (as given by Naylor - see the beginning of Section 8) and therefore include both types of syntax, computer simulation offers a new tool for science: theoretical model experiments of a scope and richness far exceeding anything available before. It is thus reasonable to conclude that we are at the threshold of an era of new scientific methodology.

Notes

[1] A fine exposition in nontechnical language of Poincarés work and the subsequent development of this subject is given by Ekeland (1988).

[2] An excellent but technical review of the different kinds of motion was presented by Berry (1983). As will be apparent from that paper, the actual situation is more complicated than can be presented here.

[3] This became known as the "butterfly effect": even a passing butterfly in the initial state can influence the predictions.

[4] For the case of physics, Redhead (1980) discusses many features of models.

[5] My emphasis.

References

Binnig, G., Quate, C.F., and Gerber, Ch. (1986), *Physical Review Letters*, 56: 930.

Bunge, M. (1964), "Phenomenological Theories", in *The Critical Approach to Science and Philosophy*, M. Bunge (ed.). New York: The Free Press, pp. 234-254.

Burks, A.W. (1966), *Theory of Self-Reproducing Automata* by John von Neumann. Urbana: University of Illinois Press.

Ekeland, I. (1988), Mathematics and the Unexpected. Chicago: University of Chicago Press.

Gaponov-Grekhov, A.V. and Rabinovich, M.I. (1990), "Disorder, Dynamical Chaos, and Structures." *Physics Today*, 43: July, 30-38.

Humphreys, P. (1990), "The Rational Use of Scientific Thought Experiments", *Colloquium in Honor of Adolf Grünbaum*, Pittsburgh, October.

_ _ _ _ _ _ _ . (1991) "Computer Simulations:, *PSA 1990*, Volume II, pp. 497-506.

Kuhn, T.S. (1964/77), "A Function for Thought Experiments", in *L'aventure de la science,* Mélanges Alexandre Koyré. Paris: Hermann, pp.307-334. Reprinted in T.S. Kuhn (1977), *The Essential Tension*, Chicago: University of Chicago Press, pp. 240-265.

Landman, U. et. al. (1990), "Atomistic Mechanisms and Dynamics of Adhesion, Nanoindentation, and Fracture", *Science* 248: 454-461.

Miller, A.I. (1984), *Imagery in Scientific Thought*. Boston: Birkhauser.

Naylor, T.H. (1966), *Computer Simulation Techniques*. New York: Wiley.

Popper, K.R. (1962), *Conjectures and Refutations*. New York: Basic Books.

Popper, K.R. (1962), *Conjectures and Refutations*. New York: Basic Books.

Redhead, M. (1980), "Models in Physics". *British Journal for the Philosophy of Science* 31: 145-163.

Rugar, D. and Hansma, P. (1990), "Atomic Force Microscopy". *Physics Today October*: 23-30.

Seiden, P.E. and Gerola, H. (1982), "Propagating Star Formation and the Structure and Evolution of Galaxies". *Fundamentals of Cosmic Rays* 7: 241-311.

Seiden, P.E. and Schulman, L.S. (1990), "Percolation Model of Galactic Structure". *Advances in Physics* 39: 1-54.

Toffoli, T. (1984), "Cellular Automata as an Alternative to (Rather than an Approximation of) Differential Equations in Modeling Physics". *Physica* 10D: 117-127.

Ulam, S.M. (1960), A Collection of Mathematical Problems. Interscience Publishers:NY.

Vishniak, G.Y. (1984), "Simulating Physics with Cellular Automata", *Physica* 10D: 96-116.

Wolfram, S. (1984), "Universality and Complexity in Cellular Automata", *Physica* 10D: 1-35.

Computer Simulations, Idealizations and Approximations[1]

Ronald Laymon

The Ohio State University

1. Introduction

It's uncontroversial that notions of idealization and approximation are central to understanding computer simulations and their rationale. So, for example, one common form of computer simulation is to abandon a realistic approach that is computationally non-tractable for a more idealized but computationally tractable approach. Many simulations of systems of interacting members can be understood this way. In such simulations, realistic descriptions of individual members are replaced with less realistic descriptions which have the virtue of making interactions computationally tractable. Such simulations can be supplemented with empirically determined correction factors which render the output produced by means of the idealizations more in accord, one hopes, with what the more realistic approach would have produced had it been computationally tractable. Another way to utilize computers is to replace an idealized but analytically tractable account of some phenomenon with a less idealized account which has no closed form or analytical solutions but where the computer can be used generate approximations to the desired solutions. It seems obvious that computer simulations such as these are correctly and naturally described in terms of idealizations and approximation. As I've said, that's uncontroversial. What's not so clear is what exactly these terms refer to. How in fact does talk of idealization and approximation serve to clarify and explain the use of computer simulations?

My plan is to take a stab at giving an analysis of the notions of idealization and approximation. I shall then propose several theses about the interrelations between these notions and the role they play in fixing or defining what the acceptance of a scientific theory is or should be. I shall then tie idealizations and approximation more directly to calculational programs that can be implemented and run on (digital) computing machines. My basic tool here will be the sort of denotational semantics for computer programs developed by Christopher Strachey and Dana Scott. The use of denotational semantics will add clarity to the proposed analysis of idealization and approximation and will also provide an improved perspective on questions of the truth and confirmation of scientific theories.

2. Approximation and Idealization

We begin with a set of ideal or paradigm cases of approximation: Archimedes' polygonal method of calculating π, Taylor expansions of differentiable continuous functions, Euclid's method for calculating roots, Newton's method for calculating roots, Euler's method of calculating solutions to initial value problems. What are the characteristics of such methods (and their calculationally more efficient brethren) and the numbers thereby generated? I shall give a deeper theory below, but for the moment it is enough to observe that these methods are constructive procedures for the production of values (the approximations) along with associated error bounds (assuming that certain conditions are met) on those values. I shall assume that the examples and this rough characterization serve to mark off a natural kind that I shall refer to as *calculational approximations*. Something is a calculational approximation then if it is the value returned by a process of the above sort. But I shall often be careless and refer to the constructive or calculational procedure itself as a calculational approximation.

Now consider this pair of examples. The governing equation for the ideal pendulum is:

$$ml\,d^2\phi/dt^2 = -mg\,\sin\phi$$

If ϕ is substituted for $\sin\phi$, this equation is *transformed* to,

$$ml\,d^2\phi/dt^2 = -mg\,\phi$$

And this equation represents a system whose period is easily calculated to be,

$$P = 2\pi\,(l/g)^{1/2}$$

The governing equation and initial conditions for a mass projected upward from the surface of the earth with velocity V is:

$$d^2x/dt^2 = -gR^2/(x+R)^2, \quad x(0) = 0, \quad dx/dt = V$$

This equation can be transformed by substituting 1 for $R^2/(x+R)^2$, which yields:

$$d^2x/dt^2 = -g, \quad x(0) = 0, \quad dx/dt = V$$

This set has the easy solution,

$$dx/dt = -gt + V, \quad x = -(1/2)gt^2 + Vt$$

What's going on in cases such as these? In general terms, we can say that there is the substitution of some term or expression X' for X in an *existing* mathematical function or equation, where it is hoped that the function is *stable* with respect to the transformation. And here I use *stable* in the ordinary mathematical sense of meaning that if X' and X are "close" then functional performance differences will be "small." Such transformations are commonplace in science. Let me emphasize that it is a necessary condition of such transformations that they be made to *already existing* theories or equations. The motivation for such transformations is usually the desire to achieve computational tractability, although on occasion such transformations may be introduced for explanatory or pedagogical purposes. Such transformations come in two extreme flavors, namely, those where mathematically derived error bounds are available and those where such error bounds are not available. In addition, there is something of a continuum between these two extremes. On the mathematically short side, the best one can do is to check for "local consistency." Roughly speaking, terms assumed small in comparison with others should remain so once the transformed equation is solved or the transformed function is allowed to run its course. Such local consistency is generally a necessary but by no means sufficient condition for actual stability. (See Lin and Segel 1974, pp. 188-189.) In between we should expect to find cases that mix physical and mathematical reasoning, where the basic move is to assume, as a working hypothesis, that the stability virtues of the phenomenon studied are possessed of the mathematical treatment as well. Perturbational analysis provides excellent examples of such mixed reasoning. Perhaps there is more than one natural kind involved in what I am calling transformational approximations. But my working assumption will be that the above characterization and examples pick out a single kind.

We now move to a consideration of *idealization*. What comes to mind here are treatments or analyses that utilize point masses, ignore medium resistance, treat finitely sized plates as if they were infinitely extended, assume the universe has two or even only one massy body. Other idealizations include treating continuous items as if they were discrete and treating discrete distributions as if they were continuous. How are treatments or analyses that make use of such devices to be characterized? To begin I'd like to insist that idealized treatments or theories *not* be modifications or transformations of existing theories or treatments. There are two reasons for this insistence. First, it captures a core of actual usage. Second, and actually more importantly, we shall need a concept that stands in contrast with transformational approximations. Assuming then the postulated necessary condition, we ask what are (or should be) the stereotypic properties of such idealizations. First and foremost, they ignore factors known to be causally relevant. More generally, there is a misstatement of the facts (or what are taken to be the facts) where this misstatement is essential to *launching* or *initiating* a formal analysis or treatment. One simply doesn't know what else to do. Given the absence then of a more realistic treatment, there will be no mathematical bounds on the effect of the idealization assumption. This absence may be mitigated somewhat by experimental evidence that supports the stability of the phenomenon with respect to the idealized or ignored causal factor. In addition, insofar as something is known about the general form of the unavailable superior account, this knowledge may be used (perhaps along with experimental evidence) to support claims about the reliability of the idealized treatment. But in general when things go wrong it will be difficult to determine what should be changed in the analysis.

We now ask the question: what are the connections between idealizations, transformational approximations and calculational approximations? *A priori* we should expect the existence of treatments that contain *both* idealizations and transformational approximations. In fact, the differential equation for the ideal pendulum, once the $\sin \phi = \phi$ substitution has been made, is exactly such an example. We should also expect that equations or accounts which are idealizations (in the sense sketched above) may be interpretable as transformational approximations with the development of more realistic or complete accounts. A simple example is that of the ideal pendulum immersed in an ideal fluid. The ideal pendulum *simpliciter* results when the hydrostatic correction factors are transformed away. (See Laymon 1989a for the details and other more realistic pendulum analyses.) In fact, it seems fair to say that there is a natural tendency in science to *convert* idealizations into transformational approximations. Such conversions provide one way of justifying commonly made claims in science that one set of idealizations or approximations is more realistic or accurate than another set. A consideration of exactly how such argumentation goes and what its limits are will have to be postponed to another occasion.[2] For now we are content to note that the claim that science aims to convert its idealizations into transformational approximations can be expanded to incorporate notions of belief and acceptance.

> Thesis 1: To accept or believe a scientific theory is to believe that every idealization can be converted into a transformational approximation.

This is, in essence, a translation of Thomas Kuhn into the concepts of idealization and approximation developed above. Another thesis I wish to put up for consideration connects the notions of transformational and computational approximations.

> Thesis 2. The mathematical structure of transformational approximations is the same as that of calculational approximations.

Exactly what this thesis comes to will be discussed below. From these theses it would seem to follow that:

> To accept or believe a scientific theory is to believe that every idealization has the structure of a calculational approximation.

On the basis of admittedly limited experience with scientists, I believe that something like this is in fact true of actual scientific practice. But I shall not press for this consequence here. Instead I shall focus on the two theses themselves. Making a case for the first is complicated by the fact that, as is typically the case in the philosophy of science, the thesis admits of both descriptive and normative readings. As a descriptive thesis its virtue is that it makes sense of those historical episodes that Kuhn called "normal science." But establishing its bona fides as a normative thesis will require an analysis of its consequences (in conjunction with thesis 2) for theory confirmation. I shall briefly discuss these consequences in a concluding section of this paper. Establishing the second thesis requires having a firmer grip on exactly what the structure of calculational approximations is supposed to be. And it is to this question of structure that I now turn.

3. Computer Program Semantics

What exactly are the approximation structures that my theses refer to? How are such structures best conceived? As earlier announced, my strategy here will be to utilize the denotational semantics for computer programs developed by Strachey and Scott. This will tie my analyses of idealization and approximation to machine implementable computations. If we think of computer simulations as machine implemented proxies for scientific theories then a natural question to ask is, how should we understand the connections between theory, program and the world. For example, does the theory attach to the world only through its implemented computations? We may also wonder what role theory plays over and above what can be programmed. Can we, as suggested by anti-realists, kick away theory? Would restriction to what's programmable unduly restrict our science? I don't propose to answer these questions here, at least not in their full generality. Instead I shall focus on the question of isolating the sense in which a computer *program* can be said to be a *proxy* for a scientific theory. More generally, we can ask, what *are* computer programs, what do they mean, and how do they get this meaning? There is great tension when we try to answer such questions. For on the one hand, *we* interpret our programs to be exactly about what we want them to be about. In the case of a physics simulation, just those functions and equations that we are interested in. But this is, of course, too glib, since we all know that what appear to be the desired mathematical functions really consist of sub-programs that change the internal memory states of the computer. In actual practice, computer languages and their associated programs are presented and learned in a mixed mode, one that combines syntax and a naive or operational semantics. The naive semantics consists in describing the effects that programs constructed in terms of the language have on the structure of some *ideal* machine. So, for example, *ld 16/c23* serves to load sixteen into the twenty-third counter of the ideal machine. While this procedure works tolerably well in practice, it is awkward and less than satisfactory in a number of deeper more theoretical respects.

To begin, there is something of a dilemma in the specification of the ideal machine. On the one hand, an ideal machine can be specified in a way that is convenient for the presentation of the language. But if so then this specification appears to be objectively arbitrary. On the other hand, problems of over-specificity arise if one attempts to make the description of the ideal machine objective by relating it to the actual hardware configuration used in some particular machine. (And one should remember that actual computers are not ideal Turing machines.) On this approach the meaning of a program just is its particular machine implementation. But if this is so then comparing different machine implementations of the same language and associated programs becomes problematic. In fact, questions of correct machine implementation of a language become meaningless. Furthermore, on this approach it is very difficult to connect the machine implementation with ordinary mathematics. (See Scott 1970 for further discussion.)

An alternative strategy is provide a *machine independent* semantics for a program that is independent of both actual and ideal machines. Implementations then are to be judged successful according to how well they respect this semantics. But how is such a machine independent semantics to be conceived? Scott-Strachey semantics seeks to circumvent the ideal or real machine implementation dilemma by temporarily sidestepping the implementation question and specifying the meaning of programming constructs in terms of a restricted class of mathematical functions. Questions of implementation are respected by requiring that certain aspects of the preferred mathematical functions be recursively effective. I shall now introduce some of the basic concepts and structures of Scott-Strachey semantics in terms of some elementary motivating examples. As we shall see, because the semantics is expressed in terms of ordinary mathematics, connecting a program in a principled way with its natural or intended meaning is considerably simplified.

4. Interval Analysis

Computers of course do not perform exact arithmetic. Because of memory limitations only finitely many numbers can be constructed and manipulated. Rounding and normalization techniques need to be employed in order to keep calculations within the restrictions imposed by finite machine memory. In a sense, then, machine numbers represent *intervals* of numbers. There's a rough correspondence here with our ordinary notions about the accuracy of numerically expressed experimental data. Such data are, even ignoring experimental error, at best only *partial* since they represent intervals of exact total values. Exactly how machine numbers represent intervals and exactly how these interval manipulations are to be best conceived is the subject of a type of numerical analysis known as *interval analysis*.

Interval analysis begins with the usual set theoretic definition of a real interval. If a and b are real numbers then the interval $[a, b]$ is defined as:

$[a, b] =_{df} \{x \mid a \leq x \leq b\}$

The elementary operations of *interval arithmetic* can be defined in analogous set theoretic fashion:

$$[a, b] * [c, d] =_{df} \{x * y \mid a \leq x \leq b, c \leq y \leq d\}$$

where * is any one of the standard arithmetic operations of addition, subtraction, multiplication or division, and where suitable adjustment of the definition is made to exclude the case of division by zero. Let me note that this definition is in terms of an *exact* underlying arithmetic on numbers. In order to apply interval analysis to computer arithmetic one needs to incorporate the fact that the exact output intervals of the above operations can only be approximated in terms of machine available numbers and operations. But I shall forgo this aspect of interval analysis and illustrate only the exact version of the theory. This will be sufficient for our purposes which are to motivate and give some understanding of Scott-Strachey semantics.

It is easy to prove that the elementary interval operations just defined are continuous in the sense that a decrease in the width of an input interval yields a decrease or no change in the width of the output interval. More precisely, the elementary operations of interval arithmetic are *inclusion monotonic*, that is, if $K \supseteq I$ and $L \supseteq J$, then $K+L \supseteq I+J, K-L \supseteq I-J, KL \supseteq IJ, K/L \supseteq I/J$. This means that more accurate input will lead to either more accurate output, or to output no worse than before. In fact, from the transitivity of interval inclusion and the fact that the elementary operations are inclusion monotonic, it can be readily shown that:

If $F(X_1, X_2, ..., X_n)$ is a *rational* expression in the interval variables $X_1, X_2, ..., X_n$, i.e., a finite combination of $X_1, X_2, ..., X_n$ and a finite set of constant intervals with interval arithmetic operations, then $X_1 \supseteq X'_1, ..., X_n \supseteq X'_n$ implies $F(X_1, X_2, ..., X_n) \supseteq F(X'_1, X'_2, ..., X'_n)$ for every set of interval numbers for which the interval arithmetic operations in F are defined. (Moore 1966, p. 11.)

None of this should be very surprising, and furthermore, once a metric is imposed on intervals one can easily restate this theorem in terms of the *continuity* of rational interval expressions. The following distance measure on our intervals is sufficient to establish a metric.

$$d([a, b], [c, d]) = \max(|a - c|, |b - d|)$$

And given this definition, it is obvious that our interval operations are continuous interval functions. In fact, it is not hard to show somewhat more, namely, that all finite combinations of interval arithmetical operations are continuous as well. (Moore 1966, theorem 4.2, p. 19.)

We shall need the notion of the *united extension* of a function. If f is some real function defined on the real interval I, then the united extension f_u of f is a function from real intervals to real intervals where, if X is an interval,

$$f_u(X) = \{f(x) \mid x \in X\}$$

That is, the f_u image of an approximation X is the union of the f images of the items approximated by X. Therefore $f_u(X)$ can be conceived as an approximation of $f(x)$. (The situation can be reversed if a filter space construction is employed, since in this case $f(x)$ can be construed as equal to the sum of all its approximations. Conceived this way, we can say that functional images do not get larger than, that is, do not exceed, their approximations.) By way of illustrating how these various notions and constructs fit together consider the function $f(x) = x/(x-2)$. The united extension of this function for the input interval [10, 12] is:

$$f_u([10, 12]) = \{ y/(y-2) \mid y \in [10, 12]\}$$

Since this function is monotonic decreasing, the endpoints by themselves are sufficient to give us the united extension,

$$f_u[10\ 12] = [1.2, 1.25]$$

Now consider the function from the point of view of interval arithmetic, that is, substitute intervals for reals and interval subtraction and division for ordinary subtraction and division in the specification of the function f. To explicitly note the fact that interval arithmetic is being used we shall denote the function computed by interval arithmetic as F. So for the input interval [10, 12] we have,

$$F([10, 12]) = [10, 12]/([10, 12] - [2, 2]) = [1, 1.5]$$

This interval contains the united extension, i.e., the exact value, but is strictly larger, as shown in figure 1. Generalizing by considering the result of applying F to the interval domain of f, we can say that F is an *approximation* of f. More in the style of Scott-Strachey semantics we can say that the function (conceived extensionally) computed by the program F is an approximation of the function computed by f. We can continue in this vein if we next divide the interval [10, 12] into the component intervals [10, 11] and [11, 12], and then compute using interval arithmetic $F[10, 11]$ and $F[11, 12]$, and finally add these values ([1.10, 1.33] and [1.11, 1.38]). In this way we get a better approximation (namely, [1.10, 1.38]) of the united extension $f_u[10, 11]$. (See figure 1.) In fact, subdividing the input interval [10, 12] in this way will lead to successively better approximations of the united extension. This convergence reinforces the idea that we view interval functions *as approximations of ordinary functions*.[4] Now if arithmetical functions can be conceived as being composed of and approximated by their interval cousins, then functions as well as data (i.e., functional input and output) can be represented as having similar interval-like structures. This is a central insight that is incorporated in an exact and general way in Scott-Strachey semantics.

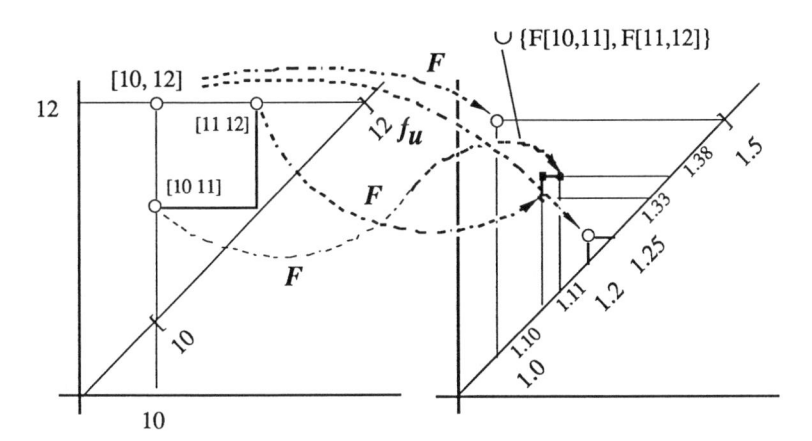

Figure 1

5. The Semantics of Recursive Calculation

Let Ψ be a function whose domain is the class of all partial functions from **N** (the natural numbers) to **N** and whose codomain is the same class, i.e., $\Psi: \text{Pfn}(N,N) \to \text{Pfn}(N,N)$. In particular let the effect of Ψ on *any* function h be defined as,

$\Psi(h(0)) = 1$

$\Psi(h(n)) = n \cdot h(n-1)$

where n is a natural number. This definition provides a recursive specification of that function which is the fixed-point of Ψ, namely, that function p such that,

$\Psi(p) = p$

The fixed-point *solution* to this equation is $\lim \Psi^n(\bot)$ as $n \to \infty$, where $\bot$ means the totally undefined function, and where $\Psi^n(\bot)$ means n successive applications of Ψ to $\bot$. E.g., $\Psi^3(\bot) = \Psi(\Psi(\Psi(\bot)))$.

	Computation - executed in some syntactic system	Meaning (function)	Kleene Semantics for Ψ
first call	$\Psi(h(0)) = 1$	$0 \to 1$	$\Psi(\bot)$
	else: $\bot$	$\begin{cases} 1 \\ \cdots \\ n \\ \cdots \end{cases} \to \bot$	
second call	$\Psi(h(1)) =$		$\Psi(\Psi(\bot))$
	$(1 \cdot \Psi(h(0)) =$	$0 \to 1$	
	$(1 \cdot 1) = 1$	$1 \to 1$	
	else: $\bot$	$\begin{cases} 2 \\ \cdots \\ n \\ \cdots \end{cases} \to \bot$	
third call	$\Psi(h(2)) =$		$\Psi(\Psi(\Psi(\bot)))$
	$(2 \cdot \Psi(h(1)) =$	$0 \to 1$	
	$2 \cdot (1 \cdot \Psi(h(0))) =$	$1 \to 1$	
	$2 \cdot (1 \cdot 1) = 2$	$2 \to 2$	
	else: $\bot$	$\begin{cases} 3 \\ \cdots \\ n \\ \cdots \end{cases} \to \bot$	

Figure 2

We now ask what constitutes a *computation* of the desired fixed-point, and what constitutes the corresponding *semantics* of the computation. The basic idea of the "all calls" semantics can be indicated by means of the table given in figure 2. In the middle column $\bot$ is used to mean undefined. The functions indicated in the middle column are the *meanings* of each stage of the computation. These functions can themselves be calculated by the Kleene sequence, which is the repeated application of Ψ to the totally undefined function. The sequence of increasingly more complete functions (move down the middle column) can be given an interpretation as being increasingly better approximations of the fixed-point solution. In fact, this sequence *converges* (in a well defined topological sense) on the recursively specified function, the fixed-point. Hence, the totally undefined function ⊥, can be interpreted as an *initial* as well as the worst approximation of the fixed-point solution.

If space permitted I would next show how Newton's method of root approximation, an example of what I've called a calculational approximation, could be reconstructed in interval form and interpreted as a fixed-point solution of a recursive equation. (See Moore 1966, pp. 58-65.)

6. Scott-Strachey Semantics

It's now time to abstract and collect the fundamental features of the examples so far considered. The basic idea behind Scott-Strachey program semantics is to interpret computer programs as mathematical functions drawn from some appropriately *restricted* class. The restrictions suggested by the examples considered are these.

(1) Function domains or possible-data spaces should be at least partially ordered sets where the ordering relation, to be denoted ≤, is to be understood in the following way: to say that $x \leq y$ means that y is consistent with x and is (possibly) more accurate than y. In short, x *approximates* y. In the examples considered, the various data spaces had an interval interpretation where the intervals were partially ordered by the containment relation. (If X and Y are intervals that contain or approximate the target value x_0, then $X \leq Y \Leftrightarrow X \supseteq Y$.)

(2) Functions between data spaces should be monotonic. So if f is a function from X to Y, then,

$x \leq y \Rightarrow f(x) \leq f(y)$

This captures a basic intuition about computation, namely, that more or better input into a computation should yield better output. We have already seen that finite combinations of interval operations satisfy this requirement. Ordinary approximation procedures, such as Newton's method of determining roots, convert into monotonic convergent procedures when intervals are used as functional input and output.

(3) The next restriction is more subtle: data spaces should have appropriate limits. Such limits are required in order to allow for principled reference to the infinite limits of sets of finite (machine implemented) approximations. Among other things, such reference is required for the semantics of *compositions* of approximations on limit constructions. (E.g., improper integrals.) Appropriate limits will be provided if our data spaces are required to be complete lattices under their partial orderings. Limits therefore will be the least upper bounds or joins of sets of data elements. If X is a data set then the lub or limit will be denoted $\bigsqcup X$.

(4) All the approximation procedures considered were continuous in the sense that input limits mapped to appropriate output limits. So if X is a set of approximations the requirement is that,

$$f(\bigsqcup X) = \bigsqcup \{f(x): x \in X\}$$

In words, a limit maps to the limit of the functional images of its approximations.

(5) Since we want our programming semantics to represent real computational and implementation possibilities, a finiteness restriction needs to be placed on data spaces. "The problem is to restrict attention to exactly those data types where the elements can be approximated by 'finite configurations' representable in machines ...." (Scott 1970, p. 12.) Roughly speaking, one needs to develop a suitable notion of a basis for generating the space and then to require that the basis be effectively given. There are several ways to do this including one that defines a topology for data spaces and uses the corresponding topological notion of a basis. We note in passing that this restriction does not mean that such data spaces are restricted to those having only countably many total or limit elements.

Assuming these five restrictions, it can be shown that sufficient functions are available for constructing a semantics adequate for computer programming. Some of the specific features of the resulting semantics are these. First, all the usual functor constructs (e.g., finite and infinite sum and product) are closed with respect to the category type. That is, sums and products of data spaces or functions are themselves data spaces or functions. Second, as suggested by our simple example from interval analysis, function spaces have the same structure as data spaces. Therefore, the same notions of approximation as used for data can be applied to functions. It also follows that functions can be used as data for other functions. That is, not only can we form compositions of functions by functor application, we can also form hierarchies of functions, by using functions as data for higher type functions, and still remain within within the initial semantic category. Given all this, a semantics can be constructed for the typed lambda calculus construed as a computer language. Semantics have also been provided for Pascal and other computer languages. (For more on Scott-Strachey semantics see Manes and Arbib 1986, Tennant 1981, Scott 1970, 1981, 1982.)

7. Comparisons with Logical Positivism and the "Semantic" View of Theories

Our problem was to isolate the sense in which a computer program could be said to be a proxy for a scientific theory. We are now in a position to meaningfully assert that a program attaches to a scientific theory by means of its programming semantics. This, of course, is at best only a partial solution to our problem. We wonder, for example, whether the program semantics is a proper subset of the theory semantics, or whether it is to be identified with the semantics of the underlying theory. Either way, the actual implementation of the program, because it itself is an approximation, can be seen as the real world representation of a subset of the theory's semantics. I shall not consider the question of whether the subset relation between programming semantics and theory semantics is proper, but shall simply adopt, as a working hypothesis, that theory and program semantics are identical. I do not mean to be assuming thereby that the semantics of talk *about* scientific theories is to be restricted to programming semantics. I mean only to be restricting the semantics of the theory itself. Given the obscurities about what should count as theories, this working hypothesis may be viewed as part of a stipulative or normative definition of *theory*.

My proposed identification of program and theory semantics suggests a parallel with logical positivism. This is because the positivists can be naturally interpreted as having been concerned with questions of the implementation of scientific theories conceived *as programmed* in ordinary first order predicate calculus. This concern is, to my mind, a significant plus for positivism. The negative side, of course, is that their models of scientific theories were not so much wrong as hopelessly simplified. One of the basic problems was the restriction to finite linguistic techniques. And it was in part this restriction that lead to Patrick Suppes' call for the wholesale use of ordinary mathematics in the philosophical representation and characterization of scientific theories. In effect, Suppes introduced a new conception of the philosophy of science and proposed that philosophers of science ignore the implementation questions of concern to the positivists. Ordinary mathematics, embedded in a set theoretic definition, was to be the canonical language for the philosophical expression of scientific theories. (For a review of the Suppes program see Moulines and Sneed 1979.) While Suppes' reform minded proposal may have been beneficial in many respects, the existence and widespread use of computer simulations dictates that some of our philosophical effort be redirected back toward those syntactic and implementational problems of interest to the logical positivists.

But there are also similarities between my identification of theory and program semantics and positions held by current "semantic" theorists. Suppes' proposal that ordinary mathematics be used in expressing theories has been seen by some as too generous and open-ended for philosophical purposes. Two leading candidates for a suitably regimented canonical form are: (1) phase space accounts (Suppe 1989, and van Fraassen 1970); (2) Bourbaki uniformities (Balzer et al. 1987). I want to add Scott-Strachey semantics to this list of candidates. It has the significant advantage of uniting the philosophical programs of the logical positivists and current semantic theorists. First of all, it captures in mathematical terms the infinitary constructs needed in science and thereby overcomes a serious limitation of logical positivism and satisfies the reasonable demands of semantic theorists for such constructs. Second, since it is a semantics of machine implementable languages it deals with the computational concerns of the logical positivists. Thirdly, it is regimented in a way that should satisfy those semantic theorists who prefer operating within preferred or canonical forms of representation. Finally, we note that the basic mathematical constructs employed (continuous lattices) are stronger than Bourbaki uniformities and that the overall approach is more regimented than phase space accounts.

8. Clarification of the Analysis of Idealization and Approximation

The reader has by now probably forgotten the original motivation for the excursus into programming semantics. It was to provide a structural account of what I called calculational approximations that would make more precise the theses I proposed about approximations and idealizations. Namely, that to accept or believe a theory is to believe that all idealizations can be made to appear as transformational approximations (thesis 1), and that transformational approximations have the same structure as calculational approximations (thesis 2). I shall now sketch an argument for thesis 2.

Assume that we have a Scott-Strachey functional space that includes some original analysis w^* (construed as a function) and all its transformational approximations. Let D be the initial (effectively given) set of finitely expressible data or functions, and let $|D|$ be the associated functional space or domain. (In essence $|D|$ consists of D and its limit points.) The limit or total elements in this domain will correspond to, or be explications of, completely specified possible functions or data. One can think of the set of total elements of $|D|$ as a set of possible worlds; partial (finite) elements then can be construed as approximate descriptions (i.e., as calculational approximations) of these possible worlds. We shall now construct another domain, call it $|W^*|$, which will order these total elements (which include w^* and its transformational approximations) into spaces that have the same basic (continuous lattice) structure as the original spaces. The basic idea is to order total elements in terms of the shared partial elements. So if *w'* and *w"* are transformational approximations of *w**, then *w** ≥ *w'* ≥ *w"* just in case every partial element consistent with *w** and *w"* is also consistent with *w'*. (Informally: *w** and *w'* share more partial elements than do *w** and *w"*.) Two simple examples of the construction are given in figure 3. In the first example, the total elements are just (1), (2) and (3). The pair (1,2), for example, is a partial element and can be read as meaning *1 or 2*. Ordering by inclusion of descriptive tokens yields the lattice on the left. If (1) is our distinguished element then the proposed construction yields the lattice on the right. So, for example, (1) and (2) share the partial descriptions (1,2) and (1,2,3), whereas (1) and (3) share the partial descriptions (1,3) and (1,2,3). Hence, the pairs (1,2) and (1,3) are both ordered below (1) and are incomparable with respect to one another. The second example is just a fragment of the binary tree with (01) as distinguished element.[3] Note that my claim is not that the set consisting of an analysis, its transformational approximations, and all calculational approximations is isomorphic to the set consisting of just the analysis and its transformational approximations. The claim is simply that these two sets can both be structured in a way that satisfies the requirements sketched earlier in section 6.

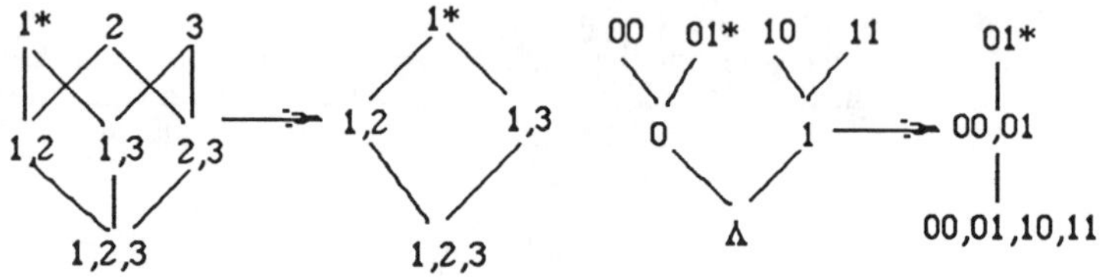

Figure 3

For the sake of completeness I shall sketch some of the formal details. The desired construction can be achieved by first identifying the tokens of the new basis or neighborhood system as being the total elements of |*D*|. (See Scott 1981 for the appropriate sense of neighborhood system; a brief primer is given in Laymon 1987.) We then let partial descriptions be defined as sets of total elements consistent with the intersection of *w** and other worlds, i.e., for any total element *w'*, let neighborhoods just be sets of the form,

$$[w^* \cap w'] =_{df} \{x \in |D| \mid x \text{ total}, x \supseteq w^* \cap w'\}$$

The intersections are well defined because |*D*| is the filter space associated with the finite basis or neighborhood system *D*. These new neighborhoods will always be non-empty since *w** ∩ *w'* is an element of the original domain, and since every partial element can be extended to a total element. Actually, we need do a slight bit more to generate a neighborhood system, namely, to identify the neighborhoods as all intersections of the [*w** ∩ *w'*]'s. That is,

$$\text{a neighborhood} = \cap_{w \in I} [w^* \cap w] \text{ for all } I \text{ where } I \supseteq \{x \mid x \text{ total}, x \in |D|\}$$

We shall have to leave it as an exercise for the reader to apply the above construction to the initial examples given in this paper of transformational approximations. The trick is to order the transformational approximations in terms of those partial interval functions (the calculational approximations) that are consistent with the exact and complete functions that correspond to the transformational approximations.

9 On the Truth of Scientific Theories

Thesis 1, that to believe a scientific theory is to believe that all its idealizations can be made to appear as transformational approximations is really a thesis about appropriate attitudes to take toward theories as products of scientific activity. As such this thesis depends in crucial ways on our notions of truth and confirmation. I want in this concluding section to consider some of the consequences of the programming semantics point of view for the understanding of these notions. One rough way to describe the relation between idealizations and transformational approximations is this: a transformational approximation is an idealization with an error bound. The practical benefits of having reliable error bounds are obvious. But exactly how such error bounds relate to the truth of a theory and its confirmation is, so far as I know, not a question ordinarily considered by philosophers. Rather than proceed abstractly here, I shall develop my analysis in terms of a pair of simple examples that have been co-opted from chaos theory. (For more on the examples and their source see Feigenbaum 1980, and Ford 1989.)

We first assume the existence of a phenomenon that operates in discrete time according to the following program, which we shall call B for benevolent:

(B) $x_{n+1} = x_n + \beta \pmod{1}$

Here x_n represents the value of the the phenomenon at time n; b is some constant. The sequence of generated values is called the trajectory. The notation *mod 1* means that the integer part of $x_n + \beta$ is to be dropped. In order to use this equation as a simple model of science, let us assume that x_0 is the true exact but unknown initial condition, and that primed versions represent idealizations, what we in fact use as input to the equation. And in order to mimic some of the features of idealizations, let us assume that while we cannot make appraisals of the distances from the truth of our idealizations we can determine their relative realism. Relative rankings such as $x_{0'}$ *is more realistic than* $x_{0''}$ will correspond therefore to the interval comparison $|x_0 - x_{0'}| \le |x_0 - x_{0''}|$. Similarly for output values. Now given these interpretative conventions, how might a program such as B be confirmed or disconfirmed? B has this important feature: if x_0 and $x_{0'}$ are two initial state descriptions then for all n (i.e., for all future states), $|x_0 - x_{0'}| = |x_n - x_{n'}|$. Hence, it would be a benevolent Nature that followed this program since the use of B would be *rewarded* in sense that better input yields better output. So if the world were as described by B, then a sensible form of scientific activity would be to attempt the development of idealizations that were more realistic and that preserved effective computability of output. In fact, successful activity of this sort would count toward the confirmation of the truth of B. Similarly, if it were found that use of more realistic input idealizations yielded less accurate predictions then B would be disconfirmed. I shall call this simple test the *monotonicity test* for confirmation: better input yields better output. But this simple connection between improvability of output and confirmation depends on the rather special error trajectory properties of B. In order to see this consider a malevolent Nature, one which runs the program, call it M,

(M) $x_{n+1} = 2x_n \pmod{1}$

This equation (equivalent to the logistic equation $y_{n+1} = 4y_n(1 - y_n)$) has the unique solution:

$x_n = 2^n x_0 \pmod{1}$

If we express x_0 in binary form, the trajectory of the phenomenon can be read off x_0 by simply moving one place to the right for each increment of time. So, for example, if $x_0 = 0.110011$, then, $x_1 = 0.10011$, $x_2 = 0.0011$, $x_3 = 0.011$, $x_4 = 0.11$, $x_5 = 0.1$, $x_6 = \perp$, …. As can be seen the trajectory can be read forward only $(k-1)$ increments, where k is the number of specified binary digits in the fractional part of the initial condition description. Each successive specification of a future state loses a digit in its representation. M is malevolent in this sense: If $x_{0'}$ and $x_{0''}$ are two initial state descriptions (assumed to be less than one and greater than zero) then, $|x_0 - x_{0'}| \leq |x_0 - x_{0''}|$ does *not* imply that $|x_n - x_{n'}| \leq |x_n - x_{n''}|$. That is, better input need not yield better output. A simple example illustrates the point. Let the true exact initial condition be $x_0 = 0.111111\ldots$. And let our two idealizations be $x_{0'} = 0.100000$ (better), and $x_{0''} = 0.001111$ (worse), where we assume that we can determine that $x_{0'}$ is better or more accurate than $x_{0''}$, though not by how much. Applying M to these values we obtain: $x_3 = 0.111111\ldots$, $x_{3'} = 0.000$, $x_{3''} = 0.111$. As can be seen $x_{3'}$ is worse than $x_{3''}$. Therefore, use of M is *not* rewarded in the sense that better input yields better output. Hence, the simple connection between improvability of prediction and confirmation does not hold in such a world. This corresponds to the commonly held intuition that a scientific theory is not automatically disconfirmed if better idealizations do not yield better predictions. (See Laymon 1985, p. 158, and Laymon 1989b, pp. 370-372.)

We are now in a position to illustrate the confirmational utility that accrues when idealizations are converted into transformational approximations with associated error bounds. We must first remind ourselves of the definition given earlier of the united extension of a function. In terms of our examples, if $x_{n+1} = f(x_n)$ is the (difference) equation in question, and if I is an interval of values in the domain of f, then the united extension for any future state x_n is,

$$f_u^n(I) = \{f^n(x_{0'}) \mid x_{0'} \in I\}$$

where superscript n means n successive applications of the function. Now if $I' \subseteq I''$ then, by a theorem discussed earlier, the following monotonicity relation holds for *both M and B*.

$$\forall n, I' \subseteq I'' \Rightarrow f_u^n(I') \subseteq f_u^n(I'')$$

Hence the monotonicity test for confirmation is restored in a variant (interval) form, even in the case of malevolent Nature, if we add error bounds to our idealizations. Some readers may know that the logistic equation M is an example of a deterministic process that is chaotic. One might wonder how this chaotic status affects the possibilities of confirmation by means of the interval version of the monotonicity test. Roughly speaking, the chaotic nature of M gets reflected in the fact that the inclusion of the united extensions ceases to be proper and degenerates into equality. So, for $n = 3$ and continuing with the values used in our earlier example,

$$f_u^3([0.100000, 0.111111\ldots]) = [0.000000\ldots, 0.111111\ldots]$$

$$f_u^3([0.001111, 0.111111\ldots]) = [0.000000\ldots, 0.111111\ldots]$$

As suggested by their titles, the two programs B and M represent two extremes of cooperative and non-cooperative Nature.[5] We should expect actual science to behave in a more modest way. That is, we should expect occasional non-monotonicity when improved idealizations are used, and by and large strict monotonicity when bounded approximations are used.

Where does all of this leave us? I think the principal virtue of the above way of proceeding is not so much in what it shows about bounded approximations (which is hardly new) but the way it sets the stage for a problem not ordinarily examined by philosophers of science, namely, that of determining how error bounds are fixed or estimated for idealizations. In short, how idealizations are converted into what I have called transformational approximations. It is a principal aim of this paper to encourage the philosophical investigation of this hitherto ignored problem.

Notes

[1] I would like to express my gratitude to the National Science Foundation (DIR-8920699) and the Ohio State University for providing funding for the project of which this paper forms a part.

[2] A basic approach is sketched in Laymon 1985, pp. 166-168. But this account must be modified so as to accommodate the complications discussed in the final section of this paper.

[3] It can be shown that the nice mathematical behavior of our example can be expected to hold in general, that is, that interval approximations never diverge from their united extension targets. See Moore 1967, theorem 4.4, p. 21.

[4] One of the benefits of the construction is that Lewis' axioms (Lewis 1973, p.14) for the semantics of counterfactuals are satisfied along all paths in the resulting lattices, i.e., Lewis spheres are automatically formed along paths. To wit, the system is centered since $[w^* \cap w^*] = \{w^*\}$. It is nested since we are restricting attention to paths. It is closed under unions since all domains are closed for unions of path elements. Finally, the system is closed under intersections since it is closed for all intersections and since intersections of path elements give elements in the path.

[5] More strictly speaking, the extreme in cooperation would be x_n = some constant for all $n > 0$. Less extreme cooperation would have $|x_{n'} - x_{n''}| < |x_{n+1'} - x_{n+1''}|$ for all n. For the importance of this form of cooperative Nature in the design of successful experiments see Laymon 1985, pp. 161-65.

References

Balzer, W., Moulines, U. and Sneed, J. (1987), *An Architectonic for Science*. Dordrecht: Reidel.

Feigenbaum, M. J. (1980), "Universal Behavior in Nonlinear Systems", *Los Alamos Science* 1: 4-27. Reprinted in P. Civtanovic (ed.) (1984), *Universality in Chaos*. Bristol: Adam-Hilger, pp. 49-84.

Ford, J. (1989), "What is Chaos, that We should be Mindful of It?", in *The New Physics*, Paul Davies (ed.). Cambridge: Cambridge University Press, pp. 348-372.

van Fraassen, B. (1970), "On the Extension of Beth's Semantics of Physical Theories", *Philosophy of Science* 37: 325-39.

Laymon, R. (1985), "Idealizations and the Testing of Theories by Experimentation", in *Experiment and Observation in Modern Science*, P. Achinstein and O. Hannaway (eds.). Boston: MIT Press and Bradford Books, pp. 147-73.

Laymon, R. (1987), "Using Scott Domains to Explicate the Notions of Approximate and Idealized Data", *Philosophy of Science*, 54: 194-221.

_____. (1989a), "The Application of Idealized Scientific Theories to Engineering", *Synthese*, 81: 353-71.

_____. (1989b), "Cartwright and the Lying Laws of Physics", *Journal of Philosophy*, 136: 353-372.

Lewis, .K. (1973), *Counterfactuals*. Cambridge: Harvard University Press.

Lin, C.C. and Segel, L.A. (1974), *Mathematics Applied to Deterministic Problems in the Natural Sciences*. New York: Macmillan.

Manes, E.G. and Arbib, M.A. (1986), *Algebraic Approaches to Program Semantics*. New York: Springer-Verlag.

Moulines, C. and Sneed, J.D. (1979), "Suppes' Philosophy of Physics", in *Patrick Suppes*, Radu J. Bogdan (ed.). Dordrecht: D. Reidel.

Moore, R.E. (1966), *Interval Analysis*. Englewood Cliffs: Prentice-Hall.

Scott, D. (1970), *Outline of a Mathematical Theory of Computation*. Technical Monograph PRG-2, Oxford University Computing Laboratory.

_____. (1981), *Lectures on a Mathematical Theory of Computation*. Technical Monograph PRG-19, Oxford University Computing Laboratory.

_____. (1982), "Domains for Denotational Semantics", in *Automata, Languages and Programming*. M. Nielson and E. M. Schmidt (eds.). New York: Springer-Verlag, pp. 577-613.

_____. and Strachey, C. (1971), *Toward a Mathematical Semantics for Computer Languages*. Technical Monograph PRG-6, Oxford University Computing Laboratory.

Suppe, F. (1989), *The Semantic Conception of Theories and Scientific Realism*. Urbana: University of Illinois Press.

Tennent, R.D. (1981), *Principles of Programming Languages*. Englewood Cliffs: Prentice-Hall.

Part XIV

LAWS, CONDITIONS AND DETERMINISM

Determinism in Deterministic Chaos

Roger Jones

University of Kentucky

1. Introduction

In a paper fifteen years ago about the meaning and the possibility of the beginning and end of time, our redoubtable session chair, John Earman, ended up like this:

> ...[T]he answers to the questions posed at the outset lie somewhere in a thicket of problems growing out of the intersection of mathematics, physics, and metaphysics. This paper has only located the thicket and engaged in a little initial bush beating. This is not much progress, but knowing which bushes to beat is a necessary first step.
>
> Some philosophers will be disappointed that the thicket is populated by so many problems of a technical and scientific nature. On the contrary, I am encouraged by this result because it shows that a long-standing philosophical problem has a non-trivial and, indeed, a surprisingly large content. Moreover, this result is a good illustration of the artificiality and danger of trying to separate philosophy from science. (Earman 1977, p. 131-2)

These remarks are a kind of manifesto of an approach to philosophical questions about space and time dating from the late 1960's. Instead of debating these questions endlessly in the sort of conceptual or abstract way in which they had been debated ever since (at least) Newton and Leibniz, largely independent of actual thinking in the physical sciences, this new approach began by taking advantage of the latest mathematical representations of space-time structure within physics. By posing philosophical questions as conjectures in this mathematical language of space-time structure, certain types of moderately precise answers could be given. These answers generally took the form of pointing out that the conjecture was true in certain models of space-time structure and false in others. There was a kind of "you pays your money; you takes your choice" atmosphere about these answers.

Now I don't want to give the impession that the answers to all philosophically interesting questions about space and time can simply be read off models of space-time structure, though there are some beautiful results. I know, for instance, that Larry Sklar would object strenuously, as he amply demonstrates in the volume of his collected essays on space-time (Sklar 1985). Often, the very interpretation of "the physics" one invokes in some philosophical discussion of space-time structure turns on the philosophical issues one is discussing. One musn't forget the metaphysical bushes in John Earman's thicket.

In any case, all I want to claim is that this historical course of philosophical discussions about space and time structure — a course moving from exclusively conceptual and metaphysical discussions to bush beating at this intersection of mathematics, physics, amd metaphysics, from searches for in-principle canonical clarity to catalogues of moderately precise "results" associated with particular mathematical structures — is characteristic of philosophical discussions about determinism as well.

In fact, there has been substantial overlap in discussions of space-time structure and those of determinism. Much such overlap is in evidence in what must be regarded as the state-of-the-art philosophical discussion of determinism in this new vein, John Earman's *Primer on Determinism* (Earman 1986), which gave Jeremy Butterfield the idea of organizing this symposium.

2. Laplacian Determinism

The notion of determinism that Earman dedicates himself to, for the most part, is what he calls Laplacian determinism. Motivated by the famous passage from Laplace, but cleansing its spirit of epistemological references to a "predictor", Earman casts his condition in terms of physically possible worlds — "worlds that satisfy the natural laws obtaining in the actual world" (Earman 1986, p. 7). A physically possible world is Laplacian deterministic just in case, given any other physically possible world, when the two worlds agree on all relevant physical properties at a given time, then they agree for all times (Earman 1986, p. 13).

Obviously, the $64 question is whether our own world is Laplacian deterministic. But, in the best spirit of the sort of space-time discussions I have just described, one is not going to get an answer from John Earman — or of course anybody — but rather a heavily annotated catalog. That is, one is going to get a whole lot of candidates for "physically possible worlds", and tests of the defining condition for determinism for them. The annotations come in various discussions of what constitutes "natural laws" and how they are distinguished from, for instance, boundary conditions; they come in discussions of "relevant physical properties", of difficulties with the concept of "a given time", of specifying physical properties at particular times, and such.

In his *Primer*, Earman's first catalog entries are for "classical" worlds. After a careful analysis of the importance of spacetime structure itself for the basic casting of the question of determinism, he turns to traditional Newtonian worlds. And very quickly, as he says, "the Gestalt of determinism safely and smoothly at work in Newtonian worlds" is switched "to puzzlement about how Laplacian determinism could possibly be true" (Earman 1986, p. 33). Many of his considerations have to do with effects imploding from and exploding to spatial infinity. Some have to do with very cleverly contrived circumstances of point particles. Throughout the difficulty lies in the status of supplementary conditions that must be imposed on the basic differential equations of motion in the particular worlds to guarantee uniqueness of evolution.

Mark Wilson, in his review of Earman's *Primer* for *Philosophy of Science* (Wilson 1989), is a little impatient with Earman's emphasis on spacetime structure. He is also a little impatient about some of the clever system/worlds designed as prima facie affronts to determinism, chiding Earman for "cast[ing] his net widely enough to sample cheerfully some fairly dubious arguments for indeterminism" (Wilson 1989, p. 527). Wilson chooses instead to concentrate on more homely systems, "ordinary machines" — two wheels linked by a connecting rod, three intermeshed gear wheels, real, inelastic billiard balls. The problems with these systems are generally problems of *instability*: in certain configurations their large-scale future behavior is extremely sensitive to minute changes in their initial conditions, rather like a pencil, balanced precariously on its point. Now the instability one finds in these systems exists only at a point, or perhaps a few points, in their entire space of states. And though even the presence of one such unstable configuration makes these homely systems, as Wilson says, adequate to "provide a tolerable first picture of how 'determinism' goes awry in many of

Earman's examples" (Wilson 1989, p. 510), both Earman and Wilson are aware of the existence of an important class of systems for which such instability is a ubiquitous aspect of their state spaces, at least asymptotically. These are the systems studied by what is popularly known as "chaos theory". And they are thought by some to introduce important and special challenges to determinism. At least they are a new class of "physically possible worlds" against which to test the defining conditions.

3. Dripping Faucets

In the best spirit of Wilson's homely machines, I want to consider a particularly homely example. I want to talk about dripping faucets. There are a couple of reasons for choosing this example to talk about determinism in chaos theory. First is its very homeliness: it's a simple phenomenon about which everyone has intuitions born of personal experience. Second, it exhibits in a simple way some of the techniques used in chaos theory, and indeed, some of the feel of research in chaos theory, which unfortunately remains an area in which there seem to be only skeptics and impassioned evangelists. The world owes largely to Robert Shaw, then a graduate student at UC Santa Cruz, the analysis of the systematics of faucet dripping (Shaw 1984). And I owe my appreciation of his work to a recent paper by Steven Kellert, Mark Stone, and Arthur Fine (Kellert et al. 1990).[1]

Dripping faucets are a part of everyone's experience, particularly in the middle of the night. What one tends to notice about them is the interval between drips. It is the waiting for the next drip that tends to produce agony. And I don't know which produces more agony, inexorably regular dripping, or subtly random dripping. In any case the drop interval is surely one of the phenomenologicaly significant aspects of a dripping faucet, and most folks know that as the flow rate of water through the pipe is adjusted upward by the processes of corruption and decay that degrade faucet washers, the drop interval becomes shorter and more and more irregular until a turbulent trickle is produced. The particular region of dripping faucet phenomenology I want to focus on is that in which the drop interval is irregular.[2] (Fig. 1)

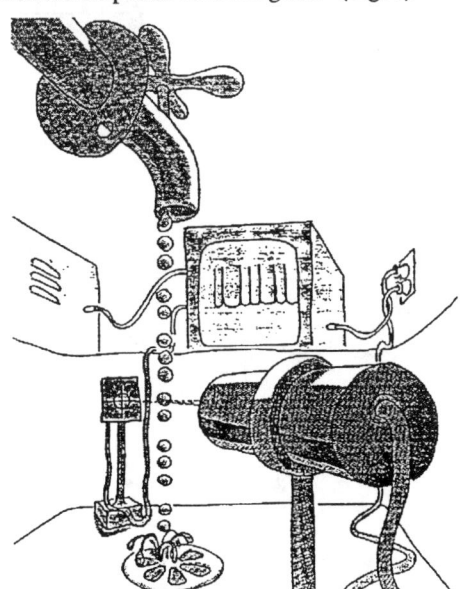

Figure 1

To understand the dynamics of this phenomenon of fluid flow using traditional hydrodynamic models would be a very imposing task. To understand the changing shapes of the water droplets as they form and detach in full detail would involve an analysis with many (even infinitely many) degrees of freedom and several force laws. But Shaw and his colleagues did not go about the task of analyzing the dripping faucet in this way. Their first tool of analysis was a display of the time series of drop intervals on a two dimensional plot in which the nth drop interval is plotted on the x axis and the n-plus-first interval is plotted on the y axis. Now when dripping is pretty regular, the nth drop interval is just equal to the n-plus-first, and the graph of all such pairs is approximately a point. (Fig. 2a) What would the appearance of such a graph be when the water flow rate is such that the drop interval is irregular? Well, if it really is irregular, and there is no correlation at all between one drop interval and the drop interval that comes after it, then the graph would simply be a random scatter of points. (Fig. 2b)

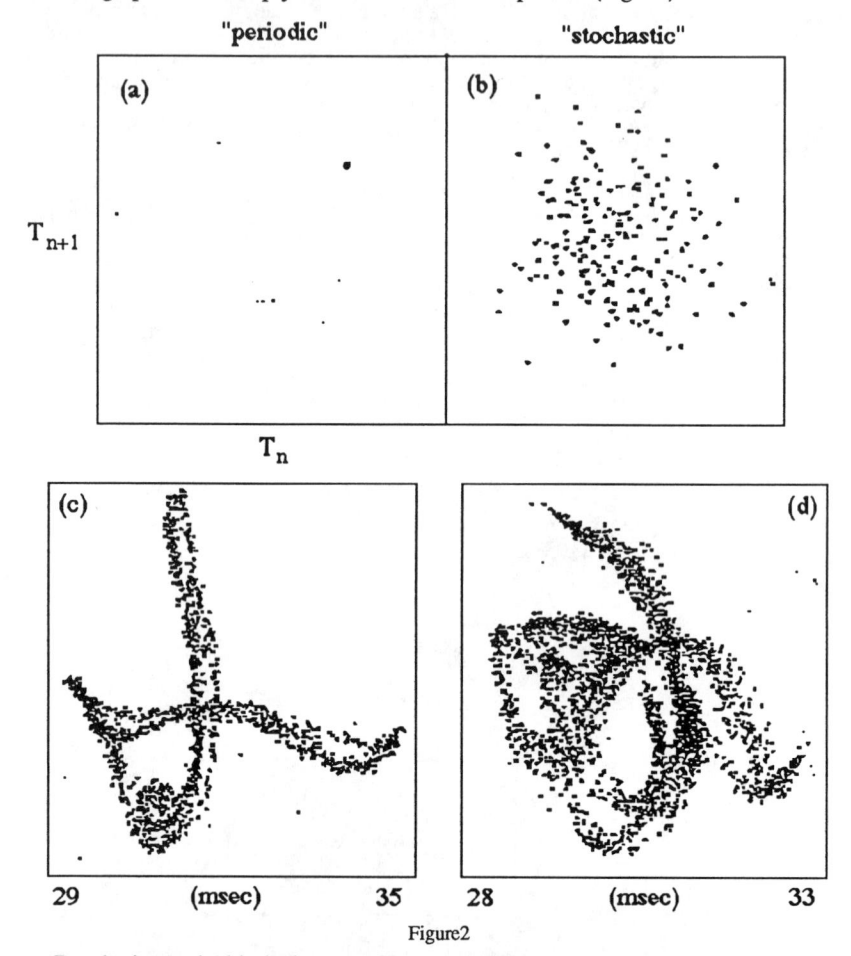

Figure 2

But that's not the kind of pattern Shaw got. With very good, minutely regulable faucets that dripped even in the daytime, and precise measurements of drop intervals, he and his buddies took data that graphed up like this. (Fig. 2b,c)

Shaw and Co. had seen such shapes before.[3] In fact, of course, they were on the lookout for them. And they knew that such shapes were often produced from remarkably simple, but non-linear differential equations, of the sort that characterize, for instance, forced and damped pendulums. So Shaw set out to find such an equation that might model the behavior of the drop intervals in this irregular region.

The equation he found was based on an analogy between a filling water drop and a body of increasing mass suspended from a spring. (Fig. 3) As the mass of the body increases, the spring stretches, and the body accelerates downwards. Now Shaw supposed that at some critical stretch, the body separated into two bodies, the one attached to the spring remaining with the same mass as at the start of the process and springing back up to its original position, and the other body falling away freely. This is a cute and wonderfully simple model, on the analogy that the viscosity of water acts like a spring coefficient, and the water flow rate as a rate of mass change.

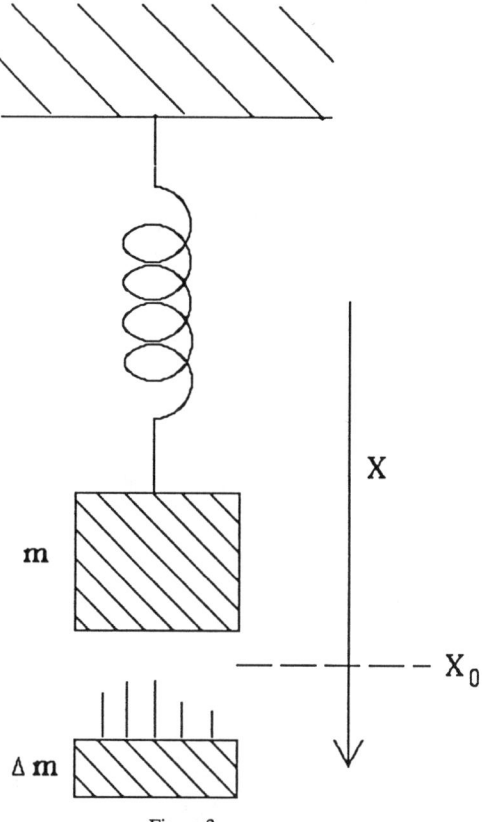

Figure 3

Well, Shaw wrote down the differential equations that characterize this spring model, and proceeded to play with them on an analog computer. Lo and behold, he discovered that, as he says, "physical faucet data can be found which closely resemble time vs. time maps obtained from the analog simulation..." (Shaw 1984, p. 16; quoted in Kellert *et al.* 1990, p. 6). (Fig. 4) The resemblance he regarded as good evidence for the success of the variable mass pendulum analogy.

data | analog model

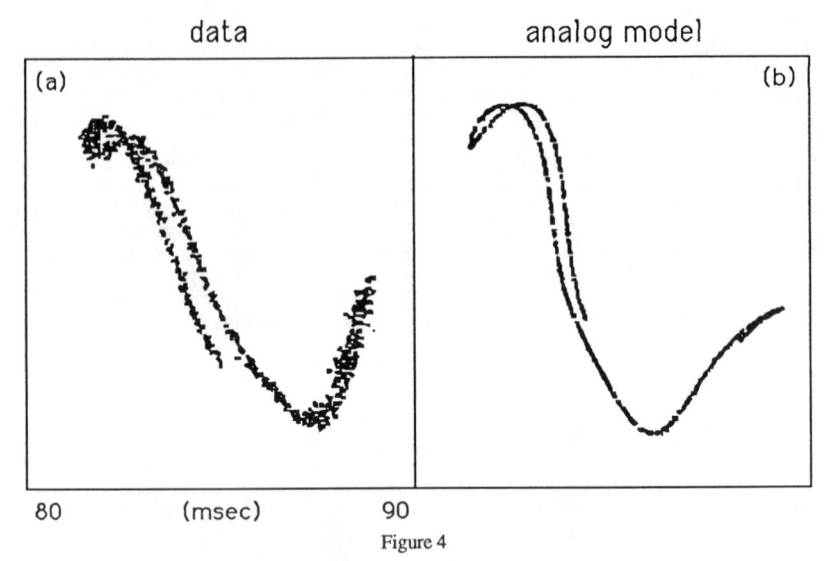

(a) | (b)

80   (msec)   90

Figure 4

I am not going to discuss here the status of the sense of "confirmation" at stake here. This is what Kellert, Stone, and Fine talk about in their paper. But it is clear that Shaw took quite seriously the functions describing the mass and spring as indicative of the underlying dynamics of the dripping faucet. And there is no doubt of the power of the mathematics of the mathematical treatment. For Shaw used it to discover an *attractor* in the state space of the variable mass pendulum system, an attractor he recognized as a two-dimensional projection of "a *Rössler attractor* in its `screw type' or `funnel' parameter regime." (Fig. 5) And, as Shaw puts it, "The close correspondence of model and experiment ... argues that such a structure is embedded in the infinite-dimensional state space of the fluid system" (Shaw 1984, p. 17; Kellert *et al.*, p. 7).

Now it is this notion of an attractor that I am particularly concerned with here. For it is the character of such attractors as Shaw identified for his pendulum model system, strange attractors, as they are called, that gives rise to much of the discussion regarding determinism in chaotic systems.

The notion of an attractor in a space describing states of a physical system is not really a very exotic one. The basic idea is simply that of a region in the state space into which nearby dynamical trajectories converge. Energy conserving Hamiltonian systems do not have attractors in their state spaces, but systems in which energy is dissipated almost always do. An ordinary real-world pendulum, for instance, will sooner or later come to rest at zero deflection from the vertical, and trajectories on a two dimensional state space coordinatized in values of its angular deflection and angular velocity will spiral in to the origin, a fixed point attractor. A stabilized real-life pendulum, such as one in a grandfather clock, will have a roughly circular "limit cycle" in state space, and trajectories will spiral out to the limit cycle from inside it, and in to the limit cycle from outside it, all inevitably "attracted".

The strange attractor Robert Shaw identified for the variable mass pendulum model is an attractor in just this way. All state trajectories (within some limits) for the system end up "on it." But its shape and structure are certainly strange. In the first place, though all nearby state trajectories converge onto it, once on it, nearby trajectories rapidly, exponentially in fact, diverge. (Fig. 6) They are able to do so, and still remain within a bounded region of state space, because of the peculiar geometric na-

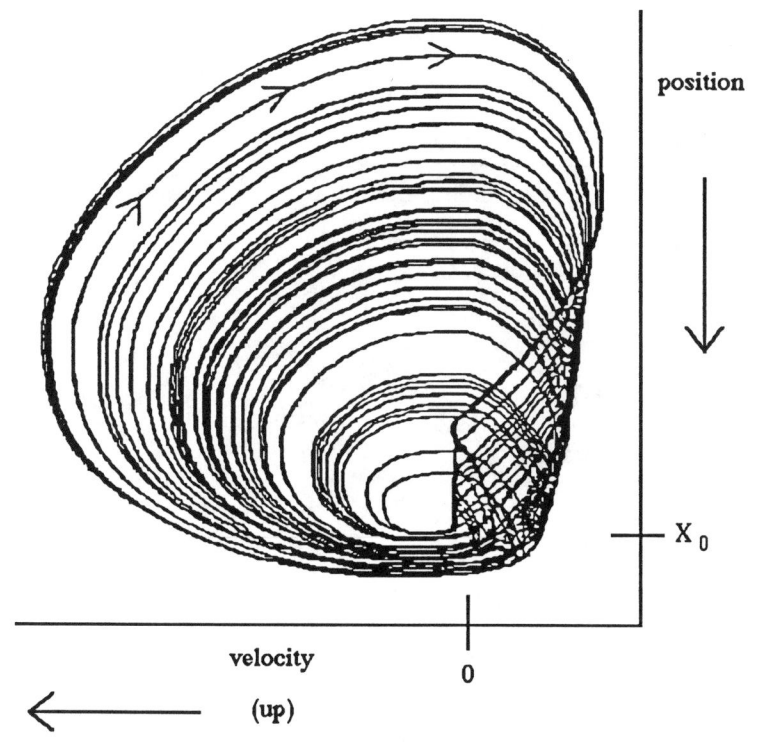

Figure 5

ture of the attractor. It is as though the state space is stretched strongly in one dimension — creating the rapid divergence of nearby state trajectories, and then folded back on itself in another dimension — creating a reconvergence of trajectories, of course in a different direction: the trajectories never cross.

It is this structure that gives rise to both the good news and the bad news about determinism in systems whose state spaces feature such attractors — chaotic systems, as I will call them from now on. Among model systems, these include oscillators such as Shaw's variable mass pendulum, compound pendulums, and various electronic analogs; there are also several easy-to-write-down one and two dimensional iterative mappings. Among modeled systems, real world systems, one finds the dripping faucet, but also controlled fluid turbulence cells, chemical reactions, biological populations, heart cells, economic structures, and, fans of such analysis insist, many, many more.

The good news is that apparently chaotic behavior, such as that of the dripping faucet at particular flow rates, can be modeled by simple, mathematically deterministic processes. These processes — like that of the variable mass pendulum — are simple in that they involve a small number of degrees of freedom. And they are deterministic, in that their evolution is described by differential equations — non-linear differential equations, to be sure — but still equations for which the existence and uniqueness of solutions can be guaranteed. So one is guaranteed that for every set of (allowable) initial conditions, the system evolves in a unique way for all time; trajectories in state space never cross. There is no need to model such systems using theories with stochastic elements; their behavior need not be seen as random, in that sense. Nor

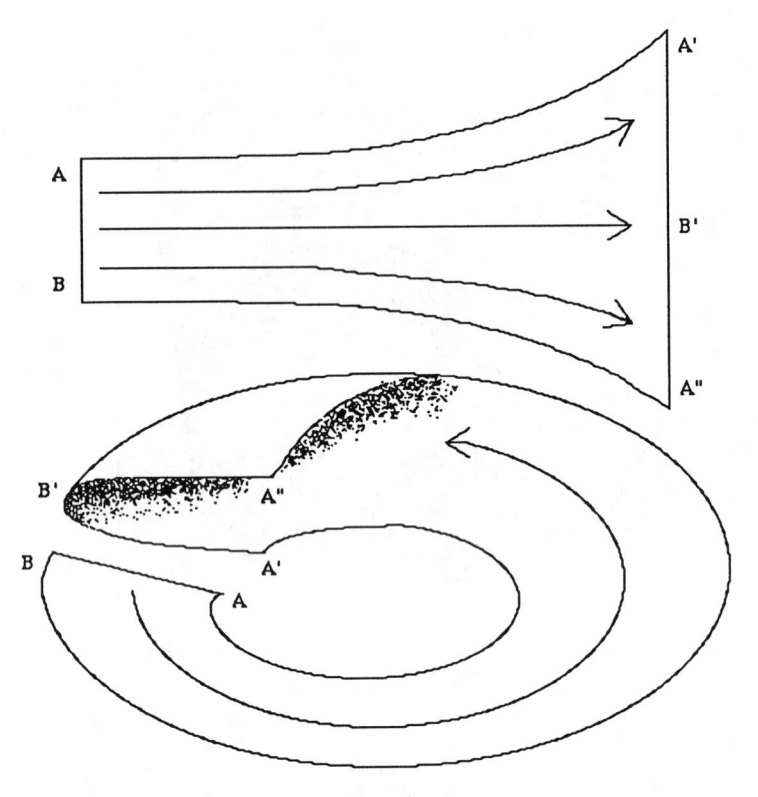

Figure 6

need the complexity in their behavior be considered to arise from the action of many competing processes, many degrees of freedom. The appearance of chaos, or random behavior, simply comes from the fact that nearby trajectories diverge so dramatically on the attractor that time series behavior, operationally obtained by sampling the state of the system at times that are long compared to the rate of divergence of the trajectories on the attractor, and even continuously modeled, is just all over the place. (Fig. 7)

But that is an important part of the bad news about determinism in "deterministically chaotic" systems. Arbitrarily nearby trajectories on the attractor diverge exponentially. Unless the state of the system is known exactly (in the real number sense), the future state of the system on the attractor is essentially wholly unpredictable. Strange attractors display, in the fairly standard parlance, sensitive dependence on initial conditions. And I do mean *sensitive*. This sensitive dependence on initial conditions of deterministically chaotic systems throws down a kind of verificationist gauntlet to some classical ideas of determinism.

For it is a fact of a very serious nature that the outcomes of measurements and the input to calculations are restricted to finitely statable numbers. And yet for every finitely statable set of initial conditions for a chaotic system there exists a multiplicity of state trajectories with those conditions, trajectories which diverge on the attractor so explosively that all predictability is lost, and one would do as well to regard the evolution of the system as random. In practical terms, such minute details as a computer's round-off algorithm will be crucial in what future states it calculates.

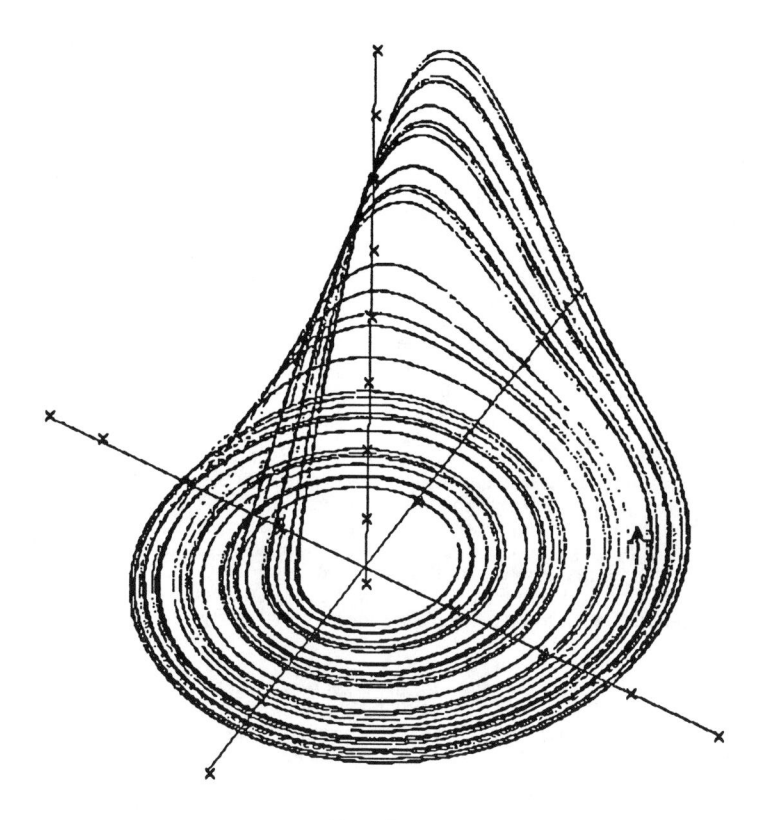

Figure 7

4. Responses

Now let me consider some responses one might make to this essential failure of predictivity. One is simply to say, "So what?" This point is persuasively made by G. M. K. Hunt (1987) and Mark Stone (1989) in short papers on just this topic, and by John Earman repeatedly in his *Primer*. In the best tradition that I mentioned at the beginning of the paper of answering questions about determinism only in the context of precise theoretical specifications, the theory of chaotic systems — non-linear dynamics — is certainly Laplacian deterministic. I mean, everybody just calls it "deterministic chaos". The mathematics of the theory guarantees the existence and uniqueness of dynamical trajectories for physical systems. Laplacian determinism requires only that two physically possible worlds that agree on the values — point values in this case — of their physically relevant properties agree at all times. That is, if they are located at identical points on identical state-space trajectories, their total histories will "trajectorially coincide". This is simply guaranteed by the mathematics of the theory of chaotic systems.

If one wishes to import this entirely straightforward discussion into this "real world", Earman, Hunt, and Stone chorus, then certain philosophical duties appropriate to such importations must be paid. One is to recognize the distinction between metaphysics and epistemology. As a metaphysical doctrine about the world, Earman says, "whether [Laplacian determinism] is fulfilled or not depends only on the structure of the world, independently of what we could do or could know of it" (1986, p. 7). But such metaphysical Laplacian determinism is not threatened in any way by chaotic systems. In accord with all I have been saying, such systems, insofar as they are part of the world, evolve along unique dynamical trajectories, associated uniquely with a set of dynamical properties at each instant. Of course, we can't identify these unique dynamical properties and follow these trajectories, because our measuring and computational abilities are ineliminably limited. And any lack of precision in specification, because of the structure of the state space of these systems, makes trajectory prediction impossible. But that's a problem of epistemology. The fact that we can never use the deterministic equations of chaos theory to find the single trajectory which the world/system follows certainly does not threaten the existence of a trajectory, which again, is guaranteed by the mathematics of the theory.

Now this seems to me to be a proper philosophical response. But I am always nervous when perfectly clear results about theoretical systems are imported into "the world", and distinctions between metaphysics and epistemology are made. For there is actually some discussion among people in this chaos business about just what kind of dynamical properties "the world" deals in. What I'm talking about here is not just some kind of basic verificationism, which would distinguish between theories in formulation and theories in use, and simply come down on the side of "observable specifications" as against "unobservable idealizations". No indeed. The discussion swirls around the fundamental appropriateness of real number specifications of state in dynamics, and it has led to proposals for whole new approaches to dynamics. The loudest voices in these discussions are those of Ilya Prigogine (e.g., Prigogine and Stengers 1984) and Joseph Ford (e.g., Ford 1983, 1986), neither of whom ever mentions the other. I want to talk about their views very briefly, and then end with a general point.

Both Prigogine and Ford argue for the inappropriateness of real number characterizations of state, and they both explicitly wish to avoid the charge of naive verificationism by appealing to "principle". The sort of principles they both mention as analogies are those associated with the speed of light and with Heisenbergian uncertainty. They both find a kind of "logical incompatibility" in the theory of "deterministic chaos", and they both embrace the chaos part — indeterminism, fundamental randomness — as the genuine way of the world. They reject the doctrine of Laplacian determinism for our world, even metaphysically, and want to construct a dynamics to reflect this indeterminism they see as fundamentally "necessary". The spirit of their positions is thus very much the same. But the details are certainly different.

Prigogine's arguments against real valued state specifications in classical (non-linear) dynamics are sometimes just discouraging conflations of "impossible in principle in our world" with "theoretically inconsistent". But he also offers much more suggestive invocations of a kind of rigorous complementarity in chaotic systems between real-valued, external-time-ordered state descriptions and distributional descriptions into which a kind of internal "system time" is built (Prigogine and Stengers 1984, pp. 272-290; Misra, Courbage, and Prigogine 1979; Batterman 1991). At present his programme is mathematically promissory, but if he can define such novel state descriptions rigorously, he assures us that choosing them as fundamental, on the basis of explicit second law considerations, holds out the prospect of making a seamless whole of "dynamics and thermodynamics, the physics of being and the physics of becoming" (Prigogine and Stengers 1984, p. 277).

Joe Ford's case against real valued state descriptions for non-linear systems is considerably more elaborate. He takes very seriously the information-based theory of algorithmic complexity due (independently) to Solomonov, Kolmogorov, and Chaitin, (see, e.g., Chaitin 1975). Variously, he points out that:

i) chaotic trajectories cannot be computed by any algorithmic rule simpler than a computer program that simply copies the state values at various times;

ii) computations of chaotic trajectories really don't deserve to be called computations at all (and certainly not predictions), because the input information for the computations must, rigorously, be equivalent to a copy of the output;

iii) the time evolution of chaotic systems is computationally incompressible: it takes just as long to compute the evolution of the system as for the system itself to evolve (e.g., Ford 1986, pp. 350-1).

For all these reasons, he finds something galling, a glaring conceptual mismatch, if not an out-and-out logical incompatibility, between the "determinism" and the "chaos" in the theory of deterministic chaos. And the charlatan is the "determinism". For what kind of "determinism" is it when the product of its action is a chaotic trajectory that is "random and incalculable; its information content ... both infinite and imcompressible" (Ford 1983, p. 46)?

Ford places much of the blame for this mismatch on the assumed input for such deterministic equations — the randomly digited, incalculable, informationally infinite and incompressible real numbers. Chaos comes out of many such deterministic calculations only because chaos is put into them. So it's kind of an artificial chaos, resulting from a bogus determinism based on treating such entities as perfectly well-defined.

Ford's suggestion is for a "humanly meaningful number system that does not involve the assumption of infinite precision" (Ford 1983, p. 47). He proposes one, isomorphic to a finite set of integers. (See Winnie 1991 for a critique of this proposal.) Other researchers are making practical efforts to develop what are called "cellular automata models" for physical processes in general.[4] Such models are defined only on discrete sets of numbers; thus a physics based on them would automatically involve a truncation of real numbers, and the kind of resultant coarse-graining would guarantee the *genuine* indeterminacy of chaotic systems, and with it an absolute irreversibility that would provide, as another fan has said, "a complete justification of classical statistical mechanics" (Jensen, p. 180), the same result that Prigogine seeks.

There is one cloud in this rosy vision. For as Ford admits, there is nothing in algorithmic complexity theory, or any other similarly deep and abstract theory, which has anything to say about a "natural bound on observational precision" (Ford 1983, p. 47). Roderick Jensen spells this out:

[I]t is possible that the scale at which the trunction of real numbers occurs may be so small that no practical consequences of the distinction between continuum and discrete theories can be deduced or verified. In that case, the issue of the ultimate discretization of the real world will pass from the domain of physics to that of philosophy". (Jensen 1987, p. 180)

The real world giveth; and the real world taketh away. Still, the domain of philosophy is not such a bad place for a doctrine such as Laplacian determinism to reside in. As I have mentioned several times, the most productive philosophical tradition of discussion of such issues is one in which various theoretical systems are carefully described and the issue decided for them. In this way we learn valuable lessons about the notion of a scientific theory, of laws of nature, of boundary and initial conditions, of experimental data, and other issues of concern to philosophers. As long as we can keep our annotated catalogs growing, we are making, in philosophy, a kind of Kuhnian progress.

Notes

[1] I learned a lot also from Steven Kellert's recent (1990) dissertation, *Philosophical Questions About Chaos Theory*. This dissertation, as far as I know, is the best large-scale study of its topic.

[2] Figures 1-5 and figure 7 are taken from Shaw 1984. Kellert *et al.* (1990) reproduce figures 2-5 as well. Figure 6 comes from Crutchfield *et al.* (1986).

[3] For a readable history of chaos theory, including the early study of attractors by a meteorologist and a population biologist, see Gleick 1987.

[4] See, for example, the articles in "Cellular Automata", *Physica* 10D (1984).

References

Batterman, R. (1991), "Randomness and Probability in Dynamical Theories: On the Proposals of the Prigogine School.", *Philosophy of Science* 58: 241-263.

Chaitin, G. (1975), "Randomness and Mathematical Proof", *Scientific American* 232 (May): 47-52.

Crutchfield, J., Farmer, D., Packard, N., and Shaw, R. (1986), "Chaos", *Scientific American* 255: 46-57.

Earman, J. (1977), "Till the End of Time", in *Foundations of Space-Time Theories*, J. Earman, C.Glymour, and J. Stachel (eds.),(Minnesota Studies in the Philosophy of Science, Vol. VIII). Minneapolis: University of Minnesota Press, pp. 109-134.

_____ . (1986), *A Primer on Determinism*. Dordrecht: Reidel.

Ford, J. (1983), "How Random Is A Coin Toss?", *Physics Today* 36: (April) 40-47.

_____ . (1986), "What Is Chaos, That We Should Be Mindful of It?", in *The New Physics*, P. C. W. Davies (ed.). Cambridge: Cambridge University Press, pp. 348-371.

Hunt, G. M. K. (1989), "Determinism, Predictability, and Chaos", *Analysis* 47: 129-133.

Jensen, R. (1987), "Classical Chaos", *American Scientist* 75: 168-180.

Kellert, S. (1990), *Philosophical Questions About Chaos Theory* (dissertation, Northwestern University)

Kellert, S., Stone, M., and Fine, A. (1990), "Models, Chaos, and Goodness of Fit", forthcoming in *Philosophical Topics*.

Misra, B., Prigogine, I., and Courbage, M. (1979), "From Deterministic Dynamics to Probabilistic Description", *Physica* 98A: 1-26.

Prigogine, I. and Stengers, I. (1984), *Order out of Chaos*. Boulder: Shambhala Publications.

Shaw, R. (1984), *The Dripping Faucet as a Model Chaotic System*. Santa Cruz: Aerial Press.

Sklar, L. (1985), *Philosophy and Spacetime Physics*. Berkeley: University of California Press.

Stone, M. (1989), "Chaos, Prediction, and Laplacian Determinism", *American Philosophical Quarterly* 26: 123-131.

Wilson, M. (1989), "Critical Notice: John Earman's A Primer on Determinism", *Philosophy of Science* 56: 502-531.

Winnie, J. (1991), "Computable Chaos", unpublished manuscript.

How Free are Initial Conditions?

Lawrence Sklar

University of Michigan

Some of what is true about the world is thought, by some, to be true of necessity. Other truths about the world are merely contingently true, it is said. Next we get a familiar distinguishing of necessity into its various kinds. Anything whose contrary would contradict the laws of logic is logically necessary. Anything compatible with these laws is logically contingent. There are, of course, grave problems in finding a principled way of discriminating the logical truths from all the others. Some propositions are not logically necessary but are metaphysically necessary. For Kant, I suppose, the alleged facts that every event had a cause, and that in any change a substance remained unchanged would be such facts. So, I suppose, would be the necessary but synthetic truths of geometry and arithmetic. For more contemporary philosophers genuine identity statements, either singular or general, might have such a status. Finally, even among the logically and metaphysically contingent remainder of assertions, there are those possesed of a more limited kind of necessity. These are the assertions that are physically necessary. Frequently these truths are identified with all that is entailed by the genuine laws of nature. All of these truths not only are the case, but "must be the case as matter of physical necessity". Other assertions state truths about nature that are merely contingent. Both they and their contradictories are compatible with the laws of nature. So while they say, truly, what is the case, they affirm only what is but might be otherwise, as far as the contraints of physical necessity are concerned.

There is, to put it mildly, much that is obscure in the notion of the "physically" necessary." Presumably this is all and only that which is contrained by the true laws of nature. But what are these? They are, of course, generalizations, but not all true generalizations are physical laws. Resort to the picturesque notion of the realm of possible worlds doesn't help much. Among all the logically possible worlds there is the proper subclass of those that are physically possible. The laws are the generalizations true not only in this world, our actual one, but in all the physically possible worlds, while the "accidental" generalizations fail to hold in some physically possible world or other. This is all well and good but if we all have to go on in characterizing the class of physically possible worlds is that it is the class of all and only those possible worlds in which all of the laws of nature hold true, the philosophical circle we go around in is so small as to almost degenerate into a point.

PSA 1990, Volume 2, pp. 551-564
Copyright © 1991 by the Philosophy of Science Association

Nor is there any obvious way of clearly delimiting the generalizations that are to count as laws of nature from those that are merely "contingent" generalizations. At least in the case of logical necessity we have a systematic means of delimiting the realm of the logical, even if we do not have a clear philosophical understanding of the ground on which the familiar characterization by the stipulation of what is to count as logical form rests. In the case of generalizations about nature no simple "syntactical" means of distinguishing the physically necessary from the physically contingent will do, for all the familiar reasons that show that such grounds as "excluding reference to particular things," and so on, just won't work. We can rely on some primitive notions of what predicates are to count as delimiting natural kinds, restricting laws to be generalizations only over such "genuine natural kinds," but any means of discriminating them from non-natural classifications of things seems as obscure as the original problem of discriminating genuine laws from mere accidental conditional generalizations in the first place.

One approach offers promise, at least to some, of simultaneously explaining what the necessity of natural necessity comes down to and of explicating the principles by which we discriminate genuine laws from merely physically contingent generalizations. This is to take a kind of "deflationary" stance to the very notion of natural necessity. It is an approach that reaches back to Hume's psycholization of the notion of necessary connection as mere habit or irresistable expectation on the part of the agent, although it need not be thought of as making natural necessity a "subjective" notion. The idea here is to recognize that while all generalizations are "mere" generalizations, some are less mere than others. Some generalizations have a status in the broad, profound and hierarchical structure of generalizations that constitutes the body of scientific theory used to explain, predict and control the world in its broadest aspects. Other generalizations, true as they may be, have no significant place in this overall structure. It is, then, place in this hierarchical structure reaching up at its top to the basic generalizations true of space and time, of the fundamental quantum kinematics of matter and to the most general dynamical principles of field theory and its concomitant theory of elementary particles, that gives a generalization the honorific of being "lawlike."

And it is this special place, characterized by the "pragmatics" of assertion, that gives the generalization those features that lead us to think of them as characterizing the physically necessary. All the apparatus of saying that the generalizations are true in other physically possible worlds as well as this one, of letting them be used as licenses to infer counter-factuals and of thinking of them as capturing not only what is in general but what "must be" in general, is asserted to follow from the fundamental characteristic they have of being central to our hierarchial scheme of generalizations describing the world. Those who want to think of physical necessity as something much "deeper" than this will be, of course, dissatisfied with such a pragmatic "deconstruction" of the notion of physical necessity but the neo-Humeian will continue to insist that if they have something more profound and more metaphysical in mind, it is about time they tell the empiricist-actualist just what this additional element of necessity comes down to.

Be all of that as it may, it isn't my primary concern in this paper. I want to focus, instead, on the frequently felt intuition that the realms of the physically necessary is the realm, at most, of the purely general. Perhaps it is a matter of necessity, of some sort, that if I let go of this piece of chalk it *must* fall. But, surely, it is purely a contingent matter that I do, in fact, let go of the chalk. While the inter-relation of states of the world at one time and another, an inter-relation governed by the physical laws that connect states at one time to states at another time, may be a matter of physical necessity, isn't it the case that "initial conditions" are a purely contingent matter?

Historically there have been those who have thought not. I suppose Leibniz would have us believe that were we smart enough we would come to see that even the most apparently contingent particular fact occurred as a matter of necessity in this best of all possible worlds with its pre-established monadical harmony. But modern intuition is that the particular is the realm of what is but of that which could have been otherwise.

What I wish to explore here are some cases from physics that might make us think twice about the idea that even if the lawlike is the realm of the necessary, the particular, the intitial state of the world at a given time, is purely contingent. The claim will be that there may be more "lawlikeness" in the initial conditions of the world than we are willing to admit. I will look at three cases, two drawn from spacetime theories and one from statistical mechanics. Other examples may exist. For example, the theory of particle interaction that dispenses with intermediate fields in favor of a combination of retarded *and* advanced direct particle interactions presents issues similar to some I will discuss. But the three cases I will look at show, I think, some diverse ways in which the claim might be supported that initial states are not as "freely choosable" as one might think.

The usual formation of the special theory of relativity will contain a characterization of the spacetime as the standard Minkowski spacetime. This might be derived from some postulation of a relativity principle, especially concerning the velocity of light as constant and isotropic in all inertial frames, and a linearity postulate to generate the full spacetime metric structure, or it might just be posited directly. But the usual formulation of the theory will also contain an assertion to the effect that the velocity of light is the limiting velocity of propagation of a causal signal to and from any particular spacetime event location. The non-existence of causal signals propagation "outside the light-cone" is taken as an empirical fact essential to the theory.

But, of course, that postulation of light as limiting causal signal is independent, logically, of the postulation of the Minkowski spacetime structure itself. So it becomes amusing to ask what a theory would look like in which there did exist superluminal causal signals, the "tachyons" of such great notoriety. Of course tolerating them requires patches here and there in physical theory. Issues of energy-momentum content, redescribability of the propagation "backward in time" of a quantum particle as its anti-particle propaging "forward in time," and so on, become interesting speculative physics.

But the real problem with tolerating tachyons is the possibility they generate of causal paradox, that is of characterizations of the state of affairs on some spacelike hypersurface that leads to the conclusion that that very state of affairs could not exist. How this occurs is easy to see. Let a faster-than-light signal propagate from a given event. There will then be, on the basis of the usual relativistic stipulations for the determination of simultaneity for events at a distance from one another, some observer (some inertial reference frame that is) relative to which the tachyonic signal will be determined to be propagating from its origin event to distant events earlier in time, relative to the chosen frame, than that origin event. This must be so since in Minkowski spacetime any event outside the light-cones of a given event will be in every time order (to the past of, simultaneous with and future with respect to) the given event for some inertial observer or other. Since the trajectory events of the tachyon are outside the light cone of the origin event, they must describe causal propagation into the past of the origin event for at least some inertial observers.

But if causal signals of arbitrarily high velocity are permitted, there will then be some causal signal propagable from the reception of the tachyonic signal at event o'

that causally influences what goes on at the origin event, *o*. For relative to some observer *o'* is before *o* and can serve as an origin of a causal signal propagated forward in time to o relative to the new observer. The net result is that postulating tachyons in conjunction with retaining Minkowski spacetime results in the possibility of the notorious closed causal loops in spacetime.

States of the world on a given spacelike hypersurface will, then, be able to causally influence their own condition by means of such closed causal loops. It is easy then, allowing oneself the postulation of arbitrary initial states on any one spacelike hypersurface, to construct a situation where an event at a given place at a given time causally determines that that very event not take place. The usual arrangements of a device that at o sends a signal to o' to intitiate a signal that prevents the original signal emission at o will do the trick. Any old variant of the time travellers paradoxical murdering of his ancestor will do.

We can eliminate the apparently paradoxical nature of tachyon theory, however, by some stipulation to the effect that all states of the world at a given time (relative to any chosen intertial frame) be self-consistent. That is, that they must be such that the causal influences they do propagate into their futures, subliminal or tachyonic, be such as to generate events that never lead to closed causal loops except those that give rise to the state we started off with. We can save tachyon theory from paradox by simply giving up our idea that any old stipulation of initial conditions on a single spacelike hypersurface is a legitimate stipulation.

My guess is, however, that if we found super-luminal causal signals to exist we would not adopt such a radical new stance to the arbitrariness of initial conditions. We would probably, instead, drop the idea that spacetime was Minkowskian, looking, instead for some alternative theory that could still save the fundamental phenomenal result that led us to special relativistic spacetime in the first place, the fact of the null results for round-trip experiments irrespective of the inertial frame in which the laboratory was at rest, but one that dropped the standard Einsteinian specification for distant simultaneity. Perhaps a resurrected aether compensatory theory would do the trick.

In any case, there is not the slightest reason to believe that tachyons do exist. So we can serenely hold to the lawlike status of the truth that there are no causal signals that propagate outside the light-cone. Given this it is then a law of nature that guarantees that we will not be hindered in our desire to be free to choose initial conditions at a single time as freely as we like. But at least there has been a hint that some real physical possibilities might lead us, not to full Leibnizianism that tells us that one and only one universe is possible, singular conditions and all, but at least to some contraints on our idea that the particular is fully the realm of the contingent with physical necessity restricted to the lawlike and hence to the particular only as conditional on other particular stipulations. If a stipulation of the particular facts at one time is lawlike connected to itself as self-determiner, some such specifications of the situation at a given time are ruled out by the lawlike connections of intial conditions to themselves.

Even if super-luminal propagation is impossible relative to physical law, however, the possibility of closed causal curves that loop from an event back around to that same initiating event still must be taken at least semi-seriously. For adding gravity to our relativistic picture of the world, at least in the favorite form for doing so, the general theory of relativity, takes us into the realm of generally curved spacetimes. And these can suffer hosts of causal pathologies that generate closed causal loops, even while obediently confining propagation into the local absolute future, that is into the union of the future light-cone from the initiating event and its lightlike boundary.

The most famous of these possible general relativistic spacetimes is Gödel's important world model. This cosmological solution to the general relativistic field equations has, among other features, a smoothed out mass distribution which is plausibly spoken of as being in uniform rotation everywhere. Rotation, that is, with respect to the compass of inertia formed by "freely moving" particles or light rays travelling only under the influence of the overall spacetime structure determined by the mass distribution. But it also has the notorious feature of containing closed timelike loops, allowing the propagation of a causal signal, always into the local future light-cone interior, that eventually loops back to the originating event. (Gödel 1949a.)

Other such causally pathological worlds can be constructed. Some, like Godel's are physically interesting, being associated with a not too unreasonable cosmological distribution of the averaged out matter of the universe. There are, for example, those models constructed to avoid, as Godel's does not, the need for a cosmological constant. Others can be constructed in a more unphysical and artificial way. One could, for example, simply take flat Minkowski spacetime with no curvature at all. Slice it along two spacelike hypersurfaces at a finite time from one another relative to a given inertial observer, and identify the two spacelike hypersurfaces with one another. The resulting "time cylindrical" spacetime is intrinsically flat everywhere, compatible with the field equations since the solution of flat empty Minkowski spacetime is compatible with them, but infected with closed causal loop.

More recently, Morris, Thorne and Yurtsever proposed a "time machine" constructible in a general relativistic world, a proposal followed up by explorations of Novikov. Here one adds to a near flat ordinary spacetime a "wormhole" that provides a handle attached to the ordinary spacetime. This results in a complete spacetime that is multiply connected. The two ends of the wormhole are labeled A and B. The B end of the wormhole is forced to follow a round-trip accelerated journey (as viewed in the ordinary spacetime) away from the other, A, end and back to its near vicinity, its clock "lagging" a clock fixed at the A end after the return due to the familiar "twin paradox" effect. But the wormhole connecting the two opening in ordinary spacetime is kept unchanged and short throughout the motion of the B end relative to the A end in ordinary spacetime. After the journey has been completed an observer entering the wormhole through the B end and passing through the wormhole will emerge out of the A opening "in his own past." Once again closed causal loops become generable in the spacetime. The physical possibility, at least in principle, of constructing such a model has been discussed by its inventors. (Morris, Thorne and Yurtsever 1988; Morris and Thorne 1987; Novikov 1989.)

But, now in any of these worlds the issue of the compatibility of a specification of the events at a spacetime location with the events at that location as causally determined, at least in part, by the very occurence of these events arises. Once more we cannot assume that any old specification of events at one time is compatible with the causal laws. Once such closed causal loops are tolerated as possibilities for the world, some specifications of the initial states become themselves impossibilities. Even now, I understand, Thorne is working on the construction of specific world models designed to show that at least some consistent "wormhole time travellers" can be constructed.

Now it is quite unlikely that the world we live in is Gödel's cosmological world, or that any of its physically "reasonable" variants. Nor is it likely to be one of the "scissors and paste" worlds with closed causal loops that provide the ingenious and amusing counter-examples to plausible but false generalities about spacetime which are found in the standard texts on global features of spacetime. It isn't likely, I suppose, that wormhole time machines exist either, although given the possibilities inherent in structures around singular regions of spacetime (which likely do exist) and of the features that can arise out of quantization of spacetime at the very small scales compatible with the Planck length, it might, I suppose, very well be that somewhere and sometime some closed causal loops do, in fact, exist in the actual world.

But even if there are, in fact, no such closed causal loops, their possibility, their physical possibility if that means compatibility with all the known standard lawlike constraints on the nature of the world, does seem to have its philosophical consequences. Of course we could make such situations impossible by just declaring it to be a "law" of nature that no closed causal loops may exist. But such a declaration of impossibility by fiat would seem unmotivated by the evidence by which we normally come to posit theory, and certainly quite unlike the denial of trans-luminal propagation that forms the supplementary law in the grounding of the usual versions of special relativity.

Gödel argued that the very possibility of closed timelike lines in general relativity showed that time as ordinarily understood (and as ordinarily experienced?) was an "ideal", a structure of immediate experience not to be identified with time as parameter as it appeared in our physical theories of spacetime. Even if such closed causal loops did not, in fact, exist, that they could exist showed that time order as understood in physics was nothing like time order as it appeared in our experience. From our perspective, it is the possibility of such closed causal loops that tells us that our ideas of the pure contingency of initial conditions has come under a cloud. For if it is even physically possible that such closed causal loops exist, then it is at least possible that some initial conditions are not merely not the case, but are themselves impossible as they entail, given the usual laws of causality, their very own non-existence. Determinism tells us that not every pair of specifications of spacelike surfaces is a possible specification. But given as closed timelike loops, a single surface must also need be compatible with itself. (Gödel 1949b.)

The existence of cosmological and other models with closed timelike or lightlike paths in the general theory of relativity has suggested to some members of the physics community that a new model of "causal explanation" is now in order. Instead of thinking of physics, in its task of constructing possible worlds, as stipulating arbitrary conditions at one time (or one one spacelike hypersurface) and then determining the inevitable development of those initial conditions as time goes on (and also as one goes back in time for theories deterministic in both time directions), the game of constructing the possible worlds is to specify a spacetime complete with its possibly causally pathological topology and to simultaneously specify a mass-energy distribution within that spacetime. The field equations then determine the overall consistency with the theory of general relativity of the total world-model so constructed. Hints that some kind of revision is necessary in what we take a theoretical explanation to be in general relativity actually antedate consideration of worlds with closed causal paths. The very nature of the field equations, connecting as they do the metric of the spacetime with the stress-energy tensor of the mass-energy that is not gravitational (i.e. not that of the spacetime field itself) already demands a notion of the construction of possible worlds as including the production of self-consistency checks. We cannot, as many have observed, naively think of specifying a mass-energy distribution and having it "cause" the spacetime to have the structure it does, for the mass-energy distribution requires an antecedent specification of the metric structure of the spacetime in which the mass-energy is to be distributed. To construct a physically possible world we must, instead, specify a spacetime structure and a distribution of mass-energy in it and then check to see if the joint specification of spacetime structure and mass- energy distribution is in concordance with the field equations.

The constraints on the possibilities allowed for initial conditions that I want to focus most attention on, though, are rather different in kind from those noted above. The elimination of some initial conditions as possible initial conditions that has been discussed above follows directly from the assumed laws of nature. Given the special relativistic lawlike structure of spacetime, the existence of tachyons makes some initial states self-incompatible and, hence, impossible relative to the assumed laws. Given the laws of general relativity, the possibilities for spacetime structures containing closed causal loops makes some imaginable initial structures for a spacetime, once again, self-incompatible and hence, actually, impossible. The constraints on initial conditions I will be concerned with now, though, don't seem to follow from any accepted laws of nature. Indeed, their peculiarity is that they seem to be constraints we must assume nature imposes on initial conditions if we are to explain important structural features of the world we find ourselves in, even though the very existence of such constraints doesn't seem to find any natural place in our conception of the world as structured in its general nature by the hierarchy of normal physical laws.

While the restrictions imposed on initial states discussed above worked by showing that, relative to assumed laws, certain initial states were impossible, the constraints I am interested in now don't seem to be framable in terms of an impossibility of any particular initial state of the world at all, if, that is, possibilities are judged relative to the constraints placed upon nature by the "ordinary" laws of nature. The constraints on initial states I have in mind now are those that tell us in any sufficiently large collection of such initial conditions of sub-systems of the world, initial states will be found to be distributed with a definite "probability." While any of a vast number of possible initial states for a system is a possible initial state, the relative frequency with which such initial states will be found must be given, subject to the usual looseness necessary between actual proportions and the probabilistic models used to generate them, by one of the familiar probability distributions for initial states found in statistical mechanics and related theories in physics.

Such stipulated probability distributions over initial conditions are a necessary component of the recent fascinating work on systems that show apparently random or chaotic behavior on the macroscopic level, even if they are described as purely deterministic model systems, the famous examples of turbulence, meterological instability, strange attractors and the like. An older physical discipline, and one with even broader importance in the description of the physical world, is the statistical mechanical theory used to underpin the results of thermodynamics.

In its application to the grounding of the macroscopic thermal behavior of systems, that being understood in its fullest generality to include all the facts in the world about the existence of macroscopic equilibrium states for systems with many microscopic degrees of freedom, the uniform, lawlike approach to such equilibrium states in the one time direction called the future and not in the reverse time direction, and the ability to characterize both the equilibrium states and the approach to them in terms of a small number of macroscopic thermodynamic parameters, the postulation of a probability distribution over microstates of the system plays different roles, with different rationales, in different portions of the theory.

In equilibrium statistical mechanics the aim is to derive from probabilistic postulates over the microstates of a system and the identification of macroscopic features of the system with appropriate averages of microscopic conditions of the components of the system, the equations of state describing the inter-relationships among these macroscopic features when the system in question is in equilibrium. The results can be obtained as one wishes using the familiar standard probability assumptions (summed up in the various Gibbs' ensemble distributions) over the microscopic conditions of the systems.

But what rationalizes or justifies our choice of such proability distributions? Why should we believe that they hold of systems in the world? And if they do, why do they? In the equilibrium theory some of these questions can be evaded to a degree by offering a "transcendental" deduction of the probability distributions, that is by showing using the fundamental dynamical laws governing the microscopic states that only the standard probability distribution over the microstate of a system will have the properties desired of a probability distribution appropriate to describe an equilibrium state. Crudely one argues from equilibrium as a temporally invariant macroscopic state of the system to the appropriate probability distribution being one that is also fully invariant in time. Then one uses the basic facts of the underlying dynamics to show that only one probability distribution will have this temporally invariant character. Actually one cannot show that, but only that there will be a unique such probability measure that also assigns probability zero to classes of micro-states given measure zero by the standard probability distribution. Even in the equilibrium case, then, with its peculiar "transcendental" model of explanation and rationalization, additional "primitive" probability assumptions over the micro-states of the system must be made to get the results one wants from the laws of the underlying dynamics.

The case of the non-equilibrium system is the one, though, that makes the role of these probability assumptions most apparent and their peculiarity in the light of our usual thoughts about scientific explanation clearest. A system is prepared in a non-equilibrium state and it then evolves toward equilibrium following a predictable dynamical path describable by the evolution of a small number of macroscopic parameters. Why does this occur? The answers given invoke, of course, the dynamical laws governing the interaction through time of the micro-components of the system. But they invoke also a probability distribution over the microstates available to the system at the moment its evolution begins. Interestingly, quite different raionals for the lawlike, marcoscopic non-equilibrium behavior are to be found in non-equilibrium statistical mechanics. Some of the rationales rest upon "mixing" properties of the dynamics, properties that hold because of basic facts about the constitution of the system and are not dependent upon its great number of degrees of freedom (its great number of micro-components). Other rationales do rest upon the fact that the system has innumerable components and, frequently, upon such idealizations as a limit of zero density (as in the Lanford derivation of the Boltzmann equation describing the non-equilibrium behavior of a rare gas). Indeed, the rationales, while not flat out contradicting one another, due to their disparate modeling assumptions and disparately limited conclusions, are sometimes at conceptual odds with one another in the way they try to account for the macroscopic non-equilibrium processes.

But one thing is clear. Each explanation will, at some point, require that one invoke in its explanatory resources a particular probability distribution over the initial microconditions possible for a system. Without such an invocation, for example, a "mixing" type rationalization can give conclusions only in infinite time limits. Deriving the finite and uniform "relaxation times" for a system is impossible without some restriction on the "initial ensemble" whose evolution models the evolution of the physical system. The need for such an initial probability distribution is made transparantly clear by the problem of time irreversibility. The underlying dynamical laws are (almost always) taken as time-reversal invariant. The only way the theory could predict the observed thermodynamical behavior of systems forward in time, and not predict the not observed "anti-thermodynamical" behavior in the reverse time direction, is to break the symmetry of what goes into the explanatory account by making a probabilistic assumption over micro-conditions that is to hold of initial but not final states of the system in question. The constraints imposed on initial conditions by statistical mechanics do not, then, demand that any particular initial micro-condition hold or fail to hold for a given system. They do insist, however, that any collection of systems prepared in the identical way as far as macroscopic conditions are concerned, the initial conditons in which the systems will be found to originate their evolution occur with a stable and repeatable frequency distribution.

It would be tempting to go further and say that statistical mechanics claims that this distribution over initial micro-states is "inevitable" or "inviolable." But that would be going too far. We do know of exceptional cases where collections of systems prepared in "anti-thermodynamic" initial states are possible, for example in the classical spin-echo experiments. These systems are, to be sure "anomalous," and prepared in a curious and tricky way, that is, by pulling a Loschmidt reversal trick on the micro-states obtained from a system that has apparently gone to equilibrium so that it evolves back to its original non-equilibrium condition, but they are possible. Even using the word "anomalous" here is revealing, with its implication of being "contrary to law." The whole puzzle is that the demanded initial probability distribution over micro-states for the well behaved systems doesn't seem to have its origin in any law of nature, if the laws are restricted to the familiar dynamical laws that form the backbone of scientific lawlike regularity.

There are, to be sure, ways in which one might attempt to recontrue the probability distribution over initial micro-states in such a way as to eliminated the very need for some contraint on initial conditions being posited. One way of doing this would be to argue that there are not any such initial micro-states to begin with, at least no micro-states that can be represented as points in an appropriate phase-space. There are those who suggest that the probability distributions over regions of phase-space ought to be taken as representing the actual initial states of individual systems, and not merely as distributions over point like micro-states that describe only ensembles. The idea here, of course, is to draw a parallel with the arguments in quantum mechanics to the effect that the wave-function is to be viewed as the irreducible state of an individual system and not as a mere probability distribution over some hidden micro-states of determinate exact position and momentum. But the arguments for this line in quantum mechanics rest on "proofs of the impossibility of hidden variables," proofs we clearly cannot apply in the statistical mechanical case, given the very definition of the statistical mechanical ensembles as just such distributions over deeper dynamical point-like state parameters. In any case, even if one took this line about the most precise states of individual systems being ensemble probability distributions, with the traditional point-like micro-states as a dispensible false idealization not representing any real physical state of systems, one would still have the problem of understanding why certain such initial states were "allowed" and other "forbidden," even though the latter forbidden initial states were permissable as irreducible probabilistic states as far as the laws of dynamics governing the evolution of such probability distributions were concerned. One would still, that is, have the problem of a constraint on initial conditions existing and of explaining why such constraints existed.

Another familar way of trying to avoid the necessity of positing a general, probabilistic constraint over initial micro-states is to try and account for this general distribution by tracing its origin to the overall cosmological structure of the universe. For all the well known reasons, reference to the expansion of the universe in one but not the other direction of time will not be, by itself, sufficient to account for the time-asymmetric entropic behavior of the word on the underlying time asymmetric probability posit. Instead reference will have to be made to the special conditions that characterized the "initial state" of the universe as a whole, that is to the state of the world an infinitesimal moment of time after the "big bang" initial singularity (assuming, of course, that there was indeed such an initial singular state). If we could account for the general, time-asymmetric, probability distribution over initial as opposed to final micro-state of systems in this way, we might hope to at least avoid the appearance in our theory of this curious element of our basic posits that doesn't fit in otherwise with our general explanatory scheme of fixed laws and arbitrary initial conditions.

But several reasons make the cosmological way out less promising than might appear at first glance. Even if one could make it work, it would itself require an explanation as to why the "big bang" initial state itself had exactly the special nature needed to give us the right results for our later "branch systems" temporarily separated from the main universal evolution. The most promising models for doing the job, such as that posited by Penrose with its "thermalized" randomness over material initial conditions combined with smoothness of the spacetime structure, the latter providing the initial low entropy on which the later agglomeration of matter into low entropy hot stars and cold space from its initial high entropy equilibrium condition feeds, itself requires an explanation (or seems to so require an explanation) to account for its occurence given its "a priori improbability". (If, indeed, talking about probability of universes and their initial states makes any sense at all.) Penrose himself suggests a "law of nature" to do the trick, a law to the effect that white-holes (of which the "big bang" may be the only example) obey a law of having the Weyl tensor or their pure spacetime structure initially equal to zero (i.e. of being conformally flat). Here, again, we see, I think, the temptation to deny that a mere initial condition by itself can ground a kind of "necessity" like that of the asymmetric probability distribution needed for statistical mechanics and the desire to found any such necessity on a "law", artificial as the process seems in this particular case.

In any case, it is dubious that one really can trace the general probability posit over initial micro-states of temporarily isolated systems back to some "natural" single initial micro-condition at the "big bang." Any plausible stipulation of that initial state, say as with smooth spacetime structure of low entropy but high entropy equilibriated matter, although it will give rise to the right entropy increase of the universe as a whole as time goes on, can only generate the appropriate probabilistic distribution over the initial micro-states of temporarily isolated systems by an additional posit that can always be matched by one that, by parity of reasoning, is applied to final micro-states of temporarily isolated systems and leads us to make incorrect anti-thermodynamic predictions about their past behavior. Without some bald-faced posit like this: "The initial micro-state of the system was just such as to generate the appropriate probability distribution over the micro-states that are initial (but not final) for temporarily isolated systems branched off from it," it is hard to see how the general posit about branch systems needed to get the parallelism in time of their entropic increase is going to follow from any "natural" stipulation about the initial micro-state. Of course if we believe in determinism there has to be some truth to the claim that the initial micro-states of all of the branch systems are as they are because the initial state of the universe as a whole is as it was. But that is different from offering any explanatory account of why the general, time-asymmetric probability posit holds and not, for example, its false time reflected anti-thermodynamic twin. At the present time it looks more as though the entropic increase of the universe as a whole is more an instance of the general rule that entropy increases in all systems in a parallel time direction (that we call the future) and not itself an explanation of the general rule.

Just what could explain the existence of time-asymmetric probabilistic constraint on initial micro-states of systems is, I believe, far from clear. The explanations that have been given so far seem to me to smuggle in that which is to be derived at some point in their arguments. Saying that the distribution is the result of our "preparation" of a system, as Krylov does, assumes that we are able to characterize preparing a system, as opposed to observing it, in such a way as to explain why all the appropriate interactions with a system are in the same time order from one another in each system. Other putative explanations rely on a presupposed time-asymmetric notion of causality, so that, for example, it is appropriate to infer the later distribution of the micro-states of an isolated system from a posited initial distribution but not to make such inferences in the reverse order, to generate the needed time-asymmetry.

But I do not intend to pursue the issue of the explanation of the time-asymmetric probability constraint on intial states here, but only to emphasize that it exists and that positing it is a vital component of our overall description of the world. But what kind of posit is it? In particular, is it just an observation about how "as a matter of fact" the micro-states of systems "happen to be" distributed in the world? Or is it, rather, the postulation of a new generalization over and above the standard dynamical laws, but one that shares with them lawlike status?

Mehlberg and Grünbaum, among others, repeatedly emphasize the allegedly "de facto" nature of the probability distribution and the consequent Second "Law" of Thermodynamics. Here they focus on the fact that the needed probability distributions over initial micro-states certainly cannot be derived from those general propositions about the regularities of evolution in general that we term dynamics, nor from these combined with the more special regularities that describe the particular structures of things that characterize the "forces" needed, along with the general dynamics, to be able to fully determine the temporal evolution of systems. But don't these two sets of generalizations fully exhaust the hierarchical struture of generalizations of fundamental depth and breadth that are sufficient for us to honor a generalization with the name of "law of nature?" And isn't it the case that relative to all of these generalizations initial conditions remain "free" or "open" in their specifications? If that is so, how could an assertion as to the proportions with which these initial conditions happen to occur be anything but a "mere de facto" truth about the world, no matter how general and important it is? (Mehlberg 1980; Grünbaum 1973.)

Others, Krylov and Prigogine to cite two noteworthy examples, insist tht the probabilistic posits of statistical mechanics must be viewed as having lawlike status. Here again the motivation is clear. How could any generalization as universal, as fundamental and (modulo peculiarly contrived special exception cases) as unevadable as the Second Law, with its constraints upon the entropic behavior of systems, be thought of as some kind of "mere accident" in the world? If the generalizations that lie at the heart of pure dynamics and at the heart of the fundamental constitutive relations governing matter are thought of as "physically necessary," as "supporting counterfactuals," and, in general, as "laws of nature and not mere generalizations," how can the Second Law, and the statistical generalization over initial micro-states of systems that grounds it, be thought of as having any lesser "modal" status? (Krylov 1979; Prigogine 1980 and 1984.)

The issues here are often connected with the issue of whether we ought to think of systems as existing in a time which is itself asymmetric or, instead, only as one in which while time is symmetric, systems behave asymmetrically in time. Even among the "de factoists" there is a split. Grünbaum claims that the "mere de facto" asymmetry in time of systems is enough to ground a notion of "asymmetry of time," Mehlberg, and more recently Horwich, argue that only an asymmetry in time of a law is enough to establish asymmetry of time. The entropic asymmetries being "merely de facto" only establish the asymmetric behavior of systems in time. (Horwich 1987.)

I actually rather doubt that any coherent issue is really at stake here. The argument of Mehlberg and Horwich is, crudely, that if we had a time asymmetric law, only the asymmetry of time itself could explain that asymmetry in the lawlike behavior of nature. But a mere de facto asymmetry in the phenomena of the world is to be explained, if at all, in terms of other merely contingent de facto conditions, and requires no positing of an asymmetry of "time itself." Presumably those who think the de facto asymmetry sufficient to establish the "asymmetry of time" would argue, rather, that such a grand and prevasive de facto asymmetry is itself constitutive of time being asymmetric. At least it is if it can "ground" (perhaps) all of our intuitive notions of what constitutes the asymmetry of time (asymmetry of causation, records of the past and not the future, and so on). I think we must first distinguish two issues: (1) What feature in the world would require an asymmetry of time as explainer?; (2) What would it take to ground the intuitions we have about the differences of past and future? If we make that distinction, then, at least, the two sides would not be just arguing past one another.

I think that the former issue is one where, on reflection, we will find that the idea that "asymmetric time itself" is both a necessary and sufficient basis for "explaining" time asymmetric laws but is not needed for explaining time asymmetries that are "merely de facto" puts a lot more weight on the idea of spacetime structure as "explainer" of symmetry and dissymmetry among phenomena than it can bear. When all is said and done I think there will remain profound issues of whether or not the asymmetries of phenomena summed up in entropic increase can really ground all the intuitive asymmetries we normally think of as distinguishing past and future. But I doubt very much if there is any real distinction between a pervasive asymmetry of phenomena in time, whether grounded in asymmetric dynamical laws or in asymmetric probability distributions over initial conditions, and some existent or non-existent asymmetry of time itself, as Mehlberg and Horwich intend that notion.

At a more profound level, I am not even sure that any coherent sense can be made of the dispute as to whether the temporal asymmetries found in thermodynamics and statistical mechanics ought to be considered "lawlike" or "merely de facto." If one takes the line that any generalization that is broad enough and deep enough to play a fundamental role in our descriptive and explanatory structure for scientifically grasping the world is a generalization that we honor with lawlike status, then surely the generalizations summed up in the Second Law and grounded in the general probability posits for non-equilibrium statistical mechanics have that status.

But whether one takes these generalizations as "lawlike" (statistically lawlike, of course) or not, they do present a feature that makes them stand out as "queer" among the other "laws of nature." Thermodynamics and statistical mechanics put aside, the remainder of fundamental physics has a simple explanatory structure. There are the basic dynamical laws that govern the interrelations of states over time of all fundamental physical systems, now, presumably, our latest refinement of quantum field theory taken as a general dynamical theory of fundamental systems. Then there are the constitutive laws that tell us what the basic components of nature are and what their fundamental inter-relationships are that are the contemporary version of the Newtonian force laws. (All of this becomes a little more complicated if one introduces gravity as curved spacetime which mixes basic kinematics and constitutive dynamics in a subtle way, but the basic claim still holds.)

But thermodynamics, and its grounding in statistical mechanics, just doesn't fit into this picture in any easy way. Universal in their applicability across all kinds of physical systems, these theories cannot be placed in a simple derivative manner into the remaining hierarchy of physical explanation. We can go a long way toward rationalizing statistical mechanics from dynamical and constitutive principles alone, but we cannot give a complete derivation of the theory from these other principles of fundamental physics. Something else must be added.

What must be added is some fundamental, independent, probabilistic posit over the distribution of intial conditions in temporarily isolated systems. Here is where the peculiarity of statistical mechanics becomes apparent. From the point of view of the other dynamical and constitutive laws, initial conditions are "free." They can be assigned to individual systems as we choose in our idealizations of real systems, and, in principle, real systems in the world can be prepared so as to have those initial states as we choose. But we know that this isn't so, at least in the probabilistic sense given us by statistical mechanics. When we prepare a collection of systems we cannot distribute their micro-initial conditions as we choose, for, a few peculiar exceptions excluded, nature will, at least probabilistically, choose the distribution for us.

So if the laws of nature do happen to allow for causal propagation outside the light cone, or if the actual topology of spacetime has the pathological structure that allows for closed causal loops, then some initial conditions are not possible at all. And as the world stands, so far as we know, some distributions of such initial conditions are "probabilistically excluded" from the realm of possibility in a manner that is generated by principles that while seeming themselves to have a kind of "lawlike" status, don't fit into the normal lawlike structure of other familiar theories in any simple way. While most of our physical theories seem to confine physical "necessity," whatever that is, to the realm of the lawlike connections of states at one time to states at some other time, leaving the state at one time in the realm of the "purely contingent," there do seem to be some aspects of physics that cast doubt on claims that the stipulation of initial states is a purely "contingent" matter as one might first think. The Leibnizian ideal of one and only one world being necessarily the case seems to be too strong. But if necessity is kept to "physical necessity" it does seem that more worlds are physically impossible (at least "probabilistically impossible") than only those excluded from possibility by the usual laws of nature alone.

References

Gödel, K. (1949a), "An Example of a New Type of Cosmological Solution of Einstein's Field Equations of Gravitation," *Review of Modern Physics* 21:447-450.

_____. (1949b), "A Remark About the Relationship Between Relativity Theory and Idealist Philosophy," in P. Schillp, ed., *Albert Einstein: Philosopher Scientist* New York: Tudor.

Grünbaum, A. (1973), *Philosophical Problems of Space and Time*, Dordrecht: Reidel.

Horwich, P. (1987), *Asymmetries in Time*, Cambridge MA: MIT Press.

Krylov, N. (1979), *Works on the Foundations of Statistical Physics*, Princeton: Princeton University Press.

Mehlberg, H. (1980), Time, *Causality, and the Quantum Theory*, Dordrecht: Reidel.

Morris, M. and Thorne, K. (1987), *Caltech Preprint* GRP-067.

_____. and Yurtsever, U. (1988), *Caltech Preprint* GRP-164.

Novikov, A (1989), "An Analysis of the Operation of a Time Machine," *Soviet Physics JETP* 68: 439443.

Prigogine, I. (198O), *From Being to Becoming,* San Francisco: Freeman.

_____. (1984), *Order Out of Chaos,* New York: Bantam Books.

Law Along the Frontier:
Differential Equations and Their Boundary Conditions

Mark Wilson

The Ohio State University

1. Introduction

This essay will survey various considerations that arise when a branch of physics requires formulation in terms of partial differential equations (or some facsimile thereof). My examples will derive almost exclusively from classical continuum (=smeared out matter) mechanics. Although the relevant formal facts are well known, it is difficult to find coherent discussions of how the underlying phenomena ought to be viewed. In this paper, I will give an introduction to some of the issues, although I will confess at the outset that I am not certain how these matters should be finally adjudicated. Certainly it would be helpful if philosophy paid more attention to the underlying issues.

Some of the points I will discuss bear upon some of Professor Sklar's observations. Although my attention will be largely directed towards the status of boundary conditions (and related forms of singular surface), many of my examples can be adapted, *mutatis mutandis*, to initial conditions as well. Professor Sklar has supplied various cases that show that the set of "initial conditions" for a physics may not be "freely chooseable". Although the relativistic cases he describes are interesting and important, they are considerably more exotic than required to make the point about "free chooseability". Even the most humdrum equations of continuum physics make it difficult to find any places in the universe where either allowed initial or boundary conditions can be plausibly viewed as "freely chooseable". Indeed, although it is relatively clear what "freely chooseable" should mean for ordinary differential equations, the matter is much less obvious when partial differential equations are concerned.

I here employ the phrase "boundary conditions" in its more or less customary physical sense. In contrast, standard philosophy of science texts encourage a rather misty and misleading understanding of this term. This faulty picture arises, as stereotypes often do, from inattention. Indeed, the standard philosophy texts say virtually nothing about boundary conditions—they are scarcely mentioned before they are packed off in an undifferentiated crate labeled "initial and boundary conditions" (usually pronounced as one word). The salient fact about "initialandboundaryconditions" is that, whatever they else they may be, they are not laws and can be safely ignored. I find that many philosophers, insofar as they entertain a separate conception of "boundary conditions", tend to think of them as "standing" or "background" conditions. But this isn't right at all. Boundary conditions, roughly characterized, represent claims about how a certain portion of the universe interacts with its surroundings along their mutual boundary. A stock example of such a boundary condition is the requirement that the two endpoints of a plucked string remain fixed. The static nature of this particular "fixed end" requirement probably suggests the confusion with "standing conditions". But consider a block of ice melting into water. The "Stefan type" boundary condition for this problem requires that the latent heat in the ice is converted to active energy at a constant rate at temperature 0°. Here the boundary moves through the ice and does not count as a "standing condition" at all. On the other hand, the statement that a substance is immersed in a constant gravitational field may count as a "standing condition", but it is certainly not a boundary condition.

Philosophers have been negligent about trying to align their rather optimistic philosophic conceptions of "law"/"nonlaw" with the concrete stuff found in physics. At a bare minimum, one will find that the linkages must operate in a much more complicated way than normally presumed. In this vein, it should be recognized that "boundary condition" in its normal physical use cannot provide a suitable place where law can be demarcated from non-law. Indeed, that Stefan type boundary condition for melting ice counts, by normal standards, as a straightforward scientific law. The fixed end condition for the string has a more complicated status, which I won't try to unravel here.

One of the quarrels that I—and this relates to Professor Sklar's concern as well—have with the stock philosophy of science picture of "initialandboundaryconditions" stems from the presumption that boundary conditions play an essentially inert role in science—it is the *laws* that serve as the real engine driving a D-N explanation forward, etc. But the phenomena I will discuss suggests that "laws" and "boundary conditions" collaborate in a more equal partnership than the inert view suggests. Accordingly, I will survey a variety of ways in which boundary conditions actively contribute to the operation of the physics within their dominion.

Unquestioning acceptance of the inert picture can effect real damage in concrete philosophy of science. For example, Hartry Field in (Field 1980) attempts to provide a nominalization of what he calls "Newtonian gravitational theory" (it really isn't that). Field treats the whole of that "theory" as contained essentially in its central "law", Poisson's equation. Field ignores any consideration of whether that equation, as he construes its mathematical content, can be properly harmonized with its usual family of boundary conditions. In fact, the answer is "no", as mathematicians in the nineteenth century recognized (at the cost of considerable distress). Indeed, the rigorous study of how Poisson's equation manages to accommodate its expected boundary conditions represents an important part of the history behind the attitudes that we will survey.

2. The clash between laws and their boundary conditions

Let us begin with a surprising fact emphasized by Jacques Hadamard around the turn of the century (although examples of the phenomena reach back to the earliest uses of partial differential equations). I have in mind the frequent mismatch that arises between "laws" formulated as partial differential equations and their "proper" boundary conditions. Suppose one has a region R occupied by a certain kind of material, say, iron. Relying upon physical principles and approximations of various sorts, one sets up a differential equation L appropriate to the region R (or set of equations—I will write in the singular for convenience). If no "forcing terms" appear in the equation, it is probably appropriate to regard this differential equation as a "law" governing the behavior inside R. As one moves out to the boundary of R, which I'll label ∂R, our "laws" L seem to place certain requirements on the iron at the edge represented by ∂R. If the iron is treated as an elastic material, one naively expects that the displacement should be twice differentiable along ∂R, for the relevant elastic law requires second spatial derivatives. But when one looks through the entire family of boundaries for regions in which one expects that the iron could be distributed, one finds many mismatches—the family includes all sorts of boundaries that are incompatible with the requirements that L seem to place upon them. In rough terms, it seems as two kinds of requirements can be placed upon the boundary ∂R: (i) *External* requirements deriving from our prior knowledge of the regions in which the underlying material can be distributed; (ii) *Internal* requirements stemming from L itself. The surprise is that these two requirements can sometimes clash.

A simple yet typical clash between boundary condition requirements and internal equation can be seen in the case of Fourier's heat equation, which governs the diffusion of heat through our piece of iron. Naively, one expects that one should be able to take a block of metal and hold its opposite sides at different temperatures, say, 0° and 100°. Such an assumption about the kind of conditions that can conceivably surround a block of material governed by the heat equation represents an example of what was dubbed an "external" requirement on the possible boundaries that the material might occupy. Collect all such possibilities into a set and call it the "externally natural class of boundary conditions" for the heat equation. Under steady state conditions, the heat equation degenerates into Laplace's equation whose only bounded solutions, in two dimensions, are constant everywhere. To accomodate our belief that boundaries can be maintained at different temperatures, the governing equation must lose validity somewhere. A natural place to expect some kind of breakdown is along the edges of the material—the sort of "boundary" we tend to select as "natural" to the problem. From the earliest times, the library of standard solutions utilized by workers in potential theory embraced solutions where the temperature jumped from one value to another along, e.g., a circular boundary. In this sense, the "internal" requirements set on boundaries have always clashed with the expected set of "external" requirements.

Prima facie one might expect that physicists would give the "internal" requirements greater credence, but, in fact, they often favor the external requirements:

> [P]rematurely ossified mathematical formulations often flounder on such examples...Thus an unenlightened mathematical formulation might lead to the assertion that no solution exists, while an enlightened formulation (or a reformulation prodded by the physics) simply widens the class of functions to which, from the outset, any eventual solutions are required to belong. (Barton 1989, p.161).

In the heat equation case, for example, one frequently encounters the claim that "analytic boundary conditions are insufficient or overly rigid to suit applications". In consequence, physics must engage in some sort of adjustment that will permit a match to be struck between the exterior conditions and the internal differential equation. With the heat equation, the adjustment needed is very simple, but in other circumstances the repair may call upon subtle constructions drawn from functional analysis.

Very similar considerations show that the set of initial conditions commonly studied in connection with the heat equation also share this "not compatible with the equation's internal solutions" character. On the other hand, if left solely to its own devices, the heat equation accepts mathematical solutions that we discard as "unphysical" precisely because these solutions deposit implausible boundary conditions upon their frontier. So we have found a clue suggesting that boundary conditions might not be quite so "inert" as typically presumed.

As I wrote, the standard adjustment between differential equation and the expected boundary conditions in the case of the heat equations is easy to obtain: one demands that the heat equation be satisfied only in the interior of R and that this interior solution approach the prescribed boundary values merely as a limit. The action of the heat equation immediately smooths our disobedient boundary data into a nice analytic function, so the heat equation can retain its validity throughout the entire interior of R. In this particular case, then, elaborate machinery is not required to repair the mismatch between the equation and its range of expected boundary conditions. But, although this "repair" has been standard practice for many, many years, it nevertheless raises the questions: What right do we have to excuse the heat equation from needing to hold upon the material's frontier? Why shouldn't we ask the heat equation to accept only the boundary conditions that suit the equation's internal demands?

Before trying to address these questions, it should be noted that the accommodation between "internal" and "external" requirements can prove a more complicated matter than this simple "solution should approach the boundary values in a `limit'" suggests. For example, if the boundary is allowed to contains corners or junctions where the boundary temperature undergoes a sudden jump, such limits may not be defined—one may get completely different values according to the direction from which the boundary is approached. Or the boundary limits may blow up to infinity, as stress will do at the inside corners of a notched elastic solid. Any of these behaviors would be viewed as physically unacceptable were they to occur within R's interior, but they have been long permitted to stand if they are confined to R's boundary.

Accordingly, if one wants to formulate the family of boundary conditions generously enough to include these standard simple cases, the requirement that boundary values obtain as a limit from the interior solution must be relaxed, to allow places where such limits fail. In complicated cases, getting this pardoning of lapses to work properly requires some rather subtle machinery, for the connections between interior solution and boundary cannot be severed in a totally cavalier fashion (we'll see later that portions of the interior must be sometimes exonerated from the demands of law as well). But which considerations excuse the frontier from its prima facie obligation to obey the internal equation? In my experience, one finds a good deal of conceptual schizophrenia when one tries to extract an answer from the scientific literature. In the sequel, I will categorize some of the answers I have found.

3. External Boundary Requirements as Mere Conveniences

Sometimes one encounters the opinion that the problematic boundary conditions are to be tolerated merely as "conveniences". In truth, the argument runs, one can never construct a box or string that has a perfectly sharp corner or maintains a strict 0° temperature at a boundary. From this point of view, the differential equation is always right in its clash with the expected family of boundary conditions—one tolerates the idealized cases because of their mathematical convenience.

Although this conservative response is sometimes warranted, in other cases it seems quite implausible. *Prima facie*, why should it be any harder to construct a box with sharp sides than a box with curved sides? Why should it be easier to form a string with a curved shape than a sharp bend, especially when we remember that, in deriving the wave equation for the string, the assumption that the string is "perfectly flexible" (=exerts no bending moment) is strongly evoked? In fact, the "mere convenience" approach clearly blurs the indisputable fact that it is very hard to construct a string meeting exact specifications, whether they be curved or sharply bent, with the more dubious claim that curved strings can readily occur in nature but not bent strings. Indeed, the usual treatments of box and string display no mechanism that would smooth out a sharp boundary once it has formed.

Likewise, what prevents a metal from being held at a specified temperature—why can't we force the metal to stay at exactly 0° by abutting it with a block of melting ice? Insofar as sudden jumps in temperature along a border go, one may be tempted to think that temperature distributions can only change smoothly in nature. But consider an elastic body partially embedded in a wall. Here one expects that its boundary conditions abruptly change at the places where the body leaves the wall. Walls, after all, terminate at their boundaries; they don't fade away in the gradual fashion required if smooth boundary conditions are to be assigned to the embedded body.

Considerations of these sorts move one towards Hadamard's claim that we sometimes know what kind of boundary conditions a material will accept, quite independently of what the internal equations for the material apparently demand. The physical world is a mosaic of regions governed by distinct differential equations and it appears that we can know facts about how the entire jigsaw fits together which are not recorded within the equations for the internal regions.

In any case, the claim that problematic boundary conditions simply represent convenient but unreal "idealizations" must be supplemented by some rather different considerations if the position is to appear at all plausible. Among other reasons, there are a wide class of circumstances where the relevant equations can meet the expected assigned boundary values only if certain portions of the interior are also excused from the requirement that the differential equation be strictly obeyed there. These cases tend to arise no matter how narrowly one tries to circumscribe ones class of allowed boundary conditions. For example, put an airplane wing in a wind tunnel and blow the air across the opening of the tunnel at high speed. The speed of this input air constitutes a boundary condition that we "externally" know can be arranged for the air inside the tunnel. But no strict solution of the usual equations for compressible flow satisfying these boundary conditions exists. The best we can do is allow the air inside to break the laws of compressible flow inside the tunnel—to allow regions where the air pressure suddenly jumps from one value to another (the laws of compressible flow, understood in their usual sense, do not permit such jumps). In fact, the regions where the lapses are tolerated are singular surfaces of zero thickness and can be regarded as new boundaries that spring up within the region that was initially regarded as the "interior". Such surfaces are familiar to most of us as the shock waves that bow out in front of a supersonic airplane wing. Analogous "shocks" can arise inside an elastic metal under the most benign of boundary conditions.

Here is a humbler example: Let water flow around an irregular obstacle. In most cases, it is impossible that the water arriving from the two sides of the obstacle will share the same velocity. To obtain a solution one must allow the water to contain a surface of "contact discontinuity" in the wake of the obstacle. Strictly speaking, this surface counts as an internal boundary where the governing hydrodynamic equations partially fail (the equations require velocity changes to be smooth).

In such cases, our question about boundaries enlarges to: how can these internal lapses from the governing differential equation be rationally justified? Since the locations of the needed internal surfaces can't be foreseen a priori, more radical repairs are generally needed to resolve the clash between internal and external demands than the various "allow lapses at the boundary" provisos we examined above. In particular, it is common practice to reinterpret the internal differential equation in some generalized way, e.g., distributionally or weakly. In some cases, how the generalization proceeds is easy to understand: suppose one has a boundary condition ∂R to which L has no classical solution. Perhaps there is a series of other boundary conditions ∂R_1, ∂R_2, ∂R_3,... that approach ∂R in some limit. Suppose further that ∂R_1, ∂R_2, ∂R_3,... have L solutions S_1, S_2, S_3,... that approach S in another limit. Then S can be called a "generalized solution" of L for boundary condition ∂R even if, strictly speaking, the mathematical operations utilized in L don't make sense along the "internal boundaries" that arise inside S.

I hasten to add that the "generalized solutions" used for shock waves display a more complicated character than this—the shocked solutions don't stay close in their behavior to that of the non-generalized solutions. But, even in the simplest cases, the family of expected "external boundary conditions" plays an important role in defining the notion of "generalized solution". Specifically, the topology selected to define the S_1, S_2, S_3,.. limit must be designed to match the expected ∂R_1, ∂R_2, ∂R_3,... topology. Such procedures cast additional doubt on the inert view of "boundary conditions", for, in allowing generalized solutions, one redefines the meaning of the differential equation with which one originally begun. The structure of the "external" family of boundary conditions thus plays an important role in determining how this redefinition should proceed. Accordingly, the clash between boundary conditions and laws cannot be resolved simply by regarding the problematic boundary conditions as "idealized conveniences".

4. Effacement and Idealization

All the same, if we look at examples, we find a large degree of manifest "idealization". Consider the usual fixed end condition for the vibrating string. The net effect of this constraint is to rechannel all energy flowing towards the endpoints back into the string's interior. Such a condition describes an impossible isolation for our string—fortunately, or we'd never hear our violins. Nor can one expect the mechanism that isolates our string can be wholly confined to the singular points marked in the standard fixed end boundary conditions. Presumably, real strings are clamped along finite portions of their length, not merely at the two singular boundary points evoked in the stock end point condition. Moreover, Newton's Third Law shouldn't allow any force, no matter how strong, to keep the endpoints truly fixed, independently of how much energy is fed into the string.

I want to argue that this kind of "boundary condition idealization" should be seen as playing an important positive role in the physics. In a very useful survey of gravitational problems, T. Damour discusses the need to "efface" a local problem from the global context in which it arises (Damour 1987). The idea is that a clean mathematical formulation of the physics of a group of local objects often cannot be set up until certain joints in nature are located along which the local problem can be severed from its surroundings. For example, one might hope to "efface" the behavior of the earth, moon and sun from the influences of the other heavenly bodies. One sets up a "cut" between inside and outside that appears as a distinction within ones mathematical formalism. In this instance, the gravitational potentials created between the earth, sun and moon will be treated one way (expressed as functions over interparticle separation), whereas the contributions of the stars and planets will appear as an averaged potential $V(x)$ (expressed as a function of position). But one can't regard this second kind of potential as wholly accurate—indeed, it is inconsistent with the usual understanding of the Third Law that the external bodies could influence the earth, sun and moon without the latter effecting, however minutely, the positions of the external bodies. But if the external bodies are so affected, it is impossible that their contributions can be accurately summarized by a position-dependent potential. Accordingly, the $V(x)$ appearing in the Hamiltonian for our system must be regarded as somewhat "idealized". Note that this division between "internal" and "external" potentials simply reflects the policy of "effacement" discussed above—one places ones "cut" in some place where the behavior of a local group of particles can be treated with reasonable accuracy using an "averaged" potential to represent the contributions of the external group. How well this "cut" will work depends in part on the accuracy demanded in the resulting calculations. For predictions extending over a longer term, our initial selection of the "cut" between internal and "external" groups may prove inadequate and the "cut" must be reassigned to encompass a larger "local" group.

In such a situation, the mathematical formulae one writes down always carry an inherent degree of "idealization" reflected in the "averaged" term $V(x)$. If the universe happens to only contain a finite number of particles, one can theoretically escape this inherent idealization by putting all particles within the chosen local group. But if the universe contains an infinite number of particles—an awkward but self-consistent proposition within classical particle physics—, every formula that one writes down will carry some degree of "idealization" within its makeup.

The recent philosophical literature has witnessed a lot of roughly hewn claims to the effect that "physics has an inherent need to treat objects in an idealized fashion". Unfortunately, little care is devoted to tracing out the source of the supposed "idealization" (few of the cases discussed in the literature involve anything that I would consider "inherent" idealization). However, in the scenario just sketched, the necessity for a partially idealized cut is "inherent" in the sense that concrete formulae cannot be written down until some such "cut" has been made. Nonetheless,—and this is the crucial observation—there is no need to suppose that a physics which treats its formulas in this manner should be regarded as "underdescribing" or "misrepresenting" the materials under its dominion. Everything that happens in an infinite point particle universe will be properly treated within the suggested physics, although at any particular stage some further facts are dealt with only in a circumscribed or approximate fashion. Thus, although no formula written down in such a physics escapes some measure of "idealization", the overall physics shouldn't be accused, point blank, of describing the universe in a "false", "idealized" or "anti-realistic" fashion.

In continuum physics the need for idealized cuts is more pressing than in a point particle universe, largely because a given hunk of material generally lies in direct contact with other materials. Furthermore, the need also arises to "efface" a material from its microscopic complexities. The standard choices of boundary condition families studied in the textbooks reflect some of the strategies for "effacement" that have been found to work for continua.

5. Boundary Conditions as Idealized Effacements of Local Microscopic Physics

These considerations suggest another way to resolve the boundary/ differential equations clashes that is frequently encountered in the literature. Specifically: such mismatches occur when some additional physics operates within the boundary region, but the policy of "effacement" finds it convenient to keep this physics partially suppressed. A paradigm of this kind of situation is the following. Water displays an almost negligible measurable viscosity. Most early physicists presumed that water's movements could be adequately approximated by Euler's equations for non-viscous flow. On the other hand, experiments also show that water remains at rest adjacent to its boundaries. In many circumstances, there are no classical solutions of Euler's equations that are compatible with these "externally" dictated boundary conditions. At the turn of the century, Ludwig Prandl observed that the neglected viscosity of water can convert the Euler-like flow found in the middle of a pipe into a viscous flow accommodating the expected boundary conditions. Accordingly, Prandl's resolution of our hydrodynamic boundary/equation clash traces to the fact that Euler's equations omit terms governing the physics of viscosity that is prominent along the pipe's boundary.

Such appeals to "missing physics" help explain some of the boundary clashes we discussed earlier. Thus in the 0°/100° heat equation case, we appealed to an adjacent block of melting ice as a way of holding one boundary firmly at 0°. Implicitly this assumption requires that a phase transition is occurring along the boundary, a process clearly not modeled in the physics formulated within the heat equation. Similar missing terms correlate with the rise of shock waves in gas dynamics. Likewise,—although we witnessed no boundary clash there—, none of the physics responsible for holding the end points of a vibrating string fixed is formulated within the governing equation for the string (which deals only with the Hookean tension force acting along the string's length). Cornelius Lanczos writes:

> And here we have first of all to record the fact that from the physical standpoint a "boundary condition" is always a simplified description of an unknown mechanism which acts upon our system from the outside. A completely isolated system would not be subjected to any boundary conditions. Imposed boundary conditions are merely circumscribed interventions from outside which express in simplified language the coupling which in fact exists between the given system and the outer world. (Lanczos 1960, pp. 504-5).

In short, one might hope that all clashes between expected boundary conditions and internal equations will trace to the way the problem has been effaced from details of its external bindings and its microscopic behavior. If one thinks through a case like the vibrating string, one readily appreciates how wonderfully effective the "effacement" policy can be; more realistic approaches to the endpoints enmeshes one in physics of a considerably messier variety, involving "cuts" extending far into the universe beyond the string.

In the next section, we'll consider a more radical resolution of boundary/equation clashes. Let us pause to underscore a point of Sklar's. In the standard picture encouraged in the philosophy of science texts, boundary conditions are regarded simply as the physical data that happen to lie along some arbitrarily chosen timelike slice through the universe ("initial conditions", likewise, represent arbitrary spacelike cuts). But the stock choices of boundary surfaces for, say, the heat equation represent locations where we expect the equation not to apply—other thermal processes will be present there. We select such boundaries because such "off-line" surfaces may be the only places where a tractable handle can be gained on the equation under study. This connects with "free chooseability"; it is relatively easy to determine that two boundaries are approximately heated to 0° and 100°, but it is absurd to expect that an exact analytic function can be precisely determined along a boundary.

6. "Complementarity" between Boundary Condition and Interior

There is no doubt that the resolution of boundary clashes suggested in the last section is often completely appropriate. Nonetheless, sketches of a more radical resolution of boundary clashes can be found in the literature. I write "sketches", for the viewpoint I'll describe is generally expressed only in the fuzziest possible manner. In consequence, I won't be able to offer a very crisp characterization of the viewpoint myself. But the idea is interesting and ought to be pursued.

Roughly speaking, the idea traces to the suspicion that the point-valued functions utilized in the classical mathematics of internal solutions are too "definitely localized" to be true to nature. Perhaps the portions of a domain marked as "interior" and "boundary" respectively reflect a somewhat conventional division within the true physics of the material, a conventional division selected to provide an effective "effacement" of the problem, but which may need to be reconsidered within a more exacting treatment. The special "excuses" permitted to boundary surfaces should be viewed as an important control on the validity of the physics recorded in the differential equations—a way of keeping the internal results honest, as it were. From this point of view, the boundary conditions turn out to be no more "idealized" than is the basic assignment of "smoothness" to the internal region. Accordingly, we might say that a sort of "complementarity" persists between assignments of "smoothness" and "nonsmoothness".

One finds some measure of such skepticism concerning the absolute status of "smoothness" assumptions in standard claims that Laurent Schwartz' distributions represent a more natural choice for modeling physical phenomena than do standard point-valued functions:

> [The] defects [of modeling by functions] are a certain rigidity and the fact that it does not correspond to physical reality. Consider, for example, a chemist studying some property of a substance at temperature T_0. He or she cannot achieve the temperature T_0 exactly, or even a uniform temperature: almost certainly, the temperatures throughout the system will vary over some range $a \leq T \leq b$. This suggests that the correct mathematical model might involve averaging over $a \leq T \leq b$, [using an "averaging" function $\emptyset(T)$]... If the (averaging) function $\emptyset(T)$ is infinitely differentiable, a reasonable hypothesis, then it is a test function... The chemist wants to test the properties of some substance at temperature T_0. So he brings together a batch of the stuff with temperatures distributed in a pulse $\emptyset(T)$ near $t=T_0$. The objective is to find some scientific law and the batch of chemicals (the pulse $\emptyset(T)$) tests the law. In a mathematical model, the scientific law will normally be represented by a function, say f(T). Now, just like the chemist, we are interested in f, not $\emptyset$! (Richards and Youn 1990).

This rather standard apology for "distributions" does not seem altogether coherent to me; it seems to suggest that the "stuff" in question is simultaneously smeared out in a pointlike fashion ("he prepares a batch of the stuff") and that it is not (such functions "do not correspond to physical reality"). Nonetheless, the passage clearly suggests a skepticism about the meaningfulness of assertions that localize physical processes to low dimensional regions, such as a boundary ∂R.

Such skepticism is sometimes evoked when a workable technical adjustment between expected boundary conditions and internal equations has been found—utilizing, say, distributional solutions—, but where it is not obvious that any hidden physics is missing from the problem. For example, structural engineers use calculations on elastic materials that may be supported only by weak solutions involving some sort of internal surfaces where the differential equation fails. These calculations seem to supply very trustworthy results, but it seems dubious that the singular surfaces described in the weak solutions can directly report any missing localized physics (e.g., like the regions of viscous balancing to which shock waves in a gas correspond). The problem traces to the fact that the stresses described in the elastic equations represent averages over moderately large groups of molecules. But the singular structures in the weak solutions seem to attribute significant detail to lengths below the scale of averaging. Perhaps these microscopic singular surfaces somehow collect or encode large scale facts about the material that was missed in the set up of the differential equations. The calculations work only because the lapses along the internal conditions systematically correct for the distortions inherent in the internal differential equations.

In fact, the question of what actually happens in those elastic calculations is still a matter of considerable controversy (Poston and Stewart 1978, p. 300). But simpler cases exist where a least some degree of "complementarity" in the assignment of interior/boundary can be plausibly motivated. Consider the following: Take a knife and cut down through the surface of water. Obviously, the water will close up behind the knife. But unless one relaxes the governing equations somewhere, one must maintain that the original upper surface of the water becomes subsequently stretched and dragged along in the water behind the knife, no matter how far or deep the knife travels. That is, the so-called "boundary" must be treated as now extending downward into the water—let us dub this dragged boundary a "wake surface". This internal boundary should not be immediately dismissed as unreal—initially the wake's presence will considerably modify the behavior of the surrounding water (the equation-violating jumps across the wake allow the water to carry a non-vanishing circulation). But after a long period of time, one expects that conditions along the wake surface will relax so that the boundary vanishes—we do believe, after all, that the movements of ocean water can be tracked without modeling the web of internal surfaces deposited by every wave that has ever crested. Unfortunately, this boundary disappearance doesn't occur in the mathematics—the conditions on the two sides of the wake may relax, but the wake surface remains permanently present (albeit greatly distorted by instabilities). The only way, mathematically, to get rid of the wake surface is to start over and redescribe the region within its vicinity as smooth. When and where this can be done seems like as a quasi-pragmatic decision—it seems theoretically possible that the jumps across an apparently weakened wake might occasionally grow strong again. But until the decision to describe an area of water as "internal" (=smooth) or "boundary" has been made, one cannot set up the physics of the problem.

Likewise, in the theory of elasticity internal surfaces are used to model cracks in the material, which, when once tolerated, mathematically never disappear. This example conveniently illustrates a second theme that contributes to skepticism about the absoluteness of the boundary/smooth region dichotomy. Namely, the internal equations predict that the stresses at the ends of the cracks will be infinite. At first blush, this indicates that the material ought to break or flow. But if you integrate (in an "improper" way, of course) the stored stress energy over a finite region, one can obtain a finite number which can be used to predict how the crack will grow. In this context, one decides not to worry about the infinite stress embodied in the crack, but concentrates upon the ability of the region containing the crack to release energy to other parts of the material.

The realization that physics and mathematics which "looks bad" regarded in terms of its pointwise behavior may "look good" inside a "smeared out" integral is a very old theme that frequently emerges in discussions of our problem. The fact that energy, which sometimes does not tolerate overly specific localization happily, is usually intimately connected with the relevant integrals contributes to doubts that physics' treatment of boundaries is necessarily more idealized than its treatment of the smooth areas. However, such themes cannot be pursued further here.

7. Closing Remarks

We have traced a variety of patterns, in an increasingly radical sequence, for resolving conflicts between differential equations and their expected boundary conditions. When such clashes are discussed in classical contexts, it is usually rather hard to determine which of these varying attitudes a given author intends to endorse. As long as one deals with classical physics, one has a (partial) excuse for tolerating such slushy resolutions of the conflict. Namely, if one tries to scrutinize the boundary conditions, or the pointwise behavior, of classical solutions in much depth of detail, one is led quickly into the realms of quantum behavior, whose own shaky foundations raise a much richer set of confusing issues. Nevertheless, our discussion of the boundary/law dichotomy in purely classical cases may have some indirect relevance to the quantum case. Very much the same tools from functional analysis utilized in classical boundary value clashes are used to construct a "well defined dynamics" in quantum mechanics. Indeed, whenever one assumes that a given quantum system lives in a Hilbert space, this implicitly assumes that a lot of delicate carpentry involving expected boundary conditions, extensions of differential operators, etc., have been expended upon a naive Schrodinger equation (or the like). We have seen that the considerations needed to set many classical systems within their proper Hilbert spaces may not represent assumptions of permanent longevity—the operative cut between "boundary" and "smooth interior" may require eventual reassessment. It therefore bothers me that, in the usual discussions of the measurement problem, the breakdown in validity of the treatment is invariably blamed on a "non-dynamical" collapse of the wave packet, without much consideration of the background decisions needed to set the problem within a proper Hilbert space in the first place. Such simple considerations hardly promise a resolution of the quantum measurement problem, but, at the same time, one would appreciate an argument that shows why these factors can be safely ignored.

References

Barton, G. (1989), *Elements of Green's Functions and Propagation*. Oxford: Clarendon Press.

Damour, T. (1987), "The Problem of Motion in Newtonian and Einsteinian Gravity" in S.W. Hawking and W. Israel (eds.), *300 Years of Gravitation*. Cambridge: Cambridge University Press.

Field, Hartry (1980), *Science Without Numbers*. Princeton: Princeton University Press.

Lanczos, Cornelius (1960), *Linear Differential Operators*. London: Van Nostrand.

Poston, Tim and Ian Stewart (1978), *Catastrophe Theory and its Applications*. London: Pitman.

Richards, J.I. and H.K. Youn (1990), *Theory of Distributions*. Cambridge: Cambridge University Press.